Palaeolimnological Proxies as Tools of Environmental Reconstruction in Fresh Water

Developments in Hydrobiology 208

Series editor

K. Martens

Palaeolimnological Proxies as Tools of Environmental Reconstruction in Fresh Water

Editors

Krisztina Buczkó[1], János Korponai[2], Judit Padisák[3] &
Scott W. Starratt[4]

[1]*Hungarian Natural History Museum, Department of Botany, 1476 Budapest, P.O. Box 222, Hungary*
[2]*West-Transdanubian District Water Authority, H-8360 Keszthely, Csik F. str. 1, Hungary; Department of Chemistry and Environmental Sciences, University of West Hungary, Szombathely, Hungary*
[3]*Department of Limnology, University of Pannonia, Egyetem u. 10. 8200 Veszprém, Hungary*
[4]*U.S. Geological Survey, 345 Middlefield Road, Menlo Park, CA 94025, USA*

Previously published in *Hydrobiologia,* Volume 631, 2009

Springer

Editors

Krisztina Buczkó
Hungarian Natural History Museum
Department of Botany
1476 Budapest
P.O. Box 222
Hungary

Judit Padisák
Department of Limnology
University of Pannonia
Egyetem u. 10. 8200 Veszprém
Hungary

János Korponai
West-Transdanubian District Water Authority
H-8360 Keszthely
Csik F. str. 1
Hungary;
Department of Chemistry and Environmental Sciences
University of West Hungary
Szombathely
Hungary

Scott W. Starratt
U.S. Geological Survey
345 Middlefield Road
Menlo Park
CA 94025
USA

Cover illustration: Lake Saint Anna, the only remaining crater lake in the Carpathian Mountains. Other crater lakes were filled and are now covered by peat-bogs. Photo: Enikő Magyari.

DOI: 10.1007/978-90-481-3387-1

ISBN: 978-94-007-3041-0 e-ISBN: 978-90-481-3387-1

Printed on acid-free paper.

Softcover reprint of the hardcover 1st edition 2009

springer.com

Contents

Foreword: A virtual congress on palaeolimnology—palaeolimnological proxies as tools for environmental reconstruction in fresh water

Krisztina Buczkó · János Korponai · Judit Padisák · Scott W. Starratt

Originally published in the journal Hydrobiologia, Volume 631, No. 1, 1–2.
DOI: 10.1007/s10750-009-9805-x

The motivation for collecting recent knowledge in a special issue of *Hydrobiologia* derives from the recognition of the importance and applicability of palaeolimnological tools to help in defining "reference conditions" as designated within the Water Framework Directives and estimating influence of global climate change on surface waters. This volume was developed by inviting contributions from prominent experts in their respective fields. The compilation not only presents papers on palaeolimnological studies, focusing mostly on Eastern and Central Europe but also includes results from other regions.

Guest editors: K. Buczkó, J. Korponai, J. Padisák & S. W. Starratt
Palaeolimnological Proxies as Tools of Environmental Reconstruction in Fresh Water

Dedicated to Olga Sebestyén (1891–1986), key scientist at the First Palaeolimnological Symposium, Vice-president of the SIL (1962–1986).

K. Buczkó (✉)
Department of Botany, Hungarian Natural History Museum, P.O. Box 222, 1476 Budapest, Hungary
e-mail: krisztina@buczko.eu

J. Korponai
West-Transdanubian District Water Authority, Csik F. str. 1, 8360 Keszthely, Hungary

J. Korponai
Department of Chemistry and Environmental Sciences, University of West Hungary, Károlyi Gáspár square 2, 9700 Szombathely, Hungary

J. Padisák
Department of Limnology, University of Pannonia, Egyetem u. 10, 8200 Veszprém, Hungary

S. W. Starratt
U.S. Geological Survey, 345 Middlefield Road, Menlo Park, CA 94025, USA

The use of palaeoecological analyses of sediments has a long tradition in Central Europe. In the nineteenth century, Lajos Lóczy (1849–1920) organized a systematic scientific research on Lake Balaton, the largest shallow lake of the region, and published the series of booklets (Lóczy 1897–1920) which are considered milestones in the development of limnology as a separate branch of science and which are comparable to Forel's (1841–1912) seminal scientific heritage (Forel, 1892). Another milestone in the development of palaeolimnology was the first palaeolimnologial meeting which was held in Hungary in 1967. This meeting included some of the most prominent limnologists in the world including G.E. Hutchinson, D.G. Frey (Chairman), Nina V. Korde, D.A. Livingstone, O. Sebestyén, and W. Tutin who together formed the organizing committee. This committee decided to hold the symposium at the Biological Research Institute, Tihany, Hungary. The meeting profited from the excellent facilities and made the attendance of scientists from socialist countries possible. This location was also appropriate in celebrating the long tradition of geological and limnological studies on the lake including the early effort of L. Lóczy. Eighty-nine individuals representing 20 different countries were registered at this symposium (Frey, 1969).

In spring 2008, a large group of scientists expressed an interest in the present special issue and 22 manuscripts were submitted, 18 of which are included in this volume. The primary objective of this special issue is to present new palaeolimnological findings from Eastern and Central Europe, as well as important findings from other regions. Although this area has sometimes received less attention than other areas of Europe, the lakes and mires, coupled with the variability in landscape and the local differences in climate, provide unique opportunities for studying palaeolimnology. A review on the Late-Quaternary records in the Carpathian region provides new results on the history of a crater lake, Lake Saint Ana, glacial lakes in the Tatra Mountains and Lake Bled. The sediments of these lakes, as well as peat bogs, also provide valuable evidence for studying climate change.

In the present issue, the various papers provide new insights on the development of lakes and bogs during the late-glacial and Holocene, using a wide range of palaeolimnological proxies, including diatoms, pollen, macrofossils, pigments, Cladocera, and Chironomidae as well as geochemistry. New results are also provided from Spain, Finland, Russia, North America and South America.

The editors express their thanks to Enikő Magyari who helped with the editorial work in all of its phases. Thanks are also due to all the referees for their efforts in evaluating and improving the manuscripts that were submitted for publication in this volume.

The guest editors

References

Frey, D. G. (ed.), 1969. Symposium on palaeolimnology. Internationale Vereinigung für Theoretische und Angewandte Limnologie Mitteilungen 17. E. Schweizerbartshe Verlagsbuchandlung, Stuttgart: 448 pp.

Forel, F. A., 1892. Monographie Limnologique. Geneve.

Lóczy, L., 1897–1920. Ergebnisse der wissenschaftlichen Untersuchungen des Balaton Sees. Magyar Királyi Természettudományi Társulat, Budapest.

Review of dated Late Quaternary palaeolimnological records in the Carpathian Region, east-central Europe

Krisztina Buczkó · Enikő Katalin Magyari ·
Peter Bitušík · Agnieszka Wacnik

Originally published in the journal Hydrobiologia, Volume 631, No. 1, 3–28.
DOI: 10.1007/s10750-009-9800-2

Abstract The Carpathian Region (including mountains and plains) has for a long time been lacking good palaeoenvironmental and especially palaeolimnological records, particularly for the Late Quaternary. In the last two decades, many new sedimentary sequences were obtained and studied using a wide range of palaeoproxies. This article reviews results from 123 sequences in the Carpathian Region, all dated by radiometric methods. Our aim was to pay attention to the existence of these data; many of them published in national periodicals and journals. Palaeoenvironmental records with at least two proxies and with palaeolimnological interpretation were compiled in both tabular form and on maps. Inspite of the density of examined sites, an assessment of the dataset led us to the following conclusions: (1) very few provide firm hydrological–limnological interpretation, such as lake level and mire water-depth fluctuation, lake productivity changes and pH changes; (2) only 47 of them are real multi-proxy studies (have at least two proxies employed on the same sediment core); (3) glacial lakes in Slovakia and Romania as well as in Ukraine are seriously under-investigated although they would be ideal objects of palaeolimnological works with the many proxies applicable on them; (4) the Hungarian lowland areas are dominated by shallow tectonic lakes or palaeochannels, often with unsatisfactory preservation of certain biological proxies (e.g. diatoms, chironomids, cladocerans). Consequently, palaeolimnological studies from this region have to apply a different combination of proxies and approach than mountain lake studies.

Keywords Palaeolimnological records · Multi-proxy · Carpathians · Pannonian Plain · Late Glacial · Holocene

Guest editors: K. Buczkó, J. Korponai, J. Padisák & S. W. Starratt
Palaeolimnological Proxies as Tools of Environmental Reconstruction in Fresh Water

K. Buczkó (✉)
Department of Botany, Hungarian Natural History Museum, P.O. Box 222, 1476 Budapest, Hungary
e-mail: krisztina@buczko.eu

E. K. Magyari
Hungarian Academy of Sciences, Palaeonthological Research Group, Hungarian Natural History Museum, P.O. Box 222, 1476 Budapest, Hungary

P. Bitušík
Research Institute & Faculty of Science, Matthias Belius University, 974 01 Banska Bystrica, Slovakia

A. Wacnik
W. Szafer Institute of Botany, Polish Academy of Sciences, Lubicz 46, 31-512 Kraków, Poland

Introduction

In recent years, significant progress has been made on continental scale quantitative climate reconstructions for the Last Glacial Maximum, Late Glacial and Holocene periods (Harrison et al., 1996; Peyron et al., 1998; Davis et al., 2003; Magny et al., 2003). These European reconstructions using pollen and lake-level records provided a basis for (1) model-data comparison, (2) assisted in our understanding of ecosystem response to orbital forcing and (3) highlighted important latitudinal and longitudinal differences in the direction of climate change at short-lived abrupt climatic oscillations, such as the perturbations of the North Atlantic thermohaline circulation (Alley et al., 1997) and changes in past solar activity (Bond et al., 2001). Looking at the data point distribution of these continental scale reconstructions, a salient feature is the scarcity of data points in the east-central European sector, and particularly in the Carpathian Region. One reason for this data shortage is the general scarcity of potential sedimentary sequences suitable for multi-proxy analyses in the Carpathian Basin. The other reason is the former political–social division of Europe that resulted in the publication of several east-central European Late Quaternary palaeoenvironmental and palaeolimnological studies in national periodicals and journals sometimes of lower scientific quality. As a consequence, the data of the local scientific groups have been buried and disappeared like 'fossils in the sediment'.

In order to the increase the awareness of existing Late Quaternary palaeoenvironmental/palaeolimnological studies from the Carpathian Region as well as to turn scientists' interest towards potential sites, an attempt is made in this study to collect, tabulate and map dated palaeolimnological sequences. Late Quaternary records with at least two proxies (of which at least one is biological), and multiple ^{14}C or ^{237}Cs isotope dates were included. Records with an exclusive emphasis on terrestrial vegetation development were excluded, as these cannot be used to infer the status and changes of water bodies. Our aim was to summarize the available proxy records from this region to make them available for regional and European-scale Late Quaternary hydrological and palaeoclimate reconstructions.

The need for the construction of European scale palaeolimnological databases has long been stressed (e.g. Batterbee et al., 2007, 2009), and several databases came into existence as outcomes of large European projects. The first attempt goes back to 1976, when IGCP Project 158 ('Palaeohydrology of the Temperate Zone During the Last 15,000 Years') was launched (Berglund, 1986). This was followed by the ELSDB (European Lake Status Data Base; Yu & Harrison, 1995), ELDP (European Drilling Programme, Brauer & Negendank, 2004) databases, and most recently, by the compilation of the Meta-database (Battarbee et al., 2007, 2009). In addition, a valuable data source is the European Pollen Database (EPD). These databases include some of the records described in this study, but the majority of them are summarized in this study for the first time.

Study sites

Situated in east-central Europe, the Carpathian Region is a well-defined geographical unit bordered by the curved ranges of the Carpathians to the north and east (Fig. 1). They encompass a vast lowland area called the Pannonian Plain that is the second major part of the geographical unit (Fig. 1). Apart from the highest peaks, this area has never been glaciated. During the last glacial period (Würm or Weichselian), the Carpathian Region fell into the periglacial zone with southward diminishing influence of the European ice sheet. The lack of glaciation means that lakes and mires are relatively few in this region, especially in the lowland. A concentration of lakes, however, appears in the glaciated mountain chains—e.g. the Retezat Mts have 58 (Clarke et al., 2005), the Tatra Mts have 138 major lakes (>0.01 ha) and 11 lakes are located in the Ukrainian part of the Carpathians (Pokynčereda, pers. com.).

Methods

The data were collected from all available literature. In addition, the EPD and the Meta-database (Battarbee et al., 2007) were used. Where research results were published in pieces or at different level of completeness, only the most comprehensive studies were cited. Sediment chronologies are mainly based on ^{14}C measurements, but different authors treat these dates in various ways; for example, some provide calibrated radiocarbon timescale, while others present the data along depth or uncalibrated radiocarbon timescale. In order to make these records comparable, we decided to present the depth of the lowermost ^{14}C

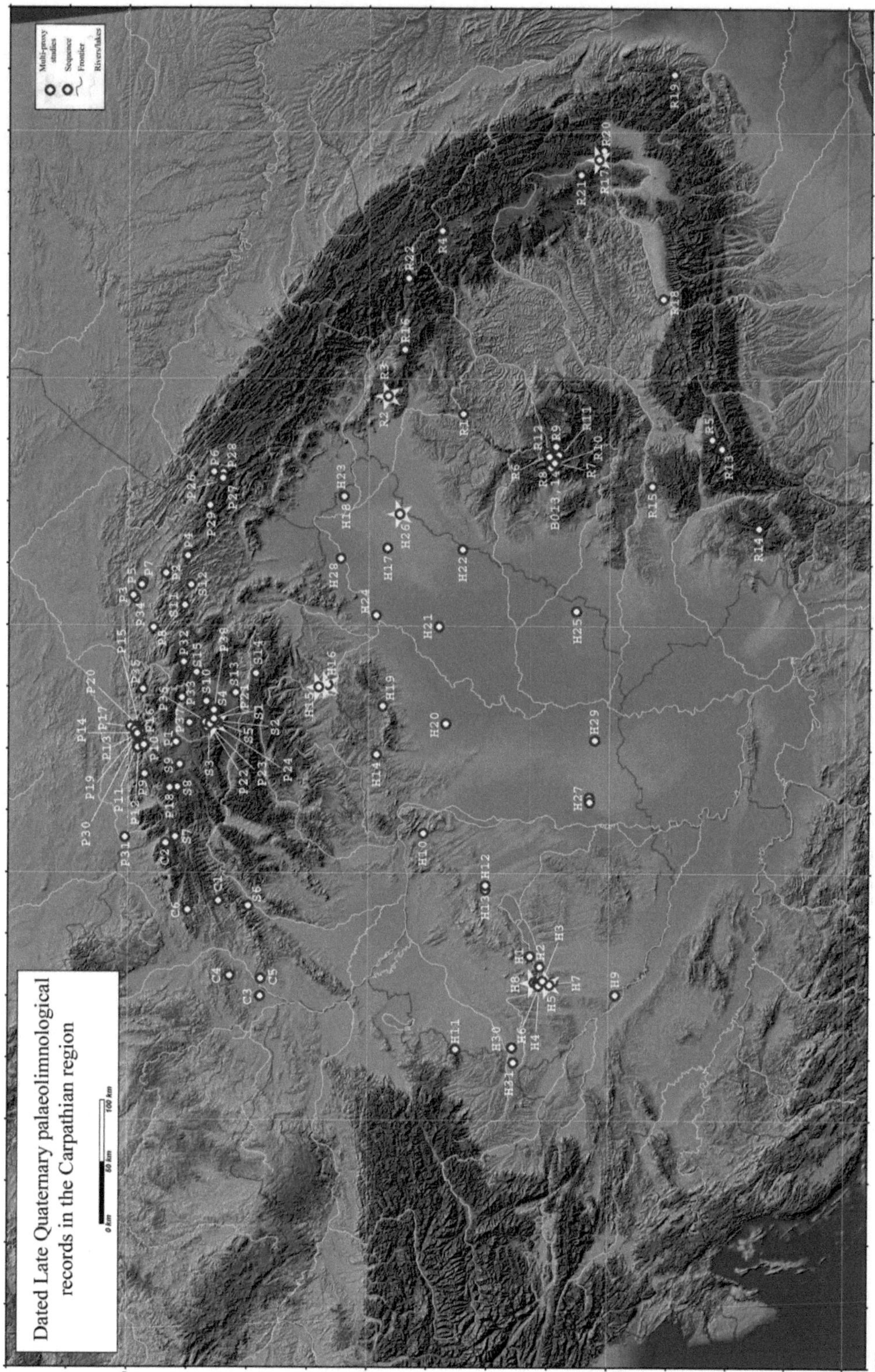

Fig. 1 Dated Late Quaternary palaeolimnological records in the Carpathian region, east-central Europe. The first letter of the locality refers to the country, C as Czech Republic, P as Poland, S as Slovakia, H as Hungary and R as Romania. For location details, see Table 1

date and the uncalibrated ^{14}C value. In addition, where possible, calibrated age ranges of the sedimentary sequences were also displayed according to the original publications. If the original publication contained only uncalibrated age ranges, then the CalPal-2007$_{Hulu}$ calibration dataset (Weninger & Jöris, 2008) and the CalPal-2007^{Online} program (Danzeglocke et al., 2008) were used to obtain the calibrated age ranges. An asterisk marks these records after the calibrated age range. This way, all the records have calibrated age ranges that make them easily comparable.

The criteria for inclusion of a site/record were as follows:

(1) the sediment is of Late Quaternary age;
(2) it has been dated by radiometric methods;
(3) published studies include palaeolimnological inferences;
(4) at least two proxies were analyzed.

Even though national and international journals were screened carefully for Late Quaternary studies, and experts were also asked to check the database, some sites fulfilling these criteria may have been omitted. We apologize authors whose data were not found.

Results

Altogether, 110 localities are listed in Table 1 including 123 sequences. Six sequences can be found in the Czech Republic, 15 in Slovakia, 38 in Poland, 37 in Hungary and 24 in Romania. The distribution of dated Late Quaternary sequences is presented on the relief map of the Carpathian Region (Horváth & Bada, 2008).

The time resolution and the number of palaeoproxies applied on these sediments vary considerably between sites (Table 2). Most of them cover the Holocene (63), but the number of Late and Full Glacial records is also significant (55). The longest continuous sequence comes from the Romanian Carpathians (Iezerul Caliman); here, pollen and lithological analyses extend back to ca. 17,700 cal year BP (Fărcaş et al., 1999, 2003, 2006a, b; Feurdean et al., 2007a). Most of the sediments analyzed from this region are peat, and only 15 lake sites are known with continuous lake deposits all over the sequence (Taul dintre Brazi, Taul Zanogutii, Pesteana, Saint Ana, Lake Balaton, Baláta-tó, Vyšné Temnosmrečinské pleso, Vyšné Wahlenbergovo pleso, Nižné Terianske pleso, Ľadové pleso, Zielony Staw Gąsienicowy, Czarny Staw Gąsienicowy, Toporowy Staw Wyżni, Długi Staw, Przedni Staw).

Sediment lithology was published from 104 sites; pollen analysis was done on 115, plant macrofossil on 33 and diatom analysis on 13 sequences. Twenty-one sites have malacological records, cladocera were analyzed at seven sites, and merely three sites have chironomid or testate amoebae records. Geochemical records are available from 17 sediment sequences, LOI was measured in 20 cases, and archaeological data supplemented the palaeoenvironmental reconstruction at 28 localities. Late Quaternary ostracod records were not found in the region.

In some cases, information can be found on the geomorphology, granulometric indexes, grain size distribution of mainly minerogenic deposits, green algae and bryophyte components of the sediment. Chrysophycean cysts were recorded only occasionally (two sites), as well as pigments and spherical carbonaceous particles (SCP).

The highest number of proxies (eight) was analyzed on the Holocene sediment sequence of Lake Saint Ana (Eastern Carpathians; Magyari et al., 2006, 2009a). In addition, the best studied lake deposit of the Western Carpathians is Nižné Terianske pleso, where altogether eight different proxies were studied, however, on different occasions and sediment cores (Šporka et al., 2002; Appleby & Piliposian, 2006; Marková et al., 2006; Bitušík et al., 2009). The number of radiocarbon dates and time resolution of the various proxies are also the highest at Nižné Terianske pleso (Appleby & Piliposian, 2006; Bitušík et al., 2009). Localities that were studied by multi-proxy methods and obtained useful evidence for hydrological changes are indicated by asterisk on the map (Fig. 1).

Discussion

The study of Late Quaternary vegetation dynamics have a long tradition in the Carpathian Region, but only recent studies provided radiocarbon-dated and high-resolution vegetation records using pollen and plant macrofossil analyses (e.g. Fărcaş et al., 1999,

Table 1 Locality details and studied proxies of dated Late Quaternary palaeolimnological records in the Carpathian region, east-central Europe

Map code	Site name Details the locality of core, Country	Latitude	Longitude	Alt (m)	Record length (cm)	Proxies	References
C1	Královec Bilé Karpaty, Czech Republic	49°07′55″	18°01′40″	560	125	Pollen, macrofossils, malacology	Rybníčková et al. (2005)
C2	Horní Lomná Západné Beskydy, Czech Republic	49°31′14″	18°37′53″	615	100	Pollen, macrofossil	Horsák & Hájková (2005), Rybníčková et al. (2005)
C3	Machová Moravskoslezské Beskydy, Czech Republic	48°49′30″	17°02′25″	460	100	Pollen, macrofossil	Rybníčková et al. (2005)
C4	Vracov Pannonian lowland, Czech Republic	48°58′40″	17°12′10″	192	270	Pollen, macrofossil	Rybníčková & Rybníček (1972)
C5	Svatobořice-Mistřín Pannonian lowland, Czech Republic	48°50′0″	17°10′0″	175	235	Pollen, archaeology	Svobodová (1989)
C6	Jablunka Moravskoslezské Beskydy, Czech Republic	49°23′	17°57′	350	28 hiatus	Pollen	Jankovská & Pokorný (2008), Kuneš et al. (2008)
P1	Puścizna Rękowiańska Orawa-Nowy Targ Basin, PUREK_N, Poland	49°29′0″	19°49′0″	656	732	Lithology, pollen	Obidowicz (1990, 1993, 1996)
P2	Cergowa Góra Lower Beskid Mts, Poland	49°32′0″	21°42′0″	495	550	Lithology, pollen, macrofossil, archaeology	Więckowski & Szczepanek (1963), Budziszewski & Skowronek (2001), Pazdur (2001), Szczepanek (2001a, b)
P3	Jasło Jasło-Sanok Depression, Poland	49°47′0″	21°28′0″	250	400	Lithology, pollen	Szafer (1948), Alexandrowicz (1984a, b), Harmata (1995)
P4	Jasiel Lower Beskid Mts, Poland	49°22′22″	21°53′13″	680	245	Lithology, pollen	Szczepanek (1987)
P5	Roztoki Jasło-Sanok Depression, ROZ14A-N, ROZ1B-N, Poland	49°43′0″	21°35′0″	230	463–585, 340–535	Lithology, pollen, mollusc	Harmata (1987)
P6	Tarnawa Wyżna Bieszczady Mts, TAR1_N, Poland	49°09′ 0″	22°49′ 0″	670	705, 265–700	Lithology, pollen, macrofossil	Ralska-Jasiewiczowa (1972, 1980)
P7	Tarnowiec Jasło-Sanok Depression, TARW-N, Poland	49°42′0″	21°37′0″	220	253	Lithology, pollen	Harmata (1987, 2008)

Table 1 continued

Map code	Site name Details the locality of core, Country	Latitude	Longitude	Alt (m)	Record length (cm)	Proxies	References
P8	Szymbark Lower Beskid Mts, SZYM_N, Poland	49°38′0″	21°06′0″	465	485	Lithology, pollen	Gil et al. (1974), Szczepanek (1989)
P9	Żurawica Beskid Makowski Mts, Poland	49°43′	19°28′	575	650 hiatus	Lithology, pollen, LOI, geomorphology, granulometric indexes, grain size	Margielewski (2006)
P10	Osieczkowa Beskid Makowski Mts, Poland	49°43′16″	19°47′08″	650	140 hiatus	Lithology, pollen, LOI, geomorphology, granulometric indexes, grain size	Margielewski (2006)
P11	Bogdanówka-Beło Beskid Makowski Mts, Poland	49°46′	19°46′	675	500 hiatus	Lithology, pollen, LOI, geomorphology, granulometric indexes, grain size	Margielewski (2006)
P12	Zięby Beskid Makowski Mts, Poland	49°46′	19°47′	535	150 hiatus	Lithology, pollen, LOI, geomorphology, granulometric indexes, grain size	Margielewski (2006)
P13	Pękalówka Beskid Makowski Mts, Poland	49°46′20″	19°54′50″	685	175 hiatus	Lithology, pollen, LOI, geomorphology, granulometric indexes, grain size	Margielewski (2006)
P14	Siódmowo Beskid Makowski Mts, Poland	49°47′19″	19°57′30″	507	150 hiatus	Lithology, pollen, LOI, geomorphology, granulometric indexes, grain size	Margielewski (2006)
P15	Mirkowo Beskid Makowski Mts, Poland	49°46′20″	19°57′30″	443	250 hiatus	Lithology, pollen, LOI, geomorphology, granulometric indexes, grain size	Margielewski (2006)
P16	Pcim Krzywica Beskid Makowski Mts, Poland	49°45′	19°57′	460	160	Lithology, LOI, geomorphology, granulometric indexes, grain size	Margielewski (2006)
P17	Kamiennik Beskid Makowski Mts, Poland	49°49′	20°00′	587	450	Lithology, pollen, LOI, geomorphology, granulometric indexes, grain size	Margielewski (2006)
P18	Pilsko Beskid Żywiecki Range, Poland	49°31′51″	19°19′13″	1,270	390 hiatus	Lithology, pollen	Obidowicz (2003)
P19	Kotoń Beskid Makowski Mts, Poland	49°46′	19°54′	740	470 hiatus	Lithology, pollen, macrofossil, LOI, geomorphology, granulometric indexes, grain size	Margielewski et al. (2003)
P20	Hajduki Beskid Średni Mts, Poland	49°47′30″	19°59′23″	475	330 hiatus	Lithology, pollen, LOI, geomorphology, granulometric indexes, grain size	Margielewski & Zernitskaya (2003), Margielewski (2006)

Table 1 continued

Map code	Site name Details the locality of core, Country	Latitude	Longitude	Alt (m)	Record length (cm)	Proxies	References
P21	Żabie Oko Tatra Mts, Poland	49°12′07″	20°04′19″	1,390	481	Lithology, pollen, macrofossil, green-algae	Obidowicz (1996), Wołowski et al. (2002)
P22	Zielony Staw Gąsienicowy Tatra Mts, Poland	49°13′43″	20°00′33″	1,671	235	Lithology, pollen	Obidowicz (1996)
P23	Czarny Staw Gąsienicowy Tatra Mts, Poland	49°13′49″	20°01′07″	1,620	200	Lithology, pollen	Obidowicz (1996)
P24	Wyżna Pańszczycka Młaka Tatra Mts, Poland	49°13′	20°02′	1,345	550	Lithology, pollen, mosses	Obidowicz (1996)
P25	Siwe Sady Tatra Mts, Poland	49°12′38″	15°50′35″	1,542	70	Lithology, pollen	Obidowicz (1996)
P26	Krzemień Bieszczady Mts, Poland	49°05′	22°44′	1,300	97	Lithology, pollen, mosses	Ralska-Jasiewiczowa et al. (2006)
P27	Tarnica Bieszczady Mts, Poland	49°04′	22°43′	1,230	5	Pollen	Ralska-Jasiewiczowa et al. (2006)
P28	Szeroki Wierch Bieszczady Mts, Poland	49°05′	22°42′	1,170	11	Pollen	Ralska-Jasiewiczowa et al. (2006)
P29	Smerek Bieszczady Mts, Smerek I, II, III, Poland	49°11′	22°26′	600	290	Lithology, pollen, macrofossil	Ralska-Jasiewiczowa (1972, 1980, 1989)
P30	Klaklowo (near Stróża) Beskid Śląski Range, Poland	49°47′	19°54′	480	360	Lithology, pollen, LOI, geomorphology, granulometric indexes, grain size	Margielewski (2001)
P31	Drogomyśl Upper Vistula valley, Poland	49°52′	18°45′	ca. 300	60 hiatus	Lithology, pollen, macrofossil, geomorphology	Niedziałkowska et al. (1985)
P32	Piwniczna-Podolik Pieniny Mts, Poland	49°25′	20°43′	360	ca. 150	Lithology, mosses, mollusc, geomorphology	Alexandrowicz (1985)
P33	Harcygrund valley (near Czorsztyn) Pieniny Mts, Poland	49°26′	20°19′	520	20	Lithology, mollusc, geomorphology	Alexandrowicz (1984a, b)
P34	Jasło Bryły Jasło-Sanok Depression, Poland	49°45′41″	21°26′15″	230	ca. 120	Lithology, pollen, mollusc, geomorphology	Alexandrowicz et al. (1985), Mamakowa and Wójcik (1987), Alexandrowicz (1988)
P35	Sowliny (near Limanowa) Beskid Wyspowy Mts, Poland	49°43′26″	20°24′52″	410	20	Lithology, pollen, macrofossil, mosses	Środoń (1987), Alexandrowicz (1988)

Table 1 continued

Map code	Site name Details the locality of core, Country	Latitude	Longitude	Alt (m)	Record length (cm)	Proxies	References
P36	Toporowy Staw Wyżni Tatra Mts, Poland	49°16′	20°02″	1,024	600	Lithology, pollen, mosses	Granoszewski et al. (2008)
P37	Długi Staw Tatra Mts, Poland	49°22′73″	20°01′70″	1,784	Kajak samples	Cladocera	Marková et al. (2006)
P38	Przedni Staw Tatra Mts, Poland	49°12′	20°02	1,668	308	Lithology, pollen, diatom, cladocera, geochemistry	Marciniak (1986), Marciniak and Cieśla (1983), Szeroczyńska (1984), Szeroczyńska & Zawisza (2007)
S1	Vyšné Temnosmrečinské pleso Tatra Mts, Slovakia	49°11′20″	20°02′22″	1,725	30	Diatom, chironomid, LOI	Bitušík et al. (2009)
S1					Kajak sampler	Cladocera	Marková et al. (2006)
S2	Vyšné Wahlenbergovo pleso Tatra Mts, Slovakia	49°09′51″	20°01′37″	2,157	35	Diatom, chironomid, LOI	Bitušík et al. (this volume)
S3	Nižné Terianske pleso Tatra Mts, Slovakia	49°10′11″	20°00′51″	1,940	30	Diatom, geochemistry, chironomid, LOI, cysts, pigment, Spheroidal Carbonaceous Particle	Bitušík et al. (this volume)
S3					Kajak sampler	Cladocera	Marková et al. (2006)
S4	Ľadové pleso Tatra Mts, Slovakia	49°11′02″	20°09′46″	2,057	30	Cladocera	Marková et al. (2006), http://www.eurolimpacs.ucl.ac.uk
S5	Starolesnianske pleso Tatra Mts, Slovakia	49°11′48″	20°10′04″	1,988	16	Cladocera, chironomid	Stuchlík et al. (2002)
S6	Tlstá Hora Biele Karpaty, Slovakia	48°53′18″	17°53′16″	460	290	Pollen, macrofossil, mollusc	Rybníčková et al. (2005)
S7	Kubríková Západné Beskydy, Slovakia	49°29′03″	18°40′20″	790	100	Pollen, macrofossil	Rybníčková et al. (2005)
S8	Dolina Zlatného potoka Podbeskydská brázda, Slovakia	49°29′	19°16′	900	350	Pollen	Rybníček & Rybníčková (1985), Rybníčková & Rybníček (1996)
S9	Bobrov Oravská kotlina, Slovakia	49°27′	19°34′	640	230	Pollen	Rybníček & Rybníčková (1985), Rybníčková & Rybníček (1996)
S10	Trojrohé pleso Tatra Mts, Slovakia	49°14′	20°15′	1,650	210	Pollen	Hüttemann & Bortenschlager (1987)

Table 1 continued

Map code	Site name Details the locality of core, Country	Latitude	Longitude	Alt (m)	Record length (cm)	Proxies	References
S11	Regetovka Busov, Slovakia	49°25′30″	21°16′45″	515	950	Lithology, pollen, archaeology	Wacnik (1995), Wacnik et al. (2001, 2006)
S12	Kružlová Ondavská vrchovina, Slovakia	49°21′	21°34′	350	250	Lithology, pollen, archaeology	Wacnik (2001), Wacnik et al. (2001, 2006)
S13	Hozelec Popradská kotlina, Slovakia	49°02′03″	20°20′48″	685	160	Pollen, mollusc	Jankovská (1988, 1998)
S14	Šafárka Volovské vrchy, Slovakia	48°52′55″	20°34′30″	600	0–120	Lithology, pollen	Jankovská et al. (2002), Kuneš et al. (2008)
S15	Sivárňa Spišká Magura, Slovakia	49°19′19″	20°35′6″	630	190	Pollen, macrofossil	Jankovská (1998), Kuneš et al. (2008)
H1	Balaton (Tó-22) Balaton, Hungary	46°49′	17°31′	104	430 hiatus	Lithology, pollen, diatom	Cserny & Nagy-Bodor (2000), Cserny (2002), Buczkó et al. (2005)
H2	Balaton (Tó-30) Balaton, Hungary	46°44′40″	17°24′03″	104	420 hiatus	Lithology, pollen, diatom, O stabile isotope	Cserny (2002), Buczkó et al. (2005), EPD
H3	Balaton (Tó-31) Balaton, Hungary	46°43′	17°15′	104	600 hiatus	Lithology, pollen, diatom, O stabile isotope	Cserny & Nagy-Bodor (2000), Cserny (2002), Buczkó et al. (2005)
H4	Alsópáhok Balaton region, Hungary	46°45′	17°12′	104	565 hiatus	Lithology, pollen, archaeology	Juhász et al. (2007)
H5	Zalavári 1 Balaton region, Hungary	46°40′00.8″	17°13′06.2″	106	380 hiatus	Lithology, pollen, archaeology	Juhász et al. (2007)
H5	Zalavári 2 Balaton region, Hungary	46°40′00.8″	17°13′06.2″	106	165	Lithology, diatom, cladocera, geochemistry	Korponai et al. (2009)
H6	Keszthely-Úsztatómajor Balaton region, Hungary	46°45′46″	17°13′30″	110	480	Lithology, pollen, archaeology	Szántó & Medzihradszky (2004), Medzihradszky (2005)
H7	Fönyed Balaton region, Hungary	46°38′22″	17°14′40″	104	480	Lithology, pollen, archaeology	Szántó & Medzihradszky (2004), Medzihradszky (2005)
H8	Balatonederics Balaton region, Hungary	46°47′	17°24′	104	520 hiatus	Lithology, pollen, macrofossil, mollusc, archaeology	Jakab et al. (2005), Juhász et al. (2007)
H9	Baláta-tó Somogy Hills, Hungary	46°18′	17°12′	154	116	Lithology, pollen, macrofossil, mollusc, archaeology, charcoal	Sümegi et al. (2007b)

Table 1 continued

Map code	Site name Details the locality of core, Country	Latitude	Longitude	Alt (m)	Record length (cm)	Proxies	References
H10	Garancsi-tó Tinnye, Hungary	47°37′25″	18°48′26″	246	44	Pollen, diatom	Nagyné-Bodor et al. (1996)
H11	Mezőlak Devecser Plain, Hungary	47°21′	16°30′	123	395	Lithology, pollen, macrofossil, mollusc, archaeology, charcoal	Ilon et al. (2006), Sümegi et al. (2007c)
H12	Sárkeszi Sárrét, Mezőség Region, Hungary	47°09′	18°16′	104	460	Lithology, pollen, mollusc, charcoal	Sümegi et al. (2007a)
H13	Nádasladány Sárrét, Mezőség Region, Hungary	47°09′18″	18°13′05″	100	390	Lithology, pollen, mollusc, archaeology	Sümegi (2003), Willis (1997)
H14	Nádas-tó Nagybárkány, N Hungarian Mid-Mts, Hungary	47°59′	19°40′	360	340 hiatus	Lithology, pollen, macrofossil, geochemistry, archaeology	Jakab et al. (2009)
H15	Kismohos Kelemér, N Hungarian Mid-Mts, Hungary	48°24′40″	20°24′30″	306	886	Lithology, pollen, geochemistry archaeology	Willis et al. (1997, 1998), Somogyi et al. (1998), Szalóki et al. (1999), Braun et al. (2005), Juhász (2005a)
H16	Nagymohos (1) Kelemér, N Hungarian Mid-Mts, Hungary	48°20′20.9″	20°25′30″	297	430 hiatus	Lithology, pollen, macrofossil, geochemistry, archaeology	Magyari et al. (1999, 2000, 2001), Sümegi et al. (2008), Zólyomi & Járainé Komlódi (2008)
H17	Nagy-mohos Kállósemjén, Nyírség, Hungary	47°52′8″	21°54′10″	123	200 hiatus	Lithology, pollen, macrofossil, mollusc, geochemistry	Braun et al. (1992)
H18	Nyíres-tó Csaroda, Bereg Plain, Hungary	48°10′61″	22°29′21″	107	420	Lithology, pollen, geochemistry, archaeology charcoal	Harrington (1995), Sümegi (1999), Juhász (2005b)
H19	Nyírjes-tó Sirok, N Hungarian Mid-Mts, Hungary	47°55′48.6″	20°11′08.4″	200	450	Lithology, pollen, geochemistry, archaeology charcoal	Gardner (2002, 2005)
H20	Meggyeserdő Jászberény, Jászság, Hungary	47°27′21″	19°59′24″	90	300	Lithology, pollen, mollusc, archaeology, sedimentology	Sümegi (2005b)
H21	Zám-Halasfenék Hortobágy, Hungary	47°30′	21°02′	83	1000	Lithology, pollen, mollusc, sedimentology	Félegyházi & Tóth (2003), Sümegi et al. (2005b)
H22	Tövises channel Pocsaj, Érmellék, Hungary	47°18′25″	21°51′14″	98	300	Lithology, pollen, mollusc, archaeology, sedimentology	Sümegi (2005a)

Table 1 continued

Map code	Site name Details the locality of core, Country	Latitude	Longitude	Alt (m)	Record length (cm)	Proxies	References
H23	Báb-tava Csaroda, Bereg Plain, Hungary	48°11′09″	22°28′59″	107	510	Lithology, pollen, macrofossil, geochemistry, archaeology, LOI	Magyari et al. (2008)
H24	Sarló-hát, Tiszagyulaháza SH-II, Middle Tisza Plain, Hungary	47°58′00″	21°10′09″	86	773	Lithology, pollen, geochemistry, archaeology, LOI	Magyari et al. (in press)
H24	Sarló-hát, Tiszagyulaháza SH-WOOD, Middle Tisza Plain, Hungary	47°58′01″	21°10′09″	86	698	Lithology, pollen, geochemistry, archaeology, LOI	Magyari et al. (in press)
H25	Fehér-tó Kardoskút, Hungarian Plain, Hungary	46°28″18″	21°09′37″	83	640	Lithology, pollen, mollusc, geochemistry, sedimentology	Sümegi et al. (1999), Sümegi (2005c)
H26	Bátorliget, littoral profile Nyírség, Hungary	47°46′	22°16′	130	150	Lithology, macrofossil, mollusc, geochemistry archaeology, O stabil isotop	Sümegi & Gulyás (2004)
H26	Bátorliget, central borehole Nyírség, Hungary	47°46′05″	22°16′04″	127	180 hiatus	Lithology, pollen, mollusc, geochemistry, archaeology, sedimentology	Willis et al. (1995), Sümegi & Gulyás (2004)
H27	Vörös-mocsár (HP-I) Császártöltés, Hungary	46°23′17″	19°09′16″	107	420	Lithology, pollen, macrofossil	Jakab et al. (2004a, b)
H27	Vörös-mocsár (CST-II) Császártöltés, Hungary	46°23′17″	19°09′16″	107	290	Lithology, macrofossil, mollusc	Jakab et al. (2004a, b)
H27	Vörös mocsár (CS-4) Császártöltés, Hajós, Hungary	46°23′38″	19°11′27″	91	390	Lithology, pollen,	Cserny (2000), EPD
H28	Tiszacsermely Bodrogköz, Hungary	48°13′	21°48′	95	800 hiatus	Lithology, pollen, sedimentology	Borsy et al. (1989)
H29	Csólyospályos Duna-Tisza Interfluve, Hungary	46°21′	19°48′	108	160	Lithology, pollen, mollusc	Sümegi et al. (2005b)
H30	Szőce Vasi Hegyhát, Hungary	46°54′	16°34′	240	120	Lithology, pollen	Cserny & Nagy-Bodor (2005)
H31	Farkasfa Vasi Hegyhát, Hungary	46°54′	16°18′	265	240	Lithology, pollen	Cserny & Nagy-Bodor (2005)
R1	Turbuta Transylvanian Basin, Romania	47°15′26.5″	23°18′42.9″	275	190 hiatus	Lithology, pollen, geochemistry, LOI	Feurdean et al. (2007b)

Table 1 continued

Map code	Site name Details the locality of core, Country	Latitude	Longitude	Alt (m)	Record length (cm)	Proxies	References
R2	Preluca Tiganului Gutaiului Mts, Romania	47°48′83″	23°31′54.6″	730	950 hiatus	Lithology, pollen, macrofossil, petrographic analysis, SIRM (saturated remnant isothermal magnetization analysis)	Wohlfarth et al. (2001), Björkman et al. (2002), Feurdean and Bennike (2004), Feurdean (2005a, b), Feurdean and Astalos (2005), Feurdean et al. (2007a, 2008a, b)
R3	Steregoiu Gutaiului Mts, Romania	47°48′48″	23°32′41″	790	592	Lithology, pollen, macrofossil	Björkman et al. (2002), Fărcaş et al. (2003, 2006a, b), Feurdean and Bennike (2004, 2008), Feurdean and Astalos (2005), Feurdean et al. (2007a, 2008a, b)
R4	Lezerul Caliman Caliman Mts, Romania	47°5′36.70″	25°15′41.47″	1,650	500	Lithology, pollen	Fărcaş et al. (1999, 2003, 2006a, b), Feurdean et al. (2007a)
R5	Taul dintre Brazi Retezat Mts, Romania	45°23′47″	22°54′06″	1,740	490	Lithology, pollen, macrofossil, diatom	Buczkó et al. (2009), Magyari et al. (2009a)
R6	Ic Ponor I Apuseni Mts, Romania	46°37′46″	22°48′24″	1,040	295 hiatus	Lithology, pollen, archaeology	Bodnariuc et al. (2002), Jalut et al. (2003) Fărcaş et al. (2003, 2006a, b)
R6	Ic Ponor II Apuseni Mts, Romania	46°37′46″	22°48′24″	1,040	165 hiatus	Lithology, pollen, archaeology	Bodnariuc et al. (2002), Jalut et al. (2003), Fărcaş et al. (2006a, b)
R7	Padis Apuseni Mts, Romania	46°35′53.2″	22°43′58.4″	1290	90	Lithology, pollen, archaeology	Bodnariuc et al. (2002), Jalut et al. (2003), Fărcaş et al. (2006a, b)
R7	Padis Plateau Apuseni Mts, Romania	46°35′53.2″	23°43′58.4″	1,290	76	Lithology, pollen, charcoal	Feurdean & Willis (2008b)
R8	Bergerie Apuseni Mts, Romania	46°37′23.1″	22°40′56.4″		230	Lithology, pollen, archaeology	Bodnariuc et al. (2002), Jalut et al. (2003), Fărcaş et al. (2006a, b)
R9	Capatana Apuseni Mts, Romania	46°30′20.88″	23° 9′5.53″	1,220	535	Lithology, pollen, testate amoebae	Fărcaş et al. (2003, 2005, 2006a, b)
R10	Molhasul Mare Apuseni Mts, Romania	46°35′24″	22°45′51″	1,224	570	Lithology, pollen, charcoal	Feurdean & Willis (2008a, b)
R11	Calineasa Apuseni Mts, Romania	46°33′47.43″	22°48′59.51″	1,360	224	Lithology, pollen, charcoal	Feurdean & Willis (2008b)
R12	Pietrele Onachii Apuseni Mts, Romania	46°38′33″	22°50′43″	1,055	185	Lithology, pollen, charcoal	Feurdean & Willis (2008b)

Table 1 continued

Map code	Site name Details the locality of core, Country	Latitude	Longitude	Alt (m)	Record length (cm)	Proxies	References
R13	Taul Zanogutii Retezat Mts, Romania	45°20′2.46″	22°48′12.53″	1,840	480 (565) hiatus	Lithology, pollen, diatom	Péterfi (1974), Fărcaş et al. (1999), Fărcaş et al. (2003, 2006a, b), Feurdean et al. (2007a)
R14	Semenic Banat Mts, Romania	45°08′23″?	21°59′00″	1,600		Lithology, pollen, macrofossil	Rösch and Fisher (2000), Fărcaş et al. (2003, 2006a, b)
R15	Pesteana Poiana Rusca Mts, Romania	45°32′36.26″	22°48′22.91″	480	510 hiatus	Lithology, pollen, testate amoebae	Fărcaş et al. (2006a, b)
R16	Varatec Lapusului Mts, Romania	47°40′	24°02′	1,340	400	Lithology, geochemistry, testate amoebae, humification	Schnithen et al. (2003, 2004, 2006)
R17	Saint Ana Harghitei Mts, Romania	46°07′30″	25°53′14″	950	420	Lithology, pollen, macrofossil, diatom, cladocera, geochemistry, testate amoebae, cysts	Magyari et al. (2006), Magyari et al. (2009a)
R18	Avrig-1 Fagaras basin, Romania	45°43′	24°23′	400	800	Lithology, pollen	Tanţău et al. (2006), Feurdean et al. (2007a)
R19	Bisoca Curbura, Romania	45°32′0.94″	26°41′58.65″	875	700 hiatus	Lithology, pollen	Tanţău et al. (2003d), Tanţău (2006), Fărcaş et al. (2006a, b)
R20	Mohos (M1) Harghitei Mts, Romania	46° 8′0.52″	25°54′13.26″	1,050	1,065	Lithology, pollen	Tanţău et al. (2003a, b), Tanţău and Fărcaş (2005), Tanţău (2006), Fărcaş et al. (2003, 2006a, b), Feurdean et al. (2007a)
R21	Luci Harghitei Mts, Romania	46°17′57.59″	25°43′13.55″	1,079	750	Lithology, pollen	Tanţău et al. (2003c), Tanţău and Fărcaş (2005), Tanţău (2006), Fărcaş et al. (2006a, b)
R22	Poiana Stiol 3 Rodnei Mts, Romania	47°35′60.00″	24°48′0.00″	1,540	315	Lithology, pollen	Tanţău and Fărcaş (2004) Tanţău (2006), Fărcaş et al. (2006a, b)

Table 2 Late Quaternary palaeolimnological records in the Carpathian region, east-central Europe

Locality and type of the core	Dating details	Estimated age cal year BP	Published deepest radiocarbon data (lab-code; sample depth; uncal age (^{14}C year BP)	Calibrated deepest data (cal year BP)
Královec/peat	2	H: 0–1,300	Gd 15176, 100 cm, 1,040 ± 110	835–1,090[a]
Horní Lomná/peat	3	H: 0–5,700	Gd 15687, 85–87 cm, 5,130 ± 95	5,763–5,990[a]
Machová/peat	1	H: 0–1,300	Gd 12647, 71–72 cm, 890 ± 100	734–909[a]
Vracov/former lake	6	LG-H: 0–15,000	Bln-1004, 264–269 cm, 11,933 ± 250 year	13,603–14,390[a]
Svatobořice-Mistřín/peat	5 bulk	H: 0–8,000	K-4229, 215–216 cm, 6,620 ± 75 year	7,459–7,567[a]
Jablunka/peat	2 AMS	>45,000 Weichselian/ Holocene		46,346–50,212[a]
Puścizna Rękowiańska/peat bog	8	H: 0–10,200	Gd-1500; 710–715 cm, 8,960 ± 80	9,942–10,190[a]
Cergowa Góra/peat bog	12	H: 0–5,500	?; 545–550 cm; 4,820 ± 70	5,475–5,619[a]
Jasło/peat bog	2	More than 12,000 Late Glacial	GD-3244; 301–304 cm, 11,890 ± 90	13,636–13,967[a]
Jasiel/peat bog	6	LG-H: 0–12,000	Gd-1846; 214–216 cm; 10,340 ± 110	11,951–12,474[a]
Roztoki/peat bog	2	LG-H: 8,200–13,700	Gd-12051; 572–575 cm; 11,740 ± 150	13,434–13,846[a]
	4	LG-H: 6,400–15,500	Gd-766; 390–395 cm; 9,920 ± 95	11,275–11,597[a]
Tarnawa Wyżna/peat bog	8 bulk	LG-H: 0–13,200	?; 680–682 cm; 11,360 ± 170	13,076–13,468[a]
Tarnowiec/peat bog	9	LG-H: 0–13,000	Gd-967; 198–203 cm; 11,190 ± 140	12,915–13,249[a]
Szymbark/peat bog	5	H: 2,500–8,900	Lv-661; 415–425 cm; 8,210 ± 150	8,974–9,368[a]
Żurawica/peat bog	21	H: ca. 1,000–6,500	Ki-10797; 543–550 cm; 5,820 ± 60	6,480–6,753
Osieczkowa/peat bog	9	LG-H: 0–12,000	Ki-11844;142–150 cm; 10,290 ± 150	11,594–12,658
Bogdanówka-Beło/peat bog	24	LG-H: ca. 0–11,000	Ki-8540; 493–495 cm; 9,480 ± 150	10,384–11,199
Zięby/peat bog	7	H: 300–6,800	Ki-11124; 101–105 cm; 5,980 ± 80	6,636–7,018
Pękalówka/peat bog	6	H: 0–10,600	Ki-11937; 112–114 cm; 9,430 ± 100	10,402–11,102
Siódmowo/peat bog	8	LGH: 0–13,000	Ki-11137; 140–147 cm; 11,530 ± 90	13,219–13,606
Mirkowo/peat bog	11	H: 500–6,800	Ki-11143; 230–232 cm; 5,980 ± 80	6,636–7,018
Pcim Krzywica/peat bog	2	H: 800–2,700	Ki 11938; 144–146 cm; 2,630 ± 60	2,690–2,869
Kamiennik/peat bog	16	H: 1,000–7,000	Ki-10146; 423–431 cm; 6,750 ± 60	5,441–5,593
Pilsko/peat bog	3	H: ca. 0–8,000	?; ?; 6,920 ± 60	7,655–7,870
Kotoń/peat bog	14	LG-H: ca. 2,000–14,000	Ki-7905; 418–425 cm; 12,140 ± 70	13,821–14,153
Hajduki/peat bog	7	LG-H: ca. 2,000–16,500	Ki-10048; 237–250 cm; 13,900 ± 90	16,152–16,972
Żabie Oko/peat bog	1	LG-H: 0–12,500	Gd-2799; ?; 8,330 ± 120	9,070–9,525
Zielony Staw Gąsienicowy/ lake	5	LG-H: 0–11,700	Ua-1446; ?; 10,040 ± 150	11,197–12,124
Czarny Staw Gąsienicowy/ lake	5	LG-H: 0–14,500	Gd-4540; ?; 12,550 ± 420	13,611–15,902
Wyżna Pańszczycka Młaka/ peat bog	3	H: 0–5,500	Gd-2323; ?; 4,570 ± 100	4,959–5,481
Siwe Sady/peat bog	3	H: 1,000–9,000	VRI-1271; ?; 4,940 ± 60	5,586–5,761

Table 2 continued

Locality and type of the core	Dating details	Estimated age cal year BP	Published deepest radiocarbon data (lab-code; sample depth; uncal age (^{14}C year BP)	Calibrated deepest data (cal year BP)
Krzemień/peat bog	2	H: 0–1,100	Poz-493; 94–97 cm; 1,090 ± 30	937–1,057
Tarnica/peat monilith	1	Modern	Poz-495; 18–22 cm; 104.6 ± 0.1 pMC	
Szeroki Wierch/peat monolith	1	Modern	Poz-494; 95–100 cm; 102.1 ± 0 pMC	
Smerek/peat bog	4	LG-H: 0–20,000	16,925 ± 325	19,439–20,824
Klaklowo/peat bog	14	LG-H: 0–15,500	Ki-8857; 350 cm; 13,150 ± 190	15,031–16,188
Drogomyśl/buried peat layer	7	LG-H: 0–13,000	Lv-966; 43–45 cm; 11,220 ± 100	12,929–13,272
Piwniczna-Podolik/ landslide	3	Modern 0–150		
Harcygrund valley/ landslide	1	H: 8,500 BP	Gd-953; ?; 7750 ± 130	8,347–8,817
Jasło Bryły/palaeochannel	3	34,000–older than 45,000 Hengelo interstadial	Older than 45,000	
Sowliny/peat layer	1	29,650 BP Denecamp interstadial	Gd-1880; ?; 29,650 ± 650	
Toporowy Staw Wyżni/lake	2	H: 0–10,000 BP	Gd-19031; 347.5–350 cm; 9,190 ± 210	7,788–8,932
Długi Staw/glacial lake	?		?	1,600
Przedni Staw/glacial lake	1	LG-H: 0–14,000	?, 200 cm, 9900 ± 120	11,095–11,827
Vyšné Temnosmrečinské pleso/glacial lake	^{210}Pb, ^{137}Cs	Holocene	29.5 cm, 957 ± 207	1,640
Vyšné Wahlenbergovo pleso/glacial lake	^{210}Pb, ^{137}Cs	Holocene	34.25 cm, 377 ± 105	317–500
Nižné Terianske pleso/ glacial lake	^{210}Pb, ^{137}Cs	Holocene	~200 ± ?	1,440 (rough estimate)
Ľadové pleso/glacial lake	^{210}Pb, ^{137}Cs	Holocene		1,680
Starolesnianske pleso/ glacial lake	^{210}Pb, ^{137}Cs	Holocene		
Tlstá Hora/peat profile	4	H: 0–4,300	Gd 12389, 265–267 cm 3,920 ± 70	4,255–4,451
Kubríková/peat profile	1	H: 0–2,200	Gd 16297, 82–84 cm 1,730 ± 210	1,442–1,898
Dolina Zlatného potoka/ peat profile	5	H: 0–8,200	Lu-2410, 330–335 cm, 7,450 ± 130	8,119–8,373
Bobrov/peat profile	13	H: 0–10,800	Lu-2219; 209–213 cm; 10,150 ± 90	11,531–11,997[a]
Trojrohé pleso/glacial lake	5	H: 0–6,800	?; 215 cm, 6050 + 110	6,783–7,088
Regetovka/peat bog	5	H: 0–8,000	Gd-11229; 725–735 cm; 6,720 + 80	7,519–7,650[a]
Kružlová/peat bog	4	H: 0–3,000	Gd-10968; 235–240 cm; 2,810 ± 80	2,759–3,084
Hozelec/mire	3	LG-H: 0–13,000	VRI-796; 146–148 cm; 11,010 ± 160	12,797–13,103[a]
Šafárka/fossile doline	4 + 2AMS	16,000–older than 52,000 Weichselian/ Holocene	ERL-4532, or 4533, 90–115 cm, older than 52,000	
Sivárňa/mire	5	LG-H: 0–14,000	13,670 ± 130	16,385–17,021

Table 2 continued

Locality and type of the core	Dating details	Estimated age cal year BP	Published deepest radiocarbon data (lab-code; sample depth; uncal age (^{14}C year BP)	Calibrated deepest data (cal year BP)
Balaton (Tó-22)/shallow lake	4	LG-H: ca. 0–16,000	350–360 cm, 12,500 ± 300	14,223–15,354
Balaton (Tó-30)/shallow lake	1	LG-H: ca. 0–14,000	390–400 cm, 10,960 ± 300	12,550–13,187
Balaton (Tó-31)/shallow lake	3	LG-H: ca. 0–17,000	360–374 cm, 12,020 ± 300	13,666–14,640
Alsópáhok/infilled peatland	4	LG-H: 0–11,000	Gif256, 540–550 cm, 10,730 ± 110	13,180–12,600
Zalavári 1/shallow lake	8	H: 1,000–11,500	Ly-11221, 350–355 cm, 9,550 ± 60	11,200–10,600
Zalavári 2/shallow lake	3	H: 0–9,000	Deb-13994; 140–148 cm; 7,935 ± 60	8,795–8,650
Keszthely-Úsztatómajor/ mire	11	H: 2,000-12,000	Deb-5060, 570–580 cm, 10,835 ± 170	13,150–12,550
Fönyed/shallow lake	12	H: 3,000–8,000	Deb-7727, 220–230 cm, 7,375 ± 70	8,150–7,950
Balatonederics/infilled peatland	6	LG-H: 4,480–17,500	509 cm 14,260 ± 150	17,194–17,738[a]
Baláta-tó/shallow lake in peatland	5 AMS	H: 0–2,700	Poz-7991, 101–105 cm, 2,425 ± 30	2,710–2,350
Garancsi-tó/shallow lake	137Cs	H: ~100 year		
Mezőlak/peat monolith	10	LG-H: 0–15,000	Poz-8093, 362 cm; 12,140 + 140	13,750–15,450
Sárkeszi/lake and peatland	6	LG-H: ca. 0–17,000	Poz-7975, 364–360 cm, 10,000 ± 50	11,950–11,800
Nádasladány/lake and peatland	10	LG-H: ca. 7000–15,000 cal BP)	AA-12394, 336 cm, 11,685 ± 90	13,405–13,711[a]
Nádas-tó/peat bog	7	LG-H: ca. 0–15,000 cal BP	Beta-224851, 325 cm, 12,543 ± 189	15,259–14,407
Kismohos/Sphagnum bog	13	LG-H: 0–14,600 cal BP	AA-11987, 868 cm, 12,495 ± 95	14,473–15,125[a]
Nagymohos (1)/Sphagnum bog	13	0–25,300 cal BP Full Glacial/ Holocene	422–426 cm, 21,756 ± 267	25,405–26,615[a]
Nagy-mohos/mire	2 bulk	H: ca. 0–9,000 cal BP	120 cm, 8,010 ± 100	8,716–9,010[a]
Nyíres-tó/infilled oxbow lake, Sphagnum bog	4	H: ca. 0–11,000 cal BP	NERC Radiocarbon Laboratory, East Kilbride, 279 cm, 7,670 ± 60	8,590–8,370
Nyírjes-tó/peat bog	5 bulk	H: ca. 0–10,000 cal BP	AA-27185, 394 cm wood, 5,805 ± 55	6,742–6,469
Meggyeserdő/ palaeochannel	4	LG-H: more than 10,000	Deb-5945, 150–140 cm, 9,055 ± 70	10,166–10,284[a]
Zám-Halasfenék/infilled riverbed	2	More than 35,000 UpperWeichselian/ Holocene	900–910 cm, >35,000	
Tövises channel/ Palaeochannel marsh	2	H: ca. 0–8,500 cal BP	135–140 cm, 3,300 ± 150	3,388–3734[a]
Báb-tava/infilled oxbow lake	4 bulk	H: 0–7,800 cal BP	Deb-6542, 500–505 cm, 6,955 ± 70	7,832–7,701
Sarló-hát/palaeochannel fen	8 AMS	H: 0–11,400	742 cm, 12,434 ± 70	(reservoir effect detected, age not used)

Table 2 continued

Locality and type of the core	Dating details	Estimated age cal year BP	Published deepest radiocarbon data (lab-code; sample depth; uncal age (^{14}C year BP)	Calibrated deepest data (cal year BP)
Sarló-hát/palaeochannel fen	14 AMS	H: 0–11,400 cal BP	612 cm, 9,400 ± 50	10,750–10,490
Fehér-tó/alkaline lake, palaeochannel	3 + 2 bulk	0–28,000 cal BP Upper Weichselian/ Holocene	ATOMKI, 620–630 cm, 23 303 ± 280	27,582–28,418
Bátorliget, littoral profile/ marsh	7	LG-H: 0–17,000	Deb-4374; 95–100 cm; 10,306 ± 122	12,649
Bátorliget, central borehole/ marsh	8	LG-H: 0–18,000	Deb-2658; 96–104 cm; 10,571 ± 110	12,259–12,666[a]
Vörös-mocsár (HP-I)/marsh	5 bulk	LG-H: 0–13,000	Deb?, 291–296 cm, 9,130 ± 130	10,180–10,460
Vörös-mocsár (CST-II)/ marsh	3 bulk	LG-H: ca. 0–14,000	260–290 cm, 11,407 ± 85	13,158—13,457[a]
Vörös mocsár (CS-4)/marsh	5	LG-H: 0–14,000	Deb-3930; 390 cm; 11,967 ± 100	13,709–14,175[a]
Tiszacsermely/ palaeochannel fen	3	LG-H: 0–>16,000	660–705 cm; >16,000	ca. 18,920–19,375[a]
Csólyospályos/infilled lake	6 bulk	LG-H: ca. 1,000–13,000	Deb-3286, 105–115 cm, 10,118 ± 81	12,350–11,300
Szőce/peat bog	2 bulk	H: ca. 0–1,500	Deb-5364, 0.70–0.80 cm, 743 ± 65	649–737[a]
Farkasfa/peat bog	1	H: <1500	?, 150–160 cm, 743 ± 50	667–725[a]
Turbuta/palaeolake	3 AMS + 11 U/Th	LG-H: 5000–13,100	Poz-16415; 163–164 cm; 10,300 ± 50	11,972–12,187
Preluca Tiganului/crater lake	14 AMS 14C	LG-H: 5000–14,400	Ua-16331; 940 cm; 12250 ± 105	15,450–14,550
Taul dintre Brazi/glacial lake	9 AMS	LG-H: 0–13,4700	Poz-27305; 578 cm; 11,590 ± 60	13,353–13,586[a]
Steregoiu/crater lake	17 AMS 14C	LG-H: 0–>14,700	Ua-16323; 5.324–5.274 m; 12,365 ± 115	15,450–14,050
Lezerul Caliman/lake	9 AMS	LG-H: 0–>17,730	450 cm; 14,800 ± 1100	16,371–19,101[a]
Ic Ponor I/peat bog	4 AMS	H: ca. 0–10,100	GifA-99669; 292–295 cm; 8,990 ± 80	10,268–9888
Ic Ponor II/peat bog	3 AMS	H: ca. 0–9900	GifA-100147; 160 cm; 8,770 ± 90	9,969–9,550
Padis/peat bog	4 AMS	H: ca. 0–5,300	Gif-11129; 85–95 cm; 4,595 ± 65	5,470–4,046
Padis Plateau/infilled dolina	3 AMS	H: 0–6,000	UBA 8511; 57–58 cm; 3,930 ± 33	4,445–4,282
Bergerie/peat bog	6 AMS	H: ca. 0–7,900	GifA-99476; 7,010 ± 182	8,177–7,562
Capatana/peat bog	7 AMS	H: ca. 0–7,000	520 cm; 5,980 ± 120	6,690–6,990[a]
Calineasa/overgrown dolina	3 AMS	H: 0–5,000	UBA-8508; 215–216 cm; 3,417 ± 36	3,700–3,650
Taul Zanogutii/glacial lake	11 AMS	LG-H: 0–14,800	465 cm; 11,140 ± 75	12,904–13,179[a]
Semenic/peat bog	4 AMS	H: 0–>7,620	146–148 cm; 6,760 ± 60	7,581–7,662[a]
Pesteana/lake	6 AMS	LG-H: ca. 0–17,000	Poz-16842; 504–506 cm; 13,670 ± 70	16,752–17,092[a]
Varatec/peat bog	5 bulk	H: 0–9,790	Deb 6805; 331–333 cm; 7,835 ± 60	8,431–8,932
Saint Ana/crater lake	7 AMS	H: 800–9,000	Poz-9976; 412–414 cm; 8,050 ± 50	8,814–9014
Avrig-1/peat bog	13 AMS 14C	LG-H: 0–17,500	Poz-741; 719–721 cm; 13,880 ± 90	16,914–17,338[a]
Bisoca/peat bog	12 AMS	H: ca. 0–12,000	670 cm; 9,680 ± 60	11,041–11,200
Mohos (M1)/peat bog	16 AMS	LG-H: 0–13,890	Gds 10615; 1,000 cm; 9,750 ± 200	10,826–11,505
Luci/peat bog	11 AMS	LG-H: 0–14,900	710 cm; 11,850 ± 60	13,481–14,100

Table 2 continued

Locality and type of the core	Dating details	Estimated age cal year BP	Published deepest radiocarbon data (lab-code; sample depth; uncal age (^{14}C year BP)	Calibrated deepest data (cal year BP)
Poiana Stiol 3/peat bog	9 AMS	H: ca. 0–11,000	290 cm; 9,220 ± 45	10,239–10,501

Details of radiocarbon dating and age ranges of the sediment sequences

H Holocene, *LG* Late Glacial, *?* Unknown data

[a] Age ranges were calibrated using CalPal-2007[Online] program on published, uncal age

2003, 2006a, b; Wohlfarth et al., 2001; Björkman et al., 2002, 2003; Tanţău et al., 2003a, b, 2006; Feurdean & Bennike, 2004), and even less studies have placed major emphasis on limnological or mire-hydrological changes (e.g. Marciniak & Cieśla, 1983; Szeroczyńska, 1984; Obidowicz, 1996; Šporka et al., 2002; Stuchlík et al., 2002; Jakab et al. 2004a, b, 2005; Schnitchen et al., 2006; Magyari et al., 2006, 2009a; Bitušík et al., 2009). Emphasis in this region has traditionally been on pollen analysis (e.g. Pop, 1932, 1960; Zólyomi, 1952; Járai-Komlódi, 1968, 1991; Ralska-Jasiewiczowa, 1980; Jankovská, 1988, 1998; Harmata, 1995; Obidowicz, 1996; Szczepanek, 2001a; Jankovská et al., 2002), and to a lesser extent on plant macrofossil analysis (Rybníčková & Rybníček, 1972, 1988; Rybníčková et al., 2005). Although the application of a classical palaeolimnological proxy, siliceous algae analysis, has long been attempted on Carpathian lake deposits, and on shallow lake deposits in the Carpathian Basin (Pantocsek, 1913; Péterfi, 1974; Marciniak, 1986), these were only occasionally interpreted together with other proxies to provide concise Late Glacial and Holocene histories of climate-induced limnological (water-depth, productivity, salinity, pH) changes (e.g. Lake Balaton: Cserny, 2002; Cserny & Nagy-Bodor, 2000). The number of more recent, multi-proxy studies is also limited, despite the richness of glacial lakes in certain mountain ranges (e.g. Tatra and Retezat Mts) offering a wide-range of palaeolimnological techniques to be applied on them. Chironomids, cladocerans and testate amoebae are abundant in these deposits, but only recently have become part of multi-proxy studies in this region (Šporka et al., 2002; Marková et al., 2006; Schnitchen et al., 2006; Bitušík et al., 2009; Korponai et al., 2009). As the main objective of the projects, in scope of which most of these recent multi-proxi palaeolimnological analyses were done (AL:PE, MOLAR and EMERGE; http://www.mountain-lakes.org/), was to assess the impact of the industrial revolution on mountain lakes, we have increasing data on Late Holocene human-induced limnological changes, but continuous Late Glacial and Holocene multi-proxy records are rare and very much needed.

The database implies that from the palaeolimnological point of view, lakes and peat bogs are seriously under-investigated in the East and South Carpathians. This is most likely due to the lack of tradition in palaeolimnological research in this region, and hence the lack of expertise. Dated Late Glacial and Holocene pollen records are, however, probably the most numerous, or most progressively increasing in this region, and sometimes with detailed interpretation of the wetland vegetation changes (Feurdean & Bennike, 2004). Given the similar characteristics of the East and South Carpathian lakes and peat bogs to the Western Carpathians, multi-proxy palaeolimnological research has a great potential in this region. One promising initiative is the CARPENVCHANGE project that has targeted multi-proxy analyses of glacial lakes in the Retezat Mts. First results from this project suggest that glacial retreat and lake formation started around 15–16 kyr cal year BP (Buczkó et al., 2008; Magyari et al., 2009a, b); thus, ecosystem responses to the Late Glacial climate fluctuations can be studied with high resolution.

The database displays several multi-proxy records from the Pannonian Plain, many of them with multiple radiocarbon dates; however, the number of continuous records is low. One drawback of the detailed and ongoing study of Lake Balaton is the calcareous nature of its deposits (Cserny & Nagy-Bodor, 2000; Cserny, 2002; Tullner & Cserny, 2003). Being the largest in east-central Europe, this lake has a great potential to detect Late Glacial and Holocene climate-induced limnological changes with regional significance (Tullner & Cserny, 2003, 2004). Plant macrofossils were, however, only recovered and radiocarbon dated from the basal peat deposits; consequently, the upper parts of the sediment sequences remained undated. Progress to solve this problem has been made recently by dating and analyzing lakeshore peat sequences (Szántó & Medzihradszky, 2004; Jakab et al., 2005; Medzihradszky, 2005) that will allow cross-correlation of the pollen records and thereby dating of the lake sediment records hopefully in the near future (e.g. oxygen isotope, diatoms, LOI).

Another problem with several multi-proxy records from this region is the application of bulk sediment ^{14}C dating. Given that most of the Pannonian Plain is covered by calcareous deposits (Rónai, 1985), this method can lead to reservoir ages; therefore, sediment chronologies must be treated with caution.

Available objects for palaeolimnological research are palaeochannel deposits (e.g. Tövises-ér, Sarló-hát, Tiszacsermely), inter-dune lakes in sandy areas (e.g. Bátorliget, Baláta-tó), or peat bogs in landslide basins and other tectonic lakes in the hill zone (e.g. Kismohos, Sirok). In general, these deposits do not or only poorly preserve biological proxies, such as chironomids, cladocera and often even diatoms. This means that the repository of research methods is limited in comparison with the Carpathian lakes; furthermore, the observed limnological changes are not directly related to climatic changes (see e.g. Magyari et al., in press). Overall, the basin provides plenty of palaeochannel sites, but limited number of lakes that accordingly has to be appreciated and studied more thoroughly in the future (e.g. Lake Balaton, Lake Velencei, Lake Fertő).

Acknowledgements We are grateful to Gábor Umann for plotting Fig. 1 and harmonizing the geographical coordinates of the presented palaeorecords, and Miklós Bálint for checking the Romanian localities. We thank the support of the Bolyai János Research Scholarship and the Hungarian Scientific Fund (OTKA F026036) under Hungarian Natural History Museum Palaeo Contribution No. 78, Hungarian Academy of Sciences. We also thank Vasil Pokynčereda for providing data on lakes of glacial origin in the Ukrainian part of the Carpathians.

References

Alexandrowicz, S. W., 1984a. Zespoły malakofauny w czwartorzędowych osadach polskich Karpat [Malacofaunal assemblages in the Quaternary deposits of the Polish Carpathians.] (in Polish). Biuletyn Instytutu Geologicznego 346: 187–205.

Alexandrowicz, S. W., 1984b. Mid-Holocene malacofauna from the Harcygrund near Czorsztyn (Pieniny Klippen Belt, Carpathians) [in Polish]. Studia Geologica Polonica 83: 95–114.

Alexandrowicz, S. W., 1985. Subfossil malacofauna from the landslide in Piwniczna (Polish Flysch Carpathians). Folia Quaternaria 56: 79–100.

Alexandrowicz, S. W., 1988. Malacofauna of Late Quaternary loess-like deposits in the Polish Carpathians. Acta Geologica Polonica 38: 85–123.

Alexandrowicz, S. W., K. Mamakowa & A. Wójcik, 1985. Jasło-Bryły; Stratygraphy of alluvial and deluvial deposits of the North Polish Glaciation. Carpatho-Balkan Geological Association XIII Congress, Cracow, Guide to excursion 5: 111–113.

Alley, R. B., P. A. Mayewski, T. Sowers, M. Stuiver, K. C. Taylor & P. U. Clark, 1997. Holocene climatic instability: a prominent widespread event 8200 yr ago. Geology 25: 438–486.

Appleby, P. G. & G. T. Piliposian, 2006. Radiometric dating of sediment records from mountain lakes in the Tatra Mountains. Biologia, Bratislava 61: 51–64.

Battarbee, R. W., D. Morley, H. Bennion & G. L. Simpson, 2007. A meta-database for recent paleolimnological studies. PAGES News 15: 23–24.

Battarbee, R. W., D. Morley, H. Bennion, G. L. Simpson, M. Hughes & V. Bauere, 2009. A palaeolimnological meta-database for assessing the ecological status of lakes. Journal of Paleolimnology (in print).

Berglund, B. E., H. J. B. Birks, M. Ralska-Jasiewiczowa & H. E.Wright (eds), 1986. Palaeoecological Events During the Last 15 000 Years – Regional Syntheses of Palaeoecological Studies of Lakes and Mires in Europe. JohnWiley and Sons Ltd., Chichester: 764 pp.

Bitušík, P., V. Kubovčík, E. Štefková, P. G. Appleby & M. Svitok, 2009. Subfossil diatom and chironomid distributions along an altitudinal gradient in the High Tatra Mountain lakes: palaeolimnological reconstruction of environmental changes. Hydrobiologia. doi:10.1007/s10750-009-9802-0.

Björkman, L., A. Feurdean, K. Cinthio, B. Wohlfarth & G. Possnert, 2002. Lateglacial and early Holocene vegetation development in the Gutaiului Mountains, NW Romania. Quaternary Science Reviews 21: 1039–1059.

Björkman, L., A. Feurdean & B. Wohlfarth, 2003. Lateglacial and Holocene forest dynamics at Steregoiu in the Gutaiului Mountains, NW Romania. Review of Palaeobotany and Palynology 124: 79–111.

Bodnariuc, A., A. Bouchette, J. J. Dedoubat, T. Otto, M. Fountugne & G. Jalut, 2002. Holocene vegetational history of the Apuseni Mountains, central Romania. Quaternary Science Reviews 21: 1465–1488.

Bond, G., B. Kromer, J. Beer, R. Muscheler, M. N. Evans, W. Showers, S. Hoffmann, R. Lotti-Bond, I. Hajdas & G. Bonani, 2001. Persistent solar influence on north Atlantic climate during the Holocene. Science 294: 2130–2136.

Borsy, Z., É. Csongor & E. Félegyházi, 1989. A Bodrogköz kialakulása és vízhálózatának változásai. [The formation and development of water regime at Bodrogköz region.] (in Hungarian). Alföldi Tanulmányok 13: 65–81.

Brauer, A. & J. F. W. Negendank, 2004. Editorial, The European Lake Drilling Programme (ELDP) from 1996–2001: high-resolution lake sediment records in climate and environment variability studies. Quaternary International 122: 1–5.

Braun, M., Gy. Lakatos, I. Mészáros, P. Sümegi, L. Szűcs & Gy. Szöőr, 1992. A kállósemjéni Nagy-Mohos láp fejlődéstörténete üledékföldtani, geokémiai, malakológiai és pollenanalitikai vizsgálatok alapján. [Geochemistry-mollusk- and pollen-based reconstruction of Nagy-Mohos bog at Kállósemjén.] (in Hungarian). In Szöőr, Gy. (ed.), Fáciesanalitikai, paleobiogeokémiai és paleoökológiai kutatások. [Studies on Facies Analyses, Paleobiochemistry and Paleoecology.]. MTA Debreceni Akadémiai Bizottság kiadványa, Debrecen: 205–245.

Braun, M., P. Sümegi, A. Tóth, K. J. Willis, I. Szalóki, Z. Margitai & A. Somogyi, 2005. Reconstruction of long-term environmental changes at Kelemér-Kis-Mohos-tó. In Gál, E., I. Juhász & P. Sümegi (eds), Environmental Archaeology in North-Eastern Hungary, Vol. 19. Varia Archaeologica Hungarica, Budapest: 25–38.

Buczkó, K., L. Vörös & T. Cserny, 2005. The diatom flora and vegetation of Lake Balaton from sediment cores according to Márta Hajós's legacy. Acta Botanica Hungarica 47: 75–115.

Buczkó, K., E. K. Magyari, J. Kédl, M. Braun & M. Bálint, 2008. Multi-fold use of siliceous microfossils in palaeoecological reconstruction: a case study from the Southern-Carpathians. In Jasprica, N., A. Car & M. Calic (eds) The 20th International Diatom Symposium Abstract book, 7–13 September, Dubrovnik, Croatia: 113.

Buczkó, K., E. K. Magyari, M. Braun & M. Bálint, 2009. Late Glacial and Holocene diatoms from glacial lake Taul dintre Brazi, Retezat Mts, Romania – Diatomededelingen 33. The joint 40th Meeting of the Dutch-Flemish Society of Diatomists (NVKD) and 3rd Central European Diatom Meeting (CeDiatoM) 26–29 March 2009, Utrecht, The Netherlands. Nederlands-Vlaamse Kring van Diatomisten: 36–39.

Budziszewski, J. & M. Skowronek, 2001. Results of the preliminary archaeological researches in the Mount Cergowa Massif, the Lower Beskid Mountains. In Machnik, J. (ed.), Archaeology and Natural Background of the Lower Beskid Mountains, Carpathians, Part I, Prace Komisji Prehistorii Karpat, Vol. II. Polska Akademia Umiejętności, Kraków: 145–164.

Clarke, G., M. Kernan, A. Marchetto, S. Sorvari & J. Catalan, 2005. Using diatoms to assess geographical patterns of biological and limnological change in high-altitude European lakes from pre-industrial times to the present day. Aquatic Sciences 67: 224–236.

Cserny, T., 2000. Komplex földtani kutatások hazai tavakon, lápokon és mocsarakon. [Complex geological survey on Hungarian lakes, mires and swamps.] (in Hungarian). In Szurdoki, E. (ed.), Tőzegmohás élőhelyek hazánkban: kutatás, kezelés, védelem. CEEWEB Munkacsoport, Miskolc: 27–41.

Cserny, T., 2002. A balatoni negyedidőszaki üledékek kutatási eredményei. [Results of an investigation into Quaternary lacustrine sediments in Lake Balaton.] (in Hungarian). Földtani Közlöny 132: 193–213.

Cserny, T. & E. Nagy-Bodor, 2000. Limnogeological investigations on Lake Balaton. In Gierlowski-Kordesch, E. & K. Kelts (eds), Lake Basins Through Space and Time. AAPG Studies in Geology 46: 605–618.

Cserny, T. & E. Nagy-Bodor, 2005. Late Holocene geohistory of the Hungarian part of the Eastern Alpine foreland in the light of recent research. Antaeus 28: 155–174.

Danzeglocke, U., O. Jöris & B. Weninger, 2008. CalPal-2007 online [available on internet at http://www.calpal-online.de]. Accessed 2008 November 19.

Davis, B. A. S., S. Brewer, A. C. Stevenson, J. Guiot & Data Contributors, 2003. The temperature of Europe during the Holocene reconstructed from pollen data. Quaternary Science Reviews 22: 1701–1716.

Fărcaş, S., J. L. de Beaulieu, M. Reille, G. Coldea, B. Diaconeasa, T. Goslar & T. Jull, 1999. First ^{14}C datings of lateglacial and Holocene pollen sequences from the Romanian Carpathians. Comptes Rendues de l'Academie des Sciences de Paris, Sciences de la Vie 322: 799–807.

Fărcaş, S., I. Tanţău & A. Bodnariuc, 2003. The Holocene human presence in Romanian Carpathians, revealed by the palynological analysis. Wurzburger Geographische Manuskripte 63: 111–128.

Fărcaş, S., J. L. de Beaulieu, I. Tanţău & I. A. Stoica, 2005. The absolute chronology aspects of the postglacial vegetation registered in the Capatana peat bog, Apuseni Mountains. Studii si Cercetari (Biologie) 10: 99–108.

Fărcaş, S., F. Popescu & I. Tanţău, 2006a. Dinamica spatiala si temporala a stejarului, frasinului si carpenului in timpul Tardi- si Postglaciarului pe teritoriul Romaniei. [Temporal and spatial dynamics of oak, ash and hornbeam in the Late- and Postglacial era.] (in Romania). Ed. Presa Universitara Clujeana, Cluj-Napoca: 214 pp. ISBN: (10) 973-610-461-3.

Fărcaş, S. I., I. Tanţău, T. M. Ursu, T. Goslar, F. Popescu & I. A. Stoica, 2006b. The study of the Late- and Postglacial dynamics of the vegetation from Pesteana, Poiana Rusca Mountains. Contributii Botanice 41: 109–118.

Félegyházi, E. & C. Tóth, 2003. A Halas-fenék lefûzödött medermaradvány üledékanyagának szedimentológiai, mikromineralógiai és palinológiai vizsgálata (Sedimentological and palynological study of the abandoned channel remnant deposits of the Halas-fenék). Acta Geographica Debrecina 36: 21–30.

Feurdean, A., 2005a. Holocene forest dynamics in northwestern Romania. The Holocene 15: 435–446.

Feurdean, A., 2005b. Tracking Lateglacial and early Holocene environmental changes: a paleolimnological study of sediment at Preluca Tiganului, NW Romania. Studia Universitatis Babes-Bolyai, Geologia 50: 3–11.

Feurdean, A. & C. Astalos, 2005. The impact of human activities in the Gutaiului Mountains Studia Universitatis Babes-Bolyai. Geologia 50: 63–72.

Feurdean, A. & O. Bennike, 2004. Late Quaternary palaeoecological and paleoclimatological reconstruction in the Gutaiului Mountains, NW Romania. Journal of Quaternary Science 19: 809–827.

Feurdean, A. & K. J. Willis, 2008a. The usefulness of a long-term perspective in assessing current forest conservation

management in the Apuseni Natural Park, Romania. Forest Ecology and Management 256: 421–430.
Feurdean, A. & K. J. Willis, 2008b. Long-term variability of Abies alba in NW Romania: implications for its conservation management. Diversity and Distributions 14: 1004–1017.
Feurdean, A., V. Mosbrugger, B. Onac, V. Polyak & D. Veres, 2007a. Younger Dryas to mid-Holocene environmental history of the lowlands of NW Transylvania, Romania. Quaternary Research 68: 364–378.
Feurdean, A., W. Wohlfarth, L. Bjorkman, I. Tanţău, O. Bennike, K. J. Willis, S. Fărcaş & A. M. Robertsson, 2007b. The influence of refugial population on Lateglacial and early Holocene vegetational changes in Romania. Review of Palaeobotany and Palynology 145: 305–320.
Feurdean, A., S. Klotz, S. Brewer, V. Mosbrugger, T. Tamas & B. Wohlfarth, 2008a. Lateglacial climate development in NW Romania – comparative results from three quantitative pollen based methods. Palaeogeography, Palaeoclimatology, Palaeoecology 265: 121–133.
Feurdean, A., S. Klotz, V. Mosbrugger & W. Wohlfarth, 2008b. Pollen-based quantitative reconstructions of Holocene climate variability in NW Romania. Palaeogeography, Palaeoclimatology, Palaeoecology 260: 494–504.
Gardner, A. R., 2002. Neolithic to Copper Age woodland impacts in northeast Hungary? Evidence from the pollen and sediment chemistry. The Holocene 12: 541–553.
Gardner, A. R., 2005. Natural environment or human impact? A palaeoecological study of two contrasting sites in northeastern Hungary. In Gál, E., I. Juhász & P. Sümegi (eds), Environmental Archaeology in North-Eastern Hungary, Vol. 19. Varia Archaeologica Hungarica, Budapest: 87–106.
Gil, E., E. Gilot, A. Kotarba, L. Starkel & K. Szczepanek, 1974. An Early Holocene landslide in the Beskid Niski and its significance for palaeogeographical reconstruction. Studia Geomorphologica Carpatho-Balcanica 8: 69–83.
Granoszewski, W., M. Derkacz & A. Wójcik, 2008. Stanowisko 6: Toporowe Stawy-geneza i wiek zagłębień wytopiskowych w północnej części Doliny Suchej Wody. [Site 6: Toporowe Stawy Lakes – development and age of the glacial lakes in the northern part of the Dolina Suchej Wody Valley.] (in Polish). XV Konferencja Stratygrafia plejstocenu Polski; Plejstocen Tatr i Podhala-zlodowacenia tatrzańskie. Państwowy Instytut Geologiczny, Warszawa: 189–192.
Harmata, K., 1987. Late-Glacial and Holocene history of vegetation at Roztoki and Tarnowiec near Jasło (Jasło-Sanok Depression). Acta Palaeobotanica 27: 43–65.
Harmata, K., 1995. Traces of human impact reflected in pollen diagram from Tarnowiec mire near Jasło (Jasło-Sanok Depression), SE Poland. Vegetation History and Archaeobotany 4: 235–243.
Harmata, K. 2008. Przekształcenia środowiska przyrodniczego pod wpływem działalności człowieka na przełomie IV i III tysiąclecia BC w świetle badań palinologicznych w Tarnowcu w dolinie Jasiołki. [Transformations of natural environment by human activities at the turn of the 4th and 3rd centuries BC in the light of palynological researches in the Jasiołka River Basin.] (in Polish). In Machnik J. (ed.), Archeologia i środowisko naturalne Beskidu Niskiego w Karpatach. Część II. Kurimská Brázda, Vol. 4. Prace Komisji Prehistorii Karpat, Kraków: 379–384.
Harrington, G. J., 1995. Anthropogenic impact upon the eastern Hungarian landscape. M.Sc. Thesis, University of Cambridge, Cambridge.
Harrison, S. P., G. Yu & P. E. Tarasov, 1996. Late Quaternary lake-level record from Northern Eurasia. Quaternary Research 45: 138–159.
Horsák, M. & P. Hájková, 2005. The historical development of the White Carpathian spring fens based on palaeomalacological data. In Poulíčková, A., M. Hájek & K. Rybníček (eds), Ecology and Palaeoecology of Spring Fens of the West Carpathians. Palacký University, Olomouc: 63–68.
Horváth, F. & G. Bada, 2008. Atlas of the present-day geodynamics of the Pannonian basin: euroconform maps with explanatory text [available on internet at http://geophysics.elte.hu/atlas/geodin_atlas.htm]. 25 October 2008.
Hüttemann, H. & S. Bortenschlager, 1987. Beitrage zur Vegetationsgeschichte Tirols VI: Riesengebirge, Hohe Tatra – Zillertal, Kühtai. Ber. Nat.-med. Verein Innsbruck 74: 81–112.
Ilon, G., I. Juhász, P. Sümegi, G. Jakab, G. Szegvári & T. Törőcsik, 2006. Mezőlak-Szélmező tőzegláp geoarcheológiai vizsgálatának eredményei. (Effects of the geoarcheological survey on the Mezőlak-Szélmező peatbog) (in Hungarian). Savaria 29: 147–215.
Jakab, G., P. Sümegi & E. Magyari, 2004a. A new paleobotanical method for the description of Late Quaternary organic sediments (Mire-development pathways and palaeoclimatic records from S Hungary). Acta Geologica Hungarica 47: 1–37.
Jakab, G., P. Sümegi & E. Magyari, 2004b. A new quantitative method for the paleobotanical description of Late Quaternary organic sediments. Antaeus 27: 181–211.
Jakab, G., P. Sümegi & Zs. Szántó, 2005. Késő-glaciális és holocén vízszintingadozások a Szigligeti-öbölben (Balaton) makrofosszília vizsgálatok eredményei alapján. [Late Glacial and Holocene water level changes in the Szigliget Bay, Lake Balaton based on macrofossil investigations.] (in Hungarian). Földtani Közlöny 135: 405–431.
Jakab, G., P. Majkut, I. Juhász, S. Gulyás, P. Sümegi & T. Törőcsik, 2009. Palaeoclimatic signals and anthropogenic disturbances from the peatbog at Nagybárkány (N Hungary). Hydrobiologia. doi:10.1007/s10750-009-9803-z.
Jalut, G., A. Bodnariuc, A. Bouchette, J. J. Dedoubat, T. Otto & M. Fontugne, 2003. Holocene vegetation and human impact in the Apuseni Mountains, Central Romania. In Tonkov, S. (ed.), Aspects of Palynology and Palaeoecology (Festschrift in honour of Elissaveta Bozilova). Pensoft Publishers, Sofia-Moscow: 137–170.
Jankovská, V., 1988. A reconstruction of the Late-Glacial and Early-Holocene evolution of forest vegetation in the Poprad basin, Czechoslovakia. Folia Geobotanica and Phytotaxonomica 23: 303–320.
Jankovská, V., 1998. Pozdní glaciál a časný holocén podtatranských kotlin – obdoba sibiřské boreální a subboreální zóny? [Late Glacial and Early Holocene of Tatra's foreground basins – an analogy of sibirian boreal and subboreal zone?] (in Czech]). In Benčaťová, B. & R. Hrivnák

(eds), Rastliny a človek. Zvolen, Technická. Univerzita.: 89–95.

Jankovská, V. & P. Pokorný, 2008. Forest vegetation of the las full-glacial period in the Western Carpathians (Slovakia and Czech Republic). Preslia 80: 307–324.

Jankovská, V., P. Chromý & M. Nižnianská, 2002. Šafárka – first palaeobotanical data of the character of Last Glacial vegetation and landscape in the West Carpathians (Slovakia). Acta Palaeobotanica 42: 39–50.

Járai-Komlódi, M., 1968. The late glacial and holocene flora of the Great Hungarian Plain. Annales Universitatis Scientiarum Budapestinensis de Rolando Eötvös Nominatae. Sectio Biologica 9–10: 199–225.

Járai-Komlódi, M., 1991. Late Pleistocene vegetation history in Hungary since the last interglacial. In Pécsi, M. & F. Schweitzer (eds), Quaternary Environments in Hungary. Studies in Geography in Hungary 26. Akadémiai Kiadó, Budapest: 35–46.

Juhász, I., 2005a. An overview of the palaeobotanical data from Kelemér-Kis-Mohos-tó. In Gál, E., I. Juhász & P. Sümegi (eds), Environmental Archaeology in North-Eastern Hungary, Vol. 19. Varia Archaeologica Hungarica, Budapest: 39–45.

Juhász, I., 2005b. Detecting antropogenic impacts in the palaeobotanical samples from Csaroda-Nyíres-tó. In Gál, E., I. Juhász & P. Sümegi (eds), Environmental Archaeology in North-Eastern Hungary, Vol. 19. Varia Archaeologica Hungarica, Budapest: 55–66.

Juhász, I., P. Sümegi, Zs. Szántó, É. Svingor, M. Molnár & G. Jakab, 2007. The Balaton region and the Balaton uplands. In Zatykó, Cs., I. Juhász & P. Sümegi (eds), Environmental Archaeology in Transdanubia. Varia Archaeologica Hungarica, Budapest: 390 pp.

Korponai, J., M. Braun, K. Buczkó, I. Gyulai, L. Forró, J. Nédli & I. Papp, 2009. Transition from shallow lake to wetland: A multi-proxy case study in Zalavári Pond, Balaton, Hungary. Hydrobiologia (submitted).

Kuneš, P., B. Pelánková, M. Chytrý, V. Jankovská, P. Pokorný & L. Petr, 2008. Interpretation of the last-glacial vegetation of eastern-central Europe using modern analogues from southern Siberia. Journal of Biogeography. doi: 10.1111/j1365-2699.2008.01974.x.

Magny, M., C. Begeot, J. Guiot & O. Peyron, 2003. Constraining patterns of hydrological changes in Europe in response to Holocene climate cooling phases. Quaternary Science Reviews 22: 1589–1596.

Magyari, E., G. Jakab, E. Rudner & P. Sümegi, 1999. Palynological and plant macrofossil data on Late Pleistocene short-term climatic oscillations in NE-Hungary. Acta Palaeobotanica. Supplement 2: 491–502.

Magyari, E., G. Jakab, P. Sümegi, E. Rudner & M. Molnár, 2000. Paleobotanikai vizsgálatok a keleméri Mohos-tavakon. [Palaeobotanical investigations on Mohos-lakes, Kelemér, N.-Hungary.] (in Hungarian). In Szurdoki, E. (ed.), Tőzegmohás élőhelyek Magyarországon: kutatás, kezelés, védelem (Peat Bog Habitats in Hungary: Research, Conservation, Management). Central and East European Working Group for Enhancement of Biodiversity, Miskolc: 101–131.

Magyari, E., P. Sümegi, M. Braun, G. Jakab & M. Molnár, 2001. Retarded wetland succession: anthropogenic and climatic signals in a Holocene peat bog profile from the NE Carpathian Basin. Journal of Ecology 89: 1019–1032.

Magyari, E. K., K. Buczkó, G. Jakab, M. Braun, Z. Szántó, M. Molnár, Z. Pál & D. Karátson, 2006. Holocene palaeohydrology and environmental history in the South Harghita Mountains, Romania. Földtani Közlöny 136: 249–284.

Magyari, E. K., G. Jakab, P. Sümegi & Gy. Szöőr, 2008. Holocene vegetation dynamics in the Bereg Plain, NE Hungary – the Báb-tava pollen and plant macrofossil record. Acta Geographica Debrecina 42: 1–16.

Magyari, E. K., K. Buczkó, G. Jakab, M. Braun, Z. Pál & D. Karátson, 2009a. Palaeolimnology of the last crater lake in the Eastern Carpathian Mountains – a multiproxy study of Holocene hydrological changes. Hydrobiologia. doi: 10.1007/s10750-009-9801-1.

Magyari, E., G. Jakab, M. Braun, K. Buczkó & M. Bálint, 2009b. High-resolution study of Late Glacial and Early Holocene tree line changes in the Southern Carpathian Mountains. European Geoscience Union General Assembly 2009, Vienna Geophysical Research Abstracts, Vol. 11. EGU2009-10549. http://meetingorganizer.copernicus.org/EGU2009/EGU2009-10549.pdf.

Magyari, E. K., J. C. Chapman, D. G. Passmore, J. R. M. Allen, J. P. Huntley & B. Huntley, in press. Holocene persistence of wooded steppe in the northern Great Hungarian Plain. Journal of Biogeography.

Mamakowa, K. & A. Wójcik, 1987. Osady organiczne środkowego Vistulianu w Jaśle-Bryłach (dolina Wisłoki). [Organic deposits of Middle Vistulian age from Jasło-Bryły (Wisłoka valley).] (in Polish). Kwartalnik Geologiczny 31: 213–214.

Marciniak, B., 1986. Late Quaternary diatoms in the sediments of Przedni Staw Lake (Polish Tatra Mountains). Hydrobiologia 143: 255–265.

Marciniak, B. & A. Cieśla, 1983. Diatomological and geochemical studies on Late Glacial and Holocene sediments from the Przedni Staw Lake in the Dolina Pięciu Stawów Polskich Valley (Tatra Mts). Kwartalnik Geologiczny 27: 123–150.

Margielewski, W., 2001. Late Glacial and Holocene climatic changes registered in forms and deposits of the Klakowo landslide (Beskid Średni Range, Outher Carpathians). Studia Geomorphologica Carpatho-Balcanica 35: 63–79.

Margielewski, W., 2006. Records of the Late Glacial-Holocene palaeoenvironmental changes in landslide forms and deposits of the Beskid Makowski and Beskid Wyspowy Mts. area (Polish Outher Carpathians). Folia Quaternaria 76: 5–149.

Margielewski, W. & V. Zernitskaya, 2003. Late Glacial-Holocene palaeoenvironmental evidence recorded in the Hajduki peat bog (Beskid Średni Mts., Outer Carpathians). In Margielewski, W. (ed.), Late Glacial-Holocene Palaeoenvironmental Changes in the Western Carpathians: Case Studies of Landslide Forms and Deposits. Folia Quaternaria 74: 57–73.

Margielewski, W., A. Obidowicz & S. Pelc, 2003. Late Glacial-Holocene peat bog on Kotoń Mt. and its significance for reconstruction of palaeoenvironment in the Western outer Carpathians (Beskid Makowski Range, South Poland). In Margielewski, W. (ed.), Late Glacial-Holocene

Palaeoenvironmental Changes in the Western Carpathians: Case Studies of Landslide Forms and Deposits. Folia Quaternaria 74: 35–56.

Marková, S., M. Černý, D. J. Rees & E. Stuchlík, 2006. Are they still viable? Physical conditions and abundance of Daphnia pulicaria resting eggs in sediment cores from lakes in the Tatra Mountains. Biologia 61: 135–146.

Medzihradszky, Zs., 2005. Holocene vegetation history and human activity in the Kis-Balaton area, Western Hungary. Studia Botanica Hungarica 36: 77–100.

Nagyné-Bodor, E., T. Cserny & M. Hajós, 1996. A Garancs-tó palynológiai és komplex földtani vizsgálata. [Palynological and complex geological investigations of Lake Garancs.] (in Hungarian). In Hably, L. (ed.), Emlékkötet Andreászky Gábor (1895-1967) születésének 100. évfordulójára. Studia Naturalia 9: 137–146.

Niedziałkowska, E., E. Gilot, M. F. Pazdur & K. Szczepanek, 1985. The upper Vistula valley near Drogomyśl in the Late Vistulian and Holocene. Folia Quaternaria 56: 101–132.

Obidowicz, A., 1990. Eine pollenanalytische d moorkundliche Studie zur Vegetationsgeschichte des Podhale-Gebietes (West-Karpaten). Acta Palaeobotanica 30: 147–219.

Obidowicz, A., 1993. Wahania górnej granicy lasu w późnym glacjale i holocenie w Tatrach [Fluctuation of the forest limit in the Tatra Mts during the last 12000 years.] (in Polish). Dokumentacja Geograficzna Instytutu Geografii i Przestrzennego Zagospodarowania PAN 4–5: 31–43.

Obidowicz, A., 1996. A Late Glacial-Holocene history of the formation of vegetation belts in the Tatra Mts. Acta Palaeobotanica 36: 159–206.

Obidowicz, A. 2003. The Holocene development of forests in the Pilsko Mt. Area (Beskid Żywiecki Range, South Poland). In Margielewski, W. (ed.), Late Glacial-Holocene Palaeoenvironmental Changes in the Western Carpathians: Case Studies of Landslide Forms and Deposits. Folia Quaternaria 74: 7–15.

Pantocsek, J., 1913. A Balaton fenékalatti mederfúrások sorozatai mélységi próbáiban talált Bacilláriák táblázatos kimutatása. [Tabular presentation of the diatoms found in the samples of the serial deep borings under the bottom of the Lake Balaton.] (in Hungarian). A Balaton Tudományos Tanulmányozásának Eredményei, Kiadja a Magyar Földrajzi Társaság Balaton Bizottsága, Budapest, Vol. 1. Part 2. Sect. 1, Supplement 4: 563.

Pazdur, A., 2001. Radiocarbon age of the sediments from the Mount Cergowa site, Poland. In Machnik, J. (ed.), Archaeology and Natural Background of the Lower Beskid Mountains, Carpathians, Part I, Vol. 2. Prace Komisji Prehistorii Karpat, Polska Akademia Umiejętności, Kraków: 183–185.

Péterfi, L. Şt., 1974. Preliminary notes on the subfossil and recent diatom flora of the Zănoguţa peat bog from the Retezat Mountains. Studia Universitatis Babeş-Bolyai, Cluj, Series Biologia 19: 5–17.

Peyron, O., J. Guiot, R. Cheddadi, P. Tarasov, M. Reille, J. L. de Beaulieu, S. Bottema & V. Andrieu, 1998. Climatic reconstruction in Europe for 18000 yr BP from pollen data. Quaternary Research 49: 183–196.

Pop, E., 1932. Contribuii la istoria vegetaiei cuaternare din Transilvania. Beitrag zur quaternären Pflanzengeschichte Siebenbürgens (Rumänien). Buletinul Gradinii Botanice Muzeul Cluj 12: 20–102.

Pop, E., 1960. Mlastinile de turba din R. S. Romina. (Die Torfmoore aus der Rumänischen volksrepublik). Editura Academiei Republicii Populare Romine: 511.

Ralska-Jasiewiczowa, M., 1972. The forests of the Polish Carpathians in the Late Glacial and Holocene. Studia Geomorphologica Carpatho-Balcanica 6: 5–19.

Ralska-Jasiewiczowa, M., 1980. Late-Glacial and Holocene vegetation of the Bieszczady Mts. (Polish Eastern Carpathians). Państwowe Wydawnictwo Naukowe, Kraków: 3–192.

Ralska-Jasiewiczowa, M., 1989. Type region P-e: Bieszczady Mts. Acta Palaeobotanica 29: 31–35.

Ralska-Jasiewiczowa, M., E. Madeyska & M. Mierzeńska, 2006. Vegetational changes in the montane grassland zone of the High Bieszczady mountains (southeast Poland) during the last millennium – pollen records from deposits in hanging peat-bogs. Vegetation History and Archaeobotany 15: 391–401.

Rónai, A., 1985. The Quaternary of the Great Hungarian Plain. Geologica Hungarica Series Geologica 21: 1–445.

Rösch, M. & E. Fischer, 2000. A radiocarbon dated Holocene profile from the Banat Mountains (Southwestern Carpathians, Romania). Flora 195: 277–286.

Rybníček, K. & E. Rybníčková, 1985. A palaeoecological reconstruction of precultural vegetation in the intermontane basins of the Western Carpathians. Ecologia Mediterranea 11: 27–31.

Rybníčková, E. & K. Rybníček, 1972. Erste Ergebnisse paläogeobotanischer Untersuchungen des Moores bei Vrackov, Südmähren. Folia Geobotanica et Phytotaxonomia 7: 285–308.

Rybníčková, E. & K. Rybníček, 1988. Holocene Palaeovegetation and Palaeoenvironment of the Kameničská kotlina Basin (Czechoslovakia). Folia Geobotanica et Phytotaxonomia 23: 285–301.

Rybníčková, E. & K. Rybníček, 1996. Czech and Slovak Republics. In Berglund, B. E., H. J. B. Birks, M. Ralskajasiewiczowa & H. E. Wright (eds), Palaeoecological Events During the Last 15000 Years. Regional Syntheses of Palaeoecological Studies of Lakes and Mires in Europe. John Wiley & Sons, Chichester: 473–505.

Rybníčková, E., P. Hájková & K. Rybníček, 2005. The origin and development of spring fen vegetations and ecosystems – palaeogeobotanical results. In Poulíčková, A., M. Hájek & K. Rybníček (eds), Ecology and Palaeoecology of Spring Fens of the West Carpathians. Palacký University, Olomouc: 29–62.

Schnitchen, C., E. Magyari, B. Tóthmérész, I. Grigorszky & M. Braun, 2003. Micropaleontological observations on a Sphagnum bog in East Carpathian region – testate amoebae (Rhizopoda: Testacea) and their potential use for reconstruction of micro- and macroclimatic changes. Hydrobiologia 506–509: 45–49.

Schnitchen, C., E. Magyari, D. Charman, I. Grigorszky, B. Tóthmérész & M. Braun, 2004. The possibility of reconstruction of past hydrological changes in a Sphagnum bog by the use of testate amoebae (Rhizopoda: Testacea) in the Carpathian Basin. In Wetzel, R. G. (ed.), Proceedings of the International Association of Theoretical and

Applied Limnology (SIL), Vol. 29, Part 4. Science Publishers, Stuttgart, Germany: 1751–1757.

Schnitchen, C., D. J. Charman, E. Magyari, M. Braun, I. Grigorszky, B. Tóthmérész, M. Molnár & Zs. Szántó, 2006. Reconstructing hydrological variability from testate amoebae analysis in Carpathian peatlands. Journal of Paleolimnology 36: 1–17.

Somogyi, A., M. Braun, A. Tóth & K. J. Willis, 1998. Speciation of elements in lake sediments investigation using X-ray fluorescence and inductively coupled plasma atomic emission spectrometry. X-Ray Spectrometry 27: 283.

Šporka, F., E. Štefková, P. Bitušík, R. Thompson, A. Agustí-Panareda, P. G. Appleby, J. A. Grytnes, C. Kamenik, I. Krno, A. Lami, N. L. Rose & E. Shilland, 2002. The paleolimnological analysis of sediments from high mountain lake Nižne Terianske pleso in the High Tatras (Slovakia). Journal of Paleolimnology 28: 95–109.

Środoń, A., 1987. Flora peryglacjalna z Sowlin koło Limanowej (Vistulian, Karpaty Zachodnie). [Periglacial flora of the Vistulian age from Sowliny near Limanowa (Western Carpathians).] (in Polish). Acta Palaeobotanica 27: 53–70.

Stuchlík, E., P. Appleby, P. Bitušík, C. Curtis, J. Fott, J. Kopáček, M. Pražáková, N. Rose, O. Strunecký & R. F. Wright, 2002. Reconstruction of long-term changes in lake water chemistry, zooplankton a nad benthos of a small, acidified high-mountain lake. MAGIC modelling and palaeolimnological analysis. Water, Air, and Soil Pollution: Focus 2: 127–138.

Sümegi, P., 1999. Reconstruction of flora, soil and landscape evolution, and human impact on the Bereg Plain from late-glacial up to the present, based on palaeoecological analysis. In Hamar, J. & A. Sárkány-Kiss (eds), In the Upper Tisza Valley. Tiscia Monograph Series, Szeged: 171–203.

Sümegi, P., 2003. New chronological and malacological data from the Quaternary of the Sárrét area, Transdanubia, Hungary. Acta Geologica Hungarica 46: 371–390.

Sümegi, P., 2005a. Palaeoenvironmental studies of the Pocsaj marsh. In Gál, E., I. Juhász & P. Sümegi (eds), Environmental Archaeology in North-Eastern Hungary, Vol. 19. Varia Archaeologica Hungarica, Budapest: 127–138.

Sümegi, P., 2005b. The environmental history of the Jászság. In Gál, E., I. Juhász & P. Sümegi (eds), Environmental Archaeology in North-Eastern Hungary, Vol. 19. Varia Archaeologica Hungarica, Budapest: 107–114.

Sümegi, P., 2005c. The Quaternary evolution of the Fehér Lake of Kardoskút. In Sümegi, P. (ed.), Loess and Upper Paleolithic Environment in Hungary. An Introduction to the Environmental History of Hungary. Aurea Kiadó, Nagykovácsi: 163–182.

Sümegi, P., & S. Gulyás (eds), 2004. The Geohistory of Bátorliget Marshland – An Example for the Reconstruction of Late Quaternary Environmental Changes and Past Human Impact from the NE Part of the Carpathian Basin. Archaeolingua 16, Budapest.

Sümegi, P., E. Magyari, P. Daniel, E. Hertelendi & E. Rudner, 1999. A kardoskúti Fehér-tó negyedidőszaki fejlődéstörténetének rekonstrukciója. (A reconstruction of the Quaternary geohistory of Fehér lake at Kardoskút) (in Hungarian). Földtani Közlöny 129: 479–519.

Sümegi, P., E. Bodor & T. Törőcsik, 2005a. The origins of alkalisation in the Hortobágy region in the light of the palaeoenvironmental studies at Zám-Halasfenék. In Gál, E., I. Juhász & P. Sümegi (eds), Environmental Archaeology in North-Eastern Hungary, Vol. 19. Varia Archaeologica Hungarica, Budapest: 115–126.

Sümegi, P., M. Mucsi, J. Fényes & S. Gulyás, 2005b. First radiocarbon dates from the freshwater carbonates of the Danube-Tisza interfluve. In Hum, L., S. Gulyás & P. Sümegi (eds), Environmental Historical Studies from the Late Tertiary and Quaternary of Hungary. University of Szeged: 103–118.

Sümegi, P., G. Persaits, K. Herbich, M. Imre, Zs. Szántó & I. Juhász, 2007a. The Mezőség region. In Zatykó, Cs., I. Juhász & P. Sümegi (eds), Environmental Archaeology in Transdanubia. Varia archaeologica Hungarica, Budapest: 361–377.

Sümegi, P., G. Persaits, Zs. Szántó, I. Juhász & G. Jakab, 2007b. Somogy Hills. In Zatykó, Cs., I. Juhász & P. Sümegi (eds), Environmental Archaeology in Transdanubia. Varia Archaeologica Hungarica, Budapest: 237–266.

Sümegi, P., Zs. Szántó, I. Juhász, G. Szegvári & G. Jakab, 2007c. The Devecser Plain. In Zatykó, Cs., I. Juhász & P. Sümegi (eds), Environmental archaeology in Transdanubia. Varia Archaeologica Hungarica, Budapest: 309–332.

Sümegi, P., I. Juhász, E. Magyari, G. Jakab, E. Rudner, Zs. Szántó & M. Molnár, 2008. A keleméri Mohos-tavak fejlődéstörténetének rekonstrukciója paleobotanikai vizsgálatok alapján. (Environmental history of the Mohos peat bogs based on paleobotanical researches.) (in Hungarian). In Doldogh, S. & T. G. Farkas (eds), A keleméri Mohos-tavak Kutatás, kezelés, védelem. The Mohos peat bogs in Kelemér Research, Conservation, Management. ANP füzetek IV. Aggteleki Nemzeti Park Igazgatóság. Jósvafő: 35–57.

Svobodová, H., 1989. Rekonstrukce přírodního prostředí a osídlení v okolí Mistřína. Palynologická studie. [A reconstruction of natural environment and settlement in the environs of Mistrin. A palynological study.] (in Czech). Památky archeologické 80: 188–206.

Szafer, W., 1948. Późny glacjał w Roztokach pod Jasłem. [Late glacial at Roztoki near Jasło.] (in Polish). Starunia 26: 1–28.

Szalóki, I., A. Somogyi, M. Braun & A. Tóth, 1999. Investigation of geochemical composition of lake sediments using ED-XRF and ICP-AES techniques. X-Ray Spectrometry 28: 399–405.

Szántó, Zs. & Zs. Medzihradszky, 2004. Holocene environmental changes in Western Hungary. Radiocarbon 46: 691–699.

Szczepanek, K., 1987. Late-Glacial and Holocene pollen diagrams from Jasiel in the Low Beskid Mts. (The Carpathians). Acta Palaeobotanica 27: 9–26.

Szczepanek, K., 1989. Type region P-c: Low Beskidy Mts. In Ralska-Jasiewiczowa (ed), Environmental changes recorded in lakes and mires of Poland during the last 13000 years. Acta Palaeobotanica 29: 17–23.

Szczepanek, K., 2001a. Late Holocene vegetation history in the Dukla Pass region (Low Beskidy, Carpathians) based on pollen and macrofossil analyses. Acta Palaeobotanica 41: 341–353.

Szczepanek, K., 2001b. Anthropogenic vegetation changes in the region of the Dukla Pass, the Low Beskid Mountains. In Machnik, J. (ed.), Archaeology and Natural Background of the Lower Beskid Mountains, Carpathians Part I. Prace Komisji Prahistorii Karpat, Vol II. Polska Akademia Umiejętności, Kraków: 171–182.

Szeroczyńska, K., 1984. Aaliza Cladocera w osadach niektórych jezior tatrzańskich. [Cladocera analyses of sediments from selected lakes in the Tatra Mts.] (in Polish). Prace i Studia Geograficzne 5: 93–102.

Szeroczyńska, K. & E. Zawisza, 2007. Paleolimnologia – historia rozwoju jezior w Polsce w świetle badań fauny wioślarek. [Palaeolimnology – history of the lake development in Poland based on Cladocera analysis.] (in Polish). Studia Limnologica et Telmatologica 1: 51–60.

Tanţău, I., 2006. Histoire de la végétation tardiglaciaire et holocène dans les Carpates Orientales (Roumanie). Ed. Presa Universitara Clujeana, Cluj-Napoca: 200 pp.

Tanţău, I. & S. Fărcaş, 2004. Chronologie de l'histoire de la végétation holocène de Monts Rodnei (Carpates Orientales). Contributii Botanice 39: 241–250.

Tanţău, I. & S. Fărcaş, 2005. Histoire de l'impact humain sur la végétation des Monts Harghita (Carpates Orientales). Studii si cercetari (Biologie) 10: 109–116.

Tanţău, I., S. Fărcaş, M. Reille & J. L. de Beaulieu, 2003a. L'analyse palinologique de la sequence de Luci: nouvelles donnees concernant l'histoire de la végétation tardiglaciaire et holocène de Monts Harghitei. Contributii Botanice 38: 155–161.

Tanţău, I., M. Reille, J. L. de Beaulieu, S. Fărcaş, S. Goslar & M. Paterne, 2003b. Vegetation History in the Eastern Romanian Carpathians: pollen analysis of two sequences from the Mohos crater. Vegetation History and Archaeobotany 12: 113–125.

Tanţău, I., M. Reille, J. L. de Beaulieu & S. Fărcaş, 2003c. Analiza palinologica a profilului turbos Mohos 1 (Muntii Harghitei). Studii si Cercetari (Biologie) 8: 33–40.

Tanţău, I., M. Reille, S. Fărcaş & J. L. de Beaulieu, 2003d. Aspects de l'histoire de la végétation tardiglaciaire et holocène dans la région des Subcarpates de la Courbure (Subcarpates de Buzau). Studia Universitatis Babes-Bolyai (Geologia) 48: 15–26.

Tanţău, I., M. Reille, J. L. de Beaulieu & S. Fărcaş, 2006. Late Glacial and Holocene vegetation history in the southern part of Transylvania (Romania): pollen analysis of two sequences from Avrig. Journal of Quaternary Science 21: 49–61.

Tullner, T. & T. Cserny, 2003. New aspects of lake-level changes: Lake Balaton, Hungary. Acta Geologica Hungarica 46: 215–238.

Tullner, T. & T. Cserny, 2004. A Balaton földtudományi adatbázisa. (Geoscientific Database of Lake Balaton). A Magyar Állami Földtani Intézet Évi Jelentése 2002: 47–53.

Wacnik, A., 1995. The vegetational history of local flora and evidences of human activities recorded in the pollen diagram from site Regetovka, NE Slovakia. Acta Palaeobotanica 35: 253–274.

Wacnik, A. 2001. Late-Holocene history of the vegetation changes based on pollen analysis of the deposits at Kružlová, Slovakia. In Machnik, J. (ed.), Archaeology and Natural Background of the Lower Beskid Mountains, Carpathians, Part I, Prace Komisji Prehistorii Karpat, Vol. 2. Polska Akademia Umiejętności, Kraków: 127–135.

Wacnik, A., K. Szczepanek & K. Harmata, 2001. Ślady działalności człowieka neolitu i brązu obserwowane w diagramach pyłkowych z okolic Przełęczy Dukielskiej i terenów przyległych. [Signs of human activity in the Neolithic and Bronze Age noted in pollen diagrams from the Dukla Pass area.] (in Polish) In Gancarski, J. (ed.). Neolit i początki epoki brązu w Karpatach Polskich. Muzeum Podkarpackie, Krosno: 207–221.

Wacnik, A., K. Szczepanek & K. Harmata, 2006. Przemiany roślinności jako zapis zjawisk osadniczych w okresie średniowiecza na terenach przyległych do Przełęczy Dukielskiej w Beskidzie Niskim. [Vegetational changes as a reflection of the Medieval settlement activity in the Dukla Pass surroundings, the Lower Beskid Mts.] (in Polish). In Gancarski, J. (ed.), Wczesne średniowiecze w Karpatach Polskich. Muzeum Podkarpackie, Krosno: 741–759.

Weninger, B. & O. Jöris, 2008. A ^{14}C age calibration curve for the last 60 ka: the Greenland-Hulu U/Th timescale and its impact on understanding the Middle to Upper Paleolithic transition in Western Eurasia. Journal of Human Evolution 55: 772–781.

Więckowski, S. & K. Szczepanek, 1963. Assimilatory pigments from subfossil fir needles (Abies alba Mill.). Acta Societatis Botanicorum Poloniae 31: 101–111.

Willis, K. J. 1997. The impact of early agriculture upon the Hungarian Landscape. In Chapman, J. & P. M. Dolukhanov (eds), In Landscapes in Flux – Central and Eastern Europe in Antiquity. Oxbow Books, Oxford: 193–207.

Willis, K. J., P. Sümegi, M. Braun & A. Tóth, 1995. The Late Quaternary environmental history of Bátorliget, N.E. Hungary. Palaeogeography, Palaeoclimatology, Palaeoecology 118: 25–47.

Willis, K. J., P. Sümegi, M. Braun & A. Tóth, 1997. Does soil change cause vegetation change or vice versa? A temporal perspective from Hungary. Ecology 78: 740–750.

Willis, K. J., P. Sümegi, M. Braun, K. D. Bennett & A. Tóth, 1998. Prehistoric land degradation in Hungary: who, how and why? Antiquity 72: 101–113.

Wohlfarth, B., G. Hannon, A. Feurdean, L. Ghergari, B. P. Onac & G. Possnert, 2001. Reconstruction of climatic and environmental changes in NW Romania during the early part of the last deglaciation (15,000–13,600 cal years BP). Quaternary Science Reviews 20: 1897–1914.

Wołowski, K., A. Obidowicz & I. Wawrzycka, 2002. *Pediastrum* species (Chlorophyceae) in Quaternary sediments of „Żabie Oko" peat bog in the Tatra Mts. Acta Palaeobotanica 42: 51–61.

Yu, G. & S. P. Harrison, 1995. Lake status records from Europe: data base documentation. Paleoclimatology Publications Series Report #3, World Data Center -A for Paleoclimatology, Boulder, CO, USA: 451 pp [available on internet at ftp://ftp.ncdc.noaa.gov/pub/data/paleo/paleolimnology/lake levels/europe/readme_europelake.txt].

Zólyomi, B., 1952. Magyarország növénytakarójának fejlődéstörténete az utolsó jégkorszaktól. [Vegetation history of Hungary since the last glacial period.] (in Hungarian). Magyar Tudományos Akadémia Biológiai Osztályának Közleményei 1: 491–544.

Zólyomi, B. & M. Járainé Komlódi, 2008. Fosszilis szubarktikus forrásláp Magyarországon. (A fossil subarctic peat bog in Hungary) (in Hungarian). In Doldogh, S. & T. G. Farkas (eds), A keleméri Mohos-tavak Kutatás, kezelés, védelem. The Mohos peat bogs in Kelemér Research, Conservation, Management. ANP füzetek IV. Aggteleki Nemzeti Park Igazgatóság. Jósvafő: 25–34.

Author Biographies

Krisztina Buczkó (b.1962) graduated at the Eötvös Loránd University of Budapest, Hungary (1985; PhD 1996), museologist, diatomist. Curator of Collection Algarum of Hungarian Natural History Museum. She has been dealing with diatom taxonomy (*Kobayasiella* genus), biomonitoring and palaeolimnology. She has experience on diatom-based palaeoecological reconstruction in shallow waters and remote alpine lakes. Secretary of the Supraindividual Jury of the Hungarian Scientific Research Fund (2007–2010). Hydrobiologia 589: 141–154.

Enikő Katalin Magyari (b. 1973) graduated at the Kossuth Lajos University of Debrecen, Hungary (1997; PhD 2002), palynologist. She has research interests in Quaternary vegetation dynamics, environmental change and human impact on the vegetation of southeast Europe. Her current work focuses on the palaeoenvironmental record of Late-glacial and Holocene landscape transformations in the Southern Carpathians and in the Pannonian Plain.

Peter Bitušík (b. 1957) graduated at the Comenius University, Bratislava, Slovakia (1981; PhD: 1991; habilitation: 2000), biologist, Professor of Ecology. He has been dealing with chironomid ecology since the beginning of his professional career. He specializes in using chironomid pupal exuviae for lake and river assessment, and using of sub-fossil chironomid remains for reconstructions of past environmental conditions. He is the president of the Slovak Limnological Society (2004–2010) and national representative of ICP Waters. He serves as a member of editorial board of the journal Biologia.

Agnieszka Wacnik (b. 1971) graduated at the Jagiellonian University in Kraków, Poland. MSc: 1995, PhD: 2003. Biologist, palaeoecologist employed at the Department of Palaeobotany, W. Szafer Institute of Botany, Polish Academy of Sciences, Kraków, Poland. She studies vegetational changes of the last glacial-interglacial cycle. Her main scientific interest concerns different aspects of early human impact on aquatic and terrestrial environments. She uses pollen as well as non-pollen palynomorphs analysis for palaeoecological reconstructions. She is engaged mainly in the interdisciplinary projects related to the area of Polish lowlands, but she has worked also on sediments of peat bogs from the Carpathian Mts. (NE Slovakia).

Palaeolimnology of the last crater lake in the Eastern Carpathian Mountains: a multiproxy study of Holocene hydrological changes

Enikő Magyari · Krisztina Buczkó · Gusztáv Jakab · Mihály Braun · Zoltán Pál · Dávid Karátson · István Pap

Originally published in the journal Hydrobiologia, Volume 631, No. 1, 29–63.
DOI: 10.1007/s10750-009-9801-1

Abstract A multi-proxy investigation (loss-on-ignition, major and trace elements, pollen, plant macrofossil and siliceous algae) was carried out on the sediment of a crater lake (Lake Saint Ana, 950 m a.s.l.) from the Eastern Carpathian Mountains. Diatom-based transfer functions were applied to estimate the lake's trophic status and pH, while reconstruction of the water-depth changes was based on the plant macrofossil and diatom records. The lowest Holocene water depths were found between 9000 and 7400 calibrated BP years, when the crater was occupied by *Sphagnum*-bog. Significant increases in water depth were found from 5350(1), 3300(2) and 2700 cal yr BP. Of these, the first two coincided with major terrestrial vegetation changes, namely (1) the establishment of *Carpinus betulus* on the crater slope and (2) the replacement of the lakeshore *Picea abies* forest by *Fagus sylvatica*. The chemical record indicated significant soil changes along with the canopy changes (from coniferous to deciduous) that led to increased in-lake productivity and pH. A further increase in water depth around 2700 cal yr BP resulted in stable thermal stratification and hypolimnetic anoxia that via P-release further increased in-lake productivity and eventually led to phytoplankton blooms with large populations of *Scenedesmus*. High productivity was depressed by

Guest editors: K. Buczkó, J. Korponai, J. Padisák & S. W. Starratt
Palaeolimnological Proxies as Tools of Environmental Reconstruction in Fresh Water

E. Magyari (✉)
Hungarian Natural History Museum Palaeonthological Research Group, Hungarian Academy of Sciences, P.O. Box 222, 1476 Budapest, Hungary
e-mail: magyari@bot.nhmus.hu

K. Buczkó
Department of Botany, Hungarian Natural History Museum, P.O. Box 222, 1476 Budapest, Hungary

G. Jakab
Institute of Environmental Sciences, Szent István University, Szabadság út 1–3, 5540 Szarvas, Hungary

M. Braun
Department of Inorganic and Analytical Chemistry, University of Debrecen, P.O. Box 21, 4010 Debrecen, Hungary

Z. Pál
Department of Physical Geography, Faculty of Geography, Babes Bolyai University of Cluj, Str. Clincilor No. 5–7, 3400 Cluj-Napoca, Romania

D. Karátson
Department of Physical Geography, Eötvös Loránd University of Budapest, Pázmány Péter sétány 1/C, 1117 Budapest, Hungary

I. Pap
Department of Mineralogy and Geology, University of Debrecen, Egyetem tér 1, 4010 Debrecen, Hungary

anthropogenic lakeshore forest clearances from ca. 1000 cal yr BP that led to the re-establishment of *P. abies* on the lakeshore and consequent acidification of the lake water. On the whole, these data suggest that Lake Saint Ana is a vulnerable ecosystem: in-lake productivity is higher under deciduous canopy and litter, and considerably repressed by coniferous canopy and litter. The lake today subsists in a managed environment that is far from its natural state. This would be a dense *F. sylvatica* forest supplying more nutrients and keeping up a more productive in-lake flora and fauna.

Keywords Lake level change · Romania · Pollen · Macrofossil · Siliceous algae · Sediment chemistry

Introduction

Situated deep in the continental interior of Europe, the Eastern and Southern Carpathian Mountains provoked special attention in the field of Quaternary palaeoecology during the last decade (Fărcaş et al., 1999, 2003; Rösch & Fischer, 2000; Wohlfarth et al., 2001; Björkman et al., 2002a, b; Bodnariuc et al., 2002; Tanţău et al., 2003, 2006; Feurdean & Bennike, 2004; Feurdean, 2005; Feurdean et al., 2007a, b, 2008a, b). Their significances in the glacial survival of many broad-leaved and coniferous tree species have been shown by pollen, plant macrofossil and population genetic studies (Fărcaş et al., 1999, 2003, 2006; Petit et al., 2001; Wohlfarth et al., 2001; Brewer et al., 2002; Grivet & Petit, 2003; Tanţău et al., 2006; Feurdean et al., 2007a, b). Pollen and speleothem δ ^{18}O-based Lateglacial and Holocene climate reconstructions showed that many of the Holocene climate oscillations known from the North-Atlantic Region are also detectable in the Eastern Carpathians, in addition to the distinctive features of its Holocene climate history (Onac et al., 2002; Feurdean & Bennike, 2004; Schnitchen et al., 2006; Constantin et al., 2007; Feurdean et al., 2008a, b). Furthermore, attempts have been made to detect century-scale hydrological changes in the Holocene using testate amoebae and plant macrofossil records (Schnitchen et al., 2006; Feurdean & Bennike, 2004, 2008).

These palaeoenvironmental studies were predominantly made on peat sediments that are numerous in the low- and mid-altitude zones of the Eastern Carpathians (Pop, 1960). Open lakes with continuous Holocene lake sediments are, however, exceptional; and Lake Saint Ana (950 m a.s.l., Fig. 1) is one of them. It was formed by the youngest volcanic eruption of the Eastern Carpathian Range some 10–30 ka years ago (Juvigné et al., 1994; Pécskay et al., 1995; Moriya et al., 1996; Vinkler et al. 2007), and given its young age, it still subsists in an open water phase, though the marginal *Sphagnum* carpet has already started its lake-ward conquest. The lake is set in a volcanic environment: dacitic pyroclastics form the bedrock making the soils, vegetation and lake water Ca^{++} poor and hence slightly acidic. The climate of the Eastern Carpathians is continental temperate with harsh winters, warm summers and considerably reduced precipitation in comparison with the Alps (Kristó, 1995). Earlier reports on the lake fauna and flora exemplified an exceptional diversity with many rare species (Schur, 1858; Nyárádi, 1929; Hortobágyi, 1942, 1943; Boros, 1943; Lepşi, 1957; Cărăus, 2002), and our reports on the lake sediment's diatom flora revealed new species and forms that have evolved and probably gone extinct in the relatively short-time period of the Holocene (Buczkó & Magyari, 2007; Buczkó & Wojtal, 2007). These features of the lake foreshadow new and exciting scientific results from the palaeoecological study its sedimentary archive.

Lake Saint Ana is a solely rain-water fed lake within the *Fagus sylvatica* forest belt and its sediment is suitable for a range of multi-proxy analyses discussed in Smol et al. (2001); Birks & Birks (2006). Here, we present the results of loss-on-ignition, total organic carbon, trace element, pollen, plant macrofossil and silicaceous algae analyses. The sedimentary sequence extends back to ca. 9300 cal yr BP and records important hydrological changes within this time window. Our first aim was to investigate the relationship between terrestrial vegetation and water-depth fluctuation inferred from several biological proxies and seek causes behind the changes, both external, i.e. climate fluctuations (Alley et al., 1997; Bond et al., 1997; Mayewski et al., 2004; Heiri et al., 2004) and internal ecosystem dynamics, such as succession of aquatic vegetation (Tallis, 1983). The second aim was to attempt quantitative reconstruction of several environmental variables including pH, total P and water depth using the diatom and macrofossil records and the transfer function approach (Birks, 2003).

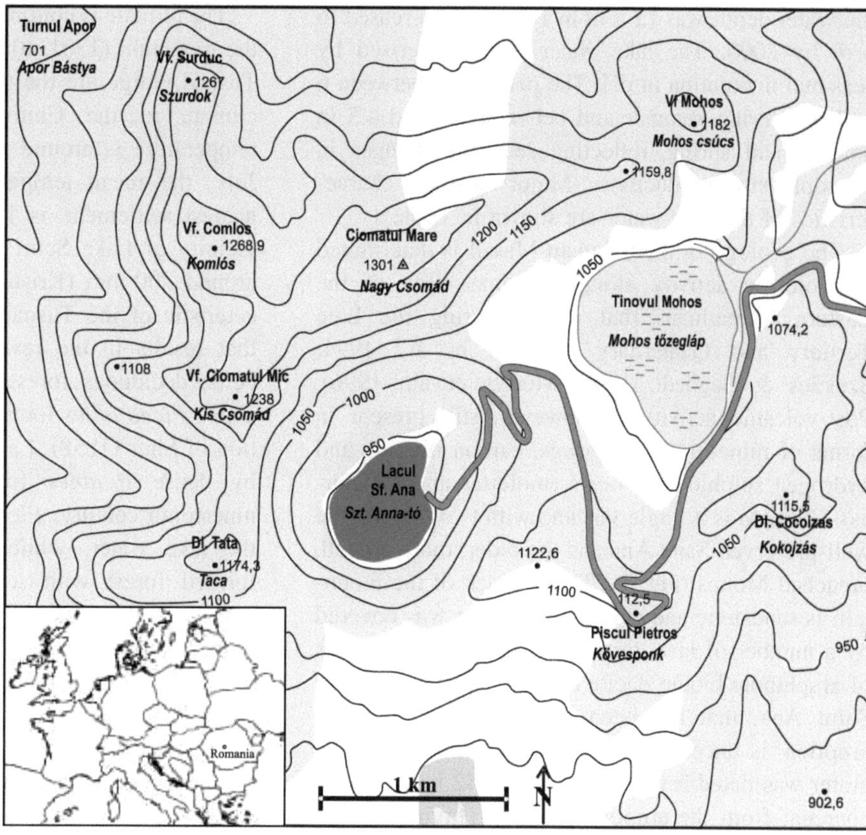

Fig. 1 Location of the study site in Europe and topographic map of the Ciomatu Massif showing the location of Lake Saint Ana. Grey-shaded area denotes closed forest and white colouring marks grazed meadows

Our preliminary results on the development of Lake Saint Ana revealed uneven sediment accumulation through the Holocene (Magyari et al., 2006). By increasing the number of AMS ^{14}C dates and applying higher stratigraphic resolution for the different proxies, this study improves the palaeoenvironmental reconstruction with a special focus on the hydrological changes.

Study area

Lake Saint Ana is a crater lake in the Ciomatu Massif of the Harghita Mountains (Fig. 1). The lake has no outlet; it is fed by rainwater and inwash from the surrounding slopes (Pál, 2000, 2001). Situated at 950 m a.s.l., the lake has an area of 0.193 km^2 (Fig. 2; Pál, 2000), while the watershed comprises 2.15 km^2. Historical documents attest to a considerable decrease in water depth and lake surface area over the last two centuries. According to Pál (2000),

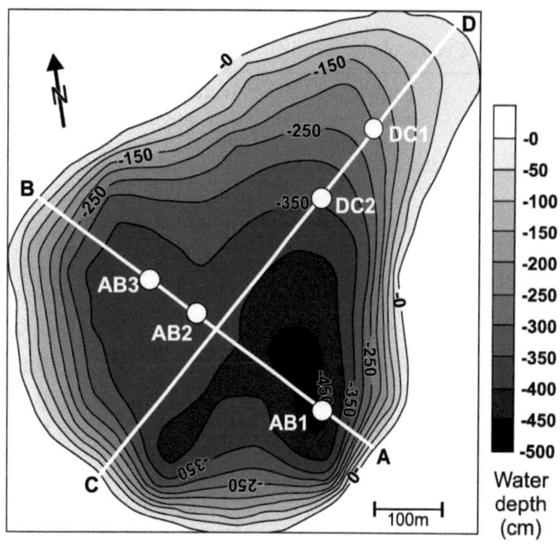

Fig. 2 Core locations and bathymetric map of Lake Saint Ana. Recorded depths on the bathymetric map are based on sondage echograms. Due to the presence of a ca. 1 m thick gelatinous stratum above the surface consolidated sediment, the echograms underestimate the water depth of the lake (Pál, 2001)

the water depth was 12.5 m in 1869 and decreased to 6 m by 2000. The lake water is characterised by seasonal fluctuation in pH. The pH ranges between 6 and 6.4 during summer and between 4.2 and 4.5 in autumn and spring, reflecting seasonal changes in phytoplankton productivity. Major chemical characteristics of the lake water are shown in Table 1.

The geology of the Ciomatu Massif is determined by volcanic activity along the inner arch of the Eastern Carpathians that abated during the Late Tertiary and Quaternary (Juvigné et al., 1994; Szakács & Seghedi, 1995; Moriya et al., 1996). Post-volcanic activity is, however, still present in forms of mineral water springs, carbon dioxide and hydrogen sulphide gas-flows (mofettas and solfataras). Ciomatu is a single volcano with two craters: the well-preserved Saint Ana and the older, more eroded, breached Mohos (Fig. 1). The bedrock of the mountain is sandstone and conglomerate that was covered by a number of lava domes and pyroclastic deposits of amphibole biotite dacite magma. The age of Lake Saint Ana, that has been formed due to the last eruption, is uncertain. The youngest eruption of the crater was dated between 10 and 35–42 ka BP using charcoal from the eruption material (Juvigné et al., 1994; Vinkler et al. 2007) and charcoal and palaeosoil (Moriya et al., 1996).

The climate is continental temperate. The height of the mountain (1301 m) and the frost retention in the Tusnad Gorge are the main factors to determine the climate in the Ciomatu Massif. January mean temperature is around -5°C. The warmest month is July; the mean temperature in Baile Tusnad, the nearest settlement, is 18°C, while it is 15°C in the vicinity of Lake Saint Ana. Annual precipitation is around 800 mm (Kristó, 1995). An important characteristic of the Tusnad Gorge is thermal inversion that results in the reverse order of the vegetation belts; deciduous forests of *F. sylvatica* are located above *Picea abies* forests. According to the observation of Shur (1858), Lake Saint Ana was surrounded by dense *P. abies* forest in the middle of the nineteenth century. *Picea* trees were in contact with the lake water, while on the slopes *F. sylvatica* formed forest with scattered occurrence of *Abies alba*. Today, the lakeshore is partly open (NE corner) and partly covered by mixed *Betula pendula*, *P. abies*, *Salix* sp. and *F. sylvatica* stands. The slopes around the lake are covered by *F. sylvatica*, and in some places *P. abies* extends upslope well into the *Fagus* forest. In the shallow NE corner of the lake, floating fen developed. Its extension increased in the last couple of years (Pál, 2000). The matrix of the floating mat is formed by *Carex rostrata* and

Table 1 Chemical characteristics of the lake water along vertical and horizontal profiles, Lake Saint Ana

	Surface water (mg/l) (03.11.2000)	Water at 2.5 mmg/l (03.11.2000)	Water at 5 m mg/l (03.11.2000)	Lake centre (mg/l) (22.08.2000)	Southern lake basin (mg/l) (22.08.2000)	Northern lake basin (mg/l) (22.08.2000)
pH	4.21	4.54	4.451	6.04	6.41	6.25
$O_{dissolved}$	9.76	10.00	10.24	6.99	5.73	4.96
Cl^-	1.37	1.30	1.41	1.84	1.51	1.73
$SO4^{-2}$	4.70	5.20	5.00	1.96	1.62	3.02
Ca	1.72	1.51	1.64	2.58	3.912	5.05
Mg	0.24	0.16	0.18	0.16	0.13	0.18
Na	0.57	0.45	0.50	0.83	0.59	0.71
K	0.60	0.59	0.58	0.60	0.56	0.82
HCO_3^-	8.54	10.98	9.76	6.71	7.32	14.64
NH_4^-	0.14	0.11	0.13	0.03	0.04	0.65
Fe	0.04	0.06	0.05	0.07	0.08	0.06
Mn	0.036	0.037	0.036	0.035	0.037	0.037
P	–	–	–	0.00	0.046	0.016
Cu	0.0017	0.0087	0.0019	0.0082	0.1426	0.0048

Note: Different dates of the vertical and horizontal measurements. Chemical analyses were carried out by the Hydrochemical Laboratory in Miercurea Ciuc

Carex lasiocarpa on which *Sphagnum* species and *Lysimachia thyrsiflora* appear. The extension of the floating mat and the occurrence of more nutrient demanding aquatic and telmatic macrophytes changed dramatically over the last two centuries based on botanical reports. For example, Shur (1858) reported the occurrence of *Pedicularis palustris*, *Utricularia vulgaris* and *Phragmites australis* that was, however, questioned by Nyárády (1929) and Boros (1943) since they failed to find these taxa on the lakeshore.

Methods

Fieldwork

The sediment of Lake Saint Ana was sampled during the summer of 2001 using a 7-cm-diameter Livingstone piston corer with a chamber length of 100 cm (Wright, 1967). Five cores were taken along two transects (Fig. 2). Here, we report the results of various analyses of the sediments obtained from the deepest part of the lake (AB3, Fig. 2). At this core location, boring started at 600 cm water depth and reached 1023 cm. The basal sediment was organic-rich gyttja that ended in a sharp boundary; the underlying pyroclast was, however, not reached as the corer could not be driven further down.

Chronology

A chronological framework for the sediments of core SZA-AB3 was established using a series of 7 AMS ^{14}C age determinations made at the Poznań Radiocarbon Laboratory using accelerator mass spectrometry (Table 3). The ^{14}C ages obtained were calibrated into calendar years using the INTCAL04 data set of Reimer et al. (2004) as implemented in the program OXCAL. An age-depth curve was constructed by linear interpolation between the mid-calibrated points of the calibrated ages and extrapolation above the near-surface dated sediment level (622.5 cm).

Sediment analysis

The sediment lithology of the core was examined and described in the laboratory. Organic matter and inorganic carbonate contents of the sediments were estimated by measuring the loss-in-weight upon ignition at 550 and 950°C for 3 h of 1-cm^3 subsamples taken at 4-cm intervals (Heiri et al., 2001). Total organic carbon (TOC) was determined on 30 samples using the Rock–Eval pyrolysis method (Espitalie et al., 1985).

Geochemistry

For the elemental analyses, 1-cm^3 dried samples taken at 2–4 cm intervals were used. A two-step wet chemical extraction technique was applied according to the recommendations of Bengtsson & Enell (1986).

Step 1: Dried samples were digested in glass beakers using 65% HNO_3 at 80°C for 3 h. Evaporated samples were treated with 2 ml 35% H_2O_2 and evaporated at 80°C. For dissolution of the extractants, 10 ml 1% HNO_3 was used. Samples were placed into ultrasonic bath to complete the dissolution process.

Step 2: Following centrifugation, the insoluble residue from Step 1 was washed into plastic reagent tubes using double de-ionised water and samples were centrifuged again. The supernatant was decanted followed by drying at 105°C. Two millitre of 38% HF was added to each sample followed by heating in closed reagent tubes at 105°C for 2 h. After cooling, 5 ml 5% boric acid was added and subsequently, sample tubes were filled up to 10 ml using double de-ionised water.

Element concentrations of both extractants were measured using inductively coupled plasma atomic emission spectrometry (ICP-AES).

Analytical studies confirm that the first step of this preparation technique takes into solution exchangeable ions, elements bound to colloids, Mn and Fe oxides, carbonates, organic matter and sulphides (Tessier et al., 1979; Somogyi et al., 1999). Silicates are not or only poorly soluble in this strong acidic medium; however, some elements bound to the surface of silicates are dissolved. Overall, the acid-soluble fraction includes the mobile element forms that broadly correspond with the elements of the endogenic/authigenic sediment fraction (Engstrom & Wright, 1984), notwithstanding a minor input from the allogenic (silicate-bound) fraction. On the other hand, the second extraction step takes into solution silicate-bound, immobile element forms, and so the

Table 2 Concentration of Al, Ca and Fe using extraction step 1 (HNO_3:H_2O_2) and step 2 (HF:HNO_3) on powdered andesite rocks taken from the vicinity of Lake Saint Ana

Sample code	Al (HNO_3:H_2O_2) (g/kg)	Al (HF:HNO_3) (g/kg)	Al (%)	Ca (HNO_3:H_2O_2) (g/kg)	Ca (HF:HNO_3) (g/kg)	Ca (%)	Fe (HNO_3:H_2O_2) (g/kg)	Fe (HF:HNO_3) (g/kg)	Fe (%)
A	6.67	84.5	7.9	1.32	28.6	4.6	7.17	27.6	26.0
B	5.40	87.3	6.2	1.16	25.3	4.6	5.65	25.8	21.9
C	7.75	83.4	9.3	0.68	23.4	2.9	10.7	24.8	43.3

Percentage share of acid-soluble (step 1) Al, Ca and Fe is shown in columns 4, 7 and 10. A–B: bedrock sample from hillslope; C: bedrock sample from the bottom of the lake

measured element concentrations will broadly correspond with the allogenic fraction as defined by Engstrom & Wright (1984).

In order to examine the proportion of silicate-derived elements in the acid-soluble fraction, andesites were collected from the lakeshore, powdered and the same two-step digestion was applied on them. Table 2 shows total and acid-soluble element concentrations and percentages in both fractions.

Macrofossil analysis

Macrofossil analyses were performed on 2–4-cm-thick sediment slices (sub-sample volume 10–21 cm^3) taken at 4-cm intervals. Between 991 and 979 cm, contiguous 2 cm-thick sediment samples were analysed. Sub-samples were wet-sieved through a 300-μm mesh. Plant macrofossils were identified to the lowest possible taxonomic level. Determinate entities, such as fruits, seeds, bract scales were counted, and estimates were made of the relative frequency of the principal components (PCs) amongst the indeterminate macrofossils, such as leaf and bryophyte fragments (ULF, UBF), unidentifiable organic matter (UOM), monocotyledons, charcoal and Cladocera using the method described in Jakab et al. (2004a). In order to obtain concentrations for the macrofossil components, known amount of marker grains (0.5 g poppy seeds, ca. 960 pieces) were added to the samples.

Pollen analysis

Sediment sub-samples of 1 cm^3 were taken at 4-cm intervals for pollen and microspore analysis. Sub-samples were processed using standard procedures for the extraction of pollen and microspores (Berglund & Ralska-Jasiewiczowa, 1986). Treatment with NaOH and HCl was followed by treatment with HF, acetolysis, staining and dehydration. Pollen and microspore concentrations were determined by adding *Lycopodium* tablets to the sub-samples prior to commencing the extraction procedure (Stockmarr, 1971). Pollen and microspore identifications were made under a Nikon Eclipse E 600 microscope (400×, 1000×) using the descriptions and identification keys in Moore et al. (1992); Beug (2004). Counting continued until at least 500 terrestrial pollen grains had been identified. Pollen percentage values were expressed relative to the sum of all terrestrial pollen taxa. The pollen-taxonomic nomenclature adopts Beug (2004).

Diatoms

Samples for siliceous microfossil analysis (1 cm^3) were taken at 4-cm intervals and treated with hot 30% hydrogen-peroxide and sulphuric-acid, followed by mounting in Zrax. For light microscope analysis, Leica DM LB2 was used (100 HCX PLAN APO). Scanning electron microscopy was performed with a Hitachi S-2600N. The diatom identification adopted the nomenclature of Krammer & Lange-Bertalot (1986–1991)and the volumes of Diatoms of Europe (Krammer, 2000; Lange-Bertalot, 2001) as well as Lange-Bertalot & Metzeltin (1996). *Eunotia* taxa were checked according to Petersen (1950) and Alles et al. (1991). A minimum of 300 diatom valves was counted at each level. The percentage ratio of diatom valves to Chrysophycean cysts was calculated using the formula of Smol (1985). Environmental variables, pH and epilimnetic total phosphorous (TP) were inferred using diatom-water chemistry transfer functions (Juggins, 2001). Training sets from the European Diatom Database (EDDI) were compared to the fossil assemblages, this way 44 species were selected for further analysis (>95% of total diatom). Environmental reconstructions were made by locally weighted weighted averaging (LWWA; Juggins, 1991). The number of common species between the best-performing training set and fossil samples was 25 for pH and 18 for TP.

Numerical methods

For the pollen, plant macrofossil and silicaceous microfossil stratigraphic sequences, local assemblage zones were distinguished using the method of optimal partitioning (Birks & Gordon, 1985) as implemented in the program psimpoll 3.00 (Bennett, 1992). The number of statistically significant assemblage zones was determined using the broken-stick model (Bennett, 1996). The element concentration diagram was zoned by clustering the principal component (PCA) sample scores of the bulk (preparation step 1 + 2) element concentrations. PCA was performed on log-transformed data using correlation matrix and varimax rotation with Kaiser normalisation. Macrofossil data were also subjected to PCA using the covariance method on normalised data. Stratigraphic diagrams were drawn using psimpoll 3.00. Multivariate analyses were performed using PSIMPOLL 3.00 (Bennett, 1992), SYN-TAX 5.0 (Podani, 1993) and SPSS (Norusis, 1990).

Results

The focus of the present study is on elucidating hydrological changes in the catchment of Lake Saint Ana during the Holocene. A more detailed description of the biostratigraphic changes and inferred terrestrial vegetation history is given in Magyari et al. (2006).

Chronology

Details of the 7 ^{14}C age determinations obtained are given in Table 3, along with the calibrated ages of the samples measured, their depths, and the materials used. All datings were made on terrestrial macrofossils to avoid contamination by older carbon. This precaution was necessary, as earlier ^{14}C datings of bulk sediment and terrestrial macrofossil samples obtained from the same sediment level revealed considerable reservoir age in the bulk samples attestable to CO_2 upwelling and uptake of the older CO_2 by the photosynthetic organisms (Magyari et al., 2006).

Results show that sediment accumulation in Lake Saint Ana started at ca. 9300 cal yr BP (Fig. 3). The extrapolated age of the uppermost sediment sample (602 cm) is 760 cal yr BP suggesting that the Livingstone corer failed to sample the youngest, unconsolidated part of the gyttja sediment.

The results of the radiocarbon dating point to very slow sediment accumulation rate (SAR) between 7900 and 5700 cal yr BP (990–961 cm), where the resolution of one proxy sample is 150–300 years. In this section, the sediment was dominated by an aquatic moss species, *Warnstorfia fluitans*, typical of nutrient-poor, rainwater-fed poor-fen pools (Fig. 5; Rybníček, 1985; Hedenäs, 1993). Similar low-sediment accumulation coupled with *W. fluitans* dominance was found in other lake/mire systems (e.g. Magyari et al., 2001) in

Table 3 Results of the AMS ^{14}C measurements, core SZA-AB3, Lake Saint Ana

Laboratory code	Dated material	Method of ^{14}C measurement	Depth below water surface (cm)	^{14}C ages BP	Calibrated BP age ranges (1σ)	Mid-point of 1σ calibrated age range
Poz-16074	*Fagus sylvatica* budscale	AMS	620–625	885 ± 30	763–885	824
Poz-9981	*Fagus sylvatica* budscale, leaf fragments	AMS	720–725	1185 ± 30	1076–1160	1118
Poz-9980	*Fagus* sp. budscale, leaf fragments	AMS	870–874	2970 ± 35	3091–3211	3151
Poz-9979	*Picea abies* seed	AMS	908–910	3475 ± 30	3712–3814	3763
Poz-9978	*Picea abies* seed	AMS	958–960	4830 ± 40	5501–5605	5553
Poz-16075	*Betula pubescens* seed	AMS	988–990	7170 ± 50	7960–8026	7993
Poz-9976	*Acer* seed	AMS	1012–1014	8050 ± 50	8814–9014	8914

All dates were converted into calendar years BP using the INTCAL04 data set of Reimer et al. (2004) and the OXCAL program

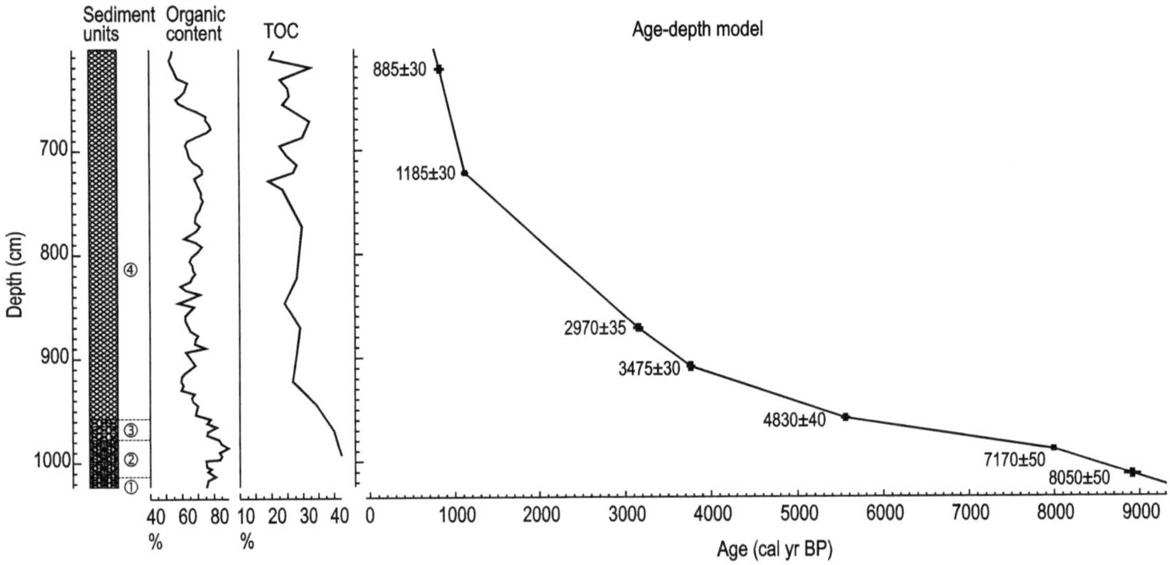

Fig. 3 Lithostratigraphy, organic content, total organic carbon (TOC) and age-depth model for SZA-AB3, Lake Saint Ana. Calibrated radiocarbon dates with 1σ age ranges are shown and the corresponding ^{14}C AMS dates are given beside the curve. Sediment units: (1) Ld2Tb1Th1Dg+ As+; (2) Tb2Ld1Th1Dg+ Dh+ As+; (3) Ld2Dg1Tb1Th+ Dh+ As+; (4) Ld3Dg1Tb+ Th+ Dh+ As+

the Carpathian region; therefore, it seems likely that the vegetation type is responsible for the low SAR. Between 9300–7900 and 5700–3800 cal yr BP, SAR was somewhat higher; 1 cm was deposited every 35–38 years resulting in a resolution of one proxy sample every 140 years. SAR increased considerably between 3800 and 1100 cal yr BP; 1 cm was deposited every 13–16 years giving an average resolution of the proxies 40–60 years. Above 1100 cal yr BP (722 cm), an extreme increase in SAR was found, each centimetre represented ca. 3 years giving one proxy sample every 6–24 years.

Overall, generally low SAR below 3800 cal yr BP (910 cm) causes lower time resolution of all proxies in the sequence, which contrasts high time resolution (6–60 years) over the last 3800 cal yr BP.

Sediment stratigraphy, organic content and TOC

Four lithostratigraphic units were recognised in the 423-cm long core of SZA-AB3 (Fig. 3). The lowermost unit (1023–1016 cm; 9300–9000 cal yr BP) is dark brown organic-rich, gyttja-rich in bryophyte (mainly *Sphagnum*) and cladoceran remains. Together with the high-organic (75%) and total organic carbon (TOC) contents (40%; see Fig. 3), this type of sediment is typical for a shallow lake mantled by *Sphagnum*-bog. Clastic inwash from the slopes was minimal, which is most likely explained by the *Sphagnum* carpet surrounding the shallow water. The transition to the overlying unit (1016–980 cm; 9000–7250 cal yr BP) is marked by an increase in organic content (80–90%) and TOC (40–45%), and bryophyte remains become more abundant at the expense of the organic detritus. These features suggest the overgrowth of the former shallow lake by a *Sphagnum*-bog in which peat accumulation was initiated. The third unit (980–960 cm; 7250–6200 cal yr BP) is dark brown organic-rich gyttja sediment with abundant plant remains. Moss leaves were still frequently encountered, but the species spectrum had changed; *W. fluitans* became abundant suggesting shallow pond environment and nutrient enrichment relative to the previous stage. Both organic content and TOC decreased gradually in this unit which indicates a gradual increase in water depth and consequent increase in detrital inwash from the slopes. The transition to the fourth unit (960–600 cm; 6200–760 cal yr BP) is very sharp, with a homogeneous mid brown gyttja characterising this uppermost unit. Cladoceran remains abundant, organic content decreases rapidly at the onset of this unit and so does TOC. Altogether, these features suggest a sudden increase in slope inwash, probably in connection with increasing water depth around 6200 cal yr BP (960 cm). Above 740 cm (1350 cal yr BP), organic content and TOC display strong fluctuation, suggesting periodic increases in clastic mineral input that may have triggered eutrophication via nutrient enrichment (Fig. 3). The sharp increases in the organic content following each erosion event support this assumption.

Regional terrestrial vegetation dynamics

Results of the main trends in the terrestrial vegetation development inferred from the pollen and plant macrofossil are given below (Figs. 4, 5). Total tree pollen (AP) was high (ca. 90%) throughout most of the profile, indicating that the landscape has been continuously forested over the last 9300 years. A diverse and dense mixed deciduous woodland dominated by *Ulmus*, *F. excelsior*, *Acer* with admixture of *Corylus*, *Quercus* and *Tilia* covered the slopes and summit around the crater between 9300 and 8800 cal yr BP. *P. abies* was abundant along the lakeshore together with *Acer*, *Betula pubescens* and *Betula pendula*. The *Picea* belt around the lake probably developed in connection with the cooler and moister microclimate of the basin as compared to the summit and hill slopes.

An important transformation in forest composition is apparent at ca. 8800 cal yr BP as rapid expansion of *Corylus* that partially replaced *Ulmus*, *F. excelsior* and *Quercus* in the forest canopy (Fig. 4). At the same time, stands of the lakeshore *P. abies* expanded, most probably upslope. *Corylus* attained maximum extension by 8200 cal yr BP.

Notable is the small-scale re-advance of *F. excelsior*, *Quercus*, *Tilia* and *Ulmus* between 7900 and 7200 cal yr BP. In this time interval, *Carpinus betulus* started to spread and partially replaced *Corylus* and *P. abies* in the forest.

By 5500 cal yr BP, *C. betulus* became the major forest tree, and between ca. 5250 and 3300 cal yr BP it formed a distinctive forest belt on the slopes and summit of the crater. The abundance of *P. abies* pollen and plant macrofossils in this phase suggests its continued presence on the lakeshore and probably also on the hill slopes. *F. sylvatica* established and showed a continuously expanding population within the *C. betulus* forest from about 4500 cal yr BP. On the lakeshore, *F. sylvatica* first appeared at 3700 cal yr BP as evidenced by its macrofossils

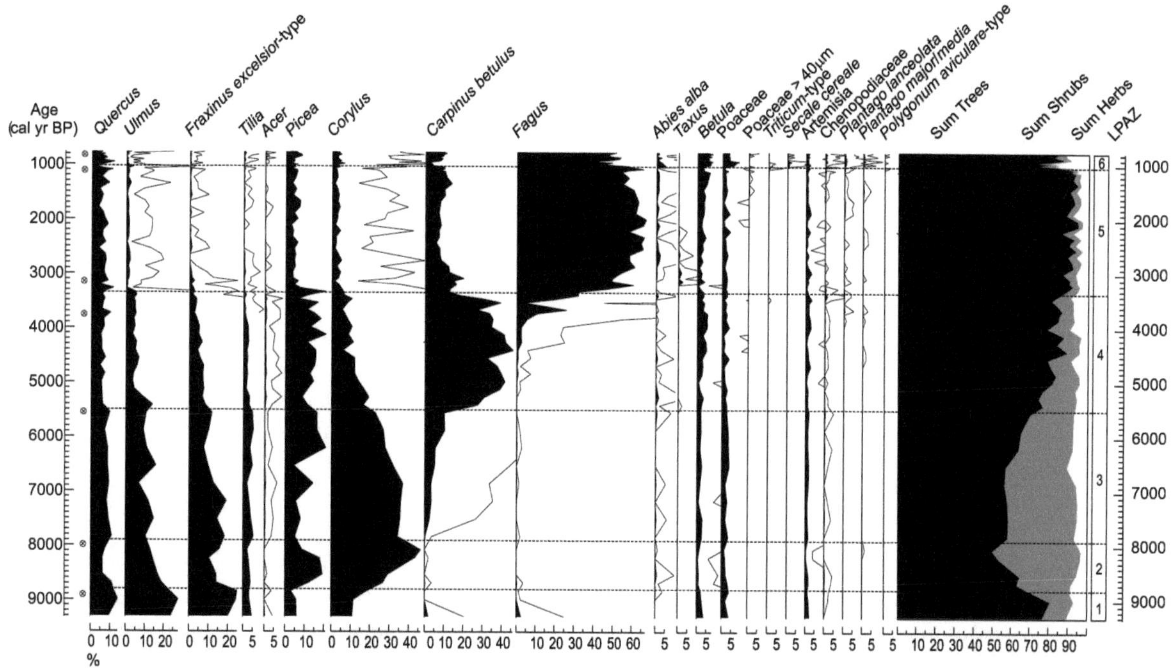

Fig. 4 Pollen percentages of selected trees, shrubs and herbaceous taxa plotted against calibrated BP timescale. Lines without filling are exaggerated (10×). LPAZ: local pollen assemblage zones

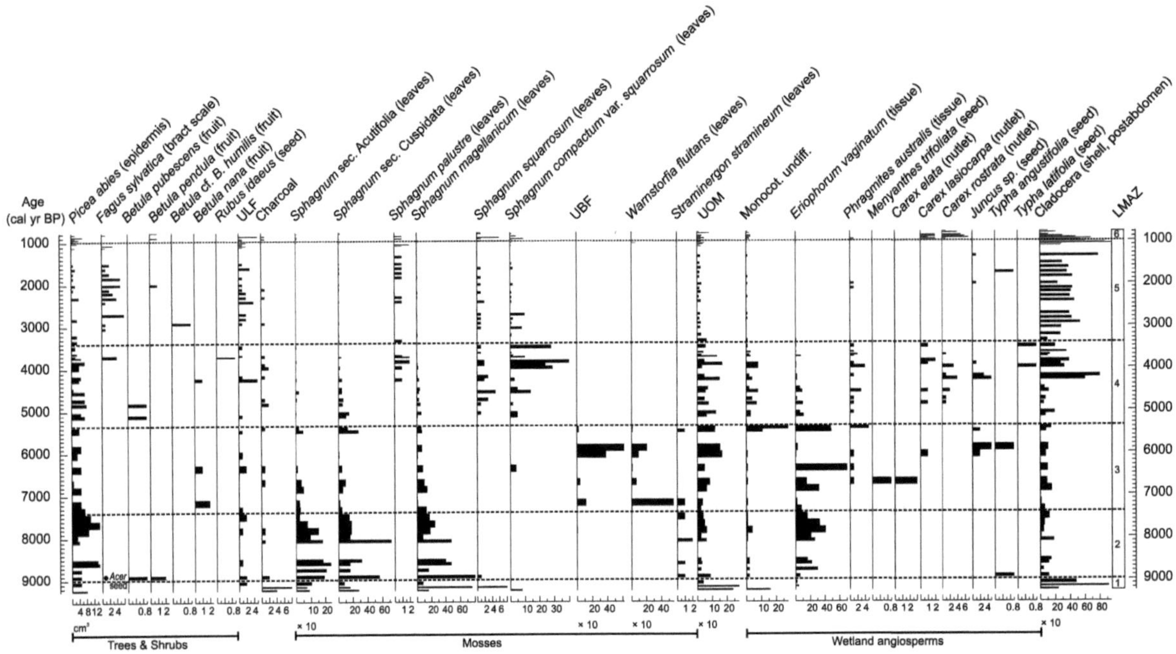

Fig. 5 Macrofossil concentration diagram for SZA-AB3, Lake Saint Ana with selected macrofossil remains presented on a calibrated BP timescale. UOM: unidentifiable organic matter; UBF: unidentifiable bryophyte fragment; ULF: unidentifiable leaf fragment; Monocot. undiff.: unidentifiable monocotyledon remains; LMAZ: local macrofossil assemblage zones. Note different scales!

(Fig. 5), and following a short reversal between 3600 and 3300 cal yr BP, *F. sylvatica* displaced *C. betulus* and partially *P. abies* to become the dominant forest tree around Lake Saint Ana. *Fagus* also out-competed the remaining populations of *Ulmus*, *F. excelsior* and *Tilia* in this period. *Fagus* retained its dominance in the forest around the lake until ca. 1100 cal yr BP, when anthropogenic forest clearance resulted in the spread of *P. abies* on the lakeshore (Fig. 5).

Local aquatic and wetland communities

Evidence of the nature of the local vegetation in and around Lake Saint Ana is provided by both the macrofossil and pollen records of wetland and aquatic plant taxa (Figs. 5, 6; Table 4). Six local macrofossil assemblage zones were distinguished, and their boundaries are shown in Fig. 5.

In order to make a quantitative assessment of the macrofossil record, PCA was performed on the 20 most abundant macrofossil components that also included cladoceran remains, the only fossil faunal element investigated. The sample scores of the second PCA axis clearly reflect the changes in water depth (Fig. 7a). Lower values along this axis were associated with open-water, lake-dwelling taxa, such as Cladocera and lake-marginal floating fen taxa (e.g. *Carex rostrata* and *Carex lasiocarpa*); while higher values along the second PC axis were associated with typical *Sphagnum*-bog macrofossil components, such as *Sphagnum magellanicum*, *Sphagnum sec. Cuspidata* and *Eriophorum vaginatum.* This second PC axis can therefore be taken as a hydrological gradient from *Sphagnum*-bog to open lake. A shift towards negative sample scores on a stratigraphic plot of this PC can therefore be interpreted as an increase in water depth and an opposite change vice versa. Figure 7b displays this stratigraphic sample score plot that will be interpreted further below.

Siliceous algae

In general, the diatom assemblages showed marked changes in species spectrum and often rapid changes in species abundances, with several species occurring only for a short period (Fig. 8; Table 5). Most diatom species identified in the sediment are acidobiontic or acidophilous and characteristic for oligo-dystrophic

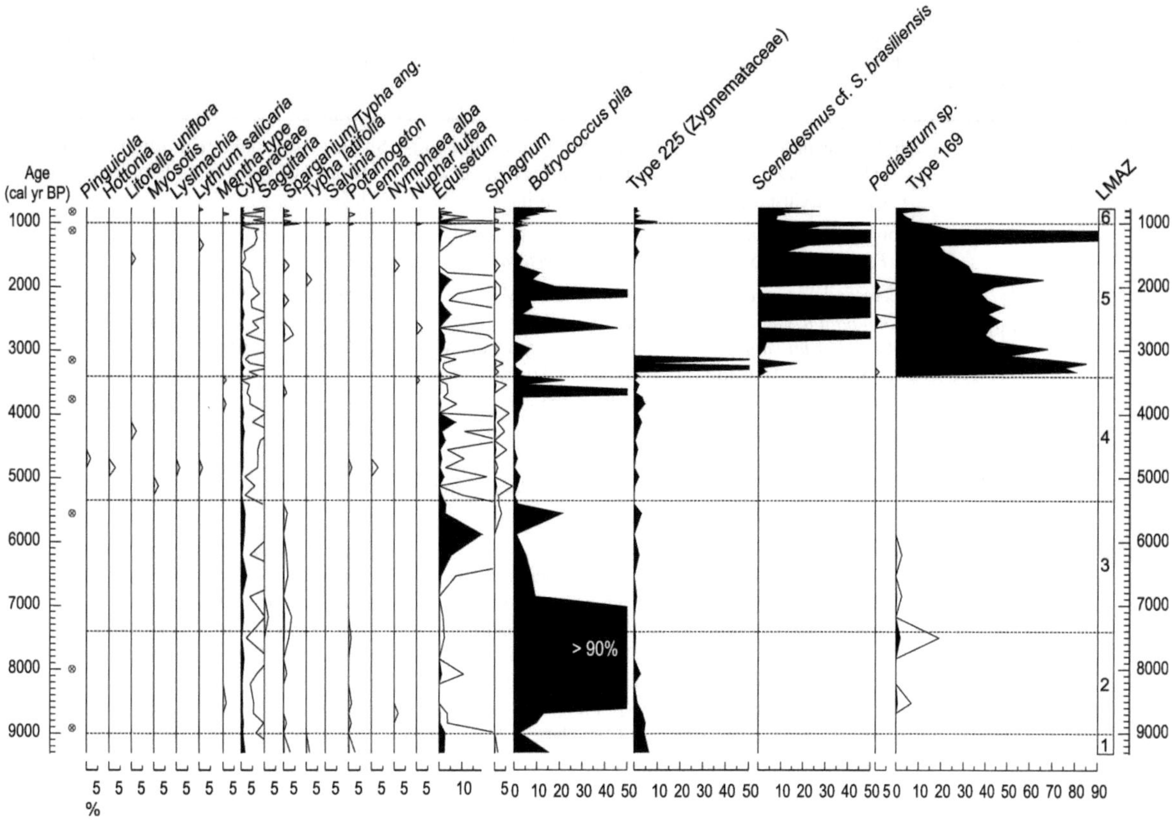

Fig. 6 Aquatic and wetland pollen types, green algae and other microspore relative frequencies from SZA-AB3, Lake Saint Ana. LMAZ: local macrofossil assemblage zones. Relative frequencies were calculated using the main pollen sum plus the count of the individual taxon. Lines without filling are exaggerated (10×)

water. Epiphitic, periphitic and benthic species were distinctly the most abundant taxa. Euplanktonic diatoms were represented by *Asterionella ralfsii* and *Stenopterobia delicatissima*. In addition, the tychoplanktonic *Tabellaria flocculosa* was abundant in several samples. This species occurs in the plankton and benthos as well. For a detailed summary of the lake's diatom flora see Buczkó & Magyari (2007).

Sediment chemistry

The chemical records of selected main elements, concentration changes of which are largely dependent on hydrological changes, are shown in Fig. 9. In order to examine the association of these elements and their changes in the sediment profile, PCA was performed separately on the two element fractions (endogenic/authigenic, allogenic) and the bulk element concentration profiles. These results are displayed on Fig. 10. In case of the allogenic fraction, the first PC explained 80% of the total variance and the stratigraphic plot of the PC values showed strong positive correlation with sediment inorganic content suggesting that changes in allogenic element concentrations mainly reflect varying input of the inorganic allochton material. The same trend is present in the stratigraphic plots of the first PC values for the endogenic/authigenic and bulk element concentrations, and association with typical clastic elements (Al, Mg) also support the interpretation of this plot in terms of varying clastic inwash. The second PC plot of the endogenic/authigenic and bulk concentrations shows similar trend and their association with Fe, Mn and Ca suggest that these curves mainly reflect endogenic/authigenic processes, such as changing oxic/anoxic conditions at the sediment–water interface, variation in the input of Fe, Mn and Ca from soil solution, and to a lesser extent, alteration of the organic sediment components (e.g. terrestrial macrophytes vs. aquatic organisms). To entangle which

Table 4 Macrofossil assemblage zones

Zone	Depth (cm) Age (cal yr BP)	Characteristics of zone	Inferred character of the wetland and lakeshore vegetation
SZA-M1	1023–1015 9300–9000	High concentrations of Cladocera (mainly *Bosmina* sp.), UOM, *Sphagnum squarrosum*, *Sph. magellanicum*, *Sph.* sec. Cuspidata & *Sph.* sec. Acutifolia. Abundant *Equisetum*; continuous presence of *Potamogeton*, *Sparganium*/*Typha angustifolia*.	Shallow open lake with oligo-mesotrophic water and *Sphagnum*-bog around the lake.
SZA-M2	1015–982 9000–7400	Decreased concentrations of Cladocera, UOM, *Sphagnum squarrosum*; increased abundance of *Eriophorum vaginatum*, *Sphagnum* sec. Cuspidata & *Sph.* sec. Acutifolia; occurrence of *Straminergon stramineum* and *Typha angustifolia* seed. Decreased abundance of *Equisetum*; *Potamogeton* occurs sparsely; single pollen grain of *Nymphaea* at 8700 cal yr BP; *Botryococcus pila* abundant, forms matrix of the sediment	Overgrowth of the shallow lake by *Sphagnum*-bog; presence of bog pools at the core location.
SZA-M3	982–950 7400–5350	Occurrence and high concentrations of *Warnstrofia exannulata*, UBF; UOM increases towards the top of the zone; temporary occurence of *Menyanthes trifoliata*, *Carex elata*, *C. lasiocarpa*, *Juncus* sp. and *Typha angustifolia*; *Phragmites australis* occurs in many samples; decreased concentration of all raised-bog *Sphagnum* species: *Sph. magellanicum*, *Sph.* sec. Cuspidata & *Sph.* sec. Acutifolia; *Potamogeton* pollen disappears; *Sparganium-Typha angustifolia* pollen present throughout; increased abundance of *Equisetum* spores from ca. 6500 cal yr BP.	Increasing water-depth, formation of poor-fen and poor-fen pools with *Warnstorfia fluitans*, *Phragmites australis*, *Typha angustifolia* and many *Carex* species; temporary decreases in water-depth inferred by *Eriophorum* peaks at 6400 and 5500 cal yr BP.
SZA-M4	950–887 5350–3400	High concentrations of heavily decomposed UOM; *Phragmites australis* remains abundant together with *Carex rostrata*, *C. lasiocapa*; decreasing abundance of *Eriophorum vaginatum*; Cladocera increase from 4300 cal yr BP; *Sphagnum compactum*, *Sph. squarrosum* & *Sph. palustre* appear and increase in this zone. *Sparganium-Typha angustifolia* pollen present throughout; temporary occurrence of *Hottonia*, *Pinguicula*, *Litorella uniflora*, *Lythrum salicaria* and *Lysimachia* pollen.	Oligo/mesotrophic shallow lake with submerged mosses (*Sph. compactum*) and floating aquatics; lakeshore tall-fen vegetation dominated by *Phragmites australis*, *Typha angustifolia* and *Carex* species. Increasing water-depth from ca. 4300 cal yr BP.
SZA-M5	887–680 3400–1000	Overall decrease in macro-remain concentrations; increased Cladocera concentration; low concentration of *Sphagum compactum* and *Sph. squarrosum*. Very low frequency of aquatic macrophyte pollen types; mass increase in Zygnemataceae at 3350 cal yr BP, followed by high relative frequencies of *Scenedesmus* cf. *S. brasiliensis*; temporary increases in *Botryococcus pila*.	Considerable increase in water-depth and volume, formation of oligotrophic open lake; sparse macrophyte vegetation. *Senedesmus* suggest seasonal nutrient enrichment, while *Botryococcus pila* peaks infer temporary water-depth declines or humic acid inwash
SZA-M6	680–600 1000–760	Increased concentration of *Carex rostrata*, *C. lasiocarpa*, Monocot. undiff., *Sphagnum compactum*, *Sph. squarrosum*; decreased concentration of Cladocera. *Sparganium*/*Typha angustifolia* pollen occurs in many samples; *Nuphar*, *Potamogeton* and *Salvinia* pollen occurs occasionally.	Decreasing water-depth, eutrophication and establishment of the lakeshore floating fen by *Carex* species and Sphagna. Temporary occurrence of nutrient-demanding aquatic plants.

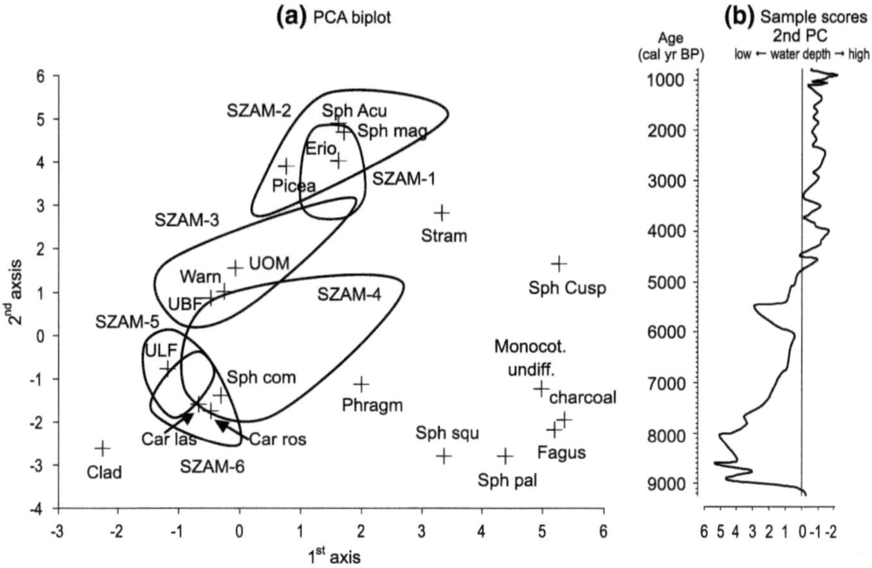

Fig. 7 PCA biplot showing the position of macrofossil zones and taxon scores together (**a**), and stratigraphic plot of the 2nd principal component scores (**b**). Macrofossil zones are mainly split along the second principal component (PC) axis that has been identified as the major hydrological gradient from peat bog (represented by high positive scores for *Sphagnum* sec. Acutifolia, *Sphagnum magellanicum* and *Eriophorum vaginatum*) to open water (represented by negative scores for Cladocera and marginal floating fen species such as *Carex rostrata* and *C. lasiocarpa*)

factor prescribes changes in these plots, they are evaluated together with the endogenic/authigenic element concentrations and inferences are given in Table 6. The stratigraphic plots of the third PCs (endogenic/autigenic, bulk) represent relatively small portions of the total variance (13% and 16%). Nonetheless, high eigenvalues for S and P suggest that the trend in these two curves can probably be interpreted in terms of endogenic/authigenic P input as S concentrations are steady all over the sequence.

Aluminium (Al) is the most abundant metal in silicate minerals and is therefore often used to infer accelerated clastic inwash usually in association with soil erosion (Engstrom & Wright, 1984). Because of the water insolubility of its minerals under circumneutral pH, catchment-derived Al usually becomes incorporated in the sediment, and hence its concentration changes mainly reflect catchment erosion processes. In the geochemical profile of Lake Saint Ana, Al concentrations, both factions, show significant inverse correlation with organic matter content ($r^2 = 0.75$, 0.82) suggesting the association of Al with inorganic sediment components, i.e. silicate minerals. The similar shape of the endogenic/authigenic and allogenic Al curves (Fig. 9a) moreover suggests that the separation between the two fractions is imperfect: some silicate-derived Al is likely present in the endogenic/authigenic fraction, and probably the same is true for Mg, Fe and Mn. This imperfect fractionation was demonstrated by applying extraction steps 1 and 2 on andesite bedrock samples taken from the lakeshore. Table 2 shows that the lakeshore andesite is rich in Al (16% Al_2O_3 content), and 7% of this is dissolved by extraction step 1. On the other hand, low Fe content (3.5–3.9%) is coupled with considerable solubility in HNO_3:H_2O_2. In order to overcome or minimise the effect of imperfect separation of the endogenic/authigenic and allogenic fractions, bulk element concentrations and Al-normalised bulk ratios were also calculated and plotted on Fig. 9b. Assuming that changes in the bulk concentration of Al are mainly attributable to changes in allogenic input, Al-normalised element ratios are expected to better represent endogenic and authigenic processes, such as redox changes in the catchment soil, changing oxic/anoxic condition in the water column and changing composition of the organic matter. Normalisation against immobile, clastic elements, such as Al, is widely applied in sediment geochemistry to aid the interpretation of the results (Brown et al., 2000; Felton et al., 2007; Schettler et al., 2007). Nonetheless, Al-solubility

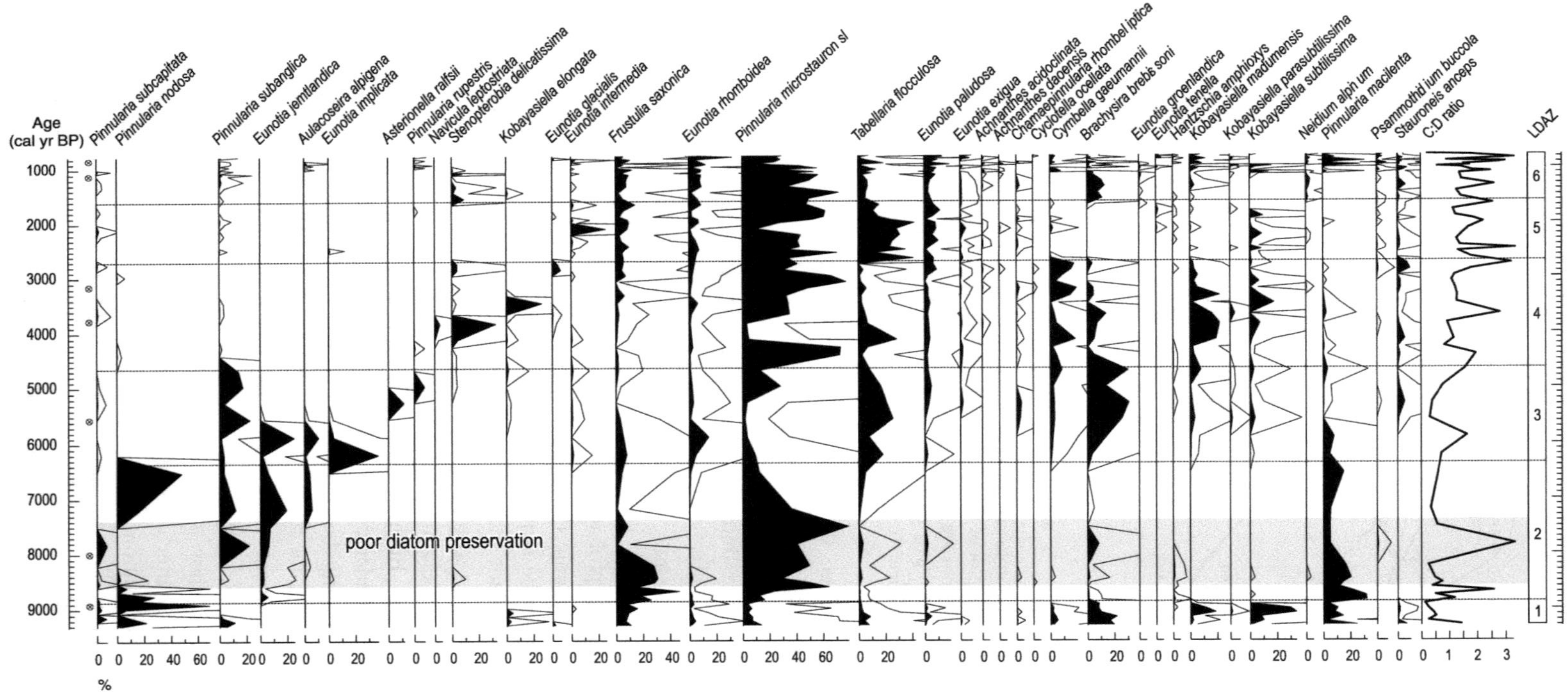

Fig. 8 Relative frequency diagram of important diatom species in SZA-AB3, Lake Saint Ana plotted against calibrated BP timescale. The grey-shaded part of the record represents poor diatom preservation and consequent uncertainty in the relative frequencies of diatom species. Diatom relative frequencies are expressed as percentage of total diatoms. Chrysophycean stomatocyst and diatom ratios (C:D) were calculated using the formula of Smol (1985). LDAZ: local diatom assemblage zones. Lines without filling are exaggerated (10×)

Table 5 Silicaceous microfossil assemblage zones

Zone	Depth (cm) Age (cal yr BP)	Characteristics of zone	Inferred character of the lake and wetland
SZA-D1	1023–1012 9300–8875	High relative frequencies *Brachysira brebissonii, Kobayasiella subtilissima*, . *madumensis, KFrustulia saxonica, Pinnularia nodosa* and *P. macilenta*. Also characteristic for this zone *Stauroneis anceps, Cymbella gaeumannii* and *Tabellaria flocculosa*; Chrysophycean cyst relative frequencies 15–20%.	Shallow lake with acidophilous diatoms suggesting the presence of lakeshore *Sphagnum*-bog, bog pools and high humic acid concentration in the water. The lake water was probably oligo/mesotrophic.
SZA-D2	1012–969 8875–6365	Increased relative frequency of *Frustulia saxonica, Pinnularia microstauron sensu lato* and *P. macilenta*. Considerably decreased frequencies of *Brachysira brebissonii, Kobayasiella subtilissima, K. madumensis, Frustulia saxonica, Pinnularia nodosa* and *Cymbella gaeumannii*; *Eunoti jemtlandica* and *Pinnularia subanglica* increase at ca. 8000 cal yr BP, while *Frustulia saxonica* decrease; high frequency of *Pinnularia nodosa* at 6450 cal yr BP; decreased frequency of *P. microstauron* after 7300 cal yr BP. Fragmented and partially dissolved diatom valves between 8600 and 7400 cal yr BP. Chrysophycean cyst relative frequencies are low, and show strong fluctuation 18–65%.	Development of *Sphagnum*-bog; dominance of acidophil diatoms until ca. 8000 cal yr BP; seasonal desiccation of the bog pools is inferred between 8600 and 7400 cal yr BP; increasing trophic level and water-depth from ca. 8000 cal yr BP.
SZA-D3	969–931 6365–4550	Increase of the benthic *Brachysira brebissonii* and tychoplanktonic *Tabellaria flocculosa* relative frequencies. Acidophilous species like *Eunotia implicata* and *E. rhomboidea* show high frequencies in the lower part of the zone (6365–5800 cal yr BP), whereas the euplanktonic *Asterionella ralfsii* occurs at 5250 cal yr BP. *Tabellaria flocculosa* relative frequencies decrease after 5300 cal yr BP. Chrysophycean cyst relative frequencies display strong fluctuation between 18 and 45%.	Gradual increase in water- depth until ca. 5250 years BP followed by temporary decrease. Fluctuation between opligotrophic and mesotrophic lake water status.
SZA-D4	931–839 4550–2700	Characterised by rapid changes in diatom frequencies; *Pinnularia microstauron* sensu lato dominates in most samples, except between 4300 and 3800 cal yr BP, when *Tabellaria flocculosa, Stenopterobia delicatissima* and *Navicula leptostriata* increase temporarily; also characteristic components of this zone *Kobayasiella madumensis, K. subtilissima* and *K. elongata*, all dwelling on *Sphagnum* species or in oligotrophic lakes; *Cymbella gauemannii*, a boreo/alpine species, also increase. Chrysophycean cyst relative frequencies increase from ca. 4600 cal yr BP to 50–65%.	Initially low water-depth and increased growth of Sphagnum-mire near core location; increasing water-depth between 4300 and 3800 cal yr BP followed by low water-depth and re-expansion of the lakeshore *Sphagnum* carpet. The lake water becomes more acidic and oligotrophic in this phase.
SZA-D5	839–759 2700–1615	Step-wise increase in the relative frequency of *Tabellaria flocculosa* (23%), increase in *Sphagnum*-mire indicator *Frustulia saxonica* and *Eunotia paludosa*; high, but strongly fluctuating frequencies of *Pinnularia microstauron sensu lato*. Decrease in the frequency of *T. flocculosa* about 2000 cal yr BP. Further increase in Chrysophycean cyst relative frequencies (55–65%).	Considerable increase in water-depth and volume, formation of oligotrophic open lake; *Sphagnum*-bog along the lakeshore; decrease in water-depth around 2000 cal yr BP.
SZA-D6	759–600 1615–760	Low relative frequencies of *Tabellaria flocculosa* (2–18%), increase in *Brachysira brebissonii* and *Stenopterobia delicatissima* between 1615 and 1200 cal yr BP; increased frequencies of *Pinnularia macilenta, Kobayasiella madumensis, Cymbella gaeumannii* after 1200 cal yr BP. Chrysophycean cyst relative frequencies remain high, but fluctuate considerably (30–65%).	Decreasing water-depth, light penetration to lake bottom favouring benthic diatoms; expansion of *Sphagnum*-dwelling epiphytic diatoms from 1200 cal yr BP suggesting expanding lakeshore *Sphagnum*-carpet; oligotrophic lake water status.

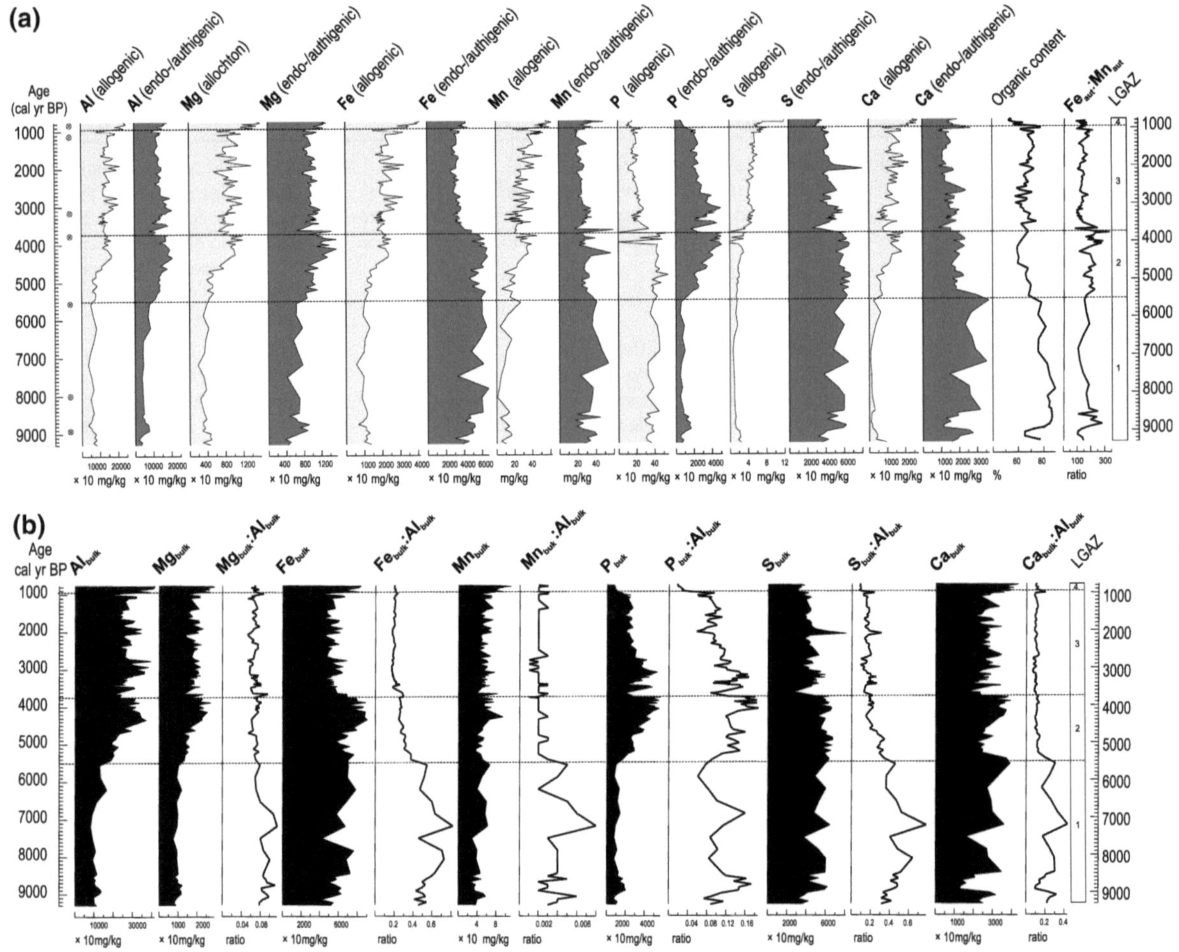

Fig. 9 Endogenic/authigenic and allogenic element concentrations (**a**); bulk element concentrations and Al-normalised bulk ratios (**b**) from Lake Saint Ana, core SZA-AB3. On the right of diagram (**a**) sediment organic content and Fe:Mn ratios are also displayed. Note different scales. LGAZ: local geochemical assemblage zones

is strongly pH dependent (Ulrich & Pöthig, 2000), and as Boyle (1994) has pointed out that this can cause increased Al-dissolution from the sediment to the water column in acidified lakes. He found that acid-liable Al concentrations in the near-surface sediments of acidified lakes were directly proportional with pH below 4.8–5.0. It is important to note, however, that acid-liable Al is only a small portion of the endogenic/authigenic element fraction of our measurements. The recent pH range of the lake is between 4.2 and 6.4 (Table 1), and the lowest diatom-inferred pH is 4.9 in the sediment profile (Fig. 11). Therefore, we cannot exclude the possibility that alternating pH had an effect on bulk sedimentary Al concentrations. Taking into account these limitations, we have chosen Al for normalisation because other weathering-resistant, immobile elements, such as Ti and Zr (Koinig et al., 2003) have not been measured. Figure 9b displays Al-normalised bulk ratios, and their interpretation is presented in Table 6.

Generally, the original element concentrations and Al-normalised ratios show considerable differences between 5500 and 3750 cal yr BP, with steeply declining Fe, Mn and S ratios that remained obscured in the concentration profiles, possibly because of the increasing input of silicate minerals. Endogenic/authigenic P shows strong positive correlation with endogenic/authigenic Al, especially in LGAZ 1–3 suggesting that inwashed P has co-precipitated with Al in the lake (Wilson et al., 2008). P is typically the

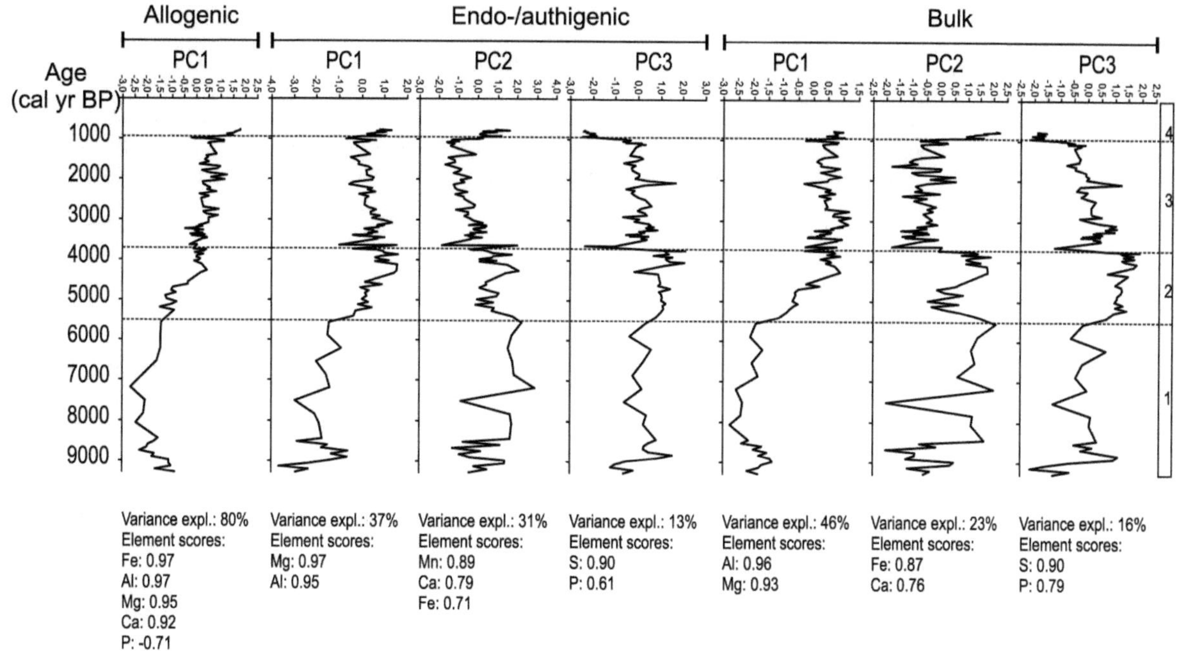

Fig. 10 Stratigraphic plots of principal component analysis (PCA) sample scores of the element concentrations. PCA was performed separately on the allogenic, endogenic/authigenic fractions and bulk element concentrations

limiting nutrient for primary productivity and, thus, controls the trophic status of a lake (Schindler, 1977). P availability in acidic/oligotrophic lakes, like Saint Ana, is controlled by many factors, but most importantly by the reductive dissolution versus oxidative precipitation of Fe and associated P during hypolimnetic anoxia versus oxia, and external Al input that removes much of the P from the water column above pH 4.8. This latter process is pH dependent but not redox dependent (Kopáček et al., 2005). It seems that during the lake's history, both of these processes influenced P availability in the water, Table 6 provides the interpretation of the record on the basis of these guidelines.

The most important indicators of changing oxic/anoxic conditions used to infer redox changes are Fe and Mn (Table 6). Their redox cycling in lakes and mires bears a relation to the availability of oxygen at the sediment–water interface. With anoxic conditions, developing usually in stratified lakes over the summer months and during winter ice cover, Fe and Mn are reduced and released into the water, while oxic condition in shallow, well-oxygenated lakes and during spring and autumn turnover of dimictic lakes favour the precipitation of Fe and Mn oxides (Mackereth, 1965; Davison, 1993; Hamilton-Taylor et al., 1996). Their absolute concentrations in the sediment are in addition affected by the redox condition of the watershed soil, with high rate of Fe and Mn mobilisation in reductive soils (such as podzol), and according to high input from soil solutions into the lake. Conversely, oxidative soils show low rate of Mn and Fe mobilisation and accordingly low input into the lake. Lake water pH also affects Mn and Fe solubility. The study of Bendell-Young and Harvey (1992), for example, revealed that naturally acid and acidified lakes (pH 5.0–5.6) contain almost twice the concentration of total Fe compared to circumneutral lakes due to the acid-induced precipitation of Fe humic substances; contrarily, acid lakes are depleted in Mn oxides as a result of increasing solubility of Mn^{2+} at low pH. Interpretation of the Fe and Mn curves therefore requires the recognition and separation of these three different processes that is aided by the examination of the Fe:Mn ratios and diatom-inferred pH changes in the sediment profile of Lake Saint Ana. These considerations were used to make hydrological inferences from the chemical profiles in Table 6.

Table 6 Sediment chemistry assemblage zones

Zone	Age (cal yr BP) Depth (cm)	Characteristics of zone	Inferred hydrological and catchment changes
SZA-G1	9300–5500 1023–953	High endogenic/authigenic concentrations and bulk Al-normalised ratios of Ca, S, Fe, Mn; low allogenic Al, Mg, Fe and Mn concentrations; note endogenic/authigenic and Al-normalised Fe and Mn minima at 7550 cal yr BP followed by peaks at 7200 cal yr BP.	High input of Fe and Mn from soil solutions in combination with oxic conditions at the sediment–water interface; high concentration of Ca and S suggest high production by wetland macrophytes and together with high organic content indicate peat accumulation. Fe and Mn concentration changes suggest bottom water anoxia around 7550 cal yr BP followed by oxic conditions and Fe, Mn precipitation around 7200 cal yr BP.
SZA-G2	5500–3750 953–908	Two-step increase in allogenic Al, Mg, Fe, Mn, Ca, S concentrations at 5500 and 4500 cal yr BP; steep decline in Al-normalised bulk Fe, Mn, Ca, S ratios; increasing endogenic/authigenic P concentrations and Al-normalised ratios. Fe:Mn ratio display small increase.	Increasing input of clastic minerals probably as a result of increasing water-depth and expansion of open lake area; reducing soils of the previous zone are replaced by more oxic soils; increasing primary production by aquatic organisms, P accumulation in the sediment under changing oxic/anoxic conditions at the sediment–water interface. Probably two-step increase in water-depth: at ca. 5500 and 4500 cal yr BP.
SZA-G3	3750–970 908–666	Rapid decrease in endogenic/authigenic Fe, Al, Mg and P concentrations at 3750 cal yr BP; endogenic/authigenic Fe remains low, while Al, Mg and P increase at 3300 cal yr BP followed by a second decline in P after 2750 cal yr BP; note that Al-normalised bulk Fe ratios show a smaller decrease than original concentrations. The endogenic/authigenic P concentration and Al-normalised ratios show similar trend; decreasing Fe:Mn ratios.	P, Al and Mn mobilisation and recycling at the sediment–water interface between 3750 and 3300 cal yr BP probably as a result of seasonal anoxia; the steady authigenic Fe decline is probably attributable to decreasing concentrations in soil solution, and so indicate more oxic soils (note coincident local *Fagus sylvatica* establishment and *Picea abies* decline); temporary return to well oxygenated sediment–water interface around 3500 cal yr BP followed by anoxia; dimictic stratification of the lake, seasonal anoxia and P recycling are inferred from ca. 2750 cal yr BP by decreasing endogenic/authigenic P, but it is also conceivable that external P input decreased after 2750 cal yr BP as P also decreased in the allogenic sediment fraction.
SZA-G4	970–760 660–600	Increased concentration of allogenic and endogenic/authigenic Al, Mg, Fe, Mn and Ca concentrations; Fe:Mn ratios decrease; decreasing endogenic/authigenic P concentrations and Al-normalised ratios.	Enhanced soil erosion, inwash of silicate minerals; probable acidification of catchment soils leading to diminished P in soil solution (fixation of P at $pH < 5.5$); note coincident return of *Picea abies* to the lakeshore and disappearance of *Fagus sylvatica*.

Inferred changes in water depth, pH and trophic status: a synthesis

For the reconstruction of changes in water depth, the relative frequencies of planktonic diatom species, chrysophycean:diatom (C:D) ratios and the macrofossil-inferred hydrological gradient (as inferred by the second PC) were used. These were reconciled with evidence of lake level and trophic change suggested by changes in sediment composition, sediment chemistry and pollen spectra. In case of the diatoms, increasing water depth was inferred from the spread of tycho- and euplanktonic species, whilst decreasing water depth was indicated by the expansion of epiphitic and littoral species (Figs. 10, 11).

A diatom-based transfer function was used to give quantitative estimates of past trends in lakewater pH and dissolved phosphorous (P), both limnological variables strongly related to catchment vegetation, soils and water depth.

Changes in the C:D ratios represent changes in the trophic status of the lake water. This relationship has been demonstrated by Smol (1985). The underlying ecological principle is the increasing ratio of Chrysophytes relative to diatoms in the epilimnion with decreasing availability of nutrients (Smol, 1985;

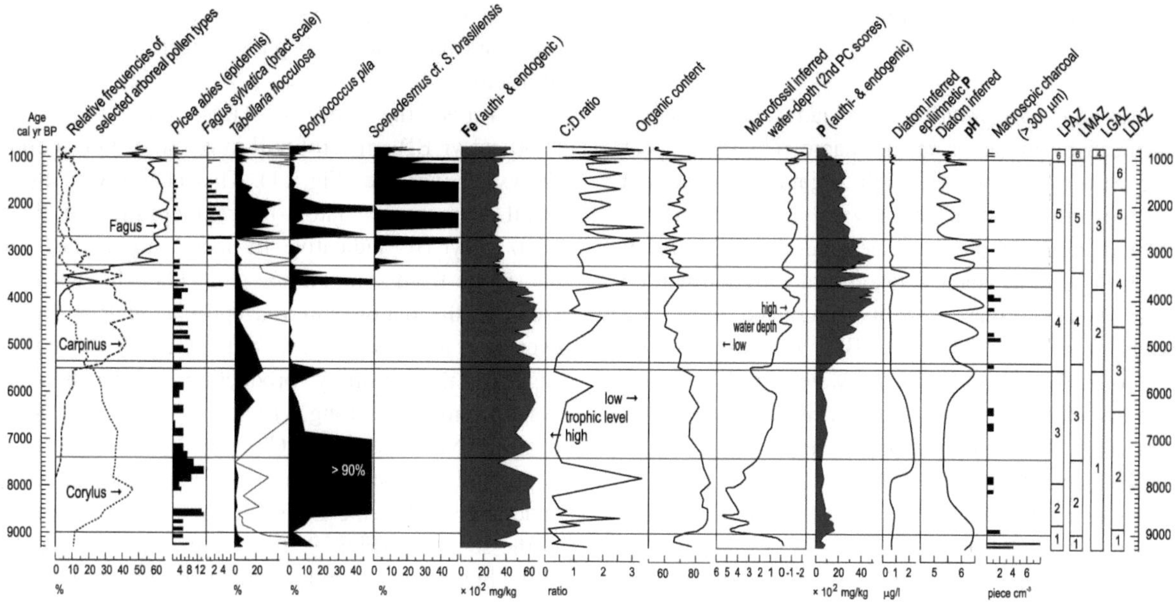

Fig. 11 Selected palaeo-proxies and assemblage zone boundaries plotted against calibrated BP timescale, SZA-AB3, Lake Saint Ana. Note different scales. For the estimation of total epilimnetic P and pH, the diatom record was used along with the training dataset of the EDDI database. Water chemistry parameters were obtained by comparing the fossil and recent diatom assemblages using transfer function (LWWA). Details of this technique are given by Birks (1998); Juggins (2001). Diatom-inferred total P and pH are smoothed using a 3 point moving average. C:D ratio abbreviates Chrysophycean stomatocyst and diatom ratios; LPAZ: local pollen assemblage zones, LMAZ: local macrofossil assemblage zones; LGAZ: local geochemical assemblage zones; LDAZ: local diatom assemblage zones

Sandgren, 1988). Therefore, increasing C:D ratios are interpreted as shifts to more nutrient-poor (oligotrophic) lake water and vice versa. In lakes set in nutrient-poor environment with small, stable catchment area, like Lake Saint Ana, available nutrients for algal growth will in a large extent be influenced by water volume via dilution of incoming nutrients (Smol et al., 2001). Therefore, increasing oligotrophy may also be connected to increasing water depth. This relationship is partially distorted in peat bogs, where periphitic species of Chrysophytes are abundant and enjoy competitive vigour over diatoms, especially in dry peat surfaces (Wilkinson et al., 1997; Rühland et al., 2000).

Changes in water depth, trophic status and associated catchment processes are discussed mainly using the macrofossil zonation scheme that showed generally good agreement with the pollen and geochemistry zones, with only small offsets in zone boundaries. Stratigraphic zone division delineated by the diatom record, however, differed markedly (Fig. 11). This implies that diatoms are tracking different environmental variables, or perhaps are more sensitive to some of the environmental changes (Jasinski et al., 1998).

9300–9000 cal yr BP: shallow lake encircled by *Sphagnum*-bog

The proxy data along with the short length of cores obtained in the NE part of the present lake basin (Fig. 2) suggest that the initial lake occupied only the deepest part of the crater, with the present shallow NE basin being exposed and probably covered by *P. abies* forest (Fig. 2). High humic acid concentration of the water and the acidophilus bog conditions suggested by the diatoms corroborate the macrofossil record that is dominated by peat bog *Sphagnum* species and cladoceran remains. The diatom-inferred pH shows values between 6 and 6.5, along with very low values of epilimnetic P (0.7 μgl^{-1}), the latter suggesting strongly P-limited primary production. All proxies therefore agree in suggesting the presence of a shallow oligo/mesotrophic lake surrounded by peat bog in the earliest phase of lake's history.

9000–7400 cal yr BP: *Sphagnum*-bog and bog pools

Decreasing concentrations of cladoceran remains, balanced by increases in the macro remains of peat bog *Sphagnum* species, high organic content and changes in the diatom flora toward the dominance of moss-dwelling epiphitic species suggest a decrease in water depth. The shallow lake became overgrown by *Sphagnum*-bog that probably alternated with bog pools from ca. 9000 cal yr BP. The diatom and macrofossil record signals the water depth decrease and associated changes in pH with an offset of ca. 400 years. The compositional change of the macrofossil assemblages at 9000 cal yr BP (Fig. 11) is followed by the damage and partial dissolution of the diatom frustules from ca. 8600 cal yr BP suggesting temporary desiccation of the peat bog surface. The driest condition is therefore inferred by the diatom record for the period 8600–7400 cal yr BP. Diatom dissolution was probably caused by organic acids under oxic conditions (Lawson et al., 1978; Bennett et al., 1991; Gell et al., 1994). This inference is corroborated by the coincident mass expansion of *Botryococcus pila* green alga in the sediment (Fig. 11). According to Komárek and Jankovská (2001) and Jankovská (2001), the presence of *B. pila* indicates dystrophic water of small peaty lakes. Its great quantity can possibly be correlated with the poor preservation of the diatoms. The oil compounds stored by these algae break up and produce organic acids during decay that, in turn, trigger the dissolution of siliceous compounds.

The diatom-inferred oxidative surface conditions, and the possible temporary drying out of the bog pools is also coincident with a major terrestrial vegetation change, i.e. the spread of *Corylus avellana* (Figs. 4, 10). Since the lake basin is fed by rainwater exclusively, the reconstructed water depth decrease can probably be connected to the decrease in available moisture in this phase, and especially between 8600 and 7400 cal yr BP.

Diatom-inferred pH displays gradual decrease around 8800 cal yr BP from 6.5 to 5.3 suggesting acid conditions. This inference is in agreement with the development of *Sphagnum*-bog in the crater, and probably also bears a relation to the expanding *P. abies* forest around the lake, which facilitated acidification in the lake basin via the acidification of the soil (Ranger & Nys, 1994). Diatom-inferred P remained very low in this phase, suggesting strong P-limitation.

Towards the end of this phase, at ca. 8000 cal yr BP, macrofossil-inferred water depth shows an increase (Fig. 11). This is followed by a small increase in diatom-inferred epilimnetic P at 7800 cal yr BP indicative of increasing inwash of phosphate from soil solution (Fig. 11). The absolute value of the diatom-inferred epilimnetic P is, however, very low; the increase from 0.6 to 1.8–2.3 $\mu g l^{-1}$ suggests that primary production in the aquatic system remained strongly limited by available phosphate P (Newbold, 1992; Cloot & Roos, 1996). This is probably the reason why neither the allogenic nor endogenic/authigenic P displays detectable increases in the sediment (Figs. 9, 10). Increasing water depth in the bog pools and more stable water levels are also suggested by the spread of benthic diatoms, such as *Eunotia jemtlandica* and *Aulacoseira alpigena*.

C:D ratios display strong fluctuation in this phase, with initially increasing values (between 9000 and 8700 cal yr BP) followed by a decline and finally a peak at 7820 cal yr BP (Fig. 11). As discussed earlier, Chrysophyceans enjoy competitive vigour over diatoms in nutrient-poor lakes and also in dry peat bogs, where periphitic forms can dominate (Wilkinson et al., 1997). Given the inferred development of a *Sphagnum*-bog in the crater, the fluctuation in C:D ratios in this phase may not entirely reflect trophic changes, but a combination of dry peat surface and small-scale changes in nutrient availability, and is therefore difficult to interpret. Dry peat surface conditions and/or low nutrient availability are inferred for 9000–8700 cal yr BP and around 7820 cal yr BP.

7400–5500 cal yr BP: poor-fen and poor-fen pools with gradual water depth increase

The gradual water depth increase that started at ca. 8000 cal yr BP continue iduring this phase. A typical poor-fen pool moss, *W. fluitans*, appears and remains abundant in this phase suggesting the transformation of the *Sphagnum*-bog into poor-fen alternating with poor-fen pools. This inference is also supported by the temporary occurrences of a set of wetland macrophytes requiring higher nutrient supply: *Menyanthes trifoliata*, *Carex elata*, *C. lasiocarpa*, *Juncus* sp., *Typha angustifolia*, *Phragmites australis* (Fig. 5; Table 4). Increasing C:D ratios also support a modest increase in nutrient availability; however, the diatom-inferred pH displays very low values (>5.5) until ca. 6200 suggesting nutrient inwash from acidic soils. This inference is in-line with the macrofossil inferred sustained dominance of *P. abies* on the lakeshore and possibly further upslope in the crater basin. Notable is that a different inference would be drawn from the pollen record alone that shows decreasing relative frequencies of *P. abies* and *C. avellana* balanced by increases in mixed oak woodland elements (*Quercus*, *Fraxinus excelsior*, *Tilia*) and *Carpinus betulus*. It seems that the pollen record captures changes on a regional scale rather than locally, and an increased load of mixed oak woodland tree pollen and *Carpinus betulus*, possibly arriving from lower elevation and from the summit zone, distort the percentage representation of the local *Picea abies* forest. Upslope transport of pollen in mountain area is common and has been demonstrated by several studies (Markgraf, 1980; Solomon & Silkworth, 1986). The pollen record is therefore rather useful in deciphering shifts in extralocal and regional vegetation that in our case are indicative of gradual warming. It is possible that nutrient availability was favoured by rising temperatures; in this respect the two proxies complement each other.

Sustained relatively high concentrations of endogenic/authigenic Ca in the sediment (Fig. 9) suggest that the organic component of the sediment is still dominated by the remains of mosses and wetland macrophytes (Kooijman, 1993). The crater probably remained under wetland macrophyte vegetation cover to a large extent in this phase, and the extent of open water and algal production was probably still limited. Only minor increases in allogenic input of clastic-associated elements are seen (Fig. 9) corroborating the macrofossil-based inference of a thick mire belt around the poor-fen pools. This probably acted as barrier against the inwash of inorganic material.

An interesting feature of this phase is the delayed decrease of *Botryococcus pila*. This alga forms the matrix of the sediment until ca. 7000 cal yr BP (Fig. 11); its withdrawal takes place only 400 years after the decrease of typical peat bog *Sphagnum* species suggesting that bog pools were only gradually transformed into poor-fen pools, and by ca. 6375 cal yr BP into a shallow lake. This latter date marks an important change in the diatom flora, namely the spread of the tychoplanktonic *T. flocculosa* and the benthic *Brachysira brebissonnii*. Although both species are acidophilous, abundant in oligo/mesotrophic lakes, poor-fens and bog pools, they replace a diatom assemblage characterised by even more acidophilous and acidobiontic species (Fig. 8). The diatom-based pH reconstruction captures this change in pH. An increase from 5.4 to 6.5 takes place between 6375 and 5500 cal yr BP. Notable is that *Sphagnum* species disappear in this higher pH period, except for *Sph. magellanicum*, that supports the diatom-based pH increase. It is unclear what might have triggered the increase in pH, as no coincident change in sediment chemistry and terrestrial vegetation is seen apart from the gradual spread of *Carpinus betulus* regionally. The most likely mechanism involves increasing soil temperatures triggering enhanced microbial activity and the transport of base ions and nutrients from soils into the pools.

5500–5350 cal yr BP: temporary decline in water depth?

The macrofossil record signals a short-term decrease in the water depth and return to increased abundance of *Sphganum sec. Cuspidata*, *Sph. sec. Acutifolia* and *Eriophorum vaginatum* between 5500 and 5350 cal yr BP (Figs. 5, 10), though statistically significant zone boundary was not found. These typical peat bog elements are, however, mixed with *Phragmites australis* and unidentifiable Monocotyledon remains, the latter probably also derived from *Phragmites* or *Carex* species, therefore the assemblage indicates a mixture of oligotrophic and more nutrient-demanding (mesotrophic) wetland habitats. *Botryococcus pila* relative frequencies also increase temporarily supporting the macrofossil inferred reversal to *Sphagnum*-bog and bog pools; however, the diatom-inferred pH and C:D ratios display increases, from which an increase in trophic level and pH are inferred. Furthermore, *T. flocculosa* and *Brachysira brebissonii* frequencies remain relatively high indicative of shallow open water or fen-pools. Overall, antagonistic changes characterise this short period taking place simultaneously with a major terrestrial vegetation change, namely the massive spread of *Carpinus betulus* in the region (Fig. 11), and very likely in the upper slope and summit area of the crater as well. *Carpinus betulus* replaced mainly *Corylus*, but at the lakeshore *P. abies* forest was not affected as evidenced by the macrofossil record (Figs. 5, 10). Increased nutrient availability in parts of the basin is related to the terrestrial vegetation change rather than wetland development. Changes in the wetland vegetation follow the rise in *C. betulus*; therefore, if a connection between the two processes is assumed, then the casual factor was probably the terrestrial vegetation change that, in turn, very likely responded to a change in macroclimate.

5350–3700 cal yr BP: mesotrophic lake phase

All hydrological proxies—diatom, D:C ratio, organic content, macrofossil—denote rapid increase in water depth at the onset of this phase. Statistically significant zone boundaries were found at 5500 cal yr BP in the pollen and chemical records, whereas the macrofossil record displayed a delayed change at ca. 5350 cal yr BP given the preceding short-term decrease in water depth. In the diatom record, the occurrence of the euplanktonic *Asterionella ralfsii* is confined to the early part of this phase (Fig. 8) further supporting increasing water depth (Figs. 10, 11). Maximal coverage of *Carpinus betulus* on the higher slopes and summit area is evidenced by the pollen record, but the lakeshore has remained under *P. abies* cover. *C. betulus* probably did not reach the lakeshore, as no macrofossil remains of this tree species were found. These changes together with increasing inorganic content (Fig. 11) suggest an increase in surface runoff probably associated with higher rainfall and/or decreased evaporation loss of the incoming water in the crater.

Al-normalised Fe and Mn ratios decreased rapidly from ca. 5500 cal yr BP (Fig. 9). Along with a small increase in Fe:Mn ratios, this was interpreted to reflect seasonal anoxia at the sediment–water interface coupled with decreased input of Fe and Mn from catchment soils (Table 6). It is well known that Fe and Mn concentrations decrease in soil solutions if soil pH increases (Mackereth, 1966). This very likely took place around Lake Saint Ana with the establishment of *C. betulus* on the slopes, and the geochemical record reveals that despite the surviving *Picea abies* forest on the lakeshore, changes in the soil further upslope have started. Endogenic/authigenic P concentrations and Al-normalised P ratios showed a massive increase in this phase, coincidently with the expansion of *Phragmites australis* (Fig. 5). Note also that diatom-inferred epilimnetic P declined at the same time along with an increase in pH (Fig. 11). Taking into account the results of several other sedimentary phosphorous studies (for summaries, see Lerman, 1978; Engstrom & Wright, 1984; Gibson, 1997), it seems likely that the increased concentration of P in the sediment can be connected to increasing inwash, bringing P in the growing lake. This facilitated the primary production and the intense accumulation of P-rich macrophytes between 5350 and 3300 cal yr BP and limited P recycling. The applied technique does not allow us to determine in what form endogenic/authigenic P is present in the sediment, but the lack of correlation with the traditionally clastic elements (Al, Mg) suggests that it is bond to organic compounds, and most likely originates from the lakeshore *Phragmites*-swamp.

Within this phase, a distinct further increase in water depth is inferred by the sudden population increase of Cladocera at 4300 cal yr BP that coincides with a decrease in organic content (Fig. 11).

3700–3300 cal yr BP: ecosystem on the move, increasing water depth punctuated by decline

This short phase is discussed separately because many short-term changes take place during this period suggesting ecosystem disturbance, and one proxy, the endogenic/authigenic Fe concentration signals irreversible change in catchment soils (Figs. 9, 10). The pollen and macrofossil records corroborate each other by signalling the spread of *F. sylvatica* on the lakeshore at 3700 cal yr BP followed by its temporary disappearance and re-establisment at 3300 cal yr BP. *Fagus* out-competed both *Carpinus* and *Picea* by 3300 cal yr BP, but this process was interrupted by external disturbances as indicted by the frequently changing profiles of Fig. 11. This figure demonstrates that the first increase in *Fagus* pollen and appearance of its macro remains at 3700 cal yr BP are concurrent with the temporary spread of *B. pila*, the latter being indicative of the inwash of humic acids, decreasing water depth and probably also colder water (Komárek & Jankovská, 2001). Note also that macrocharcoal was frequently encountered in the sediment (Fig. 11) between 4000 and 3700 cal yr BP; thus local fires may also helped the local establishment of *F. sylvatica*. Together with the macrofossil-inferred temporary water depth decline around 3700 cal yr BP, the inference of a short cool/dry event gets a firm basis. It is likely that climate change favoured the lakeshore establishment of *F. sylvatica*. The concurrent decrease in endogenic/authigenic P and Al concentrations is also indicative of a change in the catchment-aquatic system, but they are difficult to interpret. They can either indicate a pH decline in the lake water, below 4.8–5.0 (Boyle, 1994), as it leads to Al and P recycling; alternatively, the P decline may resulted from repressed primary production as conditions have changed in the lake. Supportive of the first rationale is the diatom inferred decrease in pH, but values remain around 5.5. Note, however, that these are average values, and given the large-scale fluctuation in lake water pH between seasons today, it is conceivable that during autumn and winter pH decreased well below 5.5. The second rationale is supported by the decreasing endogenic/authigenic Ca concentration around 3700 cal yr BP, being indicative of decreasing populations of waterside plants and bryophytes. On the whole, neither of these hypotheses can unambiguously be proved; therefore, the underlying processes of the temporary P and Al decline in the sediment remain uncertain.

The P and Al minima at 3700 cal yr BP are followed by peak values between 3570 and 3430 cal yr BP. Notable is the concurrent temporary disappearance of *F. sylvatica* balanced by the re-expansion of *C. betulus* (Fig. 11). The macrofossil record suggests an increase in water depth, whereas the diatom-inferred epilimnetic P displays a concurrent peak along with increasing pH, being indicative of increased production and oxic conditions in the lake. Altogether these proxies denote a reversal to warmer conditions before the final spread of *F. sylvatica*. Warmer summers and increasing water depth possibly promoted algal production, which in turn increased pH.

Endogenic/authigenic P and Al decreased again around 3430 cal yr BP and remained low until 3300 cal yr BP, probably indicating a second cycle of decreased lake level and low primary production. Note that this event is concurrent with the massive spread of *Fagus* that replaced *Picea* on the lakeshore by 3300 cal yr BP.

Another important change in this transitional period is the decreasing input of endogenic/authigenic Fe related to soil change in the catchment (Table 6), similar to the effect of spread of *C. betulus*. The *P. abies* forest present on the lakeshore since the Early Holocene has been replaced by *F. sylvatica* between 3700 and 3300 cal yr BP, and the chemical profile suggests that this was probably the cause of the further decline in endogenic/authigenic Fe input. The podzol soil typical of the *Picea* woodlands was probably altered by the litter of *F. sylvatica* via the intensification of microbial activity. This probably increased pH and thereby cut down Fe dissolution in the soil (Berger et al., 2004).

3300(2700)–1000 cal yr BP: oligotrophic deep lake phase

During the *Carpinus*/*Fagus* and *Picea*/*Fagus* forest change (ca. 3300 cal yr BP), aquatic organisms show diversified reactions. The macrofossil composition of the sediment changes concurrently with the terrestrial vegetation, but does not imply any significant increase in water depth (Fig. 11). The diatom flora, on the other hand, indicates a marked increase in the water depth at 2700 cal yr BP, i.e. 500 years later than the forest change. In the microfossil record, *Scenedesmus* cf. *S. brasiliensis* remains are first detected around 3300 cal yr BP; however, they only occur *en masse* from ca. 2700 cal yr BP (Fig. 11). *Scenedesmus* species are planktonic, the occurrence of which in the sedimentary records usually indicates increasing water depth as more pelagic habitats become available (Vorren et al., 1988), and increasing nutrient availability, especially P (Reynolds, 2006). In such strongly P-limited systems as Lake Saint Ana, the spread of *Scenedesmus* is indicative of a prominent increase in available P that is, however, inconsistent with the diatom-inferred low epilimnetic P for this phase (Fig. 11). Cell number of the *Scenedesmus* remains was generally low (2–4) that would suggest a strong nutrient limitation (Reynolds, 2006), but still experiments with internal and external P loading and phytoplankton response clearly demonstrate that increased P levels result in a greater dominance of *Scenedesmus* in natural lakes in the temperate zone (Cymbola et al., 2008). If we look at the endogenic/authigenic P concentration and Al-normalised P ratios of the sediment, they show distinct decreases at 2700 cal yr BP (Figs. 9, 10). It seems that decreasing P in the sediment and increasing *Scenedesmus* relative frequencies are inter-related. One possible mechanism of this inter-relatedness is the internal recycling of P compounds under anoxic conditions; under such circumstances the sediment releases soluble reactive phosphorous (SRP) into the hypolimnion (Steinman et al., 2004). Anoxia usually occurs during summer thermal stratification and during winter ice cover (Wetzel, 2001), both depending on water depth, light penetration and productivity. P availability in the epilimnion furthermore requires the mixing of bottom and surface water that takes place during the spring and late autumn turnover in dimictic lakes (Wetzel, 2001). It is therefore likely that the lower P concentration of the sediment after 2700 cal yr BP resulted from the enhanced internal P recycling of the sediment under anoxic conditions as the water depth increased considerably. A late-winter/early-spring anoxia under ice cover possibly occurred as a consequence of which P was released into the water, and following the spring turnover facilitated the bloom of *Scenedesmus.* The positive feedback between increased productivity and hypolimnetic anoxia has probably been gradually amplified and gives us a possible explanation for the algal blooms. The spread of the planktonic *Scenedesmus* also supports the diatom-inferred water depth increase at 2700 cal yr BP.

In case of small lakes, like Saint Ana, it is possible to calculate the water depth at which the lake becomes stratified. Using Ragotzkie's (1978) formula and the present perimeter of the lake, the position of the thermocline is estimated to 5–5.3 m. Although measurements have not yet been made on the possible existence of a thermocline, it seems unlikely that the lake is stratified today, since the largest water depth is 6 m, and this is limited to a very small area (Fig. 2). Nonetheless, if we consider the measured maximal water depth in the nineteenth century (12 m), and take into account the calculated increase in the perimeter of the lake (ca. 1.9–2 km using 3D digital basin-elevation model; see Pál, 2001), the position of the thermocline can be estimated to 5.65 m, i.e. the lake was certainly stratified at that time. Around 2700 cal yr BP, the lake level increased above the nineteenth century level; therefore both the theoretical calculation and the palaeo-record suggest that the lake was stratified and dimictic in this period.

Some studies argue that recycling of nutrients in surface water can provide almost the entire requirement for phytoplankton growth (Caraco et al., 1992), but in many cases an increase in primary production also requires an enhanced external input of nutrients on top of the internal recycling (Wetzel, 2001). This latter process is presumable in Lake Saint Ana as the forest composition and soil has changed between 3700 and 3300 cal yr BP. We argued for and increase in soil pH that could also trigger the dissolution of P compounds in the soil provided that the pH increased above pH 5.5. (Busman et al., 2002). It is therefore likely that with the establishment of *F. sylvatica* on the lakeshore, the concentration of SRP increased in the soil solutions part of which was washed into the lake. This process would also explain the sharp decrease in allogenic P concentrations around 3700 cal yr BP (Fig. 9a).

Also typical for this phase is the mass occurrence of Type 169 ascospores from ca. 3300 cal yr BP (Fig. 6) that were identified using the photo tables of Van Geel (1982/1983). These ascospores likely represent an ascomycete fungus of the Sordariales mainly living on dung and rotten waterlogged wood (Shearer et al., 2007). Its abundance is most likely connected to the local establishment of *F. sylvatica* and probably suggests the fall of dead *Fagus* trees into the lake as the water level increased.

As shown in Fig. 11, the relative abundance of *Scenedesmus* fluctuates considerably between 2700 and 1000 cal yr BP, antagonistically with *B. pila*, and is accompanied by fluctuations in the organic content. Comparing these abrupt changes with the pollen record, they cannot unequivocally be connected to anthropogenic disturbance; however, it is apparent that the aquatic environment changed frequently even within this relatively stable deep-water phase. Competition of these two algae was probably governed by changes in nutrient availability that in turn was likely dependent on oxygen availability at the sediment–water interface. If anoxia was lasting (e.g. stable thermal stratification over the summer, thick ice cover, large water depth), then internal nutrient recycling was in good working, but a disturbance in this system (e.g. lack of ice cover, low water depth, remixing by wind or soil erosion) would have facilitated other less nutrient demanding and humic acid dependent algae, e.g. *B. pila*. Detailed observations of their competition would allow tracing the underlying mechanism for these abrupt phytoplankton changes.

1000–760 cal yr BP: decreasing water depth, lake and catchment acidification

The first clearly human-induced forest clearance and soil erosion were detected in the sequence between ca. 1050 and 900 cal yr BP. This time interval concurs with the Migration Period that brought about a massive population increase in Transylvania (Makkai & Mócsy, 1986). From this date, changes in the chemical and diatom records indicate human-induced catchment erosion and acidification (see organic content, endogenic/authigenic Fe, Mn, Al-normalised Fe and Mn curves and diatom-inferred pH). The macrofossil record also signals decreasing water depth.

Catchment processes likely involved the clearance of the lakeshore *F. sylvatica* forest as inferred by the macrofossil and pollen records (Figs. 10, 11). *Fagus* was unable to rejuvenate, instead *P. abies* became established on the lakeshore, and its impact on the soil is clearly signalled by the small increases in endogenic/authigenic Fe and Mn concentrations, and sharp decline in P. Acidification of the catchment soil was probably one consequence of this vegetation change. It is unclear to what extent catchment processes are responsible for the endogenic/authigenic P decline, as increasing anoxia at the sediment–water interface could also explain it. These two processes probably acted together.

The macrofossil inferred spread of the lakeshore floating fen and *Sphagnum*-carpet (Fig. 5; Table 4) in this phase is probably also connected with the lakeshore forest clearance and subsequent spread of *P. abies*. More light combined with acidifying catchment soils probably favoured the spread of poor-fen and *Sphganum*-bog species along the lakeshore.

Unfortunately, lake and catchment processes of the last 700 years are not recorded in this sedimentary sequence, but it is very likely that *P. abies* has since dominated the lakeshore vegetation as it is the most abundant tree around the lake recently.

Discussion

Internal versus external ecosystem processes driving hydrological change

Water-depth reconstruction revealed two major increases occurring concurrently with major forest changes: around 5350; 3300 cal yr BP. As the magnitude of both water-depth increase was large, the question arises how much changes in macroclimate were responsible and how much internal factors contributed. One widely examined phenomenon that has to be considered in this respect is canopy interception defined as the precipitation that is intercepted by plant foliage and eventually evaporated back to the atmosphere rather than falling to the ground (Chang, 2003). Measurements of canopy interception under different canopy composition show that coniferous forests intercept nearly two times more water then deciduous forests, respective values being ca. 35% for pure *P. abies* stands and ca. 17.5% for pure *F. sylvatica* stands (Chang, 2003). Given this large difference between coniferous and deciduous canopy, it is likely that at least the magnitude of the water-depth increase at both forest compositional changes, from *Picea* to *Carpinus*-*Picea* dominance and from *Picea*-*Carpinus* to *Fagus* dominance, was amplified by decreased interception of the canopy as the lake receives much of its water from surface runoff. In addition, the long-lasting Early Holocene dominance of *P. abies* on the lakeshore was probably responsible to some extent for the low water depth and probably also attenuated small-scale rainfall fluctuation by intercepting a considerable amount (37%) of the rainwater.

Another important internal forcing mechanism, with profound effect on the lake water chemistry, was clearly the compositional change of the catchment soil. As we have shown in the previous section, soil solutions were rich in organic acids, Fe and Mn under *Picea* dominance, and these were readily washed into the lake, where either precipitated or became available for primary producers. They lowered pH hereby favouring the spread of *Sphagnum* species and acidobiontic diatoms, and leading to decreased in-lake productivity both during the Early and Mid-Holocene, and more recently, following man-induced alteration of the lakeshore forest. Conversely, soil changes directly following the lakeshore or upslope establishment of *F. sylvatica* and *C. betulus*, triggered increasing in-lake productivity, faster sediment accumulation and hereby enhanced infilling of the lake. These observed changes suggest a delicate linkage between lakeshore vegetation/soil and in-lake productivity that seems to be a key element of catchment–lake interactions in case of Lake Saint Ana. On the whole, these processes demonstrate the sensitivity of the lake system to vegetation changes in the catchment.

After describing the possible internal forcing mechanisms and ecosystem processes, the question still remains whether they were evoked by internal changes alone or a linkage with macroclimate has to be assumed even if acting indirectly.

It seems that all the above detailed processes were brought about by profound terrestrial vegetation changes, that themselves also require a forcing mechanism, either internal (succession) or external (e.g. macroclimate change).

The direct impact of external forcing on lake-hydrology is also visible in the Lake Saint Ana record; e.g. the water-depth increase around 2700 cal yr BP must have been due to increasing moisture availability and initiated increased phytoplankton productivity via internal P recycling.

Overall, it seems that hydrological changes in the lake primarily reflect external forcing acting both directly and indirectly. The hydrological system showed only positive feedback (e.g. via interception or internal P recycling), hereby amplifying the impact of external forcing, but still our conclusion is that water-depth changes in Lake Saint Ana were primarily forced by changes in macroclimate. The next section examines how these inferred hydrological changes relate to the regional Holocene palaeoclimate record.

Regional pattern of lake level change during the Holocene

Syntheses of Holocene lake level records by Harrison et al. (1996) and Magny et al. (2002) focus on the northwest, Baltic, Mediterranean and central zones of Europe, but no discussion is made on SE Europe, including Hungary, Romania and Bulgaria. Except for Lake Balaton in western Hungary (Cserny, 1993), this area appears blank on the lake data source maps (Magny et al., 2002) pointing to the paucity of available Holocene lake level records in this region. Several recently obtained lake level and mire surface wetness records are gathered from the Carpathian Basin, Romania and Central Europe, which are shown in Fig. 12; the discussion below uses these data for comparison.

In NW Romania, the pollen and plant macrofossil-based palaeohydrological reconstruction of Feurdean and Bennike (2004) and Feurdean et al. (2008b) suggest that around 8600 cal yr BP synchronous changes in two peat bogs (Steregoiu and Preluca

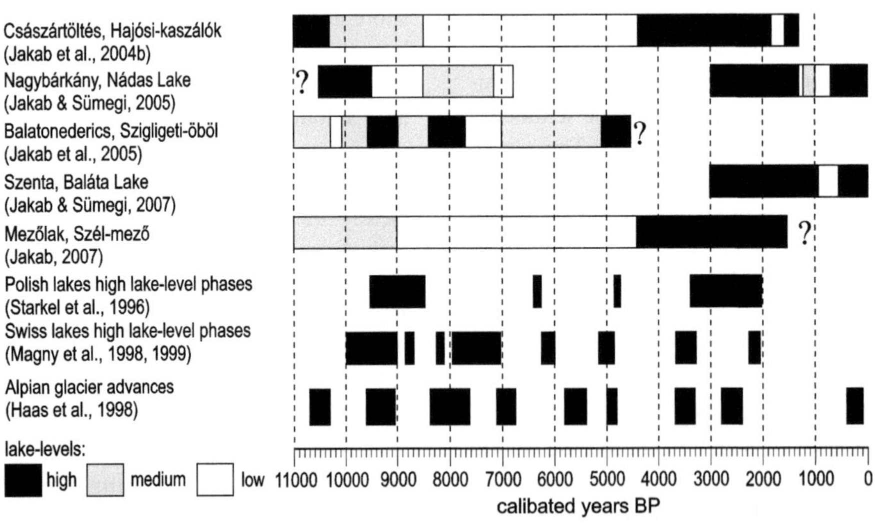

Fig. 12 Comparison of the Lake Saint Ana lake level record with other Holocene lake level records from Hungary, Romania and Central Europe

Tiganului) took place associated with drier conditions. The pollen-based palaeoclimate reconstruction moreover attested to a decrease in mean annual precipitation (MAP) by ~200 mm between 8350 and 8000 cal yr BP. The authors argued for regionally and climatically driven change, the well-known 8.2 event that is the most prominent Holocene reversal in the Greenland ice core ^{18}O record (von Grafenstein et al., 1998) and also all over the North Atlantic continental region, NW and Central Europe (Tinner & Lotter, 2001; Magny et al., 2003; Seppä et al., 2007). In Lake Saint Ana, the first decrease in the water depth was dated to ca. 9000 cal yr BP, that is about 400 years earlier, but poor diatom preservation was only detected from ca. 8600 cal yr BP suggesting that the driest period started concurrently with the NW Romanian records. In Poland, Starkel et al. (1996) found higher lake levels between 9500 and 8600 cal yr BP, in agreement with the higher lake level phase in Lake Saint Ana. In the Vistula Valley, 8600 cal yr BP marks a considerable decline in lake levels (Starkel et al., 1996). A lengthy dry phase was also established in the lowland areas of southern and northwest Hungary on the basis of the plant macrofossil records (Császártöltés, Szélmező Mire, Fig. 12). Jakab et al. (2004b) recorded drier mire conditions from 9000 cal yr BP at Szélmező and from 8500 cal yr BP at Császártöltés. In both places the dry period was followed by increasing water depths around 4400 cal yr BP. In summary, a large number of palaeo-records in the region suggest lower water depths and so drier conditions commencing around 8600 cal yr BP. These data, on the other hand, do not support hypothesis of Magny et al. (2003) that all of mid-latitude Europe became wetter during this phase as a result of a displacement and intensification of westerly circulation. It is conceivable that in parts of Central-Eastern Europe the 8.2 ka event was embedded in a longer generally drier phase, within which the period between 8200 and 8000 cal yr BP had relatively increased moisture availability, but none of the aforementioned records have yet provided evidence to support this theory.

The moderate water-depth increase from ca. 7400 cal yr BP in Saint Ana is also detectable in the testate amoebae-based hydrological record of the north-western Romanian peat bog, where a slow and gradual increase in available moisture between 8000 and 3500 cal yr BP was suggested (Schnitchen et al., 2006), interrupted by several minor wet phases peaking at 7100, 5910, 5100–4700 and 3570 cal yr BP. These wet peaks agree well with the Saint Ana hydrological record, each associated with increasing water depths; however, these not necessarily appear as peaks in our macrofossil-based water-depth reconstruction (Fig. 11), but as times with relatively high rate of water-depth increase. A reversal to lower water depth in the Saint Ana record appears between 5500 and 5350 cal yr BP. Similar decreases in mean annual precipitation (MAP) and moisture availability are suggested for this period by the pollen-based climate reconstruction of Feurdean et al. (2008b) and Schintchen et al. (2006), that is followed by a steep increase in MAP in the Preluca Tiganului record from 5300 cal yr BP (Feurdean et al., 2008b). The above mentioned regional lake level and moisture availability records suggest divergent processes occurring in East-Central Europe in comparison with the Western Mediterranean, where the Early Holocene was generally moister (Harrison et al., 1996; Magny et al., 2002). In contrast, the Carpathian Region was probably characterised by a more complex climate regime, with lower available moisture between ca. 8600 and 8000 cal yr BP, followed by gradual increase in the regional lake levels and so available moisture from ca. 8000 cal yr BP. Note, however, that in the Saint Ana record this increase in available moisture can only be detected with a delay, from ca. 7400 cal yr BP.

The moderate increase in lake level between 7400 and 5500 cal yr BP, punctuated by short-term decreases (Fig. 11), was followed by a marked increase at 5350 cal yr BP in Lake Saint Ana. Noteworthy is that Lake Balaton also displayed a considerable lake level rise at the same time and attained its largest surface area (Fig. 12; Cserny & Nagy-Bodor, 2000; Jakab et al., 2005). In the terrestrial vegetation *Fagus*, *Tilia* and *Carpinus* spread. Around Lake Saint Ana, *C. betulus* attained maximum abundance by 5200–5100 cal yr BP. In contrast with these records, in NW Romania Feurdean et al. (2008b) inferred warmer and possibly drier conditions between 5500 and 3200 cal yr BP. The accompanying terrestrial vegetation change, expansion of *C. betulus* and *F. sylvatica*, however, corresponds with the Saint Ana pollen record and suggests a regional and so probably climatically driven vegetation change that was accompanied by higher lake level in Saint Ana. As we discussed in the previous section, the canopy change probably contributed to this water depth rise around 5350 cal yr BP, however, a maximum 15% decrease in interception seems inadequate to explain the magnitude of the water depth rise. Complemented by the regional lake level data (Fig. 12), it seems likely that available moisture increased at least slightly from ca. 5350 cal yr BP.

Another inconsistency between the pollen-based MAP reconstruction of Feurdean et al. (2008b) and the Saint Ana hydrological record is the main trend of the last 3300 years; while our record suggest a prominent increase in available moisture, especially from ca. 2700 cal yr BP, the pollen-based climate reconstruction predicts the opposite, i.e. decreasing MAP accompanied by generally higher winter temperatures and constant or slightly decreasing summer temperatures. Taken into account the temperature trend, available moisture is predicted by the pollen data to decrease slightly in NW Romania, especially after 2400 years BP. A different picture emerges, however, if we look at the testate amoebae-based hydrological reconstruction of Schnitchen et al. (2006). The examined peat bog lies just 50–60 km away from Feurdean et al.'s pollen sites, and the testate amoebae-based hydrological reconstruction suggests increasing mire surface wetness from ca. 2800 cal yr BP (Schnitchen et al., 2006). In fact, 2800 cal yr BP marks the onset of the wettest phase in the peat bog's Holocene history that lasted until 2240 cal yr BP. In contrast, pollen-based climate reconstruction of Feurdean et al. (2008b) and the stalagmite isotope record of Onac et al. (2002) both suggest decreasing moisture availability around 2800 cal yr BP.

Further support to a wider extent of the Lake Saint Ana record inferred increase in moisture availability comes from the Carpathian Basin. In Hungary, several lake basins showed renewed sediment accumulation from ca. 3000 cal yr BP, following early and mid-Holocene hiatuses, e.g. Lake Baláta near Szenta (Fig. 12; Jakab & Sümegi, 2007) and Lake Nádas near Nagybárkány (Fig. 13; Jakab & Sümegi, 2005; Jakab et al., 2009). Overall these data suggest increasing moisture availability in the Carpathian basin from ca. 3400 cal yr BP with maximum moisture availability around 2700–2800 cal yr BP.

One possible explanation of the inconsistent pollen-based reconstruction is the local abundance of *P. abies* on the mires of Preluca Tiganului and Steregoiu in the Mid Holocene (Feurdean et al., 2008b). Similarly to Lake Saint Ana, *P. abies* was abundant locally, mainly on the mire surface at both sites, but retreated when *F. sylvatica* spread in the upland forests around 3400 cal yr BP. Merging the signal of the wetland and terrestrial vegetation, the Mid Holocene pollen spectra were high in *Picea* that in fact grew in the mire surface and enjoyed the cool and wet microclimate of the basin until the water depth rise submerged them to an extent they were unable to tolerate. In a modern pollen assemblage-based climate reconstruction this forest compositional change will inevitably give decreasing MAP values, even though the mire itself signals the opposite. In the first, detailed publication on Steregoiu, Björkman et al. (2002b) discussed changes in the on-site vegetation that clearly signalled a change from wooded fen vegetation with *P. abies* to sedge dominated open fen without trees around 3400 cal yr BP. The local overrepresentation would explain the divergent late Holocene moisture availability and precipitation reconstructions, but to resolve this problem, further hydrological reconstructions are needed from both peat bogs and lakes along the Eastern Carpathian Range. Another factor that probably affected the pollen-based climate reconstructions of Feurdean et al. (2008b) are human impact that has intensified over the last 3000 years in the Romanian Carpathians (Feurdean & Willis, 2008). Even the authors of this palaeoclimate study stressed the possibility of biased reconstructions for the Late Holocene due to the human-induced vegetation changes (Feurdean et al., 2008b).

Conclusions

The multi-proxy analyses of a sedimentary sequence from Lake Saint Ana provided an important record of Holocene hydrological changes in a crater lake that is the last and youngest representative of volcanic lakes in the Eastern Carpathian Mountains. Being set in a base-poor environment, the lake exemplified strong dependence of its aquatic ecosystem on lakeshore processes. Figure 13 summarises the results of the various analyses and our interpretation of the various

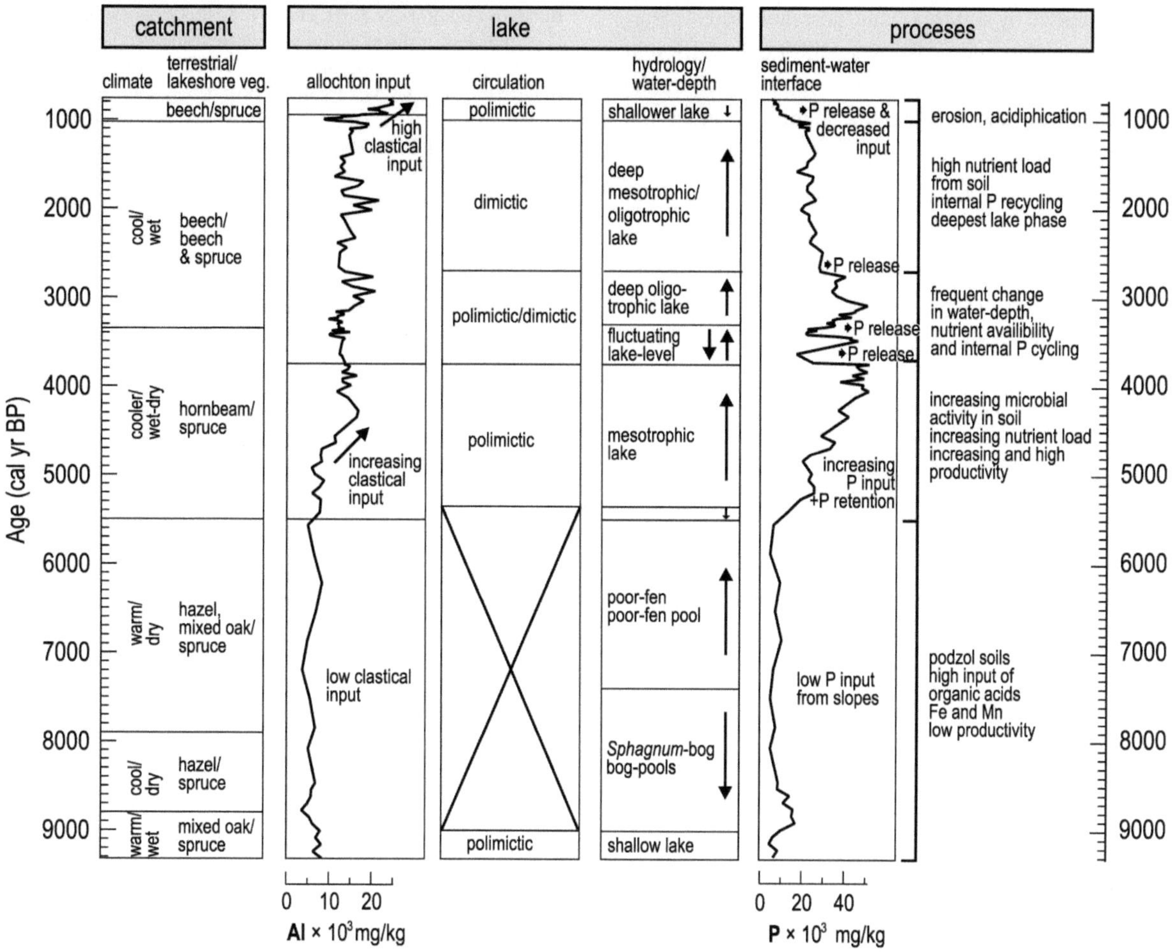

Fig. 13 Synthesis of the results of chemical, pollen, macrofossil and silicaceous algae analyses and inferred climatic and hydrological trends

limnological and terrestrial ecosystem processes. Using this figure as a guide, our interpretation of the limnological processes can be summarised as follows.

(1) The water depth of Lake Saint Ana showed considerable fluctuation during the Holocene. The lowest water depths were found between 9000 and 7400 calibrated BP years, when the crater was occupied by *Sphagnum*-bog and bog-pools. The major trend from 7400 cal yr BP was a gradual increase, but the basin was still dominated by poor-fen and poor-fen pools. Significant increases in water depth, and meso/oligotrophic lake conditions were found from 5350, 3300 and 2700 cal yr BP.

(2) Water-depth increases at 5350 and 3300 cal yr BP coincided with major terrestrial vegetation changes, namely the establishment of *C. betulus* on the crater slope, and the replacement of the lakeshore *P. abies* forest by *F. sylvatica*. The chemical record clearly indicated significant soil changes along with the canopy changes that in turn led to increased in-lake productivity and pH changes. A further increase in lake level around 2700 cal yr BP resulted in stable thermal stratification and hypolimnetic anoxia that via P-release further increased in-lake productivity and eventually led to phytoplankton blooms with massive populations of *Scenedesmus* cf. *S. brasiliesis*. High productivity was

depressed by anthropogenic lakeshore forest clearances commencing from ca. 1000 cal yr BP that led to the re-establishment of *P. abies* on the lakeshore and consequent acidification of the lake water.

On the whole, these data suggests that Lake Saint Ana is a vulnerable ecosystem; hydrological, biological and chemical processes in the lake are heavily influenced by the lakeshore forest and the soil underlying it. In-lake productivity is higher under deciduous canopy and litter, and considerably repressed by coniferous canopy and litter. The lake today subsists in a managed environment, that is, however, far from its natural state. This would likely be a dense *F. sylvatica* forest supplying more nutrients and keeping up a more productive in-lake flora and fauna.

An assessment of Holocene hydrological records in the region pinpointed divergent moisture availability trends for the last ca. 2700 years. The pollen-based palaeo-precipitation reconstruction in NW Romania showed lower precipitation (Feurdean et al., 2008b), while our, and some other records, suggested significant increase in available moisture. In order to decipher the regional patterns of moisture availability change during the Holocene and especially in the last 2700 years with inconsistent trends, more high-resolution lake level and peat bog hydrological records would be necessary. Especially the many peat bogs of the Eastern Carpathians provide the opportunity to compare and combine individual records and move to a more regional scale of long-term hydrological change during the Holocene.

Acknowledgements This research was funded by the Hungarian Research Fund through a postdoctoral fellowship held by EKM (OTKA F026036), the Bólyai Scholarship (BO/00518/07) and OTKA PD73234. This is MTA–MTM Paleo Contribution No. 77.

References

Alles, E., M. Nörpel-Schempp & H. Lange-Bertalot, 1991. Zur Systematik und Ökologie charakteristischer Eunotia-Arten (Bacillariophyceae) in elektrolyt-armren Bachoberläufen. Nova Hedwigia 53: 171–214.

Alley, R. B., P. A. Mayewski, T. Sowers, M. Stuiver, K. C. Taylor & P. U. Clark, 1997. Holocene climatic instability: a prominent widespread event 8200 yr ago. Geology 25: 438–486.

Bendell-Young, L. & H. H. Harvey, 1992. Geochemistry of Mn and Fe in lake sediments in relation to lake acidity. Limnology and Oceanography 37: 603–613.

Bengtsson, L. & M. Enell, 1986. Chemical analysis. In Berglund, B. E. (ed.), Handbook of Holocene Palaeoecology and Palaeohydrology. Wiley, Chichester: 423–451.

Bennett, K. D., 1992. PSIMPOLL: A quickBasic program that generates PostScript page description of pollen diagrams. INQUA Commission for the study of the Holocene: working group on data handling methods newsletter 8: 11–12.

Bennett, K. D., 1996. Determination of the number of zones in a biostratigraphical sequence. New Phytologist 132: 155–170.

Bennett, P. C., D. I. Siegel, B. M. Hill & P. H. Glaser, 1991. Fate of silicate in a peat bog. Geology 19: 328–331.

Berger, T. W., G. Köllensperger & R. Wimme, 2004. Plant-soil feedback in spruce (*Picea abies*) and mixed spruce-beech (*Fagus sylvatica*) stands as indicated by dendrochemistry. Plant and Soil 264: 69–83.

Berglund, B. E. & M. Ralska-Jasiewiczowa, 1986. Pollen analysis and pollen diagrams. In Berglund, B. E. (ed.), Handbook of palaeoecology and palaeohydrology. Wiley, New York: 455–479.

Beug, H. J., 2004. Leitfaden der Pollenbestimmung für Mitteleuropa und angrenzende Gebiete. Verlag Dr. Friedrich Pfeil, München.

Birks, H. J. B., 1998. Numerical tools in palaeolimnology—progress, potentialities and problems. Journal of Paleolymnology 20: 307–332.

Birks, H. J. B., 2003. Quantitative palaeoenvironmental reconstructions from Holocene biological data. In Mackay, A., R. W. Battarbee, H. J. B. Birks & F. Oldfield (eds), Global Change in the Holocene. Hodder Arnold, London: 107–123.

Birks, H. H. & H. J. B. Birks, 2006. Multi-proxi studies in palaeolimnology. Vegetation History and Archaeobotany 15: 235–251.

Birks, H. J. B. & A. D. Gordon, 1985. Numerical Methods in Quaternary Pollen Analysis. Academic Press, London.

Björkman, L., A. Feurdean, K. Cinthioa, B. Wohlfarth & G. Possnert, 2002a. Lateglacial and early Holocene vegetation development in the Gutaiului Mountains, northwestern Romania. Quaternary Science Reviews 21: 1039–1059.

Björkman, L., A. Feurdean & B. Wohlfarth, 2002b. Late-Glacial and Holocene forest dynamics at Steregoiu in the Gutaiului Mountains, Northwest Romania. Review of Palaeobotany and Palynology 25: 1–33.

Bodnariuc, A., A. Bouchette, J. J. Dedoubat, T. Otto, M. Fontugne & G. Jalut, 2002. Holocene vegetational history of the Apuseni Mountains, central Romania. Quaternary Science Reviews 21: 1465–1488.

Bond, G., W. Showers, M. Chesesby, R. Lotti, P. Almasi, P. deMenocal, P. Priore, H. Cullen, I. Hajdas & G. Bonani, 1997. A pervasive millennialscale cycle in North Atlantic Holocene and glacial climates. Science 278: 1257–1266.

Boros, Á., 1943. A Kukojszás vagy Mohos-tó és a Szentanna tó lápja (The mire of Kukojszás and Lake Saint Ana). Debreceni Szemle 12: 113–115.

Boyle, J. F., 1994. Acidification and sediment aluminium: palaeolimnological interpretation. Journal of Paleolimnology 12: 181–187.

Brewer, S., R. Cheddadi, J. L. de Beaulieu & M. Reille, 2002. The spread of deciduous Quercus throughout Europe since the last glacial period. Forest Ecology and Management 156: 27–48.

Brown, E. T., L. Le Callonnec & C. R. German, 2000. Geochemical cycling of redox-sensitive metals in sediments from Lake Malawi: a diagnostic paleotracer for episodic changes in mixing depth. Geochimica et Cosmochimica Acta 64: 3515–3523.

Buczkó, K. & E. Magyari, 2007. The Holocene diatom flora of Lake Saint Anna (Eastern Carpathians, Europe). Archiv für Hydrobiologie, Algological Studies 124: 1–28.

Buczkó, K. & A. Wojtal, 2007. A new *Kobayasiella* species (Bacillariophyceae) from Lake Saint Anna's sub-recent deposits in Eastern Carpathian Mountains, Europe. Nova Hedwigia 84: 155–166.

Busman, L., J. Lamb, G. Randall, G. Rehm & M. Schmitt, 2002. The nature of phosphorus in soils. www.extension.umn.edu/distribution/cropsystems/DC6795.html (last accessed 10/09/2008).

Caraco, N. E., J. J. Cole & G. E. Likens, 1992. New and recycled primary production in an oligotrophic lake: insights for summer phosphorus dynamics. Limnology and Oceanography 37: 590–602.

Cărăus, J., 2002. The algae of Romania. Studii si Cercetari, Biologie 73: 1–694.

Chang, M., 2003. Forest Hydrology: An Introduction to Water and Forests. CRC Press, Florida.

Cloot, A. & J. C. Roos, 1996. Modelling of a relationship between phosphorus, pH, calcium and chlorophyll-a concentration. Water SA 22: 49–61.

Constantin, S., A. W. Bojar, S. E. Lauritzen & J. Lundberg, 2007. Holocene and Late Pleistocene climate in the sub-Mediterranean continental environment: a speleothem record from Poleva Cave (Southern Carpathians, Romania). Palaeogeography, Palaeoclimatology, Palaeoecology 243: 322–338.

Cserny, T., 1993. Lake Balaton, Hungary. In Gierlowski-Kordesch, E. & K. Kelts (eds), A Global Geological Record of Lake Basins. Cambridge University Press, London: 397–401.

Cserny, T. & E. Nagy-Bodor, 2000. Limnogeology of Lake Balaton (Hungary). In Gierlowski-Kordesch, E. H. & K. R. Kelts (eds), Lake basins though space and time. AAPG Studies in Geology. American Association of Petroleum Geologists, USA: 605–618.

Cymbola, J., M. Ogdahl & A. D. Steinman, 2008. Phytoplankton response to light and internal phosphorus loading from sediment release. Freshwater Biology 53: 2530–2542.

Davison, W., 1993. Iron and manganese in lakes. Earth Science Reviews 34: 119–163.

Engstrom, D. R. & H. E. Wright, 1984. Chemical stratigraphy of lake sediments as a record of environmental change. In Haworth, E. Y. & J. W. G. Lund (eds), Lake Sediments and Environmental History. Leichester University Press/University of Minnesota Press, Leichester/Minneapolis: 11–67.

Espitalie, J., J. Deroo & F. Marquis, 1985. Rock-Eval Pyrolysis and its Applications. Institut Français du Pétrole, Report 33578.

Fărcaş, S., J. L. de Beaulieu, M. Reille, G. Coldea, B. Diaconeasa, T. Goslar & T. Jull, 1999. First ^{14}C datings of lateglacial and Holocene pollen sequences from the Romanian Carpathians. Comptes Rendues de l'Academie des Sciences de Paris, Sciences de la Vie 322: 799–807.

Fărcaş, S., I. Tanţău & A. Bodnariuc, 2003. The Holocene human presence in Romanian Carpathians, revealed by the palynological analysis. Würzburger Geographische Manuskripte 63: 113–130.

Fărcaş, S., F. Popescu & I. Tanţău, 2006. Dinamica spatiala si temporala a stejarului, frasinului si carpenului in timpul Tardi- si Postglaciarului pe teritoriul Romaniei. [Temporal and spatial dynamics of oak, ash and hornbeam in the Late- and Postglacial era in Romania.] Presa Universitara Clujeana, Cluj-Napoca.

Felton, A. A., J. M. Russell, A. S. Cohen, M. E. Baker, J. T. Chesley, K. E. Lezzar, M. M. McGlue, J. S. Pigati, J. Quade, J. C. Stager & J. J. Tiercelin, 2007. Paleolimnological evidence for the onset and termination of glacial aridity from Lake Tanganyika, Tropical East Africa. Palaeogeography, Palaeoclimatology, Palaeoecology 252: 405–423.

Feurdean, A., 2005. Holocene forest dynamics in northwestern Romania. The Holocene 15: 435–446.

Feurdean, A. & O. Bennike, 2004. Late Quaternary palaeoecological and palaeoclimatological reconstruction in the Gutaiului Mountains, northwest Romania. Journal of Quaternary Science 19: 809–827.

Feurdean, A. & O. Bennike, 2008. Plant macofossils analysis from Steregoiu, NW Romania: taphonomy, representation, and comparison with pollen analysis. Studia Universitatis Babes Bolyai, Geologia 53: 5–10.

Feurdean, A. & K. J. Willis, 2008. The usefulness of a long-term perspective in assessing current forest conservation management in the Apuseni Natural Park, Romania. Forest Ecology and Management 256: 421–430.

Feurdean, A., V. Mosbrugger, B. Onac, V. Polyak & D. Veres, 2007a. Younger Dryas to mid-Holocene environmental history of the lowlands of NW Transylvania, Romania. Quaternary Research 68: 364–378.

Feurdean, A., B. Wohlfarth, L. Björkman, I. Tanţău, O. Bennike, K. J. Willis, S. Fărcaş & A. M. Robertsson, 2007b. The influence of refugial population on Lateglacial and early Holocene vegetational changes in Romania. Review of Palaeobotany and Palynology 145: 305–320.

Feurdean, A., S. Klotz, S. Brewer, V. Mosbrugger, T. Tamas & B. Wohlfarth, 2008a. Lateglacial climate development in NW Romania—comparative results from three quantitative pollen based methods. Palaeogeography, Palaeoclimatology, Palaeoecology 265: 121–133.

Feurdean, A., S. Klotz, V. Mosbrugger & B. Wohlfarth, 2008b. Pollen-based quantitative reconstructions of Holocene climate variability in NW Romania. Palaeogeography, Palaeoclimatology, Palaeoecology 260: 494–504.

Gell, P., P. Barker, P. DeDeckker, W. Last & L. Jelicic, 1994. The Holocene history of West Basin Lake, Victoria, Australia; chemical changes based on fossil biota and sediment mineralogy. Journal of Palaeolimnology 12: 235–258.

Gibson, C. E., 1997. The dynamics of phosphorous in freshwater and marine environments. In Tunney, H., O. T. Carton, P. C. Brookes & A. E. Johnston (eds), Phosphorous Loss from Soil to Water. CAB International, Wallingford: 119–149.

Grivet, D. & R. J. Petit, 2003. Chloroplast DNA phylogeography of the hornbeam in Europe: evidence for a bottleneck at the outset of postglacial colonization. Conservation Genetics 4: 47–56.

Haas, J. N., I. Richoz, W. Tinner & L. Wick, 1998. Synchronous Holocene climatic oscillations recorded on the Swiss Plateau and at timberline in the Alps. The Holocene 8: 301–309.

Hamilton-Taylor, J., W. Davison & K. Morfett, 1996. The biogeochemical cycling of Zn, Cu, Fe, Mn, and dissolved organic C in a seasonally anoxic lake. Limnology and Oceanography 41: 408–418.

Harrison, S. P., G. Yu & P. E. Tarasov, 1996. Late quaternary lake-level record from Northern Eurasia. Quaternary Research 45: 138–159.

Hedenäs, L., 1993. Field and microscope keys to the Fennoscandian species of the Calliergon–Scorpidium–Drepanocladus complex, including some related or similar species. Biodetector AB, Märsta.

Heiri, O., A. F. Lotter & G. Lemcke, 2001. Loss on ignition as a method for estimating organic and carbonate content in sediments: reproducibility and comparability of results. Journal of Paleolimnology 25: 101–110.

Heiri, O., W. Tinner & A. F. Lotter, 2004. Evidence for cooler European summer during periods of changing meltwater flux to the North Atlantic. Proceedings of the National Academy of Sciences of the United States of America 101: 15285–15288.

Hortobágyi, T., 1942. Moszatok a Szent Anna-tóból (Algae from Lake Saint Ana). Acta Botanica 1: 102–112.

Hortobágyi, T., 1943. Beiträge zur Kenntnis des August-Phytoplanktons im St.Anna-See (in Hungarian with German summary). Botanikai Közlemények 40: 377–382.

Jakab, G., 2007. The macrobotanical remains from Mezőlak. In Zatykó, P., I. Juhász & P. Sümegi (eds), Environmental History of Transdanubia. Varia Archaeologica Hungarica 20, Budapest: 325–328.

Jakab, G. & P. Sümegi, 2005. A nagybárkányi Nádas-tó kialakulása a makrofosszília vizsgálatok alapján (Cserhát, Észak-Magyarország) (The development of Nádas Lake at Nagybárkány, Mt Cserhát, N Hungary). Kitaibelia 10: 104–114.

Jakab, G. & P. Sümegi, 2007. The macrobotanical remains from Baláta-tó. In Zatykó, P., I. Juhász & P. Sümegi (eds), Environmental History of Transdanubia. Varia Archaeologica Hungarica 20, Budapest: 247–250.

Jakab, G., P. Sümegi & E. Magyari, 2004a. A new quantitative method for the palaeobotanical description of Late Quaternary organic sediments. Antaeus 27: 181–211.

Jakab, G., P. Sümegi & E. Magyari, 2004b. A new paleobotanical method for the description of Late Quaterary organic sediments (Mire-development pathways and paleoclimatic record from S Hungary). Acta Geologica Hungarica 47: 373–409.

Jakab, G., P. Sümegi & Zs Szántó, 2005. Késő-glaciális és holocén vízszintingadozások a Szigligeti-öbölben (Balaton) makrofosszília vizsgálatok eredményei alapján (Late Glacial and Holocene water level changes in the Szigliget Bay, Lake Balaton based on macrofossil investigations). Földtani Közlöny 135: 405–431.

Jakab, G., P. Majkut, I. Juhász, S. Gulyás, P. Sümegi & T. Törőcsik, 2009. Palaeoclimatic signals and anthropogenic disturbances from the peatbog at Nagybárkány (North Hungary). Hydrobiologia. doi:10.1007/s10750-009-9803-z.

Jankovská, V., 2001. Vegetation development in the W part of the Giant Mts. during the Holocene (Pančava Rašeliniště Mire–palaeoecological research). Opera Corcontica 8: 8–17.

Jasinski, J. P. P., B. G. Warner, A. A. Andreev, R. Aravena, S. E. Gilbert, B. A. Zeeb, J. P. Smol & A. A. Velichko, 1998. Holocene environmental history of a peatland in the Lena River valley, Siberia. Canadian Journal of Earth Sciences 35: 637–648.

Juggins, S., 1991. ZONE v. 2.1. Unpublished computer program. Department of Geography, University of Newcastle, UK.

Juggins, S., 2001. The European Diatom Database, User Guide, Version 1.0. http://craticula.ncl.ac.uk/Eddi/docs/EddiGuide.pdf.

Juvigné, E., M. Gewelt, E. Gilot, C. Hurtgen, I. Seghedi, A. Szakacs, Gy. Gábris, Á. Hadnagy & E. Horváth, 1994. Une éruption vieille d'environ 10700 ans (^{14}C) dans les Carpates orientales (Roumanie). Isotopic Geochemistry. Comptes rendus de l'Académie des sciences Paris 318: 1233–1238.

Koinig, K. A., W. Shotyk, A. F. Lotter, C. Ohlendorf & M. Sturm, 2003. 9000 years of geochemical evolution of lithogenic major and trace elements in alpine lake sediments—the role of climate, vegetation and land-use history. Journal of Paleolimnology 30: 307–320.

Komárek, J. & V. Jankovská, 2001. Review of the green algal genus *Pediastrum*; implication for pollenanalytical research. Bibliotheca Phycologica 108: 1–127.

Kooijman, A. M., 1993. On the ecological amplitude of four mire bryophytes, a reciprocal transplant experiment. Lindbergia 18: 19–24.

Kopáček, J., J. Borovec, J. Hejzlar, K. U. Ulrich, S. A. Norton & A. Amirbahman, 2005. Aluminum control of phosphorus sorption by lake sediments. Environmental Science & Technology 39: 8784–8789.

Krammer, K., 2000. The genus Pinnularia. In Lange-Bertalot, H. (ed.), Diatoms of Europe 1. A.R.G. Gantner Verlag K.G., Ruggell.

Krammer, K. & H. Lange-Bertalot, 1986–1991. Süsswasserflora von Mitteleuropa. Bacillariophyceae 1–4. Gustav Fischer Verlag, Stuttgart, Jena.

Kristó, A., 1995. A Csomád hegycsoport. A Szent-Anna tó természetvédelmi területe (The Nature Reserve of Lake Saint Ana). In Papp, Á. & I. Szíjártó (eds), Kristó András emlékére (In remembrance of András Kristó). Balaton Akadémia Könyvek 13, Vörösberény, Budapest: 38–45.

Lange-Bertalot, H. 2001. Navicula sensu stricto 10 Genera separated from Navicula sensu lato Frustulia. In Lange-Bertalot, H. (ed.), Diatoms of Europe 2. A.R.G. Gantner Verlag K.G., Rugell: 526 pp.

Lange-Bertalot, H. & D. Metzeltin, 1996. Indicators of Oligotrophy. Iconographia Diatomologica 2, Koeltz Scentific Books.
Lawson, D. S., D. C. Hurd & H. S. Pankratz, 1978. Silica dissolution rates of decomposing phytoplankton assemblages at various temperatures. American Journal of Science 278: 1373–1393.
Lepşi, I., 1957. Protozoen aus dem anmoorigen Gebirgssee St.Anna in Rumänien. Travaux du Museum d'Histoire Naturelle Grigore Antipa 1: 73–110.
Lerman, A., 1978. Lakes: Chemistry, Geology, Physics. Springer, New York.
Mackereth, F. J. H., 1965. Chemical investigations of lake sediments and their interpretation. Proceedings of the Royal Society B161: 285–309.
Mackereth, F. J. H., 1966. Some chemical observations on post-glacial lake sediments. Philosophical Transactions of the Royal Society of London B 250: 165–213.
Magny, M., 1998. Reconstruction of Holocene lake-level changes in the Jura (France): methods and results. In Harrison, S. P., B. Frenzel, U. Huckried & M. Weiss (eds), Palaeohydrology as Reflected in Lake-Level Changes as Climatic Evidence for Holocene times. Paläoklimaforschung 25, Gustav Fisher, Stuttgart: 67–85.
Magny, M., 1999. Lake-level fluctuations in the Jura and French subalpine ranges associated with ice-rafting events in the North Atlantic and variations in the polar atmospheric circulation. Quaternaire 10: 61–64.
Magny, M., C. Miramot & O. Sivan, 2002. Assessment of the impact of climate and anthropogenic factors on Holocene Mediterranean vegetation in Europe on the basis of palaeohydrological records. Palaeogeography, Palaeoclimatology, Palaeoecology 186: 47–59.
Magny, M., C. Begeot, J. Guiot & O. Peyron, 2003. Constraining patterns of hydrological changes in Europe in response to Holocene climate cooling phases. Quaternary Science Reviews 22: 1589–1596.
Magyari, E., P. Sümegi, M. Braun, G. Jakab & M. Molnár, 2001. Retarded wetland succession: anthropogenic and climatic signals in a Holocene peat bog profile from north-east Hungary. Journal of Ecology 89: 1019–1032.
Magyari, E. K., K. Buczkó, G. Jakab, M. Braun, Zs. Szántó, M. Molnár, Z. Pál & D. Karátson, 2006. Holocene palaeohydrology and environmental history in the South Harghita Mountains, Romania. Földtani Közlöny 136: 249–284.
Makkai, L. & A. Mócsy, 1986. Erdély Története, Első kötet, A kezdetektől 1606-ig (The history of Transylvania, 1st volume, From the beginning to 1606). Akadémiai Kiadó, Budapest.
Markgraf, V., 1980. Pollen dispersal in a mountain area. Grana 19: 127–146.
Mayewski, P. A., E. E. Rohling, J. C. Stager, W. Karlen, K. A. Maasch, D. L. Meeker, E. A. Meyerson, F. Gasse, S. van Kreveldg, K. Holmgren, J. Lee-Thorph, G. Rosqvistd, F. Rack, M. Staubwasser, R. R. Schneider & E. R. Steig, 2004. Holocene climate variability. Quaternary Science Reviews 62: 243–255.
Moore, P. D., J. A. Webb & M. E. Collinson, 1992. Pollen Analysis, 2nd ed. Blackwell Scientific Publications, Oxford.
Moriya, I., M. Okuno, T. Nakamura, K. Ono, A. Szakács & I. Seghedi, 1996. Radiocarbon ages of charcoal fragments from the pumice flow deposit of the last eruption of Ciomadul volcano, Romania. Summaries of Researches using AMS at Nagoya University 7: 255.
Newbold, J. D., 1992. Cycles and spirals of nutrients. In Calow, P. & G. E. Petts (eds), The Rivers Handbook. Blackwell Scientific, Oxford: 379–408.
Norusis, M. J., 1990. SPSS/PC+ Advanced Statistics 4.0 for IBM PC/XT/AT and PS/2. SPSS Inc, Chicago.
Nyárádi, E. Gy., 1929. A vizek és vízben bővelkedő talajok növényzetéről a Hargitában (Vegetation of lakes and wetlands in Mt Hargita). In Csutak, V. (ed.), Emlékkönyv a Székely Nemzeti Múzeum 50 éves jubileumára (Memorandum on the 50th Anniversary of the Székely National Museum). Minerova Irodalmi és Nyomdai Műintézet Részvénytársaság, Sepsiszentgyörgy: 557–615.
Onac, B. P., S. Constantin, J. Lundberg & S. E. Lauritzen, 2002. Isotopic climate record in a Holocene stalagmite from Ursilor Cave (Romania). Journal of Quaternary Science 17: 319–327.
Pál, Z., 2000. A Szent Anna Tó: következtetések a tó mélységét és feltöltődését illetően (Lake Saint Ana: inferences regarding water-depth and lake infillment). Collegicum Geographicum 1: 65–74.
Pál, Z., 2001. A Szent Anna Tó batimetriája (Bathymetry of Lake Saint Ana). Collegicum Geographicum 2: 73–78.
Pécskay, Z., J. Lexa, A. Szakács, K. Balogh, I. Seghedi, V. Konecny, M. Kovács, E. Márton, M. Kaliciak, V. Széky-Fux, T. Póka, P. Gyarmati, O. Edelstein, E. Rosu & B. Žec, 1995. Space and time distribution of Neogene-Quaternary volcanism in the Carpatho-Pannonian Region. Acta Vulcanologica 7: 15–28.
Petersen, J. B., 1950. Observations on some small species of Eunotia. Dansk Botanisk Arkiv 14: 1–19.
Petit, R. J., S. Brewer, S. Bordács, K. Burg, R. Cheddadi, E. Coart, J. Cottrell, U. M. Csaikl, B. C. van Dam, J. D. Deans, S. Fineschi, R. Finkeldey, I. Glaz, P. G. Goicoechea, J. S. Jensen, A. O. König, A. J. Lowe, S. F. Madsen, G. Mátyás, R. C. Munro, F. Popescu, D. Slade, H. Tabbener, S. M. G. de Vries, B. Ziegenhagen, J. L. de Beaulieu & A. Kremer, 2001. Identification of refugia and postglacial colonization routes of European white oaks based on chloroplast DNA and fossil pollen evidence. Forest Ecology and Management 156: 49–74.
Podani, J., 1993. SYN-TAX 5.0: Computer programs for multivariate data analysis in ecology and systematics. Abstracta Botanica 17: 289–302.
Pop, E., 1960. Mlastinile de turba din Republica Populara Romana (Peat bogs in Romania). Academia Republica Populara Romana, Bucuresti.
Ragotzkie, R. A., 1978. Heat Budgets of Lakes. In Lerman, A. (ed.), Lakes: Chemistry, Geology, Physics. Springer, New York: 1–19.
Ranger, J. & C. Nys, 1994. The effect of spruce (Picea abies Karst.) on soil development: an analytical and experimental approach. European Journal of Soil Science 45: 193–204.
Reimer, P. J., M. G. L. Baillie, E. Bard, A. Bayliss, J. W. Beck, C. J. H. Bertrand, P. G. Blackwell, C. E. Buck, G. S. Burr,

K. B. Cutler, P. E. Damon, R. L. Edwards, R. G. Fairbanks, M. Friedrich, T. P. Guilderson, A. G. Hogg, K. A. Hughen, B. Kromer, G. McCormac, S. Manning, C. B. Ramsey, R. W. Reimer, S. Remmele, J. R. Southon, M. Stuiver, S. Talamo, F. W. Taylor, J. van der Plicht & C. E. Weyhenmeyer, 2004. IntCal04 terrestrial radiocarbon age calibration, 0–26 cal kyr BP. Radiocarbon 46: 1029–1058.

Reynolds, C. S., 2006. The Ecology of Phytoplankton. Cambridge University Press, Cambridge.

Rösch, M. & E. Fischer, 2000. A radiocarbon dated Holocene profile from the Banat mountains (Southwestern Carpathians, Romania). Flora 195: 277–286.

Rühland, K., J. P. Smol, J. P. Jasinski & B. G. Warner, 2000. Response of Diatoms and Other Siliceous Indicators to the Developmental History of a Peatland in the Tiksi Forest, Siberia, Russia. Arctic, Antarctic, and Alpine Research 32: 167–178.

Rybníček, K., 1985. A Central-European approach to the classification of mire vegetation. Aquilo Seria Botanica 21: 19–31.

Sandgren, C. D., 1988. The ecology of chrysophyte flagellates: their growth and perennation strategies as freshwater phytoplankton. In Sandgren, C. D. (ed.), Growth and Reproduction Strategies of Freshwater Phytoplankton. Cambridge University Press, Cambridge: 9–103.

Schettler, G., M. J. Schwab & M. Stebich, 2007. A 700-year record of climate change based on geochemical and palynological data from varved sediments (Lac Pavin, France). Chemical Geology 240: 11–35.

Schindler, D. W., 1977. Evolution of phosphorus limitation in lakes. Science 195: 260–262.

Schnitchen, Cs., D. J. Charman, E. Magyari, M. Braun, I. Grigorszky, B. Tóthmérész, M. Molnár & Zs. Szántó, 2006. Reconstructing hydrological variability from testate amoebae analysis in Carpathian peatlands. Journal of Paleolimnology 36: 1–17.

Schur, F., 1858. Der südliche Hochgebirgs-Zug Siebenbürgens in botanish-geographisher Beziehung. Österreichische Botanische Zeitschrift 8: 393–407.

Seppä, H., H. J. B. Birks, T. Giesecke, D. Hammarlund, T. Alenius, K. Antonsson, A. E. Bjune, M. Heikkilä, G. M. MacDonald, A. E. K. Ojala, R. J. Telford & S. Veski, 2007. Spatial structure of the 8200 cal yr BP event in northern Europe. Climate of the Past 3: 225–236.

Shearer, C. A., E. Descals, B. Kohlmeyer, J. Kohlmeyer, L. Marvanova, D. Padgett, D. Porter, H. A. Raja, J. P. Schmit, H. A. Thorton & H. Voglymayr, 2007. Fungal Biodiversity in aquatic habitats. Biodiversity and Conservation 16: 47–67.

Smol, J. P., 1985. The ratio of diatom frustules to chrysophycean statospores: a useful paleolimnological index. Hydrobiologia 123: 199–208.

Smol, J. P., J. B. Birks & W. M. Last, 2001. Tracking Environmental Change Using lake Sediments. Vol. 3. Terrestrial, Algal and Siliceous Indicators. Kluwer Academic Publishers, London.

Solomon, A. M. & A. B. Silkworth, 1986. Spatial patterns of atmospheric pollen transport in a mountain region. Quaternary Research 25: 150–162.

Somogyi, A., M. Braun, A. Tóth & K. J. Willis, 1999. Speciation of elements in lake sediments investigated using X-ray fluorescence and inductively coupled plasma atomic emission spectrometry. X-Ray Spectrometry 27: 283–287.

Starkel, L., T. Kalicki, T., M. Krąpiec, R. Soja, P. Gębica & E. Czyżowska, 1996. Hydrological changes of valley floor in the Upper Vistula Basin during the Late Vistulian and Holocene. Geographical Studies, Special Issue 9: 7–128.

Steinman, A. D., R. Rediske & K. R. Reddy, 2004. The importance of internal phosphorus loading to Spring Lake, Michigan. Journal of Environmental Quality 33: 2040–2048.

Stockmarr, J., 1971. Tablets with spores used in absolute pollen analysis. Pollen et Spores 13: 614–621.

Szakács, S. & I. Seghedi, 1995. The Calimani-Gurghiu-Harghita volcanic chain: volcanological features. Acta Vulcanologica 7: 145–153.

Tallis, J. H., 1983. Changes in wetland communities. In Gore, A. J. P. (ed.), Ecosystems of the Word. Mires: swamp, bog, fen and moor, Vol. 4B. Elsevier, Amsterdam: 311–347.

Tanţău, I., M. Reille, J. L. de Beaulieu, S. Fărcaş, T. Goslar & M. Paterne, 2003. Vegetation history in the eastern Romanian Carpathians: pollen analysis of two sequences from the Mohos crater. Vegetation History and Archaeobotany 12: 113–125.

Tanţău, I., M. Reille, J. L. de Beaulieu & S. Fărcaş, 2006. Late Glacial and Holocene vegetation history in the southern part of Transylvania (Romania): pollen analysis of two sequences from Avrig. Journal of Quaternary Science 21: 49–61.

Tessier, A., P. G. C. Cambell & M. Bisson, 1979. Sequential extraction procedure for the speciation of particulate trace metals. Analytical Chemistry 51: 844–851.

Tinner, W. & A. F. Lotter, 2001. Central European vegetation response to abrupt climate change at 8.2 ka. Geology 29: 551–554.

Ulrich, K. U. & R. Pöthig, 2000. Precipitation of aluminium and phosphate affected by acidification. Acta Hydrochimica et Hydrobiologica 28: 313–322.

Van Geel, B., D. P. Hallewas & J. P. Pals, 1982/1983. A late Holocene deposit under the Westfriese Zeedijk near Enkhuzien (Prov. of Noord-Holland, The Netherlands): palaeoecological and archaeological aspects. Review of Palaeobotany and Palynology 38: 269–335.

Vinkler, A. P., Sz. Harangi, T. Ntaflos & A. Szakács, 2007. A Csomád vulkán (Keleti-Kárpátok) horzsaköveinek kőzettani és geokémiai vizsgálata—petrogenetikai következtetések (Petrology and geochemistry of the pumices from the Ciomadul volcano (Eastern Carpathians)—implications for the petrogenetic processes). Földtani Közlöny 137: 103–128.

von Grafenstein, U., H. Erlenkeuser, J. Müller, J. Jouzel & S. Johnsen, 1998. The short cold period 8,200 years ago documented in oxygen isotope records of precipitation in Europe and Greenland. Climate Dynamics 14: 73–81.

Vorren, T., K. D. Vorren, T. Alm, S. Gulliksen & R. Løvlie, 1988. The last deglaciation (20,000 to 11,000 B.P.) on Andøya, northern Norway. Boreas 17: 41–77.

Wetzel, R. G., 2001. Limnology: Lake and River Ecosystems. Elsevier Sciences, London.

Wilkinson, A. N., B. A. Zeeb, J. J. P. Smol & M. S. V. Douglas, 1997. Chrysophyte stomatocyst assemblages associated with periphytic, high arctic pond environments. Nordic Journal of Botany 17: 95–112.
Wilson, T. A., S. A. Norton, B. A. Lake & A. Amirbahman, 2008. Sediment geochemistry of Al, Fe, and P for two historically acidic, oligotrophic Maine lakes. Science of the Total Environment 404: 269–275.
Wohlfarth, B., G. Hannon, A. Feurdean, L. Ghergaric, B. P. Onac & G. Possnert, 2001. Reconstruction of climatic and environmental changes in NW Romania during the early part of the last deglaciation (15, 000–13, 600 cal years BP). Quaternary Science Reviews 20: 1897–1914.
Wright, H. E., 1967. A square rod piston sampler for lake sediments. Journal of Sedimentary Petrology 37: 975–976.

Subfossil diatoms and chironomids along an altitudinal gradient in the High Tatra Mountain lakes: a multi-proxy record of past environmental trends

Peter Bitušík · Vladimír Kubovčík · Elena Štefková · Peter G. Appleby · Marek Svitok

Originally published in the journal Hydrobiologia, Volume 631, No. 1, 65–85.
DOI: 10.1007/s10750-009-9802-0

Abstract Multi-proxy approach was used to reconstruct the environmental conditions of remote lakes in the High Tatra Mountains (Slovakia) over the past few centuries (approximately 500–1000 years). Short sediment cores ($\sim$30 cm) taken from three morphologically similar glacial lakes distributed along altitudinal gradient (subalpine to alpine conditions) were analysed for organic matter content (LOI), diatoms and chironomids. Both descriptive and correlative approaches were used for analysing stratigraphical data. Predictive canonical correspondence analysis and co-correspondence analysis were applied to directly relate physical and biological proxies to each other. The relationship between LOI and biotic proxies was inconsistent across groups and lakes. Concordant patterns in diatom and chironomid composition were found in two non-acidified lakes. Common trends in those assemblages indicated major past environmental events such as the Little Ice Age, air pollution and lake acidification. In contrast, no relationship between the composition of diatom and chironomid assemblages was found in the formerly acidified lake, suggesting different responses of assemblages to acidification. While chironomids showed shifts that are attributable to recovery, diatoms assemblage remained relatively stable throughout the uppermost layers of the sediment record. On the other hand, climatic-driven changes in assemblages detected in the deeper layers were more pronounced in diatoms than in chironomids.

Keywords Palaeolimnology · Alpine lakes · Organic matter content · Climate · Acidification · Co-correspondence analysis

Introduction

Intensive study of remote alpine and arctic lakes in the past few decades has revealed an extremely high

Guest editors: K. Buczkó, J. Korponai, J. Padisák & S. W. Starratt
Palaeolimnological Proxies as Tools of Environmental Reconstruction in Fresh Water

P. Bitušík (✉)
Research Institute & Faculty of Science, Matthias Belius University, 974 01 Banská Bystrica, Slovakia
e-mail: bitusik@fpv.umb.sk

V. Kubovčík · M. Svitok
Faculty of Ecology and Environmental Sciences, Technical University in Zvolen, T. G. Masaryka 2117/24, 960 53 Zvolen, Slovakia

V. Kubovčík
e-mail: kubovcik@vsld.tuzvo.sk

E. Štefková
Institute of Zoology, Slovak Academy of Sciences, Dúbravská cesta 9, 845 06 Bratislava, Slovakia
e-mail: elena.stefkova@savba.sk

P. G. Appleby
Department of Mathematical Sciences, University of Liverpool, P.O. Box 147, Liverpool L69 3 BX, UK
e-mail: appleby@liverpool.ac.uk

sensitivity of these lakes to global effects such as atmospheric pollution and climatic change, as well as minor changes of regional importance (Rouse et al., 1997). Relatively subtle, short-term variability in air temperature is extremely well mirrored in epilimnion lake water temperature (Livingstone & Lotter, 1998). The complex effects of long-term climatic trends, regardless of whether natural or human induced, are potentially recorded in lake sediments, and lake sediments can thus be used as a means of reconstructing past climate over long time-scales. However, the usefulness of this approach depends on the sensitivity and accuracy of the various proxies and the extent to which the climate signal in the sediment record is obscured by noise from other influences (Battarbee et al., 2002b). Past climatic changes can be tracked by assemblages of aquatic organisms such as diatoms and chironomids that may be more sensitive indicators of climatic changes than terrestrial groups (Smol et al., 1991). They can be affected by climatic variability directly through their life cycles (Davidson, 1991; Smol et al., 1991) and indirectly through processes such as stratification, water quality and habitat changes (Anderson et al., 1996; Gaedke et al., 1998; Walker, 2001; Weckström & Korhola, 2001). Due to this sensitivity, diatoms and chironomids are used quite extensively as proxy indicators of climatic changes (e.g., Anderson, 2000; Brooks & Birks, 2004).

Lakes of glacial origin in the Tatra Mountains, situated at the Slovak-Polish border, became the focus of multidisciplinary research within the EU-funded projects AL:PE, MOLAR and EMERGE. Palaeolimnological analyses were employed to document the long-term effects of acidification in the lake communities (Cameron et al., 1999; Stuchlík et al., 2002; Clarke et al., 2005; Kubovčík & Bitušík, 2006) and to reconstruct climate patterns over the last 200 years, comparing the palaeolimnological evidence with instrumental climate data (Šporka et al., 2002). However, comparisons between various aquatic assemblages as climate proxies and instrumental climate records during the past 200 years did not show clear-cut results over a range of European mountain lakes (Battarbee et al., 2002a).

Specifically, in the Tatra Mountains, a relatively strong correlation was found between diatom species composition and reconstructed summer air temperature in Nižné Terianske pleso. In spite of decadal-scale temperature fluctuations at this lake (Agustí-Panareda & Thompson, 2002), the response of chironomids was weak (Šporka et al., 2002). An equivocal reaction of diatoms and chironomids to climatic change was observed in other alpine lakes, too (e.g., Koinig et al., 2002). In contrast, those data overall indicated that organic matter content of sediments may be a good indicator of varying mean annual temperature in lakes, especially in soft water systems where the sources of organic matter are mainly autochthonous (Battarbee et al., 2002a).

Shared responses are ecologically interesting because they suggest that taxonomically divergent groups are controlled by relatively few environmental factors (Paavola et al., 2003). Similar structures between different groups, however, may be a consequence of biotic interactions among trophic levels (e.g., Jackson & Harvey, 1993). In a bottom-up approach of biodiversity, plant species composition will first affect the herbivores that directly depend on these plants, which will in turn affect higher trophic levels (Schaffers et al., 2008). Therefore, when dealing with a range of taxonomic groups that have different trophic levels and positions in the food web, one should bear in mind various indirect ways of climatic influences and possible complex interactions (Battarbee et al., 2002a).

By examining the degree of synchronicity of the responses among biostratigraphic indicators, it is possible to assess which proxy indicator(s) has responded most sensitively to subtle Holocene environmental changes. Only few palaeolimnological studies have tackled this subject (Fallu et al., 2005). To our knowledge, no studies have explicitly tested for shared structures between subfossil assemblages.

The motivation for this study was to provide analyses of the deeper parts of sediment cores that have not yet been studied, taken from three Tatra Mountain lakes situated at different altitudes. The primary aim of this study was to discern to what extent the sediment records from these lakes could reflect environmental changes through time within the region. Our approach has been to make comparisons between the organic matter content as a useful temperature proxy and the composition of diatom and chironomid assemblages. In conjunction with this, the sensitivity and usefulness of subfossil diatoms and chironomids used as proxy indicators were evaluated. Subsequently, the relationships between assemblages of different trophic levels were directly evaluated and compared with those provided by organic matter.

Study sites

The three lakes of glacial origin are located in the High Tatra Mountains in northern Slovakia. The mountain range is characterised by steep changes in temperature and precipitation along an altitudinal gradient. The average annual air temperature decreases with elevation by 0.6°C per 100 m, being 1.6 and −3.8°C at elevations of 1,778 and 2,635 m, respectively (Konček & Orlicz, 1974). The amount of precipitation varies from ~1.0 to ~1.6 m y r^{-1} between 1,330 and 2,635 m a.s.l., but reaches >2.00 m y r^{-1} in some valleys (Chomitz & Šamaj, 1974). Snow cover usually lasts from October to June at elevations >2,000 m.

The surveyed lakes are situated above the present-day timberline and span elevations from 1,725 to 2,157 m a.s.l. (Fig. 1). Bedrock in the study area consists mainly of granitoids (biotite granodiorites to tonalites). Soils are dominated by undeveloped podsols, leptosols and regosols. The dominant vegetation of lake catchments changes from subalpine bushes with dwarf pine (*Pinus mugo*) to alpine meadows (dry tundra) with increasing percentage of rocks. The lake basins are relatively deep with small surface areas. All lakes are soft water, oligotrophic and fishless. Further details on some environmental characteristics can be found in Table 1.

The lakes encompass a gradient of climate and catchment characteristics (soil and vegetation

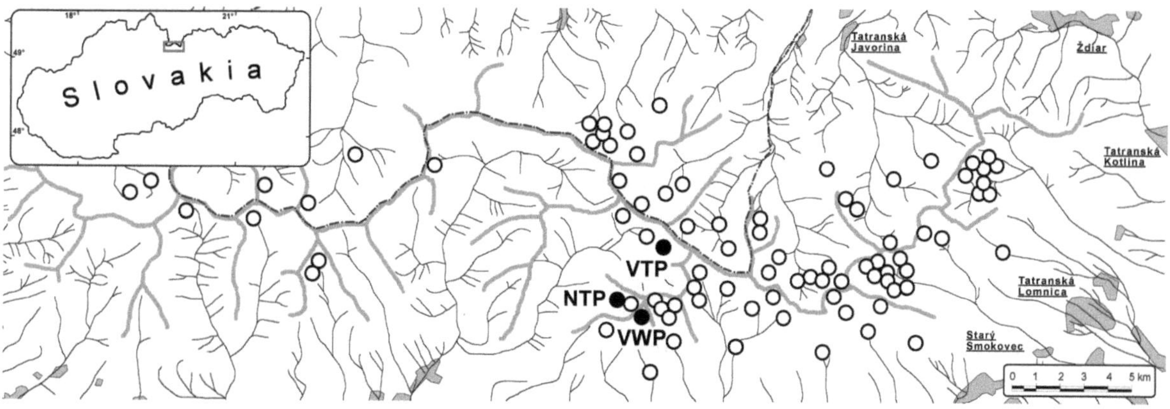

Fig. 1 Location of the most important Tatra Mountains lakes (circles). Study lakes are denoted by black circles and assigned as follows: VTP—Vyšné Temnosmrečinské pleso, NTP—Nižné Terianske pleso and VWP—Vyšné Wahlenbergovo pleso

Table 1 Locations and environmental characteristics of the study lakes

Characteristic/lake name	Vyšné Temnosmrečinské pleso	Nižné Terianske pleso	Vyšné Wahlenbergovo pleso
Latitude (N)	49°11′20″	49°10′11″	49°09′51″
Longitude (E)	20° 02′22″	20°00′51″	20°01′37″
Altitude (m a.s.l.)	1725	1940	2157
Lake area (ha)	5.56	5.56	5.17
Maximum depth (m)	20	47.3	20.6
Catchment area (ha)	112	110	32
Max. mean LSTW in 2001 (°C)	12.9	11.8	11.6
Ice-cover duration (days)	155	203	217
Catchment rocks/screes/alpine meadows (%)	40/34/26	40/32/28	37/51/12
Surface inflow/outflow	+/+	+/+	−/−

Explanations: data on altitude, area and maximum depth of lakes come from Gregor & Pacl (2005), data on LSTW (lake surface water temperature) from Šporka et al. (2006)

coverage). They span a gradient of human impacts, as well, since the negative effects of acid deposition on the lakes in the second half of the twentieth century increased with altitude (Kopáček & Stuchlík, 1994). Recently, there are no direct human activities occurring in the lake catchments. All kinds of land use have been prohibited since the 1950s when the Tatra Mountains became a national park.

Methods

Lake sediment cores were taken in August 1996 and in April 2001 using a modified Kajak corer from the deepest parts of the lakes. The cores were sectioned in the field into 0.5-cm thick layers except for Nižné Terianske pleso, which was divided into 0.2-cm slices. The samples were stored in plastic bags and kept at 4°C for later analysis.

The organic matter content (%) in the sediment cores was measured through loss-on-ignition (LOI) when dried sediment was combusted in a muffle furnace at 550°C for 2 h. The analyses were performed at the Faculty of Sciences, Charles University in Prague.

The upper parts of the cores, representing the past ca. 200 years, were analysed for ^{210}Pb, ^{226}Ra, ^{137}Cs and ^{241}Am by direct gamma assay using Ortec HPGe GWL series well-type coaxial low background intrinsic germanium detectors (see Appleby & Piliposian (2006) for more details). Radiometric dates were calculated from the ^{210}Pb and ^{137}Cs records using the procedures described in Appleby (2001). Dating of deeper sediment samples was estimated applying a mean sedimentation rate. A summary of all cores is given in Table 2.

Sediment samples for diatom analysis were treated through the standard procedures (Battarbee, 1986) using H_2O_2 and HCl and repeated washing in distilled water. Diatom samples were then mounted on slides in Naphrax. At least 400 diatom valves were counted using a LEICA DMLB microscope with 100× oil immersion and phase contrast objectives. Identification, taxonomy and basic information on ecological preferences of the diatom taxa followed Krammer & Lange-Bertalot (1986, 1988, 1991a, b), Lange-Bertalot & Krammer (1989), Charles (1985), van Dam et al. (1994), Tolotti (2001), and Schmidt et al. (2004).

Sediment samples for analysis of chironomid remains were sieved using a 233- and then a 85-μm sieve, respectively (Walker & Paterson, 1985). The fractions retained on the sieves were manually sorted at 7–40× magnifications in a counting tray. After separation, the sediments were dried at 120°C to constant weight to allow the calculation of head capsule concentrations per gram of the dry sediment, except for the sediments from Nižné Terianske pleso, where the number of chironomid remains was expressed per 10 cm^3 of wet sediment.

Chironomid head capsules were mounted, ventral side up, in Berlese solution on microscopic slides. For identification, the Wiederholm (1983), Kowalyk (1985), Schmid (1993) and Ekrem (2004) keys were used. The nomenclature of Chironomidae follows that of Sæther & Spies (2004). In order to achieve more reliable taxonomic results, larvae and pupae collected from the investigated lakes during the past two decades were compared with those in the subfossil material. Remains consisting of the right half of the head capsule or more than half of the mentum were enumerated as a whole head capsule.

Table 2 Details on the cores studied: coring, subsampling and dating

Lake name	Core code	Coring date	Length (cm)	Sampling interval (cm)	Period covered (AD)	Studied period (years)
Vyšné Temnosmrečinské pleso	TE-1	April 2001	30	0.5	2001–1828 ± 37 1828–*1044 ± 207*	*957 ± 207*
Nižné Terianske pleso	TERI96/5 (chironomids) TERI96/7 (diatoms)	August 1996	30.4	0.2	1996–1784 ± 18	156 ± 17
Vyšné Wahlenbergovo pleso	FU-1	April 2001	19	0.5	2001–1862 ± 39 1862–*1624 ± 105*	*377 ± 105*

Estimated ages are indicated by italics

Fragments that consisted of the left half or of less than half of the mentum were excluded.

Diatom and chironomid stratigraphic diagrams were produced using TILIA 2.0.b.4 and TGView 2.0.2 computer software (Grimm, 2004). Biostratigraphical zones were detected according to the major changes in composition of assemblages using constrained incremental sum of squares cluster analysis (CONISS, Grimm, 1987). Diatom and chironomid data were square root transformed for cluster analysis. All the taxa were included in the analysis.

For the purpose of ordination analyses, data on assemblage composition were summarised in the incidence (presence/absence) and abundance matrices for each taxonomic group (diatoms, chironomids) and lake. Both kinds of composition matrices were used in the analyses because each can provide different but complementary information (Heino, 2008). Species with less than five specimens were deleted from the original data matrices to improve the signal-to-noise ratio (Gauch, 1982). Altogether, 12 site-by-species matrices were constructed (two groups × two matrices × three lakes). The chironomid incidence matrix of Vyšné Wahlenbergovo pleso was excluded from the subsequent analyses due to a lack of variability (matrix of ones). Ordination methods were used to examine the relationships between organic matter and ecological assemblages and between individual assemblages.

The diatom and chironomid composition matrices were subjected to preliminary indirect ordination analysis. Detrended correspondence analysis (Hill & Gauch, 1980) with detrending by segments was used to determine whether linear- or unimodal-based ordination techniques were more appropriate for the data. A majority of the analyses yielded gradient lengths from 1.6 to 2.2 standard deviations. The features of the data suggested that both linear- and unimodal-based techniques may be appropriate (ter Braak and Prentice, 1988). However, the qualitative nature of incidence matrices and the predefined total abundance of diatoms per layer make unimodal techniques more appropriate (ter Braak & Schaffers, 2004). Thus, methods related to correspondence analysis were applied to all the composition matrices, facilitating comparability of the results across different datasets.

Canonical correspondence analysis (CCA) was used to assess the relationship between the organic matter content of sediments as a good temperature indicator (Battarbee et al., 2002b) and composition of diatom and chironomid assemblages, respectively. Predictive CCA (ter Braak & Schaffers, 2004) was applied rather than the usual exploratory version of this analysis, which enabled a direct comparison of the results with those of predictive co-correspondence analysis (CoCA; see below). Predictive CCA uses a 'leave-one-out' procedure to validate the models. This means, in our case, that an analysis was carried out as many times as there were layers, each time with a different layer left out and applying the obtained model to the omitted layer to predict its species composition from the organic matter content. Predicted taxonomic composition was compared to the observed composition on the basis of sum of squared prediction errors (ssp_a). The prediction accuracy of each model was assessed using the cross-validatory fit, calculated as $100 \times (1 - ssp_a/ssp_0)$, where ssp_0 is the sum of squared prediction errors under the null model of no relationship. Unlike explained variation in explanatory CCA, cross-validatory fit can be negative when the model fit is so poor that the null model predicts the data better. Any fit above zero, however, indicates that prediction is better than could have been expected by chance, implicitly validating the model without additional statistical tests (Schaffers et al., 2008).

The relationships between diatom and chironomid assemblages of different lakes were assessed using CoCA (ter Braak & Schaffers, 2004). This direct ordination method relates one community dataset to another by maximising the weighted covariance between the weighted averaged species scores of the communities. The CoCA attempts to identify the patterns (ecological gradients) that are common to both assemblages. Both a symmetric descriptive and an asymmetric predictive form have been developed. Predictive CoCA provides a way to use the diatom assemblage as a predictor of the chironomid assemblage and to assess the predictive value of models via a cross-validation procedure (see above). The use of the same cross-validatory fit measure in predictive CCA and predictive CoCA allowed a direct comparison of the ability of different predictors (diatom incidence matrix, diatom abundance matrix and organic matter content) to predict the composition of chironomid assemblages (defined in terms of incidence or abundance).

Schaffers et al. (2008) pointed out that the predictive models themselves are validated implicitly (when the cross-validatory fit is above zero), but there is still a need to judge whether differences between models (using different predictors) are actually significant or could just as well be attributed to random variation. The predictive power of different models was compared using a two-sided simple randomisation test (van der Voet, 1994). The difference in mean square of prediction errors (T) of the compared models was used as a test statistic. The significance of difference between models was obtained by a comparison of the observed T statistic with a distribution of this statistic generated by randomised data (999 random rearrangements of site prediction errors).

Results

Dating

The ^{210}Pb results from the Vyšné Temnosmrečinské pleso core TA0019 suggested a slow and relatively uniform sedimentation rate of 0.0064 g cm^{-2} yr^{-1} (0.058 cm yr^{-1}) during the past 150 years or so. Since this value is typical of remote undisturbed sites, it is not unreasonable to suppose that this rate has persisted over a much longer timescale. Table 3 gives dates calculated on this basis.

Results of an extrapolated chronology for the Nižné Terianske pleso core TERI96/7, down to a depth of 6 cm (dated 1784), are given in Appleby (2000). The presence of a number of layers of dense sediment at depths varying from 2 to 29 cm suggests that sedimentation rates at this site are likely to have been quite variable, with several episodes of rapid accumulation. In view of this, any further extrapolation is likely to be very unreliable unless supported by well-defined correlations with cores from more stable sites.

Due to the very slow accumulation rate in the Vyšné Wahlenbergovo pleso core FU-1, $^{210}Pb/^{226}Ra$ equilibrium being reached at a depth of just 6 cm, there were only two ^{210}Pb data points above the $^{210}Pb/^{226}Ra$ equilibrium depth. However, the results did suggest a mean sedimentation rate of 0.0057 g cm^{-2} yr^{-1} (0.043 cm yr^{-1}), a value typical of remote undisturbed sites. In the absence of evidence of earlier disturbances, it would therefore

Table 3 Extrapolated ^{210}Pb chronology of Vyšné Temnosmrečinské pleso core TA0019

Depth		Chronology			Sedimentation rate		
cm	g cm^{-2}	Date AD	Age y	±	g cm^{-2} y^{-1}	cm y^{-1}	± (%)
0.00	0.00	2001	0				
0.25	0.01	1999	2	0	0.0064	0.11	21.5
0.75	0.04	1994	7	1	0.0064	0.09	21.5
1.25	0.08	1988	13	3	0.0064	0.07	21.5
1.75	0.13	1981	20	4	0.0064	0.07	21.5
2.25	0.17	1974	27	6	0.0064	0.07	21.5
2.75	0.22	1967	34	7	0.0064	0.07	21.5
3.25	0.27	1959	42	9	0.0064	0.07	21.5
3.75	0.32	1952	49	11	0.0064	0.06	21.5
4.25	0.37	1944	57	12	0.0064	0.06	21.5
4.75	0.42	1936	65	14	0.0064	0.06	21.5
5.25	0.47	1927	74	16	0.0064	0.06	21.5
5.75	0.53	1919	82	18	0.0064	0.05	21.5
6.25	0.59	1909	92	20	0.0064	0.05	21.5
6.75	0.65	1899	102	22	0.0064	0.05	21.5
7.25	0.71	1890	111	24	0.0064	0.05	21.5
7.75	0.77	1880	121	26	0.0064	0.05	21.5
8.25	0.84	1871	130	28	0.0064	0.05	21.5
8.75	0.90	1860	141	30	0.0064	0.04	21.5
9.25	0.98	1848	153	33	0.0064	0.04	21.5
9.75	1.07	1835	166	36	0.0064	0.04	21.5
10.25	1.15	1821	180	39	0.0064	0.04	21.5
10.75	1.24	1808	193	41	0.0064	0.04	21.5
11.25	1.33	1794	207	45	0.0064	0.03	21.5
11.75	1.44	1777	224	48	0.0064	0.03	21.5
12.25	1.54	1760	241	52	0.0064	0.03	21.5
12.75	1.65	1744	257	55	0.0064	0.03	21.5
13.25	1.74	1729	272	59	0.0064	0.03	21.5
13.75	1.86	1711	290	62	0.0064	0.02	21.5
14.25	*2.01*	*1687*	*314*	*68*	*0.0064*	*0.02*	*21.5*
14.75	*2.15*	*1665*	*336*	*72*	*0.0064*	*0.02*	*21.5*
15.25	*2.29*	*1644*	*357*	*77*	*0.0064*	*0.02*	*21.5*
15.75	*2.43*	*1622*	*379*	*82*	*0.0064*	*0.02*	*21.5*
16.25	*2.55*	*1604*	*397*	*85*	*0.0064*	*0.03*	*21.5*
16.75	*2.69*	*1582*	*419*	*90*	*0.0064*	*0.02*	*21.5*
17.25	*2.88*	*1552*	*449*	*97*	*0.0064*	*0.01*	*21.5*
17.75	*3.11*	*1515*	*486*	*105*	*0.0064*	*0.02*	*21.5*
18.25	*3.28*	*1489*	*512*	*110*	*0.0064*	*0.02*	*21.5*
18.75	*3.38*	*1474*	*527*	*113*	*0.0064*	*0.03*	*21.5*
19.25	*3.52*	*1451*	*550*	*118*	*0.0064*	*0.02*	*21.5*
19.75	*3.69*	*1425*	*576*	*124*	*0.0064*	*0.02*	*21.5*

Table 3 continued

Depth		Chronology			Sedimentation rate		
cm	g cm^{-2}	Date AD	Age y	±	g cm^{-2} y^{-1}	cm y^{-1}	± (%)
20.25	*3.83*	*1403*	*598*	*129*	*0.0064*	*0.03*	*21.5*
20.75	*3.95*	*1385*	*616*	*133*	*0.0064*	*0.03*	*21.5*
21.25	*4.08*	*1365*	*636*	*137*	*0.0064*	*0.02*	*21.5*
21.75	*4.26*	*1337*	*664*	*143*	*0.0064*	*0.02*	*21.5*
22.25	*4.42*	*1311*	*690*	*149*	*0.0064*	*0.02*	*21.5*
22.75	*4.57*	*1288*	*713*	*153*	*0.0064*	*0.02*	*21.5*
23.25	*4.76*	*1259*	*742*	*160*	*0.0064*	*0.02*	*21.5*
23.75	*4.96*	*1227*	*774*	*167*	*0.0064*	*0.02*	*21.5*
24.25	*5.12*	*1203*	*798*	*172*	*0.0064*	*0.02*	*21.5*
24.75	*5.23*	*1186*	*815*	*176*	*0.0064*	*0.03*	*21.5*
25.25	*5.32*	*1171*	*830*	*179*	*0.0064*	*0.03*	*21.5*
25.75	*5.41*	*1157*	*844*	*182*	*0.0064*	*0.03*	*21.5*
26.25	*5.50*	*1143*	*858*	*185*	*0.0064*	*0.04*	*21.5*
26.75	*5.59*	*1129*	*872*	*188*	*0.0064*	*0.04*	*21.5*
27.25	*5.68*	*1115*	*886*	*191*	*0.0064*	*0.03*	*21.5*
27.75	*5.77*	*1100*	*901*	*194*	*0.0064*	*0.03*	*21.5*
28.25	*5.87*	*1085*	*916*	*197*	*0.0064*	*0.03*	*21.5*
28.75	*5.98*	*1069*	*932*	*201*	*0.0064*	*0.03*	*21.5*
29.25	*6.10*	*1050*	*951*	*205*	*0.0064*	*0.03*	*21.5*

NB: Dates in italics (for sediments below 14 cm) should be regarded with more caution due to high irregular variations in values of dry bulk density

Table 4 Extrapolated ^{210}Pb chronology of Vyšné Wahlenbergovo pleso core FU-1

Depth		Chronology			Sedimentation rate		
cm	g cm^{-2}	Date AD	Age y	±	g cm^{-2} y^{-1}	cm y^{-1}	± (%)
0.00	0.00	2001	0				
0.25	0.02	1997	4	1	0.0057	0.05	28.1
0.75	0.08	1987	14	4	0.0057	0.04	28.1
1.25	0.16	1974	27	8	0.0057	0.04	28.1
1.75	0.23	1960	41	12	0.0057	0.04	28.1
2.25	0.30	1948	53	15	0.0057	0.04	28.1
2.75	0.37	1936	65	18	0.0057	0.04	28.1
3.25	0.44	1924	77	22	0.0057	0.04	28.1
3.75	0.50	1913	88	25	0.0057	0.04	28.1
4.25	0.57	1901	100	28	0.0057	0.04	28.1
4.75	0.64	1889	112	31	0.0057	0.04	28.1
5.25	0.70	1878	123	35	0.0057	0.05	28.1
5.75	0.76	1867	134	38	0.0057	0.05	28.1
6.25	0.81	1858	143	40	0.0057	0.06	28.1
6.75	0.86	1850	151	42	0.0057	0.06	28.1
7.25	0.91	1841	160	45	0.0057	0.06	28.1
7.75	0.96	1832	169	47	0.0057	0.05	28.1
8.25	1.02	1823	178	50	0.0057	0.05	28.1
8.75	1.07	1813	188	53	0.0057	0.05	28.1
9.25	1.12	1804	197	55	0.0057	0.06	28.1
9.75	1.17	1796	205	58	0.0057	0.06	28.1
10.25	1.22	1786	215	60	0.0057	0.05	28.1
10.75	1.28	1777	224	63	0.0057	0.05	28.1
11.25	1.33	1768	233	65	0.0057	0.06	28.1
11.75	1.38	1759	242	68	0.0057	0.05	28.1
12.25	1.44	1749	252	71	0.0057	0.05	28.1
12.75	1.49	1739	262	73	0.0057	0.05	28.1
13.25	1.54	1730	271	76	0.0057	0.06	28.1
13.75	1.59	1722	279	78	0.0057	0.06	28.1
14.25	1.64	1713	288	81	0.0057	0.06	28.1
14.75	1.69	1704	297	83	0.0057	0.06	28.1
15.25	1.74	1695	306	86	0.0057	0.06	28.1
15.75	1.79	1687	314	88	0.0057	0.05	28.1
16.25	1.85	1677	324	91	0.0057	0.05	28.1
16.75	1.90	1667	334	94	0.0057	0.05	28.1
17.25	1.96	1657	344	97	0.0057	0.05	28.1
17.75	2.01	1648	353	99	0.0057	0.05	28.1
18.25	2.07	1639	362	102	0.0057	0.05	28.1
18.75	2.12	1629	372	104	0.0057	0.05	28.1
19.25	2.17	1620	381	107	0.0057	0.05	28.1
19.75	2.22	1611	390	110	0.0057	0.06	28.1

seem reasonable to suppose that sedimentation rates have been stable over a much longer period than that covered by the ^{210}Pb time span. Table 4 gives dates calculated on this basis.

Biostratigraphies

Vyšné Temnosmrečinské pleso

Diatoms A total of 100 diatom taxa were recorded, most of them benthic. Fourteen species occurred at a minimum of over 5%. Of these, three species achieved abundance higher than 30% (*Fragilaria construens* f. *venter*, *F. pinnata* and *F. pseudoconstruens*).

Based on changes in the assemblage composition, four zones can be detected (Fig. 2). Zone DTE-I (depth 30–10 cm) is dominated by *Fragilaria* species at relative abundance amounting to 81% at the 28 cm depth, except for at 14 cm where they reach only 40%. Also, *Achnanthes* species (7–17%) significantly

Table 4 continued

Depth		Chronology			Sedimentation rate		
cm	g cm^{-2}	Date	Age		g cm^{-2} y^{-1}	cm y^{-1}	± (%)
		AD	y	±			
20.25	2.28	1602	399	112	0.0057	0.06	28.1
20.75	2.32	1593	408	114	0.0057	0.06	28.1
21.25	2.37	1585	416	117	0.0057	0.06	28.1
21.75	2.42	1576	425	119	0.0057	0.06	28.1
22.25	2.47	1568	433	122	0.0057	0.06	28.1
22.75	2.52	1560	441	124	0.0057	0.06	28.1
23.25	2.57	1551	450	126	0.0057	0.06	28.1
23.75	2.62	1542	459	129	0.0057	0.06	28.1
24.25	2.67	1532	469	132	0.0057	0.05	28.1
24.75	2.73	1522	479	135	0.0057	0.04	28.1
25.25	2.80	1510	491	138	0.0057	0.04	28.1
25.75	2.86	1499	502	141	0.0057	0.05	28.1
26.25	2.91	1490	511	144	0.0057	0.05	28.1
26.75	2.97	1481	520	146	0.0057	0.05	28.1
27.25	3.02	1471	530	149	0.0057	0.05	28.1
27.75	3.08	1461	540	152	0.0057	0.05	28.1
28.25	3.13	1451	550	154	0.0057	0.05	28.1
28.75	3.19	1441	560	157	0.0057	0.05	28.1
29.25	3.26	1430	571	160	0.0057	0.05	28.1
29.75	3.32	1419	582	163	0.0057	0.04	28.1
30.25	3.38	1409	592	166	0.0057	0.05	28.1
30.75	3.44	1398	603	169	0.0057	0.05	28.1
31.25	3.50	1387	614	172	0.0057	0.04	28.1
31.75	3.56	1376	625	175	0.0057	0.05	28.1
32.25	3.62	1366	635	178	0.0057	0.05	28.1
32.75	3.69	1353	648	182	0.0057	0.04	28.1
33.25	3.77	1340	661	185	0.0057	0.04	28.1
33.75	3.83	1328	673	189	0.0057	0.04	28.1
34.25	3.91	1316	685	192	0.0057	0.04	28.1

contribute to this assemblage. *Cymbella minuta*, *Denticula tenuis*, *Navicula minuscula* and *Nitzschia paleacea* make up a stable part of the assemblage through this zone, with a relative abundance <8% whilst *Navicula minuscula* increases in the uppermost samples of the zone. In Zone DTE-II (10–7 cm), *Fragilaria* species, mainly *F. pinnata* (25–41%) and *F. pseudoconstruens* (22–28%), remain as dominant elements. *F. brevistriata*, *F. construens* f. *construens* show an increase in proportion of the assemblage. *Cyclotella stelligera* reaches its greatest proportion in the bottom of this zone. Zone DTE-III (7–4 cm) is characterised by a slight decrease of *Fragilaria pseudoconstruens* and *Achnanthes* species, while an increase in the relative abundance of *F. brevistriata* and *F. construens* f. *construens* is evident. Zone DTE-IV (4–0 cm) is dominated by *F. construens* f. *venter* and *F. pinnata*, and *F. construens* f. *construens*, *F. brevistriata*, and *F. pseudoconstruens* disappear, but the second species reappears again from 1.5 cm. Other *Fragilaria* species persist as a minor element of the assemblage. In conjunction with this trend, slight increase in *Navicula minuscula* occurs and *Denticula tenuis* reaches significant proportion (up to 12%).

Chironomids In total, 14,025 chironomid head capsules from 15 taxa were analysed. Subfossil chironomid density varied from 6 to 354 specimens g^{-1} dry sediment. The *Tanytarsus lugens* group (57.8%) and *Micropsectra* spp. (33.7%) made up 91% of the chironomid assemblages. The other taxa that achieved an average abundance of more than 1% were *Procladius* (*Holotanypus*) sp. (5.5%) and *Heterotrissocladius marcidus* (1.1%).

Changes in the chironomid record can be divided into three zones (Fig. 3). The oldest sediments of Zone ChTE-I (30–26 cm) are characterised by a high number of chironomid remains and dominance of the *Tanytarsus lugens* group. Opposite trends in the relative abundance of both taxa show a marked transition from Zone ChTE-I to Zone ChTE-II. The beginning of Zone ChTE-II (26–9 cm), especially between 24 and 17.5 cm, is defined by a significant reduction in the head capsule density. *Micropsectra* spp. remains a dominant taxon but declines in relative abundance, while the importance of the *Tanytarsus lugens* group starts to increase in the uppermost samples of the zone. *Heterotrissocladius marcidus*, *Diamesa* spp. and *Zavrelimyia* sp. reach their greatest proportions in this interval. In Zone ChTE-III (9–0 cm), the *T. lugens* group becomes dominant, while *Micropsectra* spp. gradually decreases in the most recent sediments, and *Paratanytarsus austriacus* first appears. *Procladius* (*Holotanypus*) sp. and *Cricotopus*/*Paratrichocladius* remain relatively unchanged throughout the whole sediment record.

Nižné Terianske pleso

Diatoms A total of 110 diatom taxa were identified. Twenty-four species were found at >5% abundance,

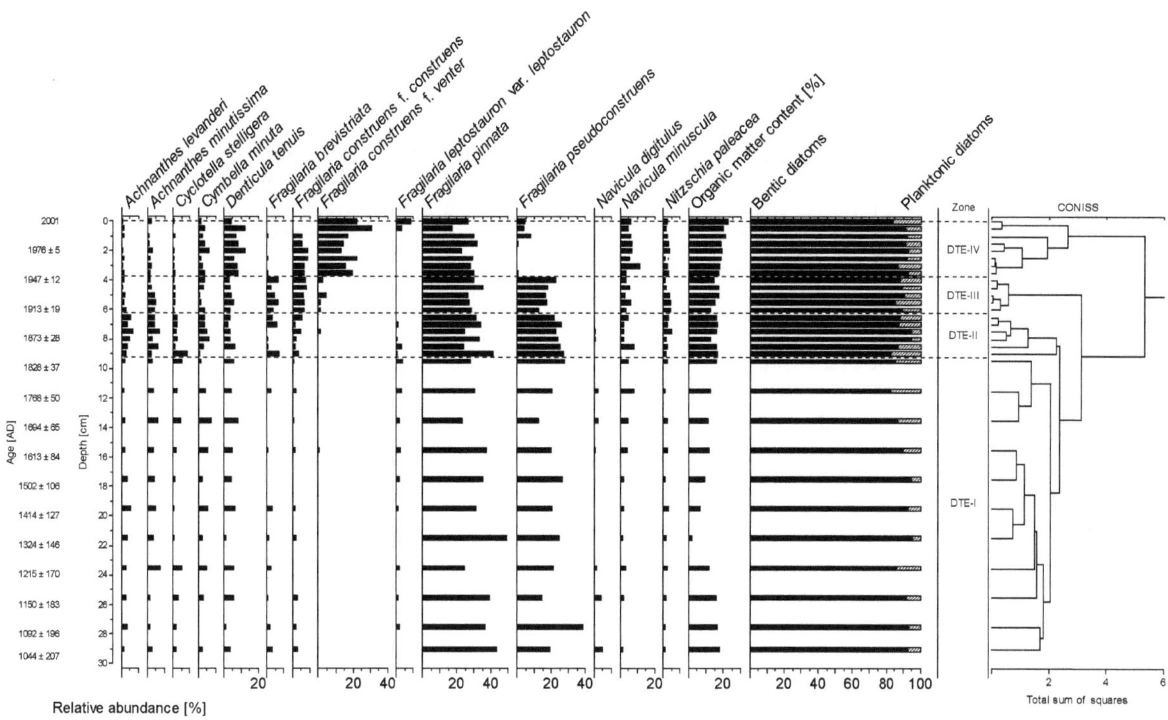

Fig. 2 Changes in percent abundances of selected diatom taxa in the sediment core from Vyšné Temnosmrečinské pleso

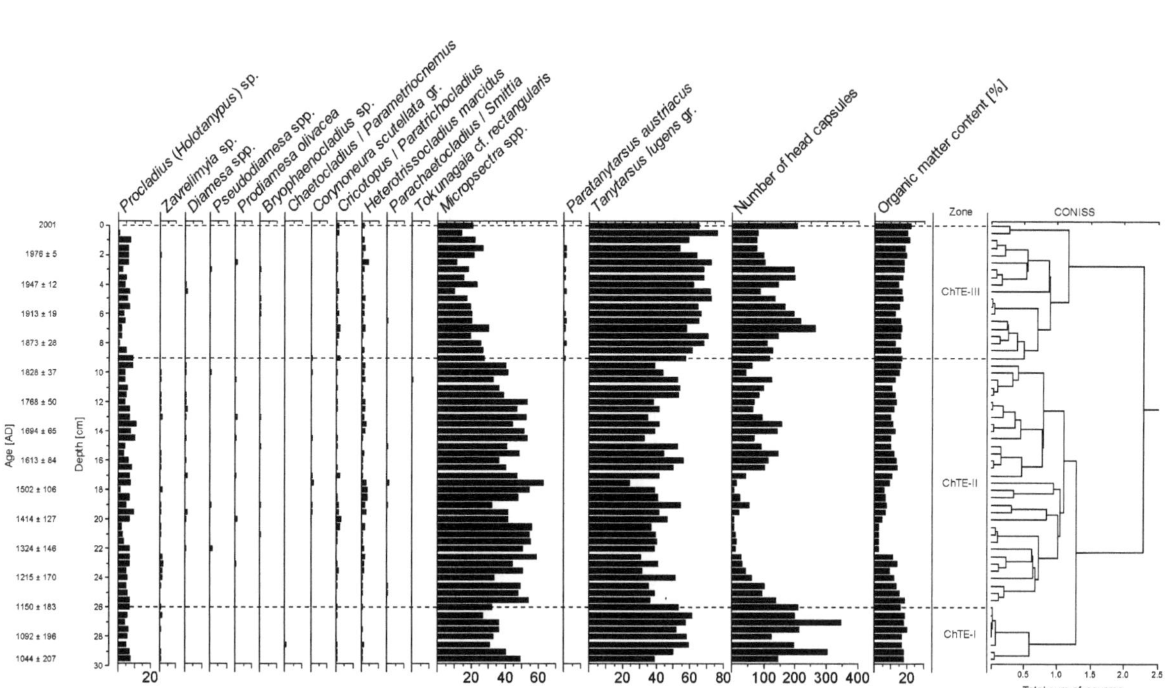

Fig. 3 Changes in percent abundances of chironomid taxa in the sediment core from Vyšné Temnosmrečinské pleso

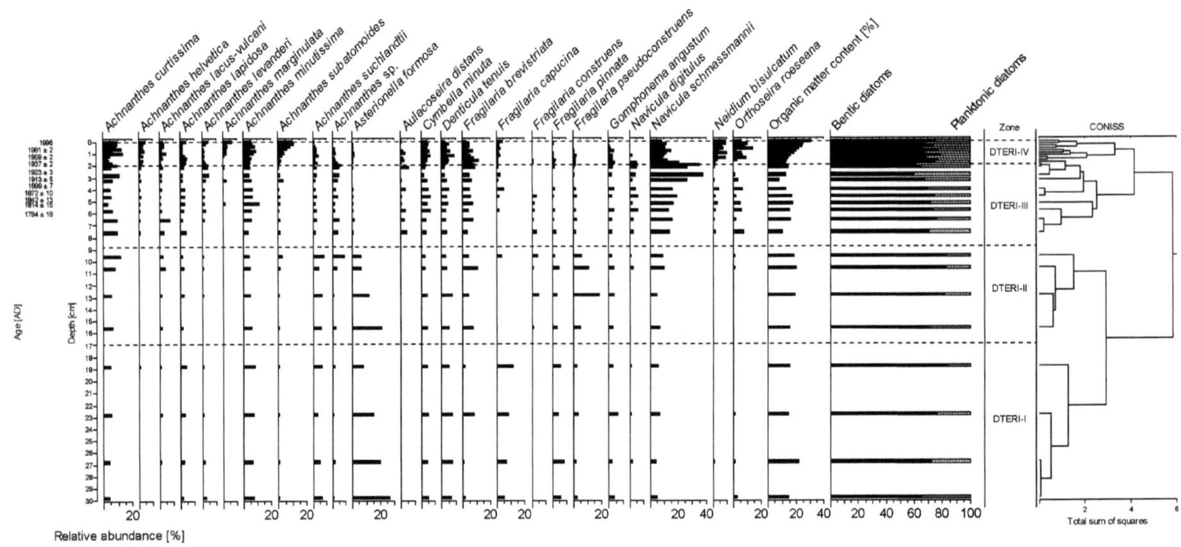

Fig. 4 Changes in percent abundances of selected diatom taxa in the sediment core from Nižné Terianske pleso

and ten of them achieved an average abundance of more than 10%.

Four diatom zones were identified through the stratigraphically constrained cluster analysis of 28 samples (Fig. 4). Zone DTERI-I (30–17 cm) is clearly dominated by planktonic *Asterionella formosa* (>20%) and together with *Fragilaria pinnata*, *F. capucina*, *Denticula tenuis* and *Achnanthes minutissima* reach their maximum extent in this zone. Zone DTERI-II (17–9 cm) is characterised by a decline in *Asterionella formosa. Fragilaria pseudoconstruens* rapidly becomes a major element (19%) and *Fragilaria brevistriata*, *Achnanthes curtissima*, *A. suchlandtii* and *Denticula tenuis* remain relatively abundant in the assemblage. *Navicula schmassmannii* increases throughout this zone and reaches approximately 10% at the top. In Zone DTERI-III (9–2 cm), a major shift in diatoms occurs. *N. schmassmannii* increases gradually and reaches its greatest proportion (up to 40%) at 3–2 cm. On the other hand, *Asterionella formosa* rapidly declines in relative abundance to near extinction. The centric diatom *Orthoseira roeseana*, which was present in very low numbers (1–2%) in the bottom part of the core, significantly increases in number at 8 cm, and becomes a significant member of the assemblage (14%).

Another change in the diatom assemblage is evident in Zone DTERI-IV (2–0 cm). Many species, mostly rare or absent in the deeper zones, appear here at their highest values. This zone is dominated by *Achnanthes* species (45%), and to a lesser extent by *Orthoseira roeseana* (15%), *Fragilaria brevistriata* (13%) and *Neidium bisulcatum* (9%).

Chironomids Altogether 2,211 chironomid head capsules were recovered, comprising 11 taxa. The number of remains was variable, ranging from 1 to 69 specimens per 10 cm^3 of sediment. The most abundant taxa comprising the subfossil record were *Micropsectra radialis* (44.8%), *Procladius* (*Holotanypus*) sp. (22.4%) and *Micropsectra* cf. *junci* (10.9%). Other important taxa, such as *Heterotrissocladius marcidus* and *Diamesa* spp., were found almost in all of the samples at >9%.

The chironomid diagram is divided into six assemblage zones (Fig. 5). Zone ChTERI-I (30.4–27.6 cm) is dominated by *Procladius* (*Holotanypus*) sp., making up 40% of the chironomid assemblages. Within Zone ChTERI-II (27.6–17.4 cm), *Procladius* (*Holotanypus*) sp. declines rapidly, and *Micropsectra radialis, Micropsectra* cf. *junci* increase, making up 40% of the chironomid assemblage, along with *Diamesa* spp. and *Heterotrissocladius marcidus.* The numbers of specimens per 10 cm^{-3} of sediment vary, with a peak at 18–17.4 cm. The two zones ChTERI-III and ChTERI-IV (17.4–12.6 cm) cover relatively short periods with a rapid, marked decline in the number of chironomid head capsules and consequently in relative abundances of chironomids.

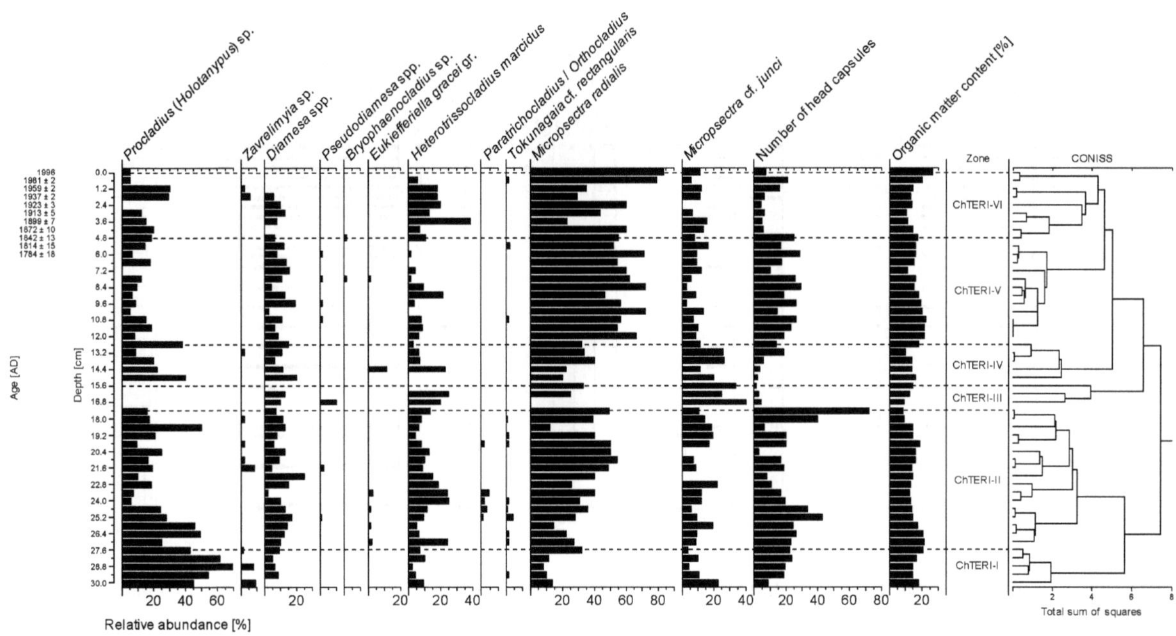

Fig. 5 Changes in percent abundances of chironomid taxa in the sediment core from Nižné Terianske pleso

Procladius (*Holotanypus*) sp. disappears from the record between 17.4 and 15.6 cm and appears again within Zone ChTERI-IV (15.6–12.6 cm), when *M. radialis* again attains significant relative abundance. In Zone ChTERI-V (12.6–4.8 cm), *M. radialis* continues to be a dominant species with a coincident decrease of *Procladius* (*Holotanypus*) sp. The decline of relative abundance of *Micropsectra radialis* in Zone ChTERI-VI (4.8–0 cm) is accompanied by a slight increase in *Procladius* (*Holotanypus*) sp. and a marked increase in *H. marcidus*. *Diamesa* head capsules are missing from the most recent samples. The number of head capsules and organic content in the sediments increase towards the top of the record.

Vyšné Wahlenbergovo pleso

Diatoms Altogether 112 diatom taxa were identified. The assemblage consisted of planktonic and benthic species, and was dominated by *Achnanthes* species. Twenty-three species occurred at abundances of greater than 5%, eleven of these were found at >10%, four species at >20% and one species at >30%.

Changes in the diatom assemblage can be divided into four zones (Fig. 6). Zone DFU-I (20–15 cm) is characterised by a dominance of *Navicula schmassmannii* (up to 24%) and *Cymbella minuta* (about 15%). *Achnanthes minutissima, A. oblongella* (both about 10%), *Fragilaria exigua*, *F. pinnata* and *F. pseudoconstruens* persist as very stable elements of the assemblage in this interval. *Aulacoseira distans* declines in relative abundance from 12% in the bottom layers to 1% in 15 cm. In Zone DFU-II (15–10 cm), *Cymbella minuta* and *Navicula schmassmannii* remain as dominant species (about 20%). A significant increase in *Achnanthes* species (*A. lacus-vulcani*, *A. minutissima*, *A. oblongella*, and *A. subatomoides*) occurs through this zone, reaching up to 30% in some samples. *Aulacoseira distans* making up 10% of the assemblage at the beginning of the zone decreases from 11.5 cm to the top. Zone DFU-III (10–3 cm) shows a gradual reduction in *Navicula schmassmannii* reaching 6% at the top of this interval. In conjunction with this trend, *Achnanthes* species—*A. subatomoides*, *A. lacus-vulcani*, *A. levanderi* and *A. minutissima* become major components of the assemblage (on average 50%). *Cymbella minuta* retains its previous abundance. In the sediments of Zone DFU-IV (3–0 cm), *Achnanthes* species attain significant proportions (50–70%) of the diatom assemblage composition. *Aulacoseira distans* reaches its maximum extent (24%) at 3 cm. *Cymbella minuta* declines in relative abundance similarly to

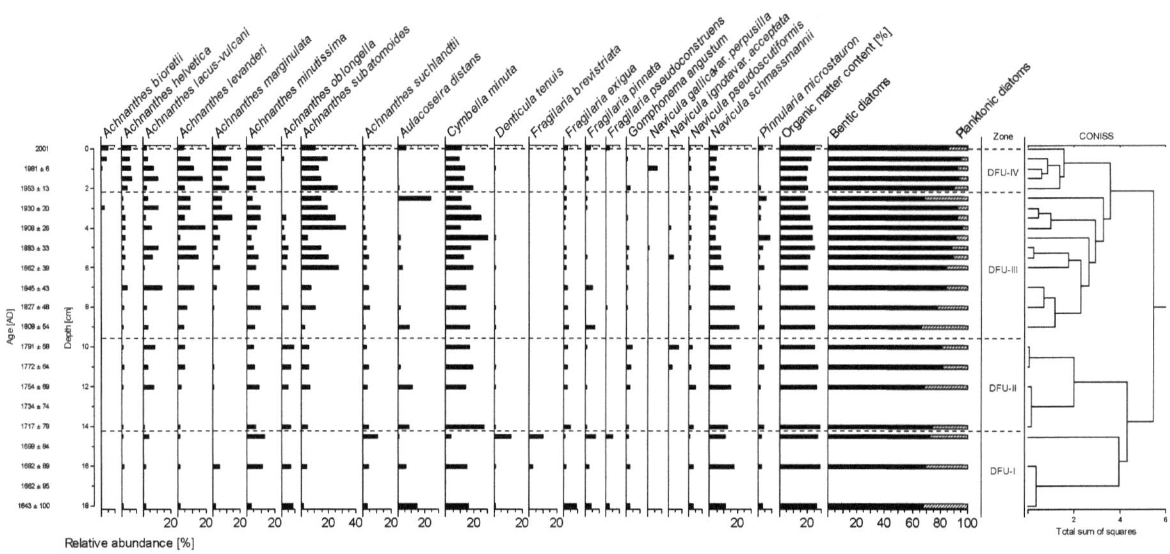

Fig. 6 Changes in percent abundances of selected diatom taxa in the sediment core from Vyšné Wahlenbergovo pleso

Pinnularia microstauron, whereas *Navicula gallica* var. *perpusilla* increases up to 7%. *N. schmassmannii* persists as a relatively small proportion of the assemblage.

Chironomids In total, the chironomid analysis provided 2,048 head capsules from five taxa. The down-core number of chironomid head capsules showed large fluctuations, ranging from 6 to 136 head capsules g^{-1} dry sediment. *Micropsectra radialis* (68.6%), *Heterotrissocladius marcidus* (14.6%) and *Pseudodiamesa arctica* (16.5%) dominate the whole subfossil record.

Three biostratigraphic zones were identified by cluster analysis (Fig. 7). Zone ChFU-I (19–3 cm) is dominated by *Micropsectra radialis* (>55%). The number of chironomid remains slightly decreased throughout this zone, but declines suddenly from 15 cm to the lowest level at 14–13 cm. In Zone ChFU-II (3–1 cm), *Heterotrissocladius marcidus* rapidly become an important faunal component, whereas *Micropsectra radialis* and *Pseudodiamesa arctica* decline in relative abundances. The upper boundary is marked by a sharp decrease in the number of head capsules. Within the most recent Zone ChFU-III (1–0 cm), *Micropsectra radialis* increases to its previous level, and the abundance of *Heterotrissocladius marcidus* is reduced. The number of chironomid head capsules remains low.

Organic matter and subfossil assemblages

The relationship between the abundance of organic matter in the sediments and composition of the subfossil assemblages depends on matrix type (Table 5). A significant predictive power of organic matter is found for all of the abundance matrices. The best prediction is recorded for chironomid assemblages in Vyšné Wahlenbergovo pleso where the proportions of *Pseudodiamesa arctica* and *Micropsectra radialis* increase with the amount of organic matter in the sediments. In general, the strength of this organic matter–assemblage relationship changes inconsistently across groups and lakes: relatively poor results in Nižné Terianske pleso for both of these groups, a higher predictive power for diatoms in Vyšné Temnosmrečinské pleso and a high predictive value for chironomids in Vyšné Wahlenbergovo pleso. The pattern common to all lakes is the increasing proportion of the diatom *Achnanthes levanderi* in the organic-poor layers.

In contrast to the abundance matrices, the amount of organic matter does not appear to be a good predictor of species occurrence in all but one assemblage. A weak relationship is shown in Vyšné Wahlenbergovo pleso, where *Denticula tenuis*, *Achnanthes oblongella*, *Pinnularia subrostrata* and a few other diatom species almost completely disappear from the sediment with lower organic matter content.

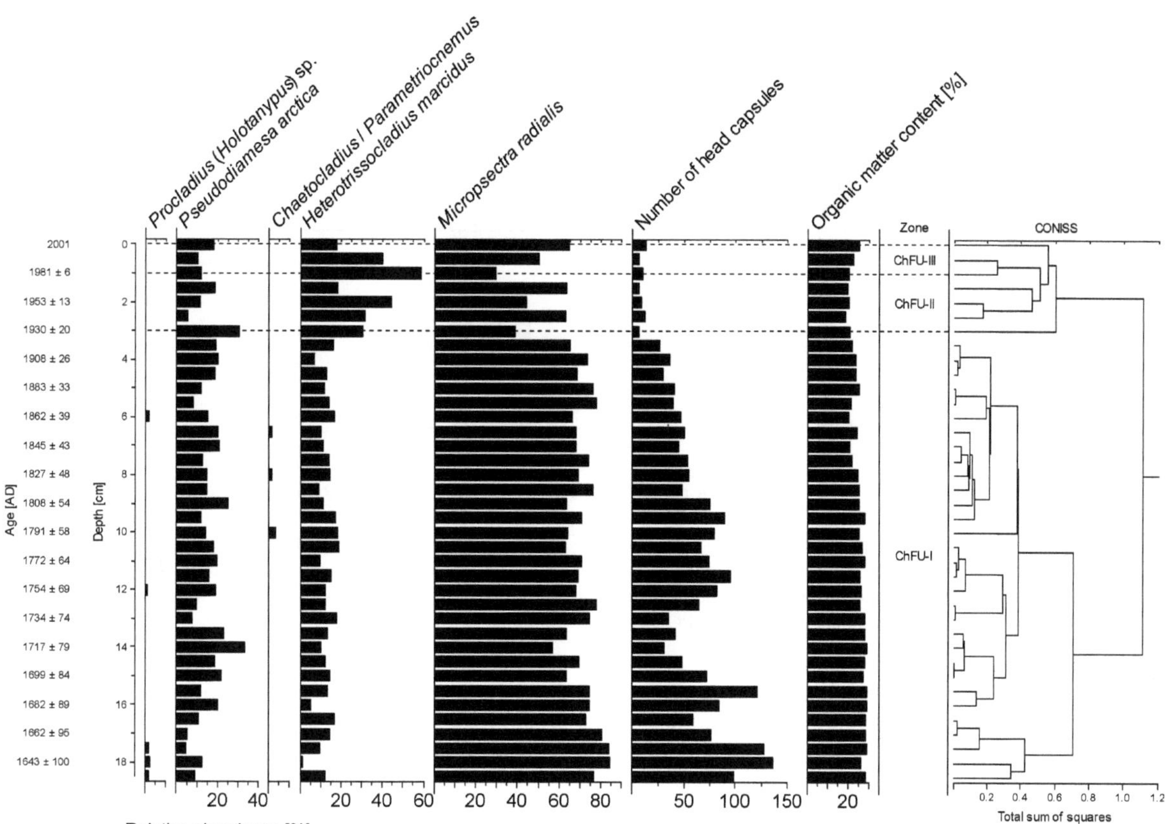

Fig. 7 Changes in percent abundances of chironomid taxa in the sediment core from Vyšné Wahlenbergovo pleso

Table 5 Percentage cross-validatory fit of predictive CCA models relating diatom and chironomid assemblages to the organic matter content of the three studied lakes

Lake	Incidence matrix		Abundance matrix	
	Diatoms	Chironomids	Diatoms	Chironomids
Vyšné Temnosmrečinské pleso	−0.5	−3.7	6.9	3.7
Nižné Terianske pleso	−1.3	−1.2	0.4	0.7
Vyšné Wahlenbergovo pleso	3.5	–	3.5	17.8

In other cases, organic matter gives negative percentages, and thus has no predictive value for the presence/absence of diatoms and chironomids, respectively.

Relationships between assemblages

In two of the three studied lakes, predictive CoCA shows that chironomid assemblages are more closely related to the species composition of diatom assemblages than to the amount of organic matter in the sediments (Fig. 8). The diatom species expressed either as abundance or as presence/absence, predict the chironomid data better than organic matter content. A notable exception is Vyšné Wahlenbergovo pleso where no relationship between subfossil assemblages is seen. The overall prediction level is higher for quantitative (abundance) than qualitative (presence/absence) chironomid data as judged by cross-validatory fit. The effect of different

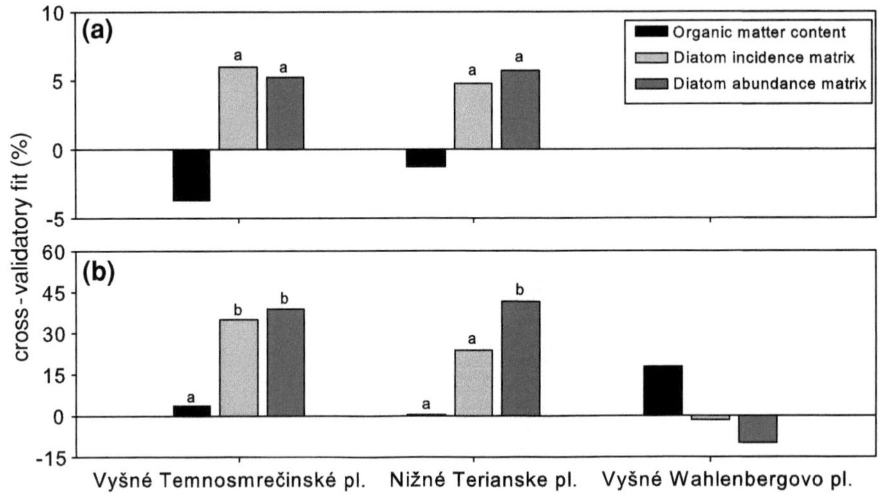

Fig. 8 Chironomid assemblage prediction using different predictors. Prediction levels are given as percentages of cross-validatory fit for chironomid composition defined in terms of incidence (**a**) and abundance (**b**). Within each combination of lake and chironomid matrix type, different letters indicate significant differences ($P < 0.05$, simple randomisation test). Only predictors with positive fit were included in the pair-wise comparison

diatom matrix types is relatively unimportant. However, the matrix of diatom abundances in Nižné Terianske pleso offers a significantly higher level of prediction than its presence/absence counterpart.

Discussion

Biostratigraphies

Vyšné Temnosmrečinské pleso

Both diatom and chironomid analyses revealed relatively stable taxonomic composition throughout the sediment record. The diatom assemblages are heavily dominated by small *Fragilaria* species reaching more than half of the total diatom abundance. Zone DTE-I (~AD 1044 ± 207 to 1828 ± 37) roughly corresponds to the chironomid zones ChTE-I, II (~AD 1044 ± 207 to 1854 ± 31), though the changes in diatom assemblages are not as marked as they are in the chironomid record. The major change in diatom assemblages started in the middle of the nineteenth century and lasted until recently, whereas the most pronounced change in chironomids took place in the second half of the twentieth century.

It is important to note that compared with other European mountain regions, the Tatra Mountains were heavily exposed to acidifying pollutants for more than a half of a century, peaking in the late 1980s (Curtis et al., 2005). Though this lake was classified as non-acidified during this acidification period (Fott et al., 1994), the adverse effect of acid conditions on littoral species cannot be excluded. Despite this, we suggest that some factors other than pH-driven changes must be invoked to explain the diatom oscillation through the sequence in this well-buffered lake.

The non-acidified status of this lake during the acidification period (Fott et al., 1994) is supported by the dominance of diatoms *Fragilaria pseudoconstruens* and *Fragilaria pinnata* (Tolotti, 2001). *Fragilaria* species are considered to be highly sensitive to climate-driven variables, too (Schmidt et al., 2004). The ratio of planktonic diatoms to *Fragilaria* species in alpine lakes has been used as an indicator of ice-cover duration, as the number and abundance of *Fragilaria* species may be the result of longer ice-cover period (Lotter et al., 2000). Clarke et al. (2005) hypothesised that changes in the ratio of planktonic to non-planktonic diatoms are associated with the effects of climate warming. They found larger proportion of planktonic diatoms over the last 150 years in many European remote lakes, although indistinct changes were discovered in the Tatra lake district.

Unlike the other two investigated lakes, where the proportion *Fragilaria* species decreased in the upper parts of the sediment records, the dominance of *Fragilaria* in this lowermost located lake makes the profile difficult to interpret. Other environmental variables might be more important than climate change in explaining the diatom variation.

Fluctuations in relative abundances of *Micropsectra* spp. and the *Tanytarsus lugens* group are crucial for the interpretation of the chironomid sediment record. *Micropsectra* spp. consists of several species including *M. radialis*. Since head capsules were mostly worn, no attempt was made to identify them to the lower taxonomic level. Thus, valuable information on the distribution of *Micropsectra* species throughout the core has been lost.

Only *Tanytarsus bathophilus* from the *T. lugens* group occurs in the modern chironomid fauna of the Tatra Mountain lakes (Bitušík et al., 2006). The species is generally absent in cold alpine lakes and can be found in lakes with higher trophic status (Ekrem, 2004). Considering this, the dominance of *T. lugens* group could indicate a warmer climate during the oldest investigated period (eleventh to thirteenth century).

The decline in the *T. lugens* group and subsequent dominance of *Micropsectra* spp. accompanied by a dramatic decrease in the number of subfossil remains in Zone ChTE-II suggest a cooler climate than was inferred for the previous interval. By extrapolation of ^{210}Pb dates, we estimated that the shift in chironomid assemblages spans the period between the thirteenth and sixteenth centuries and roughly corresponds with cold periods during the Little Ice Age (Beer et al., 2000). However, since there is a section of the core between 24 and 14 cm in which the bulk density of the sediments is significantly higher than that in the other parts of the core and varies irregularly with depth, dates of sediments below 14 cm (~AD 1700) should be regarded with more caution than those for sediments above this depth. Our interpretation of the chironomid record is consistent with observed chironomid-inferred cooling during the Little Ice Age from the sediments of other European mountain lakes (Lotter et al., 2000, Kamenik et al., 2000, Tátosová & Stuchlík, 2006).

The increase in the *T. lugens* group between the end of the nineteenth century and recent decades suggests warmer climatic conditions and increased lake productivity. This is in accordance with reconstructed air temperatures for Nižné Terianske pleso over the last ~200 years (Agustí-Panareda & Thompson, 2002). The appearance of *Paratanytarsus austriacus* during this period is particularly interesting. The subfossil record points to a relatively recent colonisation of the lake. The modern distribution of the species in the Tatra Mountains is restricted to warmer subalpine lakes. Its occurrence in higher situated alpine lakes in the future could be a signal of changes in their temperature regimes.

Nižné Terianske pleso

The analyses display a relatively stable taxonomic composition of diatom and chironomid assemblages throughout the investigated period. The important zone boundary in the diatom record is seen at 9 cm, with the most prominent change at 3 cm (~AD 1900). Assemblages dominated by *Asterionella formosa* and *Fragilaria capucina* were replaced by the assemblage with *Achnanthes* spp., *Navicula schmassmannii,* and *Orthoseira roeseana* the dominant species approximately 200 years ago. Šporka et al. (2002) found that the increase in *Achnanthes* species is linked with mean October temperature increases. Further positive relationships between diatoms and summer temperature were found as well. On the other hand, despite the high lake alkalinity, acid deposition cannot be excluded as an important driving force in the change of diatom assemblages. The diatom zones DTERI-I and DTERI-II roughly correspond to chironomid zones ChTERI-I, II, III, IV. The dominance of *Asterionella formosa* could be indicative of cold climate conditions. This planktonic species has low light requirements and is favoured by late-spring overturn after a long period of ice cover (Maberly et al., 1994).

The major changes in the chironomid fauna can be described as a general succession from an assemblage dominated by *Procladius* (*Holotanypus*) sp. (Zone ChTERI-I) to an assemblage with an increasing importance and dominance of *Micropsectra radialis.* This general pattern is interrupted twice (Zones ChTERI-III, IV, VI). This change from the *Procladius*-dominated assemblage to an assemblage with a dominance of the cold stenothermic *Micropsectra radialis* could possibly be interpreted as a signal of cooler climatic conditions which have extended until recently. The first shift in this cold-adapted assemblage occurs between 17.4 and 13.8 cm. The change involves the dramatic reduction of *Micropsectra radialis* and the disappearance of *Procladius* (*Holotanypus*) sp., *Diamesa* spp. and *Heterotrissocladius marcidus* from the sediment record. Accompanying this event is the marked decrease in chironomid density and organic matter contents in the sediments. These characteristics indicate that the in-lake production decreased at this time. Moreover, higher density layers at depths of 18 and 16 cm (Appleby, 2000) suggest episodes of rapid accumulation from increased erosion in the catchment. *Micropsectra* cf. *junci* is the only chironomid which increased during this period. As a littoral inhabitant, it may reflect favourable conditions in marginal benthic habitats during prolonged ice cover on the lake. Changes in seasonal oxygen availability during the long winter ice cover and/or during summer stratification could be a possible explanation for the strong decline in profundal taxa. The timing of this shift is unknown, but it evokes the cold climatic event observed in Vyšné Temnosmrečinské pleso.

The second important shift between 3.6 and 1.2 cm (AD 1899 ± 7–1959 ± 2) is characterised by similar characteristics as in the previous. A reduction in *Micropsectra radialis*, *Procladius* (*Holotanypus*) sp. and chironomid remains is accompanied by another episode of rapid sediment accumulation, and a decrease in organic matter content in the sediments. Percentages of *Heterotrissocladius marcidus* and *Micropsectra radialis* are more or less complementary, and a very similar distributional pattern is seen in Vyšné Wahlenbergovo pleso. These observations support Warwick's (1989) opinion on the affinity of *Heterotrissocladius* for sediments with a higher proportion of mineral components. As mentioned in the previous study (Šporka et al., 2002), chironomids as a whole showed no correlation to temperature fluctuations in the last 200 years. Only *Micropsectra radialis* was found to be significantly related to temperature trends. The relative abundance of head capsules mirrored the temperature fluctuations during the studied period. but this trend was interrupted in 1930s and 1940s when amelioration was accompanied with a fall in the abundance. The coincidence of elevated *Micropsectra radialis* abundance with higher temperature may be the result of improved food conditions and suggests the indirect response of this cold-stenothermic species to temperature change. Actually, the negative relationship with temperature observed in the subfossil record could be the response of the species to restricted food sources. Bitušík & Kubovčík (1999) pointed out the difference between abundant remains of *Diamesa* spp. and other taxa closely associated with stream conditions in the older sediment layers with the absence of them in the recent sediments. The change is considered to be an evidence of a stronger inflow in the past.

Vyšné Wahlenbergovo pleso

Unlike chironomids, more pronounced changes in diatom assemblages occurred in this lake. The notable shift in diatoms at 10 cm (∼AD 1791 ± 58) is associated with an increase in most *Achnanthes* species and a decline in *Fragilaria* species. Similar to the scheme observed in Nižné Terianske pleso, increase of temperature may be of importance in explaining this trend. Zone DFU-IV (3–0 cm), dominated by small benthic *Achnanthes* species, corresponds to chironomid Zone ChFU-III. Acidophilous species (*Achnanthes marginulata, A. subatomoides, Aulacoseira distans, and Fragilaria exigua*) reflect the acidification period of the lake (Jones et al., 1989; Guilizzoni et al., 1996).

Chironomids show a relatively constant composition throughout the older sediment samples (AD 1624 ± 105 to ∼1930s). The assemblage reflects cold, high-alpine conditions (Lotter et al., 2000; Bigler et al., 2006). The rapid decrease of subfossil remains between 17 and 13 cm (AD 1662 ± 95 to 1734 ± 74) in conjunction with a reduction of *Micropsectra radialis* and increased values of *Pseudodiamesa arctica* indicate a cooler period, perhaps corresponded to the coldest phase of the Little Ice Age (AD 1645–1715, Beer et al., 2000). The major shift in the chironomid assemblages is observed at the end of the studied period (∼AD 1930s to 2001). This period is characterised by a decrease in *Micropsectra radialis* and an increase in *Heterotrissocladius marcidus*. In conjunction with this trend, the density of subfossil remains rapidly declines and organic matter content decreases relative to the previous zone as a consequence of low in-lake production. This pattern corresponds very well with the period of acidification, when the lake became more oligotrophic (Kopáček et al., 2006). It is interesting to note that none of the chironomid taxa (including acid-sensitive *Micropsectra radialis*) present prior to acidification disappeared during the acidification stage.

This data supports the opinion that chironomids better reflect the lake trophic state rather than acidity (Brodin, 1990) provided that pH does not decrease to near or below 5 (Olander et al., 1997). In the most recent Zone ChFU-I, the decrease in *Heterotrissocladius marcidus* and the increases both in *Micropsectra radialis* and organic matter content indicate a recovery stage after acidification (Kopáček et al., 2006).

Response of assemblages to temperature

Subfossil remains in lake sediments may contain valuable information about past climatic conditions. Diatom and chironomid subfossils are widely used in environmental and climate reconstructions (reviewed by Smol et al., 1991; Walker, 1995). However, the reaction of both groups can be different across a range of sites (Battarbee et al., 2002a). We chose both descriptive and correlative approaches to compare the responses of these assemblages. Independent zonations of diatom and chironomid stratigraphies showed corresponding zonation patterns without much asynchrony (see above).

The correlative approach, based on the response of assemblages to organic matter as a temperature proxy, showed inconsistent results across groups and lakes. In general, the species inventory of the lakes was not affected by the temperature variability. However, this is unsurprising as the temperature changes in the Holocene are subtle, and even quantitative-based temperature reconstructions may be rendered insensitive by the overall predominance of common species with wide ecological tolerances (Birks & Birks, 2006). In Vyšné Wahlenbergovo pleso, the disappearance of some diatom species correlated with decrease in organic matter was rather the consequence of acidification stress than climate change (e.g., Tolotti, 2001; Štefková, 2006). The same is true for proportional changes in the chironomid composition in this lake, when dominant taxa responded to low organic matter content due to acidification and subsequently oligotrophication (Kubovčík & Bitušík, 2006).

The relationship between elevated *Achnanthes levanderi* abundance and the lower organic matter content observed in all studied lakes is unclear, and further study is needed to interpret this response.

Irrespective of altitude, the significant predictive power of organic matter reflecting overall temperature governs changes in assemblage compositions. However, relatively small differences in predictive strengths between groups and its inconsistent change along altitude are difficult to interpret. The organic matter content used here as a simple temperature proxy can be the result of complex influences and rather hard to interpret per se, especially when using single core measurements (Shuman, 2003). Moreover, response of assemblage composition to environmental changes may differ with geographical position of lake and assemblages may be influenced by different environmental factors along altitudinal gradients. Heegaard et al. (2006) showed existence of several critical altitudinal boundaries ('aquatic ecotones') where major changes in assemblages composition occurred. In contrast to those ecotones, rate of compositional change appeared to be relatively low at the centre of altitudinal gradient. This complexity might be the reason for inconsistent results that prevents a straightforward interpretation and comparison of sensitivity of different assemblages to climate variability. We can therefore concur with Birks & Birks (2006) that each proxy takes its own unique place in the ecosystem network and may be used to reconstruct different facets of the ecosystem.

Relationships between assemblages

Our results clearly show a strong relationship between diatom and chironomid assemblages in all but one lake (Fig. 7). Biotic interactions between trophic levels (e.g., Jackson & Harvey, 1993; Grenouillet et al., 2008) and similar but independent responses to major environmental gradients (e.g., Allen et al., 1999; Paszkowski & Tonn, 2000) are considered as the most obvious mechanisms behind congruent patterns in assemblages. Direct effects of diatom composition on chironomids are unlikely because grazing on algae is not a dominant feeding strategy for chironomids inhabiting the investigated lakes (Hamerlík, unpublished data). Reasons for the high covariance between assemblages are rather indirect. Diatoms can sensitively reflect different aspects of lake environmental conditions (Stoermer & Smol, 1999). Consequently, diatom assemblage composition integrates and efficiently synthesizes a number of causal factors which may also be relevant for chironomids. Still, possible direct effects cannot be excluded and a combination of both effects can make diatom composition a much better predictor of the chironomid assemblages than organic matter (cf. Schaffers et al., 2008). Interestingly, diatom abundance matrices contain a certain amount of redundant information for prediction purposes. This means that the most relevant information for chironomids is summarised in the presence/absence of diatoms, or that at least abundance matrices do not offer a higher predictive power than incidence matrices.

In Vyšné Wahlenbergovo pleso, no common trends in the assemblage compositions are observed. One might argue that a unimodal-based technique is inappropriate in the case of relatively stable chironomid assemblage of Vyšné Wahlenbergovo pleso. However, when indirect approach using linear-based technique has been applied (i.e. redundancy analysis of chironomids with the first axis of diatom correspondence analysis in the predictor role), the results remained insignificant. Thus, both the approaches suggest discordant trends in composition of diatom and chironomid assemblages in this lake. This asynchrony probably arises, at least in part, from different response of diatom and chironomid species to acidification of this lake. According to both, the chironomid and the diatom assemblages, the effect of acidification can be tracked back to the first half of the twentieth century. Recently, chironomid assemblage showed shifts that are attributable to recovery, while diatom assemblage remains relatively stable throughout the uppermost layers of core. On the other hand, climatic-driven changes in assemblages of deeper layers (see above) were more pronounced in diatoms than chironomids. Strong decline and even disappearance of some *Fragilaria* species contrasted sharply with relatively slight changes in proportion of chironomid species during the seventeenth and eighteenth centuries. These discrepancies may be a consequence of the fact that chironomid and diatom assemblages may be controlled by different environmental factors (Larocque & Bigler, 2004). Heegaard et al. (2006) showed that different assemblages of aquatic organisms (diatoms, chironomids and cladocerans) differ in rate of compositional change along the same altitudinal gradient, and both the position and strength at the 'aquatic ecotone' differ considerably among the organisms when comparing the rate of change. Thus, the interpretation of the asynchronicities between proxies is rather complex. Nevertheless, more robust insights can be gained from studies that quantify a diverse suite of geochemical and biological time series. Since many organisms exhibit contrasting responses to common environmental forcing, the use of multiple proxies of environmental change can help avoid errors associated with happenstance variation in individual indicators or hidden causal mechanisms (McGowan & Leavitt, 2009). Advances in multivariate numeric techniques such as co-correspondence analysis (or CoCA) (ter Braak & Schaffers, 2004) or co-inertia analysis (Dray et al., 2003) can help rigorously identify common trends in multivariate datasets such as biotic proxies and allow more complex hypothesis testing.

Acknowledgements The authors wish to acknowledge colleagues who have contributed to the collection and analysis of data used in this article. Special thanks are due to E. C. Grimm (Illinois State Museum, USA) for donation of TGView program, and to E. Stuchlik and J. Tatosová for data on LOI. We thank Lucia Poláková and Mária Némethová for their help during laboratory work. The manuscript benefited from the constructive comments of two anonymous reviewers. This study was co-supported in two EU founded projects: MOLAR ('MOuntain LAke Research'), and EMERGE ('European Mountain lake Ecosystems: Regionalisation, diaGnostics & socio-economic Evaluation'), grant No. 1/4334/07 of the Scientific Grant Agency of the Ministry of Education of Slovak Republic and the Slovak Academy of Sciences (VEGA), and Technical University in Zvolen (project IPA TUZVO No. 1/08).

References

Agustí-Panareda, A. & R. Thompson, 2002. Reconstructing air temperature at eleven remote alpine and arctic lakes in Europe from 1781 to 1997. Journal of Paleolimnology 28: 7–23.

Allen, A. P., T. R. Whittier, D. P. Larsen, P. R. Kaufmann, R. J. O'Connor, R. M. Hughes, R. S. Stemberger, S. S. Dixit, R. O. Brinkhurst, A. T. Herlihy & S. G. Paulsen, 1999. Concordance of taxonomic composition patterns across multiple lake assemblages: effects of scale, body size, and land use. Canadian Journal of Fisheries and Aquatic Sciences 56: 2029–2040.

Anderson, N. J., 2000. Diatoms, temperature and climate change. European Journal of Phycology 35: 307–314.

Anderson, N. J., B. Odgaard, U. Segerström & I. Renberg, 1996. Climate–lake interactions recorded in varved sediments from a Swedish boreal forest lake. Global Change Biology 2: 399–405.

Appleby, P. G., 2000. Radiometric dating of sediment records in European mountain lakes. Journal of Limnology 59(Suppl. 1): 1–14.

Appleby, P. G., 2001. Chronostratigraphic techniques in recent sediments. In Last, W. M. & J. P. Smol (eds), Developments in Paleoenvironmental Research. Volume 1, Tracking Environmental Change using Lake Sediments: Physical and Chemical Techniques. Kluwer, Dordrecht: 171–203.

Appleby, P. G. & G. T. Piliposian, 2006. Radiometric dating of sediment records from mountain lakes in the Tatra Mountains. Biologia, Bratislava 61(Suppl. 18): S51–S64.

Battarbee, R. W., 1986. Diatom analysis. In Berglund, B. E. (ed.), Handbook of Holocene Palaeoecology and Palaeohydrology. John Wiley & Sons, Chicester: 527–570.

Battarbee, R. W., J.-A. Grytnes, R. Thompson, P. G. Appleby, J. Catalan, A. Korhola, H. J. B. Birks, E. Heegaard & A. Lami, 2002a. Comparing palaeolimnological and instrumental evidence of climate change for remote mountain lakes over the last 200 years. Journal of Paleolimnology 28: 161–179.

Battarbee, R. W., R. Thompson, J. Catalan, J.-A. Grytnes & H. J. B. Birks, 2002b. Climate variability and ecosystem dynamics of remote alpine and arctic lakes: The MOLAR project. Journal of Paleolimnology 28: 1–6.

Beer, J., W. Mende & R. Stellmacher, 2000. The role of the Sun in climate forcing. Quaternary Science Reviews 10: 297–317.

Bigler, C., O. Heiri, R. Krskova, A. F. Porter & M. Sturm, 2006. Distribution of diatoms, chironomids and cladocera in surface sediments of thirty mountain lakes in southeastern Switzerland. Aquatic Sciences 68: 154–171.

Birks, H. H. & H. J. B. Birks, 2006. Multi-proxy studies in palaeolimnology. Vegetation History and Archaeobotany 15: 235–251.

Bitušík, P. & V. Kubovčík, 1999. Sub-fossil chironomids (Diptera: Chironomidae) from the sediments of the Nižné Terianske pleso (High Tatra Mts., Slovakia). Dipterologica Bohemoslovaca 9: 11–20.

Bitušík, P., M. Svitok, P. Kološta & M. Hubková, 2006. Classification of the Tatra Mountain lakes (Slovakia) using chironomids (Diptera, Chironomidae). Biologia, Bratislava 61(Suppl. 18): S191–S201.

Brodin, Y.-W., 1990. Midge fauna development in acidified lakes in northern Europe. Philosophical Transactions of the Royal Society B: Biological Sciences 327: 295–298.

Brooks, S. J. & H. J. B. Birks, 2004. The dynamics of Chironomidae (Insecta: Diptera) assemblages in response to environmental change during the past 700 years on Svalbard. Journal of Paleolimnology 31: 483–498.

Cameron, N. G., H. J. B. Birks, V. J. Jones, F. Berge, J. Catalan, R. J. Flower, J. Garcia, B. Kawecka, K. A. Koinig, A. Marchetto, P. Sánchez-Castillo, R. Schmidt, M. Šiško, E. Solovieva, E. Štefková & M. Toro, 1999. Surface-sediment and epilithic diatom pH calibration sets for remote European mountain lakes (AL:PE project) and their comparison with the Surface Waters Acidification Programme (SWAP) calibration set. Journal of Paleolimnology 22: 291–317.

Charles, D. F., 1985. Relationship between surface sediment diatom assemblages and lakewater characteristics in Adirondack lakes. Ecology 66: 994–1011.

Chomitz, K. & F. Šamaj, 1974. Zrážkové pomery (Precipitation characteristics). In Konček, M., et al. (eds), Klíma Tatier. Veda, Bratislava: 443–536 (in Slovak).

Clarke, G., M. Kernan, A. Marchetto, S. Sorvari & J. Catalan, 2005. Using diatoms to assess geographical patterns of change in high-altitude European lakes from pre-industrial times to the present day. Aquatic Sciences 67: 224–236.

Curtis, C. J., I. Botev, L. Camarero, J. Catalan, D. Cogalniceanu, M. Hughes, M. Kernan, J. Kopáček, A. Korhola, R. Psenner, M. Rogora, E. Stuchlík, M. Veronesi & R. F. Wrigh, 2005. Acidification in European mountain lake districts: a regional assessment of critical load exceedance. Aquatic Sciences 67: 237–251.

Davidson, I. R., 1991. Environmental effects on algal photosynthesis: temperature. Journal of Phycology 27: 2–8.

Dray, S., D. Chessel & J. Thioulouse, 2003. Co-inertia analysis and the linking of ecological data tables. Ecology 84: 3078–3089.

Ekrem, T., 2004. Immature stages of European Tanytarsus species I. The eminulus-, gregarius-, lugens- and mendax species groups (Diptera, Chironomidae). Deutsche entomologische Zeitschrift 51: 97–146.

Fallu, M.-A., R. Pienitz, I. R. Walker & M. Lavoie, 2005. Paleolimnology of a shrub-tundra lake and response of aquatic and terrestrial indicators to climatic change in arctic Québec, Canada. Palaeogeography, Palaeoclimatology, Palaeoecology 215: 183–203.

Fott, J., M. Pražáková, E. Stuchlík & Z. Stuchlíková, 1994. Acidification of lakes in Šumava (Bohemia) and in the High Tatra Mountains (Slovakia). Hydrobiologia 274: 37–47.

Gaedke, U., D. Ollinger, P. Kirner & E. Bäuerle, 1998. The influence of weather conditions on the lake seasonal plankton development in a large and deep lake (L. Constance). III. The impact of water column stability on spring algal development. In George, D. G., J. G. Jones, P. Punčochář, C. S. Reynolds & D. W. Sutcliffe (eds), Management of Lakes and Reservoirs During Global Climate Change. Kluwer Academic Publishers, Netherlands: 71–84.

Gauch, H. G., 1982. Multivariate Analysis in Community Ecology. Cambridge University Press, Cambridge.

Gregor, V. & J. Pacl, 2005. Hydrológia tatranských jazier. (Hydrology of the Tatra mountain lakes) (in Slovak). Acta Hydrologica Slovaca 6: 161–187.

Grenouillet, G., S. Brosse, L. Tudesque, S. Lek, Y. Baraillé & G. Loot, 2008. Concordance among stream assemblages and spatial autocorrelation along a fragmented gradient. Diversity and Distributions 14: 592–603.

Grimm, E. C., 1987. CONISS: A Fortran 77 program for stratigraphically constrained cluster analysis by the method of incremental sum of squares. Computers & Geosciences 13: 13–35.

Grimm, E. C., 2004. TGView version 2.0.2. Illinois State Museum. Research and Collections Center, Springfield, Illinois, USA.

Guilizzoni, P., A. Marchetto, A. Lami, N. G. Cameron, P. G. Appleby, N. L. Rose, Ø. A. Schnell, C. A. Belis, A. Giorgis & L. Guzzi, 1996. The environmental history of a mountain lake (Lago Paione Superiore, Central Alps, Italy) for the last c. 100 yrs: a multidisciplinary, palaeolimnological study. Journal of Paleolimnology 15: 245–264.

Heegaard, E., A. F. Lotter & H. J. B. Birks, 2006. Aquatic biota and the detection of climate change: are there consistent aquatic ecotones? Journal of Paleolimnology 35: 507–518.

Heino, J., 2008. Influence of taxonomic resolution and data transformation on biotic matrix concordance and assemblage–environment relationships in stream macroinvertebrates. Boreal Environmental Research 13: 359–369.

Hill, M. O. & H. G. Gauch, 1980. Detrended correspondence analysis: an improved ordination technique. Vegetatio 42: 47–58.

Jackson, D. A. & H. H. Harvey, 1993. Fish and benthic invertebrates: community concordance and community–environment relationships. Canadian Journal of Fisheries and Aquatic Sciences 50: 2641–2651.

Jones, V. J., A. C. Stevenson & R. W. Battarbee, 1989. Acidification of lakes in Galloway, southwest Scotland: a diatom and pollen study of the post-glacial history of the Round Loch of Glenhead. Journal of Ecology 77: 1–23.

Kamenik, C., K. A. Koinig, R. Schmidt, P. G. Appleby, J. A. Dearing, A. Lami, R. Thompson & R. Psenner, 2000. Eight hundred years of environmental changes in a high Alpine lake (Gossenköllesee, Tyrol) inferred from sediment records. In Lami, A., N. Cameron & A. Korhola, (eds), Paleolimnology and ecosystem dynamics at remote European Alpine lakes. Journal of Limnology 59(Suppl. 1): 43–52.

Končeк, M. & M. Orlicz, 1974. Teplotné pomery (Temperature characteristics). In Konček, M., et al. (eds), Klíma Tatier. Veda, Bratislava: 89–179 (in Slovak).

Koinig, K. A., C. Kamenik, R. Schmidt, A. Agustí-Panareda, P. G. Appleby, A. Lami, M. Prazakova, N. L. Rose, Ø. A. Schnell, R. Tessadri, R. Thompson & R. Psenner, 2002. Environmental changes in an alpine lake (Gossenköllesee, Austria) over the last two centuries—the influence of air temperature on biological parameters. Journal of Paleolimnology 28: 147–160.

Kopáček, J. & E. Stuchlík, 1994. Chemical characteristic of lakes in the High Tatra Mountains, Slovakia. Hydrobiologia 274: 49–56.

Kopáček, J., E. Stuchlík & D. Hardekopf, 2006. Chemical composition of the Tatra Mountain lakes: II. Recovery from acidification. Biologia, Bratislava 61(Suppl. 18): S21–S33.

Kowalyk, H. E., 1985. The larval cephalic setae in the Tanypodinae (Diptera: Chironomidae) and their importance in generic determinations. The Canadian Entomologist 117: 67–106.

Krammer, K. & H. Lange-Bertalot, 1986. Bacillariophyceae, 1. Teil: Naviculaceae. Süßwasserflora von Mitteleuropa 2/1. Gustav Fischer Verlag, Jena.

Krammer, K. & H. Lange-Bertalot, 1988. Bacillariophyceae, 2. Teil: Bacillariaceae, Epithemiaceae, Surirellaceae. Süßwasserflora von Mitteleuropa 2/2. Gustav Fischer Verlag, Jena.

Krammer, K. & H. Lange-Bertalot, 1991a. Bacillariophyceae, 3. Teil: Centrales, Fragilariaceae, Eunotiaceae. Süßwasserflora von Mitteleuropa 2/3. Gustav Fischer Verlag, Stuttgart, Jena.

Krammer, K. & H. Lange-Bertalot, 1991b. Bacillariophyceae, 4. Teil: Achnanthaceae, Kritische Erganzungen zu Navicula (Lineolate) und Gomphonema. Süßwasserflora von Mitteleuropa 2/4. Gustav Fischer Verlag, Stuttgart, Jena.

Kubovčík, V. & P. Bitušík, 2006. Subfossil chironomids (Diptera, Chironomidae) in three Tatra Mountain lakes (Slovakia) on an acidification gradient. Biologia, Bratislava 61(Suppl. 18): S213–S220.

Lange-Bertalot, H. & K. Krammer, 1989. Achnanthes, eine Monographie der Gattung mit Definition der Gatung Cocconeis. Bibliotheca Diatomologica. Band 18, J. Cramer, Berlin & Stuttgart.

Larocque, I. & C. Bigler, 2004. Similarities and discrepancies between chironomid- and diatom-inferred temperature reconstructions through the Holocene at Lake 850, northern Sweden. Quaternary International 122: 109–121.

Livingstone, D. M. & A. F. Lotter, 1998. The relationship between air and water temperature in lakes of the Swiss Plateau: a case study with palaeolimnological implications. Journal of Paleolimnology 19: 181–198.

Lotter, A. F., W. Hofmann, C. Kamenik, A. Lami, C. Ohlendorf, M. Sturm, W. O. van der Knapp & J. F. N. van Leeuwen, 2000. Sedimentological and biostratigraphical analyses of short sediment cores from Hagelseewli (2339 m a.s.l.) in the Swiss Alps. Journal of Limnology 59(Suppl. 1): 53–64.

Maberly, S. C., M. A. Hurley, C. Butterwick, J. E. Corry, S. I. Heaney, A. E. Irish, G. H. M. Jaworski, J. W. G. Lund, C. S. Reynolds & J. V. Roscoe, 1994. The rise and fall of Asterionella formosa in the South Basin of Windermere: analysis of a 45-year series of data. Freshwater Biology 31: 19–34.

McGowan, S. & P. R. Leavitt, 2009. The role of paleoecology in whole-ecosystem science. In Miao, S., S. Carstenn & M. Nungesser (eds), Real World Ecology: Large-Scale and Long-Term Case Studies and Methods. Springer, New York: 161–208.

Olander, H., A. Korhola & T. Blom, 1997. Surface sediment Chironomidae (Insecta: Diptera) distributions along an ecotonal transect in subarctic Fennoscandia: developing a tool for palaeotemperature reconstructions. Journal of Paleolimnology 18: 45–59.

Paavola, R., T. Muotka, R. Virtanen, J. Heino & P. Kreivi, 2003. Are biological classifications of headwater streams concordant across multiple taxonomic groups? Freshwater Biology 48: 1912–1923.

Paszkowski, C. A. & W. M. Tonn, 2000. Community concordance between the fish and aquatic birds of lakes in northern Alberta, Canada: the relative importance of environmental and biotic factors. Freshwater Biology 43: 421–437.

Rouse, W. R., M. S. V. Douglas, R. E. Hecky, A. E. Hershey, G. W. Kling, L. Lesack, P. Marsh, M. McDonald, B. J. Nicholson, N. T. Roulet & J. P. Smol, 1997. Effect of climate change on the freshwaters of arctic and subarctic North America. In Cushing, C. E. (ed.), Freshwater Ecosystem and Climate Change in North America. Wiley, Chichseter: 55–84.

Sæther, O. A. & M. Spies, 2004. Fauna Europaea: Chironomidae. In de Jong, H. (ed.), Fauna Europaea: Diptera, Nematocera. Fauna Europaea version 1.1, http://www.faunaeur.org.

Schaffers, A. P., I. P. Raemakers, K. V. Sýkora & C. J. F. ter Braak, 2008. Arthropod assemblages are best predicted by plant species composition. Ecology 89: 782–794.

Schmid, P. E., 1993. A key to the larval Chironomidae and their instars from the Austrian Danube region Streams and

Rivers. Part 1: Diamesinae, Prodiamesinae and Orthocladiinae. Wasser und Abwasser, Suppl. 3/93: 513 pp.
Schmidt, R., C. Kamenik, H. Lange-Bertalot & R. Klee, 2004. Fragilaria and Staurosira (Bacillariophyceae) from sediment surfaces of 40 lakes in the Austrian Alps in relation to environmental variables, and their potential for climatology. Journal of Limnology 63: 171–189.
Shuman, B., 2003. Controls on loss-on-ignition variation in cores from two shallow lakes in the north-eastern United States. Journal of Paleolimnology 30: 371–385.
Smol, J. P., I. R. Walker & P. R. Leavitt, 1991. Paleolimnology. Paleolimnology and hindcasting climatic trends. Verh Internat Verein Limnol 24: 1240–1246.
Stoermer, E. F. & J. P. Smol (eds), 1999. The Diatoms: Applications for the Environmental and Earth Sciences. Cambridge University Press, Cambridge.
Šporka, F., E. Štefková, P. Bitušík, A. R. Thompson, A. Agustí-Panareda, P. G. Appleby, J. A. Grytnes, C. Kamenik, I. Krno, A. Lami, N. Rose & N. E. Shillanmd, 2002. The paleolimnological analysis of sediments from high mountain lake Nižné Terianske Pleso in the High Tatras (Slovakia). Journal of Paleolimnology 28: 95–109.
Šporka, F., D. M. Livingstone, E. Stuchlík, J. Turek & J. Galas, 2006. Water temperature and ice cover in lakes of the Tatra Mountains. Biologia. Bratislava 61(Suppl. 18): S77–S90.
Štefková, E., 2006. Epilithic diatoms of mountain lakes of the Tatra Mountains (Slovakia). Biologia, Bratislava 61(Suppl. 18): S101–S108.
Stuchlík, E., P. Appleby, P. Bitušík, C. Curtis, J. Fott, J. Kopáček, M. Pražáková, N. Rose, O. Strunecký & R. F. Wright, 2002. Reconstruction of long-term changes in lakewater chemistry, zooplankton and benthos of a small, acidified high-mountain lake: MAGIC modelling and palaeolimnological analysis. Water, Air, and Soil Pollution: Focus 2: 127–138.
Tátosová, J. & E. Stuchlík, 2006. Seasonal dynamics of chironomids in the profundal zone of a mountain lake (Ľadové pleso, the Tatra Mountains, Slovakia). Biologia, Bratislava 61(Suppl. 18): S203–S212.
ter Braak, C. J. F. & I. C. Prentice, 1988. A theory of gradient analysis. Advances in Ecological Research 18: 271–317.
ter Braak, C. J. F. & A. P. Schaffers, 2004. Co-correspondence analysis: a new ordination method to relate two community compositions. Ecology 85: 834–846.
Tolotti, M., 2001. Littoral diatom communities in high mountain lakes of the Adamello-Brenta Regional Park (Trentino, Italy) and their relation to acidification. In Jahn, R., J. P. Kociolek, A. Witkowski & P. Compere (eds), Lange-Bertalot-Festschrift: 327- 352.
Van Dam, H., A. Mertens & J. Sinkeldam, 1994. A coded checklist and ecological indicator values of freshwater diatoms from the Netherlands. Netherlands Journal of Aquatic Ecology 28: 117–133.
van der Voet, H., 1994. Comparing the predictive accuracy of models using a simple randomization test. Chemometrics and Intelligent Laboratory Systems 25: 313–323.
Walker, I. R., 1995. Chironomids as indicators of past environmental changes. In Armitage, P. D., P. S. Cranston & L. C. V. Pinder (eds), The Chironomidae: Biology and ecology of non-biting midges. Chapman & Hall, New York & London: 405–422.
Walker, I. R., 2001. Midges: Chironomidae and related Diptera. In Smol, J. P., H. J. B. Birks & W. M. Last (eds), Tracking Environmental Change using Lake Sediments, Zoological Indicators, Vol. 4. Kluwer Academic Publishing, Dordrecht, The Netherlands: 45–66.
Walker, I. R. & C. G. Paterson, 1985. Efficient separation of subfossil Chironomidae from lake sediments. Hydrobiologia 122: 189–192.
Warwick, W. F., 1989. Chironomids, lake development and climate: a commentary. Journal of Paleolimnology 2: 15–17.
Weckström, J. & A. Korhola, 2001. Patterns in the distribution, composition and diversity of diatom assemblages in relation to ecoclimatic factors in Arctic Lapland. Journal of Biogeography 28: 31–46.
Wiederholm, T. (ed.), 1983. Chironomidae of the Holarctic region. Keys and diagnoses. Part 1—Larvae. Entomologica Scandinavica, Suppl. 19: 457 pp.

Palaeoclimatic signals and anthropogenic disturbances from the peatbog at Nagybárkány (North Hungary)

Gusztáv Jakab · Péter Majkut · Imola Juhász · Sándor Gulyás · Pál Sümegi · Tünde Törőcsik

Originally published in the journal Hydrobiologia, Volume 631, No. 1, 87–106.
DOI: 10.1007/s10750-009-9803-z

Abstract Investigation of a small peatbog in northern Hungary provides a late Quaternary record of vegetation development effected by climatic changes and anthropogenic disturbances. The aim of this study was to separate climatic signals from the development of a continental eutrophic peatland with the use of plant macrofossil analysis. The development of water catchment is reconstructed using pollen and geochemical analyses. The formation of the lakebed can be traced back to the Late Glacial. A higher water level with oligotrophic conditions can be existed from the Late Glacial to middle Holocene, when the reed beds covered a small area only. This was followed by a hiatus spanning ca. 4400 years, caused by peat cutting during the Imperial Age. The water level decreased and the water quality was more eutrophic. A reed bed evolved around the lake. Terrestrialization started with a bulrush floating mat phase at the close of the Árpádian Age, ca. cal yr AD 1400. The initiation of the *Sphagnum*-bog underwent similar phases as in the other Hungarian peatbogs. Although remarkable anthropogenic disturbances can be reconstructed in the development of the peatbog, some climatic effects and authogenic processes might be separated by palaeoecological analyses.

Keywords Wetland development · Macrofossil · Geochemistry · Pollen analysis · Holocene · Sphagnum

Introduction

Lakes, fens and bogs have different vegetation depending on water level and nutrient status, and produce characteristic sediments in these different environments (Birks & Birks, 1980). Hydrologically speaking, peatlands can be grouped into two major groups. Ombrotrophic peatlands ('bogs') heavily rely on precipitation to establish a water balance; thus, their evolution and distribution is primarily influenced by the climatic conditions. The water balance is mostly controlled by the ratio of precipitation to evaporation, while the role of the vegetation and the storage capacity of the peat is negligible. Ombrotrophic peatlands are generally restricted to the Atlantic and mountain areas of Europe. Water in the second group of minerotrophic peatlands ('fens')

Guest editors: K. Buczkó, J. Korponai, J. Padisák & S. W. Starratt
Palaeolimnological Proxies as Tools of Environmental Reconstruction in Fresh Water

G. Jakab (✉)
Environmental and Water Management Faculty, Institute of Environmental Sciences, Szent István University, Szabadság Str. 1–3, 5540 Szarvas, Hungary
e-mail: cembra@freemail.hu

P. Majkut · I. Juhász · S. Gulyás · P. Sümegi · T. Törőcsik
Department of Geology and Palaeontology, University of Szeged, Egyetem Str. 2–6, 6722 Szeged, Hungary

comes from the surface watercourses as well as the groundwater in addition to local precipitation. As such, these peatlands are not as dependent on the amount of precipitation as the previous group for maintaining a constant water balance. These two groups are further divided into various subgroups based on the morphology and the vegetation. Several classification systems are known from different parts of Europe and North America. In Atlantic Europe, the ombrotrophic peatlands are often subdivided into raised bogs and blanket bogs. Raised bogs, as the name implies, form a positive morphology in a depression. Conversely, the thinner peat layers of blanket bogs follow the natural morphology of the landscape. In continental Europe, continental peatlands are known. In the drier summer months, an advancement of the arboreal vegetation is characteristic for these peatlands. Floating mats or floating fens are another typical form of continental peatlands, generally developing in the carbonate-rich or alkaline-rich shallow littoral parts of lakes and ponds (Balogh, 2000a, b). There are two major types for the initiation of a peatland: (1) a terrestrial area is turned into a peatland through the process of paludification and (2) the advancement of a peatland into a lacustrine basin during terrestrialization. This process is especially important in central Europe (Pokorný & Jankovská, 2000).

The peatlands best suited for paleoclimatic reconstructions are those of ombrotrophic peatlands, which emerged via paludification. Climatic conditions favoring the evolution of these types of peatlands are mainly restricted to the western parts of Europe under the influence of the oceanic climatic influences. As such, most of the peatland paleoclimatic records are known from the areas of the United Kingdom, Germany, Denmark and Sweden. Here, the moisture gradient is unambiguously reflected in the distribution of certain *Sphagnum* taxa. Barber & Charman (2005) questioned the suitability of strongly continental peatlands for paleoeclimatic reconstructions. In ombrotrophic peatbogs, the layers of scarcely humified *Sphagnum* peat were taken as indicators of fast peat growth and therefore of wet climatic conditions. Layers of humified peat with wood fragments were taken to indicate a dryer mire surface and so a drier and warmer climate. The composition of aquatic plant communities is largely influenced by hydrological conditions. These communities are highly prone to centennial-scale fluctuations in the climate. The application of macrofossil analysis to peat and lacustrine deposits enables to identify long-term vegetation changes in aquatic ecosystems.

Former studies aimed at deciphering past climatic conditions via detailed analysis of peatland deposits, primarily focused on the investigation of *Sphagnum*-peats of ombrotrophic peatlands (Mauquoy & Barber, 1999; Barber et al., 2000; Blackford, 2000; Barber & Langdon, 2001; Barber, 2005). Owing to various methodological problems, high-resolution quantitative macrofossil studies on the eutrophic peatlands of continental Europe are generally lacking. A slight modification of the method proposed by Barber et al. (1994) enabled us to retrieve proxy-climate data from the eutrophic peatlands of continental areas as well (Jakab et al., 2004b).

This article discusses the development of a small eutrophic peatbog in northern Hungary. The aim of this study is to use plant macrofossil analysis to separate climatic signals from the development of a continental eutrophic. The history of the watershed was reconstructed using pollen and geochemical analyses.

In this study, we attempt to answer the following questions: (1) Were climatic fluctuations or anthropogenic disturbances the critical factors in peatland development? (2) Is there any autogenic process in the hydrosere? (3) Is there any connection between the water level changes, the landscape usage and the historical events?

Study site

Hungary enjoys a moderately continental climate modified by marine influences coming from the west and by Mediterranean influences coming from the south. The regional characteristics of the country's climate are largely determined by the geographic conditions in individual basins. There is a gradual increase in the amount (or influence) of continentality, accompanied by a drop in the annual precipitation, as we move from the side of the basin towards the centre. The lowland regions of the Great Hungarian Plains (Alföld) have an average annual rainfall of 500–550 mm. Conversely, the marginal mid-mountain areas receive an average rainfall of 600–800 mm per annum. The driest parts of the basin are found in the central region of the Alföld (e.g. in the Hortobágy region), where the rate of average annual rainfall is very often below 500 mm (Bacsó, 1959). These harsh climatic conditions are not favourable for the formation of *Sphagnum* bogs. The total number of *Sphagnum* occurrences hardly exceeds 50, and in the central driest areas of the Great Hungarian Plains Sphagna are completely absent. Only sporadic *Sphagnum*-bogs are known in the country with a total number well below 20, most of them being extremely small with an area of a mere couple ha. True raised bogs are completely missing. The peatlands are fringed by deciduous woodlands generally. The majority of *Sphagnum* bogs is restricted to the western parts of the country which is dominated by maritime climatic influences, and those of the areas of the Northern Mid-Mountains and Northern Great Hungarian Plains are controlled by the montane-type climatic influences of the Carpathians (Boros, 1968; Szurdoki & Nagy, 2002). Despite the seemingly unfavourable conditions, Hungary is covered by more than 90000 ha of wetlands. The estimated quantity of peat in these wetlands is more than 973 million tons preceding artificial regulation and drainage efforts (Dömsödi, 1988). Most of this peat is reed and sedge peat, rather than *Sphagnum* peat. These peatlands are round in the neotectonic basins of the Great Hungarian Plains, abandoned riverbeds like those of the Pleistocene Danube and along the shores of major lakes like Lake Balaton and Lake Fertő.

The peatlands of the Carpathian basin are highly prone to even minor climatic changes because of their marginal position. Following moderate desiccation of *Sphagnum* bogs, a succession of reed, different sedge taxa and/or birch is observable depending on local climatic endowments (Lájer, 1998, Borhidi & Sánta, 1999). Extensive reed beds are generally found in the peatlands of the Alföld and along the shores of larger Hungarian lakes. Lowstands are generally characterized by tussocks (mainly composed of *Carex elata*), with depressions harbouring rich hollow vegetation. Floating mats, which appear in unusual alkaline aquatic habitats, are also very important (e.g. those of the lakes Velencei and Fertő) (Borhidi & Balogh, 1970; Balogh, 2000a, b).

The Nádas-tó (360 m a.s.l.) at Nagybárkány lies on the northern side of Mt. Hármas-Határhegy, rising to a height of 516 m in the Eastern or Pásztó Cserhát Mountains (Fig. 1). The lakebed has an elongated, north-west orientation, with a very narrow extension in the south. Its length is roughly 100 m, its greatest width is 40 m and it covers an area of roughly 2000 m^2. The narrow section is about 5–10 m wide. Accumulation in the catchment basin of the Nádas-tó started in the Late Glacial when a mass movement (rotational landslide) occurred on the slope of the Miocene sandy and silty sediment covered land

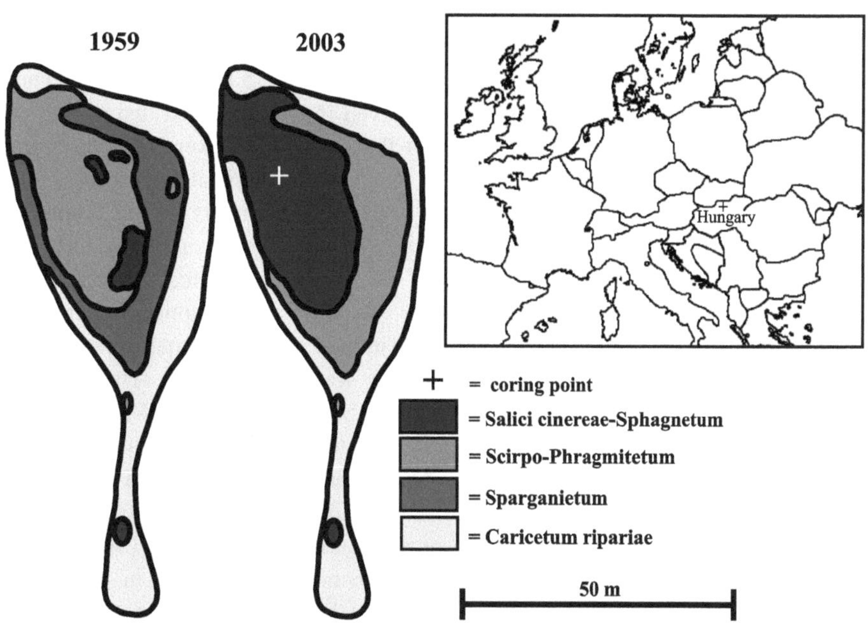

Fig. 1 Nagybárkány–Nádas-tó site location in Hungary (47°59′56.05″N, 19°40′29.50″E) and the core location on the vegetation map from 1959 (after Máthé & Kovács, 1959) and in 2003 (Gusztáv Jakab)

surface. A slump hollow formed in the source area between the landslide toe and the scarp which was filled up by water forming a small round-form lake. The annual rainfall is between 600 and 700 mm. The peatbog is solely fed by rainwater. There is not any visible watercourse in the drainage area.

The lakebed is fringed by an oak forest. Three plant communities can be distinguished in the recent bog (Fig. 1). The central part of the bog is covered with willow swamp (Salici cinereae-Sphagnetum recurvi). This community has low species diversity and is characterized by the dominance of *Salix cinerea* and a carpet of *Sphagnum squarrosum*. This association is rather rare in Hungary, occurring in the well-watered, undrained valleys of the Great Hungarian Plains and the Northern Mountain Range, as well as in smaller local hollows. The willow swamp is fringed by reed beds (Scirpo-Phragmitetum), except on the western shore. The reed beds are similarly poor in species; the presence of *Lythrum salicaria*, *Lycopus europaeus* and *Utricularia vulgaris* can be noted. Tall sedge communities (Caricetum ripariae) line the reed beds. These communities are dominated by *Carex riparia*. The peatbog underwent significant changes since the preparation of the first vegetation map in 1959 (Máthé & Kovács, 1959). The extension of the willow swamp was rather small at the time, and peat moss too covered a relatively small area only. Burr reed communities (Sparganietum erecti) are typical for the area. In 2003, the willow swamp encroached on the one-time reed beds, while the reed beds replaced the former burr reed associations (Fig. 1). Burr reed communities can no longer be observed in the area, indicating the oligotrophication of the bog caused by the expansion of peat mosses.

Methods

The sampling of the 320 cm deep, undisturbed sedimentary sequences from the Nádas-tó basin was carried out using a 5 cm diameter Russian type corer (Jowsey, 1966). Overlapping cores were extracted conforming to the general practice in Quaternary palaeoenvironmental studies (Aaby & Digerfeldt, 1986). Coring was carried out in the northwestern part of the bog, now occupied by a willow swamp (Fig. 1). After transportation to the laboratory, the cores were cut lengthwise for various analyses; the sections for palaeobotanical and geochemical analyses were stored at 4°C in accordance with the international standards. The samples submitted to lithological analyses were identical with the ones used for the pollen, macrobotanical, geochemical and radiocarbon analyses.

The main lithostratigraphic characteristics of the sedimentary sequence were identified and described using the system and symbols of Troel-Smith developed for unconsolidated sediments were adopted (Troels-Smith, 1955).

Radiocarbon dating of the sequence was obtained by both gas decay and accelerator mass spectrometry (AMS) analyses. Four bulk samples of sediment were analysed for radiocarbon ages at the Nuclear Research Centre of the Hungarian Academy of Sciences, Debrecen, Hungary, and three samples of terrestrial plant macrofossils were analysed for AMS date at the radiocarbon dating facility in Poznńan, Poland, and at the Beta Laboratory in Florida, USA. In order to allow comparison with other archaeological data, the dates were calibrated using the CalPal-2007 online calibration programme (Danzeglocke et al., 2008), using the most up-to-date CalPal-2007$_{\mathrm{Hulu}}$. calibration data set (Weninger & Jöris, 2008). The original dates (14C) are indicated as BP, while the calibrated dates are indicated as cal BC or cal BP.

The core was divided into 4 cm samples. The organic content of the core samples was measured by loss-on-ignition at 550°C for 5 h and the carbonate content by the further loss-on-ignition at 900°C for 5 h (Dean, 1974). The inorganic content was further analysed using the sequential extraction method. Mackereth (1966) was the first to recognise the potential of geochemical investigations on the sediments of the catchment basin for the purpose of environmental reconstructions in his review of bulk chemical analyses on deposits from the Lake District. The application of bulk analysis, however, is quite problematic since it does not shed light unequivocally onto the origin of the chemical constituents (Engström & Wright, 1984). Mackereth's work was later enhanced by researchers working on the combination of chemical analyses with palynological investigations. Engström (1983) subsequently devised a procedure to split the different chemical elements preserved in catchment basin deposits in an attempt to trace the origin of the chemical constituents.

A new sequential extraction method (Dániel et al., 1996; Dániel, 2004) with a long established history in the analysis of geochemical composition of lacustrine sediments was adopted in our study. From the full procedure, the step of water extraction for unseparated samples was sufficient to suit our analytical needs as it was shown by previous works (Sümegi et al., 1999; Dániel, 2004), the most important palaeohydrological and palaeoecological data originate from water extraction samples. Therefore, the geochemical results from water extraction samples are shown in this study.

Water was extracted form the samples following the methods of Dániel (1996, 2004) and Sümegi et al. (1999). Distilled water was purified using a Millipore 5 Plus Water Purification System for water extraction samples. One-hundred millilitre distilled and purified water was added to 1.0 g sample and was shaken for 1 h (Dániel, 2004), and then the water extract elements of Na, K, Ca, Mg, Fe were analysed using a Perkin-Elmer AAS spectrometer. Principal components analyses computed on correlation matrices were performed after logarithmic transformation of the geochemical data (Rollinson, 1993). The geochemical zones were identified by cluster analysis of principal components (Dowdeswell, 1982) using squared Euclidean distance and Ward aggregation method.

The sampling interval for pollen analysis was 8 cm throughout the core, with a finer interval of 4 cm for the lowermost section, between 312 and 248 cm. Samples were processed for pollen (Berglund & Ralska-Jasiewiczowa, 1986) with exotic pollen added to each sample to determine the concentration of pollen (Stockmarr, 1971). A minimum count of 300 grains (all pollen plus spores) per sample was made in order to ensure a statistically significant sample size (Maher, 1972). The pollen diagram was divided into nine Local Pollen Assemblage Zones (LPAZ) reflecting the changes of the most important taxa. The lowermost section of the core (340–312 cm) did not contain any pollen.

For the description of macrofossils, we used a modified version of the QLCMA technique (semi-quantitative quadrat and leaf-count macrofossil analysis technique) of Barber et al. (2004) and Jakab et al. (2004b). Organic remains from peat and lacustrine sediments rich in organic matter can be divided into two major groups. Some remains can be identified with lower ranking taxa (specific peat components), while others cannot be identified using this approach (non-specific peat components). The most important non-specific peat components are the following:

- *Unidentified organic material* (UOM): irregularly shaped tissue fragments, often moderately decomposed.
- *Undifferentiated monocotyledon remains* (Monocot. undiff.): opaque or slightly pigmented rhizomes and epidermal tissue fragments, with elongated or short cells.
- *Unidentified bryophyte fragments* (UBF): only the tubular, brown pigmented "stem" survives in decomposed peat with the stub of the "leaf veins".
- *Unidetifiable leaf fragments* (ULF): moderately humified deciduous tree leaf fragments. Easily recognisable by the remains of web-like veins.
- *Charcoal*: tiny, 1–3 mm large charcoal fragments (macro-charcoal), probably of allochthonous origin.
- *Wood*: lignified plant tissues can be easily recognised from their compact, thick-walled wood fibres.

The most important specific peat components are seeds, fruits, sporogons, mosses, rhizomes and epidermis (e.g. *Carex* species), leaf epidermis, other tissues and organs (hairs, tracheids, etc.), insect remains and Ostracoda shells. The identification of herbaceous plant tissues was based on the procedure described by Jakab & Sümegi (2004a). The size of subsamples for macrofossil analysis was 3 cm^3. We defined the amount of peat components on the 1 cm^3 level and the amount of seeds on the 3 cm^3 level. The samples were washed through a 300-μm sieve. Concentration levels were determined by adding a known amount of indicator grains (0.5 g poppy seed, ca. 960 pieces) and by counting the poppy seeds and the remains using a stereo microscope in ten 10 mm by 10 mm quadrates in a Petri dish. Similarly to mosses, rhizomes can only be identified with a light microscope. We removed a 100 monocotyledon remains and mounted them in water on microscopic slides for determining the percentages of individual taxa and of undivided Monocots. The values for different moss species and UBF were determined using a similar procedure. The concentration can be described with the following equation:

$$\text{macrofossil concentration} = \frac{\text{counted macrofossil (average)} \times 960 \text{ (total poppy seeds)}}{\text{counted poppy seeds (average)} \times \text{sample volume (cm}^3)}.$$

The Psimpoll (Bennett, 1992) program was used for plotting the analytical results.

The distribution of the most important peat components was evaluated using multivariate statistical methods. The method of PCA (standardised, mixed/Rohlf/biplot) was adopted following Podani (1993) using the software package SYN-TAX 5.0.

Results

Chronology and sediment stratigraphy

Peat was present to a depth of 110 cm, with an underlying water pocket down to 130 cm. Between 130 and 300 cm, we found peat and peat-mud with varying organic content. Between 300 and 340 cm, there was a silty lacustrine sediment (Table 1).

The radiocarbon dates indicate a hiatus of roughly 4400 years (from 4970 BC to ca. 330 BC) between 248 and 240 cm. The results of the radiocarbon measurements analyses of the sequence described in this study are shown in Table 2.

Geochemistry

The distinct elemental and lithological stratigraphy identified in the core can serve as a potential record of the palaeohydrological and palaeoecological history of the catchment basin of the Nádas-tó. According to the data, six geochemical zones are present in the sediment profile of the core from Nagybárkány (Fig. 2).

Table 1 The lithological description of the Nagybárkány–Nádas-tó sequence (Pál Sümegi—Gusztáv Jakab)

Depth (cm)	Troel-Smith (1955) system	Description
0–40	Tb4 (Sphag.)	*Sphagnum* peat
40–110	Dg2Th1Tb1(Sphag.)	*Sphagnum* peat mixed with limus detritus, made up mostly of *Phragmites* (40–80 cm) and *Typha* rhizomes (80–100 cm)
110–130	–	Water
130–134	Dg2Tb1Th1	Burnt, charcoal rich peat layer with *Phragmites* rhizomes
134–255	Ld3Sh1 Tb+ (Sphag.)Th+Tl+	Dark brown eutrophic lacustrine deposits (clayey silt) with varying organic content, large amount of wood fragments at 225 cm
255–277	As3Ld1 Th+Gs+(min)	Pale yellow, brownish-grey slightly laminated silty clay with yellow spots
277–295	Ld3Sh1 Tb+(Sphag.)Th+Gs+(min)	Brownish-grey and pale yellow clayey silt with yellow spots
295–300	As3Ld1Gs+(min)	Transitional layer
300–340	As3Ag1Gs+(min)	Greenish-grey, clayey silt with frost marks (oligotrophic lacustrine deposists)

Table 2 Radiocarbon data for the Nagybárkány–Nádas-tó

Sample number	Depth (cm)	Sample type	δ^{13}C(PDB) ± 0.2 [‰]	^{14}C age (BP)	Cal AD/BC (2σ)
deb-11110	NB-45	Peat	−28.02	100% ± 0.40 pM ^{14}C	1950–1960 cal AD
deb-11098	NB-100	Peat	−27.73	740 ± 60	1230–1300 cal AD
deb-11009	NB-180	Peat	−28.49	1600 ± 60	400–540 cal AD
deb-11100	NB-250	Charcoal	−27.52	6090 ± 60	4956–5146 cal BC
Beta-194559	NB-280	Charcoal	−24.90	8050 ± 40	6875–7061 cal BC
Beta-224851	NB-325	Charcoal	−25.5	12543 ± 189	13309–12457 cal BC

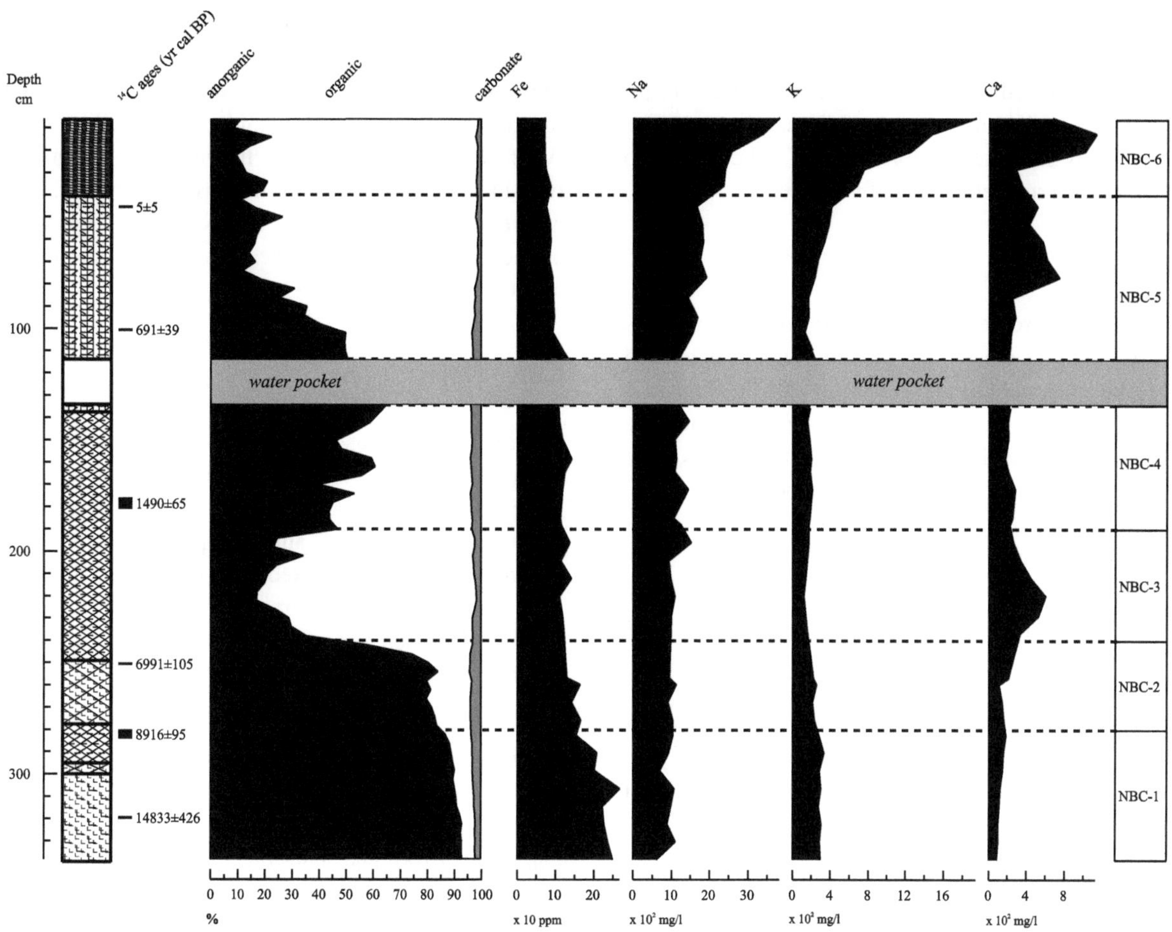

Fig. 2 The results of geochemical analyses (Péter Majkut)

NBC-1 (340–280 cm, 15260–8800 cal yr BP)

The lake is characterized by the deposition of non-calcareous and low-organic content sediment during the Late Glacial and early Holocene. This sediment was predominantly inorganic (90–95%) and contained a high concentration of the water soluble Fe and K and low concentration of water soluble Ca and Na.

NBC-2 (280–240 cm, 8800–6000 cal yr BP)

During the middle Holocene, the inorganic content (80–90%) decreased and there was a gradual increase in carbonate (to about 5%) and organic (10–15%) content. The water soluble Fe and K content decreased, while the level of water soluble Ca, Mg input prior to this increase indicates that the transformation of the vegetation continued and deciduous forest elements spread around the lake basin.

NBC-3 (240–190 cm, 2300–1500 cal yr BP)

There is a sudden decrease in the inorganic content of the deposits from the depth of 240 cm upwards, with an increase of the organic matter from the previous 10–15% to 70–80%. Elements to increase included water soluble Ca, Mg suggesting authigenic changes within the catchment (Dániel, 2004). Probably Ca and Mg acceptor water plants, such as *Typha*, *Phragmites* colonized the catchment basin and the increase in water soluble Ca and Mg originated from these plant remains.

NBC-4 (190–130 cm, 1500–700 cal yr BP)

There is a gradual decrease in the organic content and water soluble Ca and Mg content accompanied by an increase in the inorganic content with water soluble Na content between 190 and 130 cm of the core profile. Previous studies (Mackereth, 1966; Engström

& Wright, 1984; Dániel, 2004) have indicated that an increase of the abundance of these elements is indicative of both physical and chemical weathering associated with soil erosion and human impact. The increase of the water soluble Na may indicate a drop in lake level as well during this phase.

NBC-5 (110–40 cm, 700–0 cal yr BP)

A gradual increase in the organic content indicates decreasing soil erosion and human influences around the lake catchment basin. The water soluble Na, K, Ca and Mg content increased gradually in this zone. The observed composition of these elements may be due to the development of a floating mat on the water surface.

NBC-6 (40–0 cm, last 50 yrs)

There is a rapid increase in the amount of water soluble Ca, Mg, K and Na as well as the organic content in this zone, with peak values present here in the entire profile. According to the chemical composition of this zone, the emergence of a closed peat layer with mosses and the formation of a small peatbog could have been inferred for the last 50 years.

Pollen analysis

Samples taken between the depths of 312 and 40 cm yielded material suitable for evaluation. A summary of pollen analytical results is depicted on Fig. 3.

NBP-1 (312–304 cm)

During the Late Glacial, immediately following the formation of the channel, the surrounding hill slopes were covered by a mixed taiga vegetation with scattered patches of forest-steppe. The most important arboreal taxa are *Pinus* (40%), *Picea* (10%), *Betula* (8%) and *Abies* (5%), and traces of *Corylus* and *Juniperus* are also present. Significant non-arboreal pollen (NAP) include Poaceae and other steppe elements like *Artemisia*, type Asterioidae, type Cichorioideae and Chenopodiaceae.

NBP-2 (304–292 cm)

A species-rich mixed oak woodland appears early in this zone. Some patches of the former mixed taiga and forest-steppe also survived. There is a major decrease in the amount of pine at the beginning of the zone. This decline was accompanied by an increase in *Betula* to 15% and of *Corylus* to 10%. After this period, the first thermophylus elements turn up in the profile like those of *Quercus* (20%) and *Ulmus* (2%). Among the NAP, the dominant elements are Poaceae, *Artemisia*, type Asterioidae and Chenopodiaceae.

NBP-3 (292–272 cm)

A rapid spreading of *Fagus* (2%) and *Ulmus* (5%) can be traced within the newly developed mixed oak woodland accompanied by a gradual retreat of coniferous and steppe elements. The arboreal flora is dominated by *Quercus* (40%) and *Corylus* (25%) with minor amounts of conifers and *Betula*. A sudden appearance of *Fagus* and *Carpinus* can also be traced here (2–3%).

NBP-4 (272–248 cm)

The dominant elements of the flora in this zone are *Quercus* (40%) and *Corylus* (10%) together with *Fagus* (2–3%) and *Ulmus* (5%) comprising the minor component of the flora. There is a sudden drop in the amount of *Pinus sylvestris* accompanied by a decrease in the proportions of *Fagus* and *Carpinus* in the upper part of the zone. Evidence of the first human impacts was identified between the depths of 260–248 cm. Selective logging of the woodlands is responsible for the change in the forest.

NBP-5 (248–192 cm)

In this zone, a mixed oak woodland was reestablished containing elements of hazelnut and scattered stands of *Fagus* and *Carpinus*. However, this woodland is characterized by a more closed canopy than the one present in the previous zone. The undergrowth is relatively poor in species. There is a gradual decrease in the amount of *Corylus* (hazelnut) in the second part of the zone was accompanied by an appearance of *Tilia*, *Ulmus* and the reoccurrence of *Fagus* and *Carpinus*. The diversity of the NAP also increaed in this zone.

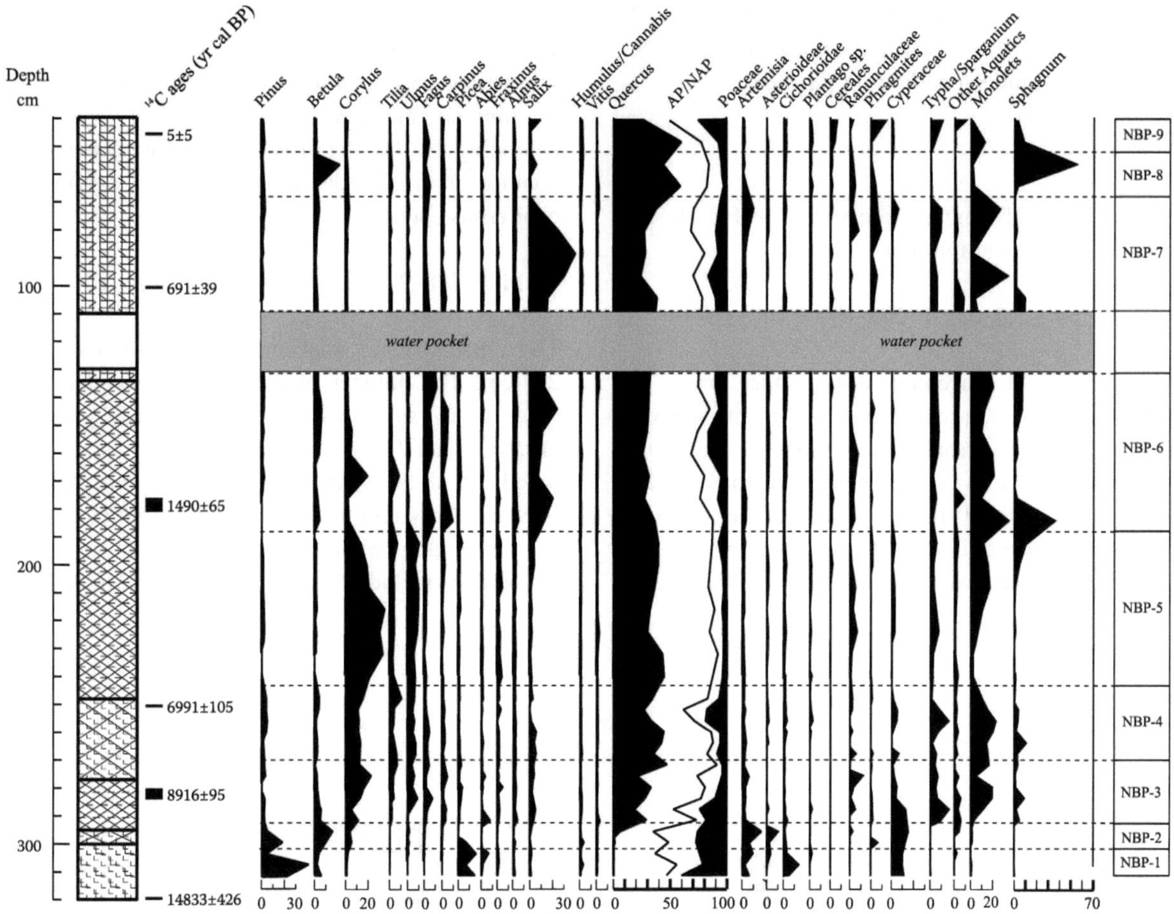

Fig. 3 Selected pollen diagram of Nagybárkány–Nádas-tó (Imola Juhász)

NBP-6 (192–136 cm)

Besides the general dominance of *Quercus*, there is a sudden increase in the proportion of *Fagus* and *Carpinus*. The increased abundance of NAPs marks the gradual opening of the canopy from about 168 cm upwards. The presence of numerous weeds (*Plantago lanceolata*, *Centaurea cyanus*) and cereals marks an intensified human activity in the area.

NBP-7 (104–72 cm)

The pollen profile points to the development of a closed canopy woodland with such dominant elements as *Salix* (40%) and *Quercus* (30–40%), together with *Fagus* and *Carpinus*. Other arboreal pollen (AP) types are rare. The presence of numerous weeds (*P. lanceolata* and *C. cyanus*) and cereals marks an intensified human activity in the area.

NBP-8 (72–56 cm)

This zone marks the gradually closing canopy of the woodland comprised *Quercus*, *Betula*, *Corylus*, *Fagus* and *Carpinus*. Very few NAP species are present here, with the dominance of cereals and weeds marking a less intensive human influence in the area (*Rumex* and *Plantago*).

NBP-9 (56–40 cm)

The top of the profile is characterized by the development of an altered closed oak woodland with a presence of such AP species as *Fagus* and

Carpinus. The presence of numerous weeds (*P. lanceolata*, *Urtica* and *Rumex*) and cereals marks an intensified human activity in the area again.

Macrofossils

The macrofossil diagrams are shown in Fig. 4. The profile was divided into nine zones.

NBM-1 (340–288 cm, 15260–10000 cal yr BP)

The macrofossil concentration was rather low in the lower silty, organic-matter poor sediment, suggesting a high oligotrophic water level, and little vegetation cover. A narrow belt or patches of reed beds probably lined the lake margin. The radiocarbon measurements indicated that this zone covered period from the Late Glacial to the Pleistocene-Holocene transition. Rough peat moss, *S. squarrosum*, a characteristic feature of the recent bog vegetation, was already present at this time, even if in minimal amounts. The presence of peat moss belonging to the *Acutifolia* section is noteworthy, coming from the damp soil of the surrounding coniferous forests.

NBM-2 (288–272 cm, 10000–8000 cal yr BP)

The macrofossil concentration increased from 288 cm. The number of *Phragmites communis* rhizomes and *S. squarrosum* leaves increased. *Typha angustifolia* and *Typha latifolia* made their appearance, together with the seeds of plants typical for reed beds (*Ranunculus sceleratus*, *Rorippa amphibia* and *Alisma plantago-aquatica*) and *Daphnia* ephippia. Bryophytes include the peatmoss *Sphagnum palustre*

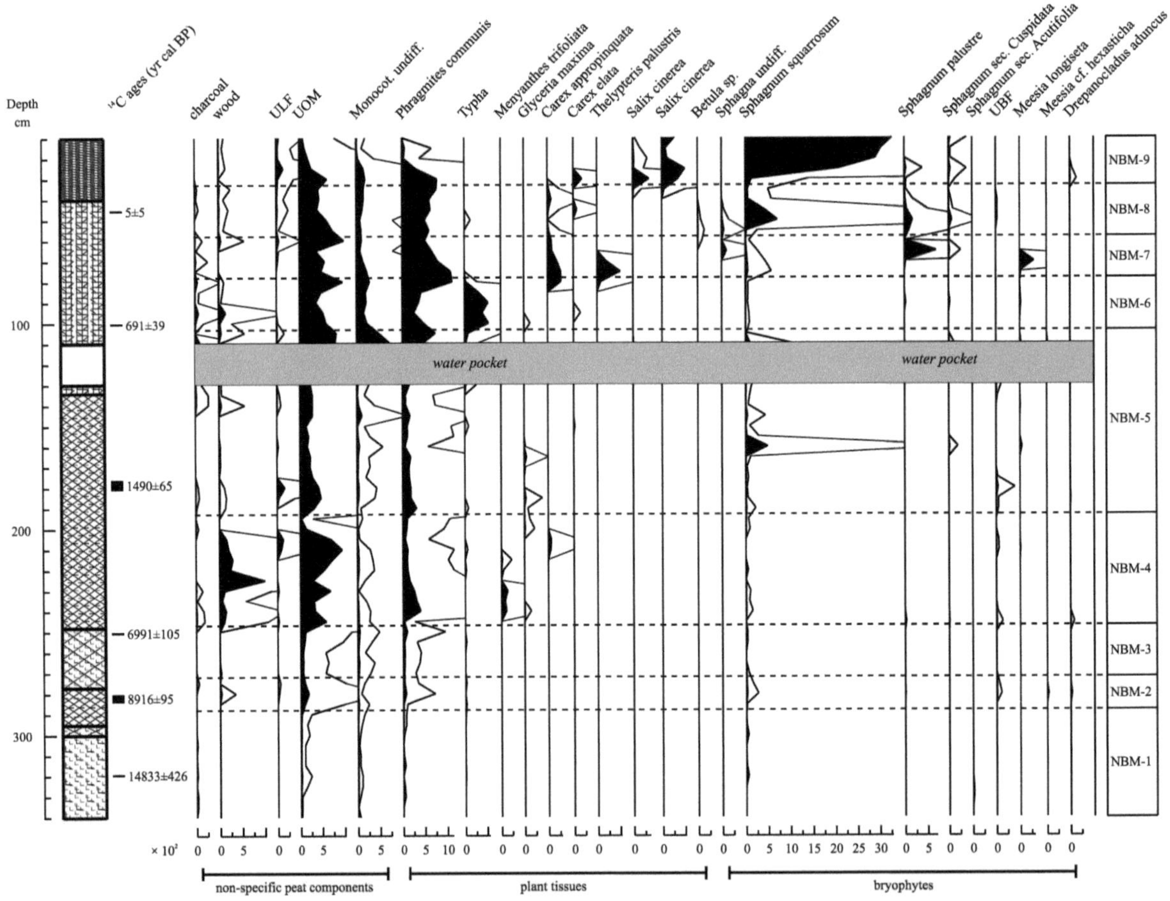

Fig. 4 **a, b** Fossil plant tissues, mosses (pc cm^{-3}) and seeds from Nagybárkány–Nádas-tó (pc cm^{-3}; seed) (Gusztáv Jakab)

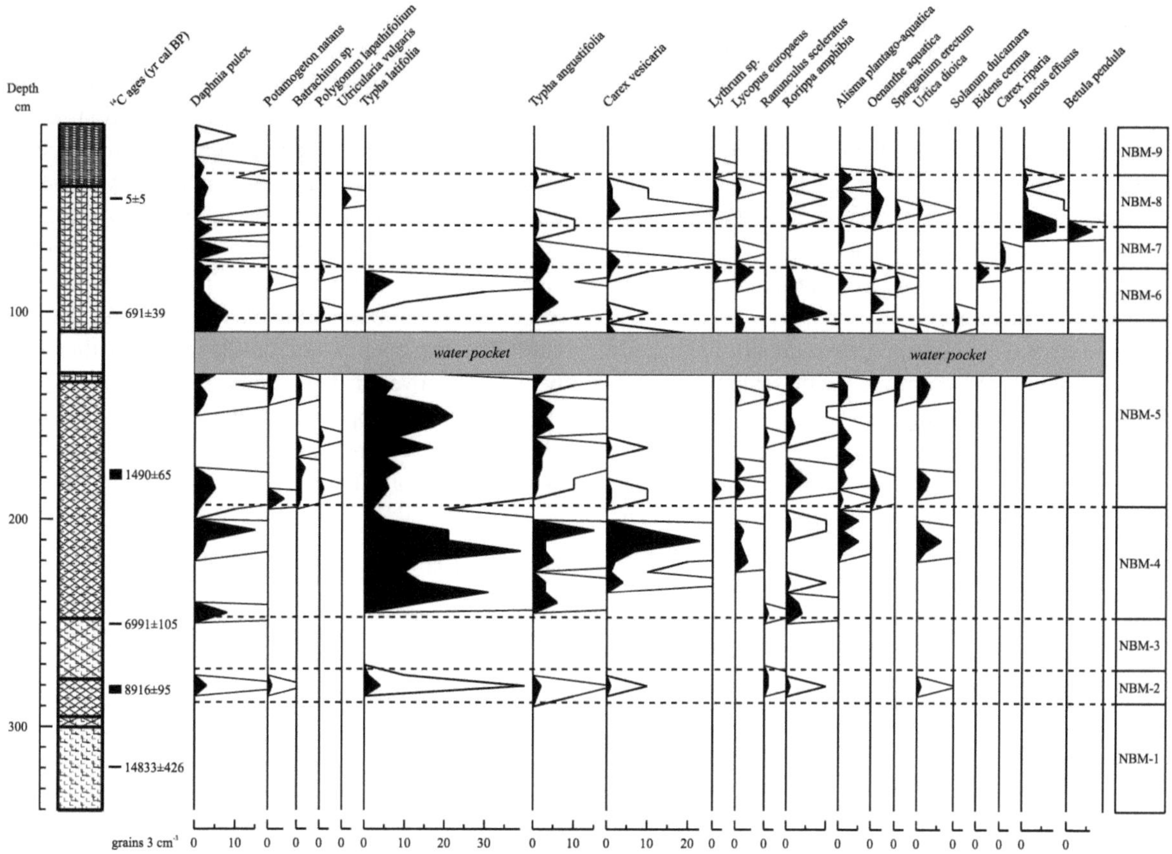

Fig. 4 continued

and various other mosses (*Drepanocladus aduncus*, *Meesia* cf. *hexasticha*). The macrofossils reflect a lower water level and a mesotrophic water quality. The reed bed at the edge of the lakebed probably formed a continuous belt by this period.

NBM-3 (272–247 cm, 8000–6000 cal yr BP)

The macrofossil concentration and the number of *Phragmites* decreased at 270 cm, and marshland and bog species disappeared, suggesting a rise in water level. In the second part of the zone, after 7000 cal yr BP, the macrofossil concentrations and the amount of *Phragmites* again increased, parallel to the renewed appearance of various Sphagna (*S. squarrosum*, *S.* sec. *Cuspidata*, and *S. palustre*) and moss species (*Drepanocladus aducus*). These changes reflect another decrease in the water level and the spread of wetland vegetation.

NBM-4 (247–193 cm, 2300–1500 cal yr BP)

The radiocarbon measurements of the sediment samples between 187 and 176 cm indicated a hiatus of roughly 4400 years at the beginning of the zone. This sediment hiatus is also confirmed by the abrupt change in the pollen composition (Juhász, 2005) and the macrofossil analyses, reflecting a marked change in the environment. The extrapolation of the measurements suggests that this sediment hiatus developed around 2300 cal yr BP, during the Imperial Age when the area was probably settled by Celtic groups who probably deepened the peatbog which had evolved by then. A similar phenomenon was observed at Nagy-Mohos-tó near Kelemér (Magyari et al., 2001). However, the pollen from this location did not include hemp (*Cannabis*) remains, suggesting that the lakebed was deepened to gain water. The huge amounts of wood found in the sediment are probably associated with the deeper lake; it is unlikely the wood originated from trees which had fallen from the edge of the lake, for considerably smaller amounts of wood accumulated in the sediment from later periods characterized by a more closed woodland in the area. Especially, high amounts of wood were found at 225 cm (ca. 1000 pieces cm^{-3}), perhaps the remains of a larger object or of some wooden structure associated with the lakebed's function.

The macrofossil concentration suddenly increased at the beginning of the zone, with strikingly high UOM values, indicating an eutrophic marshland environment. *Phragmites* cover expanded significantly over the lakebed. The zone contained high amounts of the leaf sheat epidermis of bogbean (*Menyanthes trifoliata*), which probably grew at the edge of the reed bed facing the open water or in hollows. This section of the zone contained *S. squarrosum*. Reed abundance declined in the second half of the zone, parallel to the expansion of bulrush. The presence of *T. latifolia* and *T. angustifolia* was noted. The bulrush probably grew on the southern side or along the lake's edge, because while seeds were highly represented, there were few rhizomes. This period is characterised by *Carex vesicaria* and various moss species (*Amblystegium serpens*, *Calliergonella cuspidata*, *D. aduncus*). *Daphnia* ephippia occurred in high numbers, and the spread of eutrophic marshland species, such as *L. europaeus*, *R. amphibia*, *A. plantago-aquatica* and *Urtica dioica* were noted. The macrofossils indicate a lower water level and meso-eutrophic conditions at the beginning of the zone and eutrophic conditions from ca. 2100 cal yr BP.

NBM-5 (193–103 cm; 130–110 cm: water, 1500–700 cal yr BP)

The macrofossil concentration declined slightly in this zone. The concentration of *Phragmites* rhizomes was relatively low. Various pondweed species (*Potamogeton natans*, *Batrachium* sp. and *Polygonum lapathifolium*) made their appearance, suggesting a relatively high water level. *Sparganium erectum* became typical. The reed bed was quite species rich, with species such as *L. europaeus*, *Lythrum* sp., *R. sceleratus*, *R. amphibia*, *A. plantago-aquatica*, *Oenanthe aquatica* and *U. dioica*.

The concentration of *Phragmites* rhizomes was quite high at the beginning of the zone, but declined continuously, parallel to the spread of *Typha*. The transition is marked by the brief desiccation of the lakebed at 160 cm, with the significant increase of *S. squarrosum*. The trophic condition was meso-eutrophic, changing to eutrophic from 160 cm. Between 130 and 110 cm there was a water pocket. Between 110 and 100 cm, the number of *Phragmites* rhizomes increased significantly, suggesting that the extent of the open water diminished and that reed beds also covered the sampling location. *Typha* (rhizomes) also appeared at the sampling location, although to a lesser degree. The peak of *R. amphibia* similarly indicates the decrease of the water level.

NBM-6 (103–78 cm, 700–400 cal yr BP)

The macrofossil concentration in this zone was extremely high. Many trees fell into the lake. The charcoal concentration is also high, reflecting the intensive exploitation of the environment. *Typha* abundance increases at the sampling location is noted (increase of rhizomes), and the proportion of *T. angustifolia* was higher than the previous one. *T. angustifolia* gradually replaces *T. latifolia*, indicating paludification and a higher water level. This zone can be regarded as the first stage in the development of the present-day bog, when its central part was covered by floating bulrush mat at the expense of pondweed communities. This zone thus represents the lake/bog transition. It seems likely that the reed bed broke loose from the sediment in consequence of rising water levels, leading to the formation of a floating mat. *Phragmites* were soon replaced by *T. angustifolia*.

NBM-7 (78–58 cm, 400–200 cal yr BP)

A genuine floating mat phase. As a result of intensive oligotrophication, a floating reed swamp (*Phragmitetum communis thelypteridetosum*) developed in the sampling location with *C. riparia* and *C. appropinquata*. The zone is characterized by abrupt changes, reflecting further oligotrophication. There is a large-scale increase in marsh fern (*Thelypteris palustris*), which later decreases, parallel to the expansion of a rare moss, *Meesia longiseta*.

Following the decline of the latter, the values of peat mosses, especially of *Sphagnum palustris*, increase.

NBM-8 (58–33 cm, 200–0 cal yr BP)

A *Sphagnum* bog (*Phragmiti communis-Sphagnetum*) developed in consequence of oligotrophication in the sampling area. *Typha* virtually disappeared. In addition to *S. squarrosum*, *S. palustre* and *S.* sec. *Cuspidata* are represented with significant values. The presence of *U. vulgaris* in the zone is noteworthy. The amount of autochtonous arboreal remains (wood and ULF) increases in the zone. The presence of *Betula pendula* on the bog surface seems likely. The high values of *Juncus effusus* are also characteristic for this zone.

NBM-9 (33–10 cm, last 50 yr)

The *Sphagnum*-bog is replaced by a willow swamp in the last zone. The recent expansion of *S. cinerea* could be noted in the area (see Fig. 1). The number of reed species and reed beds decreases, parallel to the increase of wood remains (wood and ULF). *S. cinerea* remains (leaves and roots) are quite frequent. The values of *S. squarrosum* increase significantly.

Multivariate statistical analysis of macrofossil data

The distribution of the 15 most important peat components in the studied samples was evaluated using multivariate statistical methods to elucidate the major ecological, hydrological gradients of the individual macrofossil zones. The ordination of variables (peat components) and objects (sediment samples) is depicted on Fig. 5. As it is clearly visible on the dual-scattergrams, samples are distributed along the first axis (PCA1) in accordance with the degree of water coverage. Lower PCA1 values correspond to samples with taxa preferring open-water conditions (e.g. *Potamogeton*) and samples of the zones NBM-1, NBM-2 and NBM-3. Higher PCA1 values refer to conditions with greater plant coverage and those characteristic of marshlands, as seen in taxa of *Phragmites*, UOM, *Typha* and samples of the zones NBM-6, NBM-7 and NBM-8. The second principal component (PCA2) seems to correspond to the trophic level. Higher values here represent taxa

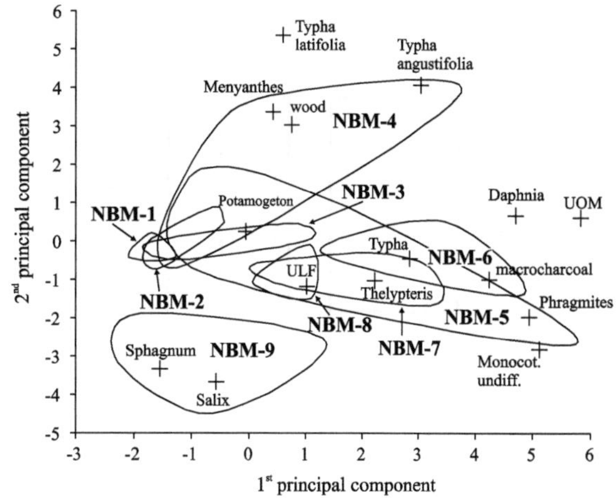

Fig. 5 PC biplot of selected macrofossil data (scale factor: 6,7666) (Gusztáv Jakab)

preferring eutrophic conditions (e.g. *Typha*, *Menyanthes* and *Daphnia*) and samples corresponding to zones NBM-4 and NBM-5. Lower PCA2 values seem to correspond to taxa preferring oligotrophic conditions (e.g. *Sphagnum*, *Phragmites*) and samples with the zone of NBM-9. The received PCA1 and PCA2 values were depicted on a separate graph in accordance with the vertical position (depth) of the studied sediment samples (Fig. 6).

Discussion

Initial pond phase (Late Glacial to Middle Holocene: 15260–6000 cal yr BP)

The first phase of the bog's development, lasting until the middle Holocene, was determined by relatively stable trophic conditions and smaller fluctuations in the water level. The water level dropped somewhat at ca. 8800 cal yr BP and the peatbog developed, probably as a consequence of the onset of a warmer and drier climate. *S. squarrosum* and *S. palustre*, which are an indication of paludification, are present. Based on pollen profiles (Juhász et al., 2004), this phase coincided with the appearance of thermophilous species-rich oak forests. The water level again rose at 8000 cal yr BP and fell after 7000 cal yr BP, causing the expansion of reed beds. The macrofossil analyses of sediments from Lake Balaton also indicated an increasing water level at 8000 cal yr BP (Jakab et al., 2005).

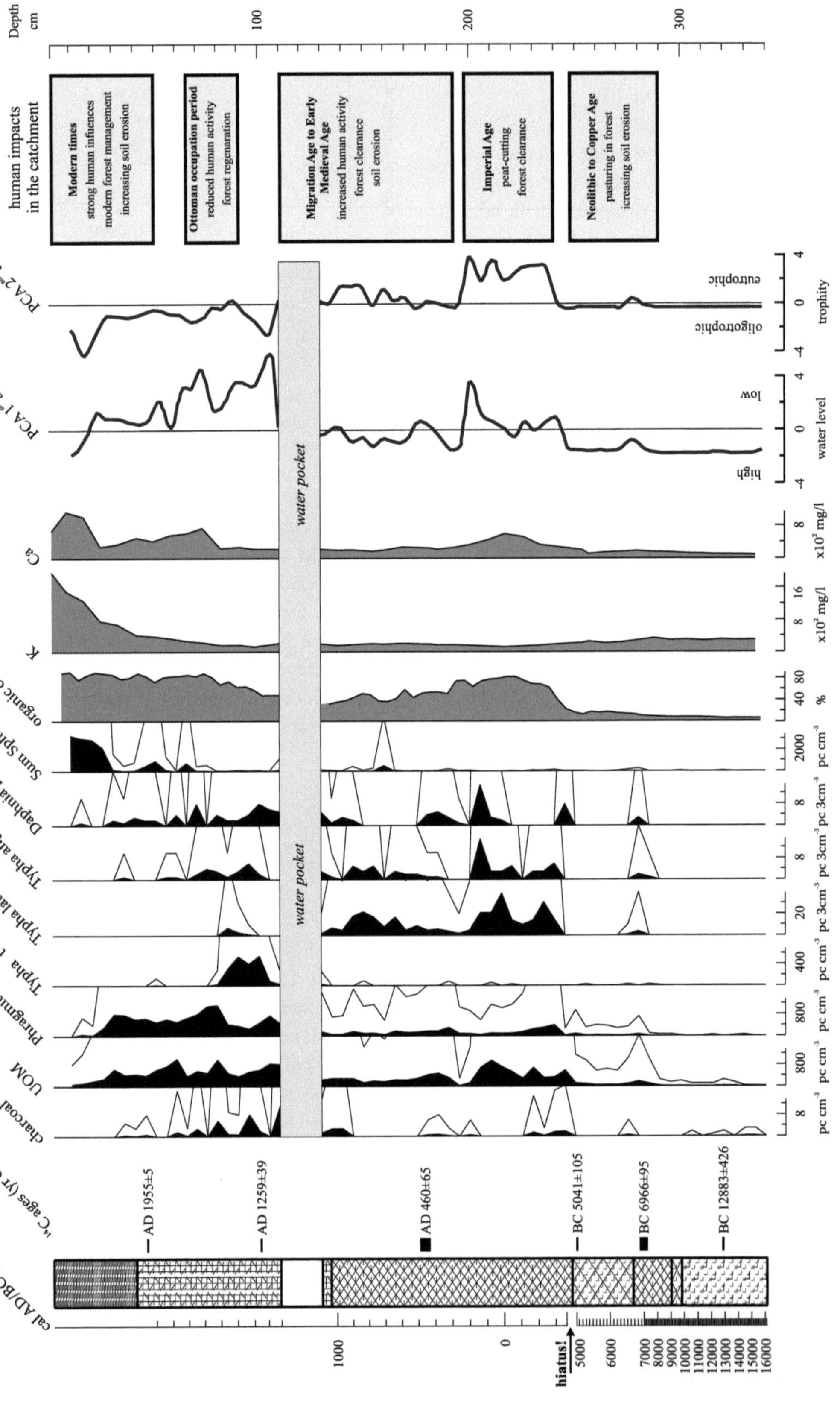

Fig. 6 Cumulative diagram of macrofossils and geochemical data from Nagybárkány–Nádas-tó (t: tissue, s: seed, e: ephippium) (Gusztáv Jakab)

Between 15260 and 8800 cal yr BP, inputs of K and Mg into the basin suggest that erosion of the slopes surrounding the lake basin was occurring, and trophic conditions of the lake were oligotrophic (Mackereth, 1966; Engström & Wright, 1984; Engström & Hansen, 1985). The high content of water soluble Fe suggest that a combination of acidic silicate-rich bedrock, the coniferous trees and the cool Late Glacial climatic conditions resulted in podzol soil formation around the catchment basin. The deposited lacustrine sediments also contain varying quantities of minor pebbles until around 6000 cal yr. These must indicate erosion of the shore, which lacks a closed reed belt.

According to geochemical analysis, the early Holocene lacustrine phase differed significantly from the previous pond stage in sedimentary, chemical composition and temperature conditions. While the earlier, Late Glacial and Early Postglacial lake environment can be characterized by sedimentation in a cold and oligotrophic water lacking Ca content and low vegetation cover, the early Holocene can be described as being relatively rich in Ca, with high carbonate and organic content and with a vegetation typical of warm water. The amount of Ca increased from about 100 to 200 ppm. Changes in the chemical composition refer to intensified erosion around the catchment basin and the transformation of the Late Glacial oligotrophic lake into an open mesotropic lake phase. The increasing level of soil erosion into the catchment basin must have developed under increasing human impacts (e.g. woodland grazing). As shown by archaeological data (Bácsmegi, 2005; Bácsmegi & Fábián, 2005, Neolithic and Copper Age communities settled around the analysed region between 7500 and 6000 cal yr BP (5500–4000 cal yr BC). These prehistoric human communities transformed their forested environment to open surface for agriculture and pasture. This type of human disturbance might trigger intensified soil erosion into the catchment basin.

Climate-driven mire phase (Late Holocene: 2300–700 cal yr BP)

The second phase of peat land development started with a short but remarkable anthropogenic disturbance. Abrupt geochemical changes indicate an hiatus in the catchment basin. It appears that the unconformity was the result of human impact. Results of the radiocarbon measurements are presented in Table 2, from which a sedimentary hiatus is apparent between 240 and 220 cm. This hiatus associated with a thin layer of macrocharcoal. This and a subsequent change in lacustrine stratigraphy from mesotrophic lake to reed peat shows ca. 4000 years difference in age between adjacent samples. According to radiocarbon and sedimentological data, this event coincides with a peat-cutting in the Imperial Age when Barbarian groups (Dacs, Celts and German tribes) lived around the lake basin (Vaday, 2005). Probably, one of these ancient tribes cleared the area around the pond around 2300 cal yr BP. After this clearing, a mass of water plants covered the artificially transformed pond surface.

This second phase lasted from ca. 300 cal BC until cal AD 1400. The water level decreased and the lake was encircled by a marshland belt. Following the anthropogenic disturbance, this period was characterized by fluctuations in the trophic conditions, which can most likely be linked to climatic causes. Figure 7 shows the changes in the values of six peat components, which were typical for this phase, together with the historically recorded climatic phases.

Figure 7 reveals that the *Phragmites* concentrations in the sediment decreased during warmer periods in the Imperial Age and the Middle Ages, parallel with an increase of *Typha* seeds and *Daphnia* ephippia, which suggest increasing eutrophication (Keller et al., 2002). This reflects a competitive situation, characterized by alternating dominances of reed and bulrush in the lakebed. Reed and bulrush are both competitive species under favourable conditions. Bulrush is a light-demanding plant. Reeds grow higher, overshadowing bulrush and easily out competes the latter. At the same time, reeds require relatively coarse sediment to anchor its roots, otherwise it collapses. Conversely, bulrush settles more easily in finer-grained sediments (Haslam, 1972; Toivonen & Bäck, 1989; Nurminen, 2003). In the pollen record, this phase is characterized by the increase of herbaceous elements, which suggests the opening up of the forest canopy. It would appear that during periods of greater solar activity, the lake received more light, in part owing to the retreat of species forming higher and more closed forest

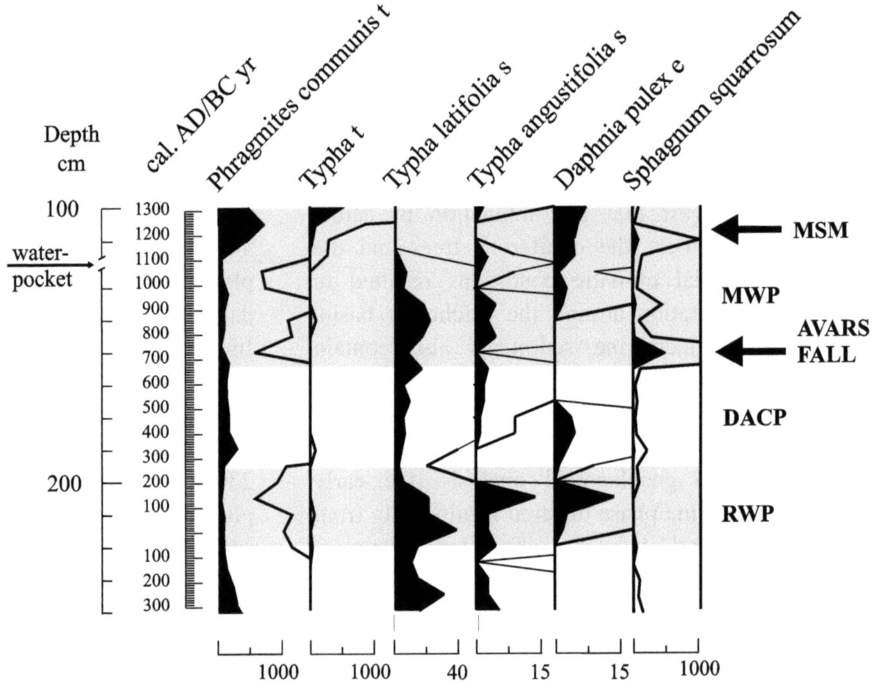

Fig. 7 Trophic fluctuations at Nagybárkány–Nádas-tó between 330 BC and 1300 AD. The grey areas indicate periods with a warmer climate (MSM: medieval solar activity maximum, MWP: medieval warm period, DACP: Dark Ages cold period, RWP: Roman Age warm period; the interbedded water layer between 130 and 110 cm has been omitted) (Gusztáv Jakab)

canopy, like *Fagus* or *Carpinus*. The lake is fringed by high, steep slopes in the south and southeast, from where the high trees cast a shadow over the greater part of the lake. The expansion of phytoplankton at the time of greater solar activity is indicated by an increase of *Daphnia* feeding on them. The expansion of phytoplankton leads to the development of looser sediments, encouraging the spread of *Typha*.

The beginning and the close of the Medieval Warm Period saw the maximum of solar activity (Bradley et al., 2003). The first maximum, dated to AD 700–800, caused significant droughts in the Carpathian basin and contributed to the decline and fall of the Avar Empire (Györffy, 1995, Győrffy & Zólyomi, 1994). The end of this warm period ca. AD 1250 was marked by the so-called medieval solar activity maximum, which caused serious droughts in Europe and North America. The sudden expansion of *S. squarrosum* coincides with the time of the two maximums. *Phragmites* and *Typha* both declined around AD 800, suggesting the brief desiccation of the bed, when peat moss temporarily covered the entire lakebed. In the pollen record at ca. AD 600, Cyperaceae and aquatic species (*Nuphar*, *Nymphea*, *Lemna* and *Butomus*) have a temporary minimum as well (Juhász, 2005). The water level decreased for a longer period of time around AD 1200–1300, enabling the expansion of the reed bed over the lake's entire surface and causing the reduction of open water. According to the geochemical data between cal AD 500 and 1300 a drier and may be a warmer phase occurred. The increase of water soluble Na shows that a decrease of the pond water level occurred during this phase.

The referred period was coeval with one of the major crisis period of medieval Hungary, the invasion of Mongolian ('Tartar') tribes dated between AD 1241 and 1242. From the written record we do know that some chronicle writers blamed the Mongolian invasion of the country on the severe cold weather, while others related it to the unusual droughts hampering Europe during the summers of the thirteenth century. Barber et al. (2000) declared this period as the driest of the past 2000 years in the history of Europe. In Hungary, the extremely cold winter of AD 1241 had a devastating effect on the political and economic fate of the country, when the river Danube was completely frozen enabling the Mongolian tribes to cross the river safely and destroy the settlements of Transdanubia as well. As Kiss (2000, 2003) clearly stated, the controversies lying in the contrast of the extremely cold winters and summer droughts can easily be resolved. A complete freezing of the river Danube was not an unusual event in Hungary prior to human regulation of the river. The summer droughts might have struck Hungary as well. The droughts must have reinforced the effects of famine attributable to the war itself. This warm and dry weather must have contributed to a complete desiccation of the Nádas-tó of Nagybárkány, when reed coverage must have extended across the entire lacustrine basin.

The geochemical changes indicate increasing soil erosion around the lake which can be attributed to a extensive local human impact during the Age of Great Migrations and the Early Medieval Age. According to the available archaeological data (Vaday, 2005; Takács, 2005; Zatykó, 2005), the area around Nagybárkány was well populated from the Dark Ages to Early Medieval Period.

Autogenous peatbog phase (Since late medieval age: after 700 cal yr BP)

The development of the present-day bog and the commencement of peat accumulation can be dated to the end of the Árpádian Age. Around AD 1400, the number of *Typha* rhizomes increases significantly, indicating a rise in the water level and the formation of a floating mat. The hydroseries of the bog was from this point on characterized by autogenic processes, with a tendency towards a gradual oligotrophication, triggered by climatic deterioration.

Based on detailed phytogeographical studies, floating mats of *Phragmites* and *Typha* are frequent components of lake shore vegetation in Hungary. The first signs of peatbog formation from these floating mats can be seen in a massive expansion of *T. palustris*. As time goes by, peat mosses also turn up on these mats (Borhidi & Balogh, 1970; Balogh, 2000a, b). Based on the palaeobotanical investigations, this bog development passed through phases characterized by *Typha* → *T. palustris* → *M. longiseta* → *Sphagnum* spp. This development shares numerous similarities with the formation of two other *Sphagnum* bogs at Csaroda–Báb-tava (Jakab & Magyari, 2000) and Kelemér–Nagy-Mohos (Magyari et al. 2000, 2001), suggesting some sort of regularity in the formation of Hungarian *Sphagnum* bogs. *M. longiseta* preceding the expansion of Sphagna is highly interesting. The taxon *M. longiseta* is a unique component in the flora of the Carpathian basin with a single occurrence recorded in Hungary from 1885 (Boros, 1968). According to Hall (1979), the appearance of *M. longiseta* can be linked to a distinct phase of wetland succession, characterized by the transformation of the brown moss-sedge floating mat into acidic *Sphagnum* bog. This species was observed during the secondary succession of abandoned lakes used for retting hemp as well. Odgaard (1988) reported the same for *Meesia triquetra*.

Representatives of *Sphagna* seemed to have appeared in similar quantities during the middle part of the seventeenth century as today, followed by a complete drop till AD 1950. Surprisingly, the taxon *S. palustre* indicating mesotrophic conditions was present in the largest numbers. In the pollen record, *Sphagnum* reaches a maximum peak at 70%, at the same time. This expansion of *Sphagnum* coincides with the coldest period of the Little Ice Age Hungary (Rácz, 2001), which was also the coldest time of the past 2000 years. The Little Ice Age dates from the middle part of the sixteenth till the middle part of the nineteenth centuries (Bradley et al., 2003). The most significant cooling is put to the terminal part of the sixteenth century, when a major drop in the average temperatures is traceable across entire Europe (Pfister, 1999; Pfister & Brázdil, 1999).

As shown by archaeological data (Zatykó, 2005), the traditional Medieval Age settlement system collapsed and was abandoned during the Ottoman occupation. Scattered farmstead-like settlements have been present from the sixteenth century onwards. The geochemical composition and the increased amount of charcoal indicate that the human disturbance decreased around the basin in the eighteenth century. According to the pollen, human impact on the vegetation during this time was more intense. There was a rapid decrease in the amount of *S. palustre*, which prefers mesotrophic conditions (Daniels & Eddy, 1985). A rapid expansion of *J. effusus* indicates eutrophication and the quick spread of weeds. Forest management, accompanied by increased soil erosion in the study area resulted in an enrichment of nutrients in the peatbog. The steady expansion of *S. squarrosum* suggests the emergence of an acidophilic, eutrophic peatland in the area. *S. squarrosum* is relatively tolerant to high Ca, bicarbonate levels, and pH (Clymo, 1973), and in a mineral-rich habitat with a high nutrient supply, grows very fast

(Kooijman, 1993). The expanding beds of *S. squarrosum* are capable of capturing and accumulating Ca via ion exchange (Anschutz & Gessner, 1954; Clymo, 1963; Kooijman, 1993). The record of the Ca content of the sediments seems to display a strong correlation with the amount of peat moss, which is a clear sign of the excellent Ca ion bonding capacity of peat moss. Reeds are also capable of accumulating Ca in rhizomes in the same way as peat mosses (Tóth & Szabó, 1958; Kovács et al., 1978; Podani et al., 1979; Penksza et al., 1994).

Conclusions

The development of the peatbog can be divided into three main phases in the light of macrofossil analysis. The first phase spanned the Late Glacial to the middle Holocene. The trophic conditions in this phase were oligo-mesotrophic and mesotrophic. The water level was high, although with minor fluctuations. A hiatus of roughly 4400 years occurred after this phase owing to peat-cutting during the Imperial Age. The water level decreased slightly, the trophic conditions became eutrophic, and the lake was fringed by macrophyte vegetation. This period was characterized by the fluctuation of the lake's trophic conditions. This phase can be dated to the late Holocene, lasting until the end of the Árpádian Age, at the end of the fourteenth century. The last phase, from the end of the fourteenth century to the present, saw the terrestrialization of the lake and the cessation of the open water surface. The hydrosere was characterized by autogenic processes, with *T. palustris* and *M. longiseta* playing a key role. Peat mosses appeared in the same quantities during the Little Ice Age dated to the middle part of the seventeenth century as today. The subsequent periods saw a temporary decrease in their amounts. Forest management, accompanied by increased soil erosion in the study area resulted in an enrichment of plant type nutrients on the peatbog during the past 200 years. The steady expansion of *S. squarrosum* refers to the emergence of an acidophilic, eutrophic peatland in the basin. Based on our findings, changes in the palaeohydrology and aquatic vegetation of the bog were mainly driven by climatic changes and autogenic processes. Recurring human influences have also significantly modified the natural succession in the region.

Acknowledgements The first author of this article is grantee of the János Bolyai Scholarship. Authors would like to express their gratitude to the Directorate of the Bükk National Park for granting access to the study area. To András Schmotzer for handing over data on the area and Sándor Molnár for his kind help in the fieldwork. This work was supported by grants Hungarian Scientific Research Fund (OTKA) T-034392 and National Research and Development Programme (NKFP) 5/0063/2002.

References

Aaby, B. & G. Digerfeldt, 1986. Sampling techniques for lakes and bogs. In Berglund, B. E. (ed.), Handbook of Holocene Palaeoecology and Palaeohydrology. John Wiley and Sons Ltd, Chichester, New York, Brisbane, Singapore: 181–194.

Anschutz, I. & F. Gessner, 1954. Der Ionenaustausch bei Torfmoosen (*Sphagnum*). Flora 141: 178–236.

Bácsmegi, G., 2005. The copper age int he Nagybárkány area. In Gál, E., I. Juhász & P. Sümegi (eds), Environmental Archaeology in North-Eastern Hungary. Varia Archaeologica Hungarica XIX, Budapest: 243–245.

Bácsmegi, G. & Sz. Fábián, 2005. The neolithic of Nagybárkány and its environs. In Gál, E., I. Juhász & P. Sümegi (eds), Environmental Archaeology in North-Eastern Hungary. Varia Archaeologica Hungarica XIX, Budapest: 237–242.

Bacsó, N., 1959. Magyarország éghajlata [Climate of Hungary]. Akadémiai Kiadó, Budapest: 302 pp.

Balogh, M., 2000a. A lápok rendszerezése [Classification of peatbogs]. In Szurdoki, E. (ed.), Tőzegmohás élőhelyek Magyarországon: kutatás, kezelés, védelem [Sphagnum-bogs in Hungary: research, conservation, management.]. CEEWEB Munkacsoport, Miskolc: 57–65.

Balogh, M., 2000b. Az úszólápi szukcesszió kérdései I (On the succession of floating bogs I.). Kitaibelia 5: 9–16.

Barber, K. E. & D. Charman, 2005. Holocene palaeoclimate records from peatlands. In Mackay, A., R. Battarbee, J. Birks & F. Oldfield (eds), Global Change in the Holocene. Hodder Arnold, London: 210–226.

Barber, K. E. & P. G. Langdon, 2001. Peat stratigraphy and climate change. In Brothwell, D. R. & A. M. Pollard (eds), Handbook of Archaeological Sciences. Wiley, Chicester: 155–166.

Barber, K. E., F. M. Chambers, D. Maddy & J. Brew, 1994. A sensitive high resolution record of the Holocene climatic change from a raised bog in northern England. The Holocene 4: 198–205.

Barber, K. E., D. Maddy, N. Rose, A. C. Stevenson, R. Stoneman & R. Thompson, 2000. Replicated proxy-climate signals over the last 2000 yr from two distant UK peat bogs: new evidence for regional palaeoclimate teleconnections. Quaternary Science Reviews 19: 481–487.

Bennett, K. D., 1992. PSIMPOLL—a quickBasic program that generates PostScript page description of pollen diagrams. INQUA Commission for the study of the Holocene: working group on data handling methods. Newsletter 8: 11–12.

Berglund, B. E. & M. Ralska-Jasiewiczowa, 1986. Pollen analysis and pollen diagrams. In Berglund, B. E. (ed.), Handbook of Holocene Palaeoecology and Palaeohydrology. John Wiley & Sons, New York: 455–479.

Birks, H. J. B. & H. H. Birks, 1980. Quarternary Palaeoecology. University Park Press, Baltimore: 289 pp.

Blackford, J. J., 2000. Palaeoclimatic records from peatbogs. Trends in Ecology and Evolution 15: 193–198.

Borhidi, A. & M. Balogh, 1970. Die Enstehung von dystrophen Schaukelmooren in einem alkalischen(szik-). Acta Botanica Hungarica 16: 13–31.

Borhidi, A. & A. Sánta (eds), 1999. Vörös könyv Magyarország növénytársulásairól 1-2 [Red Book of Plant Associations of Hungary 1–2.] Természetbúvár Alapítvány Kiadó, Budapest: pp. 362, pp. 404.

Boros, Á., 1968. Bryogeographie und Bryoflora Ungarns. Akadémiai Kiadó, Budapest: 466 pp.

Bradley, R. S., M. K. Hughes & H. F. Diaz, 2003. Climate in medieval time. Science 302: 404–405.

Clymo, R. S., 1963. Ion exchange in *Sphagnum* and its relation to bog ecology. Annals of Botany 27: 309–324.

Clymo, R. S., 1973. The grows of *Sphagnum*: some effects of environment. Journal of Ecology 61: 849–869.

Danzeglocke, U., O. Jöris & B. Weninger, 2008. CalPal-2007 online. http://www.calpal-online.de.

Dániel, P., 2004. Geochemical analysis. In Sümegi, P. & S. Gulyás (eds), The Geohistory of Bátorliget Marshland. Archaeolingua Press, Budapest: 52–57.

Dániel, P., B. Kovács, Z. Győri & P. Sümegi, 1996. A Combined Sequential Extraction Method for Analysis of Ions Bounded to Mineral Component. In 4th Soil and Sediment Contaminant Analysis Workshop. Lausanne (Switzerland).

Daniels, R. E. & A. Eddy, 1985. Handbook of European Sphagna. Institute of Terrestrial Ecology, Hundtingdon: 262 pp.

Dean, W. E., 1974. Determination of carbonate and organic matter in calcareous sediments and sedimentary rocks by loss on ignition: comparison with other methods. Journal of Sedimentary Petrology 44: 242–248.

Dömsödi, J., 1988. Lápképződés, lápmegsemmisülés [Bog Formation, Bog Destruction]. MTA Földrajztudományi Kutató Intézet, Budapest: 120 pp.

Dowdeswell, J. A., 1982. Relative dating of Late Quaternary deposits using cluster and discriminant analysis, Audubon Cirque, Colorado Front Range. Boreas 11: 151–161.

Engström, D. R., 1983. Chemical stratigraphy of lake sediments as a record of environmental change. Unpublished Ph.D Thesis, University of Minnesota, Minnesota, USA.

Engström, D. R. & B. C. S. Hansen, 1985. Postglacial vegetation change and soil development in southeastern Labrador as inferred from pollen and chemical stratigraphy. Canadian Journal of Botany 63: 543–561.

Engström, D. R. & H. E. Wright Jr., 1984. Chemical stratigraphy of lake sediments as a record of environmental change. In Haworth, E. Y. & J. W. G. Lund (eds), Lake Sediments and Environmental History. University of Minnesota Press, Minneapolis, MN: 11–68.

Györffy, Gy., 1995. Hová lettek az avarok? [Where the Avars Diappeared?]. História 17: 3–9.

Győrffy, Gy. & B. Zólyomi, 1994. A Kárpát-medence és Etelköz képe egy évezred előtt [Picture of the Carpathian basin and Etil Interfluve one thousand year before]. In Győrffy, Gy. & L. Kovács (eds), Honfoglalás és régészet [Hungarian conquest and archaeology]. Balassi Kiadó, Budapest: 13–37.

Hall, A. R., 1979. A note on the Quaternary history of *Meesia longiseta* Hedw. Journal of Bryology 10: 511–515. (in Britain).

Haslam, S. M., 1972. Biological flora of the British Isles. *Phragmites communis* Trin. Journal of Ecology 60: 585–610.

Jakab, G. & E. Magyari, 2000. Új távlatok a magyar lápkutatásban: szukcessziókutatás paleobryológiai és pollenanalitikai módszerekkel. (Progress in Hungarian mire succession studies: the use of palaeobryological and palynological techniques in the reconstruction of hydroseres.). Kitaibelia 5: 17–36.

Jakab, G. & P. Sümegi, 2004a. A lágyszárú növények tőzegben található maradványainak határozója mikroszkópikus bélyegek alapján. (Keys and descriptions for the determination of fossil plant remains occur in peat.). Kitaibelia 9: 93–129.

Jakab, G., P. Sümegi & E. Magyari, 2004b. A new paleobotanical method for the description of Late Quaternary organic sediments (Mire-development pathways and palaeoclimatic records from S Hungary). Acta Geologica Hungarica 47: 1–37.

Jakab, G., P. Sümegi & Zs. Szántó, 2005. Késő-glaciális és holocén vízszintingadozások a Szigligeti-öbölben (Balaton) makrofosszília vizsgálatok eredményei alapján. (Late Glacial and Holocene water level changes in the Szigliget Bay, Lake Balaton based on macrofossil investigations.). Földtani Közlöny 135: 405–431.

Jowsey, P. C., 1966. An improved peat sampler. New Phytologist 65: 245–248.

Juhász, I. E., 2005. Preliminary palaeobotanical results from the Nádas-tó peat bog at Nagybárkány: the Late Holocene environmental history. In Gál, E., I. Juhász & P. Sümegi (eds), Environmental Archaeology in North-Eastern Hungary. Varia Archaeologica Hungarica XIX, Budapest: 79–85.

Juhász, I. E., G. Jakab & P. Sümegi, 2004. Vegetation dynamics of a small *Sphagnum* peat bog from North Hugnary, from the Late Pleistocene until Late Holocene: palynological and macrobotanical data. In Abstracts of the 11th International Palynological Congress, Granada, Spain: 508 pp.

Kiss, A., 2000. Weather events during the first Tartar invasion in Hungary (1241–42). Acta Geographica Szegediensis 37: 149–156.

Kiss, A., 2003. Ecce, in hyemis nivis et glaciei habundantia supervenit". Időjárás, környezeti krízis és a tatárjárás [Climate, environmental crisis and the invasion of the Mongols]. In Nagy, B. (ed.), Tatárjárás [Invasion of the Mongols.] Osiris Kiadó, Budapest: 439–452.

Keller, B., H. R. Bürgi, M. Sturm & P. Spaak, 2002. Ephippia and *Daphnia* abundances under changing trophic conditions. Verhandlungen Internationale Vereinigung für Theoretische und Angewandte Limnologie 28: 851–855.

Kooijman, A. M., 1993. On the ecological amplitude of four mire bryophytes, a reciprocal transplant experiment. Lindbergia 18: 19–24.

Kovács, M., I. Précsényi & J. Podani, 1978. Anhäfung von Elementen im balatoner Schilfrohr (*Phragmites communis*). Acta Botanica Academiae Scientiarum Hungaricae 24: 99–111.

Lájer, K., 1998. Bevezetés a magyarországi lápok vegetációökológiájába [Introduction to the vegetation ecology of the Hungarian peatbogs.] Tilia 4: 84–238.

Magyari, E., G. Jakab, P. Sümegi, E. Rudner & M. Molnár, 2000. Paleobotanikai vizsgálatok a keleméri Mohostavakon [Palaeobotanica investigations on the Kelemér Mohos Lakes]. In Szurdoki, E. (ed.), Tőzegmohás élőhelyek Magyarországon: kutatás, kezelés, védelem [Sphagnum-bogs in Hungary: research, conservation, management.]. CEEWEB Munkacsoport, Miskolc: 101–131.

Magyari, E., P. Sümegi, M. Braun, G. Jakab & M. Molnár, 2001. Retarded wetland succession: anthropogenic and climatic signals in a Holocene peat bog profile from north-east Hungary. Journal of Ecology 89: 1019–1032.

Maher, L. J., 1972. Nomograms for computing 95% limits of pollen data. Review of Palaeobotany and Palynology 13: 85–93.

Mauquoy, D. & K. Barber, 1999. A replicated 3000 yr proxy-climate record from Coom Rigg Moss and Felicia Moss, The Border Mires, northern England. Journal of Quaternary Science 14: 263–275.

Máthé, I. & M. Kovács, 1959. A Cserhát tőzegmohás lápja (Das Sphagnummoor des Cserhát-Gebirges.). Botanikai Közlemények 50: 106–108.

Mackereth, F. J. H., 1966. Some chemical observations in post-glacial lake sediments. Proceedings of Royal Academy of Sciences 250: 165–213.

Nurminen, L., 2003. Macrophyte species composition reflecting water quality in adjecent water bodies of Lake Hiidenvesi, SW Finland. Annales Botanici Fennici 40: 199–208.

Odgaard, B. V., 1988. Glacial relicts—and the moss Meesia triquetra in Central and West Europe. Lindbergia 14: 73–78.

Penksza, K., G. Turcsányi & M. Kovács, 1994. A siroki Nyírjes-tó tőzegmohalápjának elemkatasztere (Element concentration cadasters of peat profiles in Nyírjes Bog near Sirok in Hungary.). Botanikai Közlemények 81: 29–41.

Pfister, C., 1999. Wetternachhersage: 500 Jahre Klimavariationen und Naturkatastrophen (1496–1995). Haupt, Bern: 304 pp.

Pfister, C. & R. Brázdil, 1999. Climatic variability in sixteenth-century Europe and its social dimension: a synthesis. Climatic Change 43: 5–53.

Podani, J., 1993. SYN-TAX 5.0: computer programs for multivariate data analysis in ecology and systematics. Abstracta Botanica 17: 289–302.

Podani, J., M. Kovács & M. Dinka, 1979. A balatoni nád elemtartalmának vizsgálata I. A nád szerveinek összehasonlítása, az elemek közötti korrelációk. Botanikai Közlemények 66: 275–284.

Pokorný, P. & V. Jankovská, 2000. Long-term vegetation dynamics and the infilling process of a former lake (Švarcenberk, Czech Republic). Folia Geobotanica 35: 433–457.

Rácz, L., 2001. Magyarország éghajlattörténete az újkor idején [Climate history of Hungary in the modern times.] Juhász Gyula Felsőoktatási Kiadó, Szeged: 303 pp.

Rollinson, H., 1993. Using Geochemical Data: Evaluation, Presentation and Interpretation. Longman, London, Essex: 352 pp.

Stockmarr, J., 1971. Tablets with spores used in absolute pollen analysis. Pollen et Spores 13: 615–621.

Sümegi, P., E. Magyari, P. Dániel, E. Hertelendi & E. Rudner, 1999. A kardoskúti Fehér-tó negyedidőszaki fejlődéstörténetének rekonstrukciója (Reconstrucion of the Quaternary geohistory of Fehér lake at Kardoskút.). Földtani Közlöny 129: 479–519.

Szurdoki, E. & J. Nagy, 2002. *Sphagnum* dominated mires and *Sphagnum* occurences of North-Hungary. Folia Historico-Naturalia Musei Matraensis 26: 67–84.

Takács, M., 2005. The 10–11th century settlement history of Nagybárkány, Sirok and Tarnabod. In Gál, E., I. Juhász & P. Sümegi (eds), Environmental Archaeology in North-Eastern Hungary. Varia Archaeologica Hungarica XIX, Budapest: 387–395.

Tóth, L. & E. Szabó, 1958. Über die chemische Zusammensetzung vershieder Schilfproben vom Balaton-See. Annales Instituti Biologici (Tihany) Hungaricae Academiae Scientarium 25: 363–374.

Toivonen, H. & S. Bäck, 1989. Changes in aquatic vegetation of a small eutrophicated and lowered lake (southern Finland). Annales Botanici Fennici 26: 27–38.

Troels-Smith, J., 1955. Karakterisering af lose jordater. Danmarks Geologiske Undersogelse 4(3): 1–73.

Vaday, A., 2005. The Nagybárkány area in the Iron Age and the Roman period and Migration period. In Gál, E., I. Juhász & P. Sümegi (eds), Environmental Archaeology in North-Eastern Hungary. Varia Archaeologica Hungarica XIX, Budapest: 345–347.

Weninger, B. & O. Jöris, 2008. A 14C age calibration curve for the last 60 ka: the Greenland-Hulu U/Th timescale and its impact on understanding the Middle to Upper Paleolithic transition in Western Eurasia. Journal of Human Evolution 55: 772–781.

Zatykó, Cs., 2005. The medieval environment and settlement history of the Nagybárkány area. In Gál, E., I. Juhász & P. Sümegi (eds), Environmental Archaeology in North-Eastern Hungary. Varia Archaeologica Hungarica XIX, Budapest: 405–409.

Late Pleistocene–early Holocene transition recorded in the sediments of a former shallow lake in the Czech Republic

Tomáš Bešta · Jitka Šafránková · Miloslav Pouzar · Jan Novák · Kateřina Nováková

Originally published in the journal Hydrobiologia, Volume 631, No. 1, 107–120.
DOI: 10.1007/s10750-009-9804-y © Springer Science+Business Media B.V. 2009

Abstract Quaternary limnic sediments from Velanská Cesta, a former lake in the Czech Republic, were investigated to document the response of an aquatic environment to global climatic shifts during the late Pleistocene–early Holocene transition. The obtained multi-proxy record, including subfossil diatoms, diatom-inferred variables (index of trophic state, halobic index, TP, and pH), plant macrofossils, and geochemical data, was compared with the δ^{18}O GISP2 curve. Diatom accumulation zones and subzones established on the basis of a ConSLink cluster analysis corresponded well with the boundaries of the Bölling, Older Dryas, Alleröd, Younger Dryas, and Preboreal periods. The diatom-inferred trophic state and salinity data showed increased values during the colder epochs, which were supported by the findings of low-temperature macrofossil indicators in the corresponding samples. The results of this study document a long-term forcing of a shallow lacustrine ecosystem in central Europe by climatic events of a global scale.

Keywords Late Pleistocene–Holocene transition · Shallow lake · Diatoms · Macrofossils · Geochemistry

Introduction

The late Pleistocene–early Holocene transition represents a period of abrupt shifts in temperature and humidity resulting in climatic conditions similar to present. Microfossils in lacustrine sediments (e.g., diatoms, cladocerans, chironomids, and pollen) provide detailed information on the prevailing conditions in the past aquatic environments and, in combination with radiocarbon dating, enable a temporal reconstruction of the development of lakes in the context of climatic shifts.

Numerous studies conducted in central Europe specified the regional aspects of the climatic changes during the Pleistocene–Holocene (P–H) transition (e.g., Amman & Lotter, 1989; Isarin & Bohncke, 1999). However, paleoecological research of the transition in the region of the Czech Republic has been scarce, with the exception of studies focused on pollen (e.g., Jankovská, 1980). Multidisciplinary studies were

Guest editors: K. Buczkó, J. Korponai, J. Padisák & S. W. Starratt
Palaeolimnological Proxies as Tools of Environmental Reconstruction in Fresh Water

T. Bešta (✉) · J. Šafránková · J. Novák · K. Nováková
Department of Botany, Faculty of Science, University of South Bohemia, Branišovská 31, 370 05 České Budějovice, Czech Republic
e-mail: Bobiz@seznam.cz

M. Pouzar
Department of Environmental and Chemical Engineering, Faculty of Chemical Technology, University of Pardubice, Studentská 573, 532 10 Pardubice, Czech Republic

performed at two localities in the Czech Republic, the former lake of Švarcenberk (Pokorný & Jankovská, 2000) and the Plešné jezero lake (e.g., Jankovská, 2006; Tátosová et al., 2006); however, diatom analyses were restricted to sediments dated outside the P–H transition (e.g., Řeháková, 1986, 1988), or they documented subrecent lake acidification (Arzet, 1987, 1993). The lack of data from lacustrine sediments is partly caused by the lack of natural lakes in the area. There are only five glacial lakes situated in the Bohemian Forest Mountains, and some Quaternary limnic sediments have been discovered in peat bogs (e.g., Břízová, 2002, Jankovská, 2007, Petr & Pokorný, 2008).

This study aims to reveal the dynamic of a shallow lake during the climatic fluctuations occurring over the course of its lifetime, from its origin in the Bölling period until the completion of the natural infilling process in the Preboreal period, recorded in diatoms, diatom-inferred variables, macrofossils, and geochemistry.

Materials and methods

Study site

The existence of the former lake of Velanská Cesta was first noted by Jankovská (1980). It is located in the south of the Czech Republic (48°46′32.96″ N; 14°53′58.51″ E), and situated in a transition area between the Třeboň Basin and the Novohradské hory Mountains at an altitude of 500 m. The bedrock is composed of Tertiary sediments of the Mydlovarské Souvrství complex of strata, with dominant sandstones and pudding stones at the base. Topological depressions are filled with Quaternary alluvial silt, gravel, and peat bogs. Limnic sediments are overlaid by peat, which formed after the natural infilling of the lake at approximately 10,260 cal year BP. The nitrogen content of the soil is low, phosphate availability is medium, calcium carbonate is often deficient, and potassium only occurs in deep soil horizons (Husák & Hejný, 1978). The locality is supplied by waters rising along geological faults and drained by an artificial melioration system. Recent vegetation is composed of a secondary (artificially planted) stand of Scots pine (*Pinus sylvestris*) with undergrowth formed by the association of *Dicranio-Pinion.*

Coring, dating, and core chronology

The main profile (240 cm) was selected as the standard profile, having the longest and most continuous record. The profile was dug manually in the summer of 2004. Samples of sediment were collected from the pit wall using 50 × 8 × 10 cm metal corers. The samples were divided into 2-cm thick slices and refrigerated at 4°C. The average depth of the sediment layer was used for the labeling of individual samples (e.g., the sample from the depth of 220–218 cm was labeled as "219 cm").

The base of the profile (240–216 cm) was composed of sands and gravelly sands with a slight admixture of organic matter in the upper part (222–214 cm). The layers between 214 and 180 cm contained algal gyttja, the layers between 180 and 160 cm were composed of algal gyttja and larger organic compounds, the layers between 160 and 130 cm contained sedge-reed peat with algal gyttja, the layers of 130–50 cm contained sedge-reed peat, the layers between 50 and 6 cm were composed of sedge reed peat with a high content of charcoals, and the topmost layer of 6–0 cm contained degraded peat.

Four AMS radiocarbon dates (Table 1) from macrofossils obtained from lithostratigraphical breakpoints were analyzed at the Poznań Radiocarbon Laboratory and calibrated using the OxCal 3.1 software (Ramsey, 2005). The dating of individual samples was based on a linear interpolation of the median values of the calibrated radiocarbon dates.

Diatom analysis

A total of 55 samples were analyzed at a magnification of 1000×. Qualitative permanent slides were prepared after Battarbee (1986) using sulfuric acid for digestion and Pleurax (Fott, 1954) as a mounting medium. At least 700 valves were counted per each slide, except for the sample 215 cm (50 valves) and samples adjacent to non-lacustrine sediment (samples 139–129 cm; a minimum of 350 valves), where low diatom concentrations prevented adequate counts. Taxonomic identifications followed the current nomenclatural system (Krammer & Lange-Bertalot, 1986, 1988, 1991a, b; Lange-Bertalot, 1996; Krammer, 1997a, b; Krammer, 2000, 2002, 2003; Houk, 2003).

Table 1 Table of AMS radiocarbon dates analyzed at the Poznań Radiocarbon Laboratory

Depth (cm)	Material	Lab. no.	^{14}C year BP	Cal year BP median value (95.4% probability)
131	*Carex* sp.	Poz-9636	9,080 ± 60	10,290 ± 140
181	*Betula* sp.	Poz-9637	11,670 ± 60	13,535 ± 165
213	*Potamogeton gramineus*	Poz-9641	11,970 ± 60	13,845 ± 135
221	*Potamogeton gramineus*	Poz-9642	12,410 ± 250	14,500 ± 750

Macrofossil analysis

Approximately 120 cm^3 of sediment was sieved per sample to obtain plant macrofossils, and 25 cm^3 was sieved for remains of charophytes and cladoceran ephippia, using pedological sieves (200–300 μm). Sediment samples with a high content of organic matter were treated with 5–10% KOH for several hours following the procedure of Wasylikowa (1986) and Latałowa (1999). Macrofossils were determined using an Olympus SZ51 stereo microscope and stored in Hantsch medium (Watts 1978). Taxonomic identifications followed Mądalski (1949), Soják (1958), Dąmbska (1964), Kac et al. (1965), Bergren (1969, 1981), Marjatta (1970), Anderberg (1981), Körber-Grohne (1991), Cappers (1994), and Haas (1994). A collection of recent plant seeds was used for direct comparison.

Geochemistry

Elemental composition of the sediments was determined using a desktop energy-dispersive XRF spectrometer (Elva X, Elvatech Ltd, Kiev, Ukraine) equipped with a Ti-anode X-ray tube. The excitation X-ray beam was pre-treated by a collimator (gap diameter 4 mm). A thermoelectrically cooled solid-state Si-pin detector (PF-770, MOXTEK, USA) was employed. Two separate energy ranges were adopted for the measurements. A range of 0–4 keV was applied for the $K_{\alpha 1}$ lines of Si, P, S, K, and Ca. The tube was operated at 10 kV and 25 μA for this range. The excitation beam was modified by a Ni filter, and a helium micro-flush of the sample site was used. The $K_{\alpha 1}$ lines of Mn, Fe, and Zn were measured in the range of 4–34 keV. The operating conditions of the tube were 45 kV and 25 μA. The spectrum collection time was 300 s per range (600 s per sample). The Elvatech commercial software package Elva X version 2.35 was used for the evaluation of the spectra. Empirical calibration according to the geological standard reference materials was adopted for the calculation of concentrations.

The standards and samples were analyzed in the form of pellets. Approximately, 500 mg of dried (105°C, 12 h) and pulverized geological material was mixed with 100 mg of boric acid. The mixture was then pressed in an evacuable KBr die set with an inner diameter of 16 mm (ICL, Garfield, USA) by force of 10 tons for 2 min.

Organic matter content in the sediment samples was measured using loss-on-ignition at 550°C for 4 h.

Data processing

Diatom and macrofossil percentage diagrams were created using PolPal (Nalepka & Walanus, 2003). Diatom accumulation zones (DAZ) were established reflecting the results of ConSLink (constrained single linkage of samples) cluster analysis. Diatoms were divided into ecological groups following Van Dam et al. (1994) and information available in the identification literature. Indexes of trophic state and the halobic index were calculated according to the formula described in Zelinka & Marvan (1963) with the species weight item deleted. Indifferent species were excluded.

Transfer functions available on EDDI (http://craticula.ncl.ac.uk/Eddi/jsp/) were used for qualitative estimates of pH, TP, and TOC. We chose this approach because the geological and climatological conditions at our locality during sediment deposition (Bölling–Alleröd) were more similar to the conditions in the present northwestern Europe than to those of the study region. The modern analog technique was used to measure the similarity between the dataset and each diatom sample.

The GISP2 δ^{18}O curve (Stuvier et al., 1995) was used to compare our paleolecological data to the

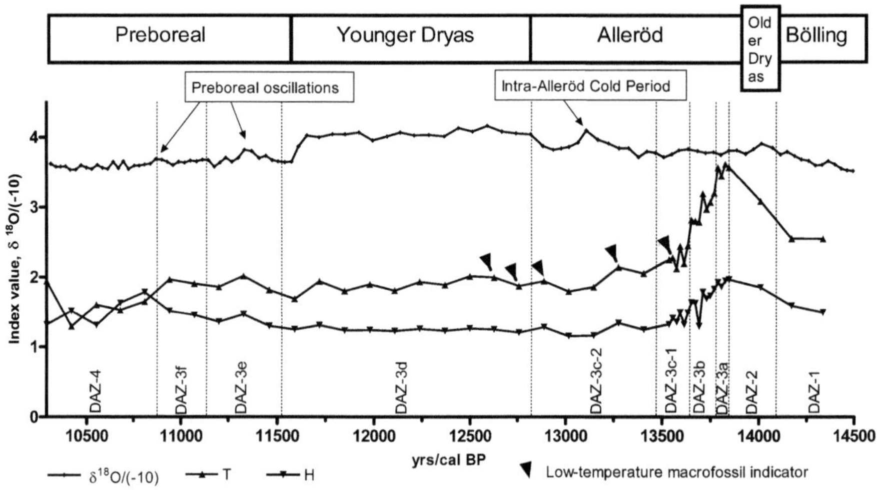

Fig. 1 Development of diatom-inferred trophic state and salinity according to δ^{18}O GISP2 curve with major late Pleistocene and early Holocene intervals. δ^{18}O GISP2 values were divided by −10 for the purpose of the graph; therefore, higher values correspond with colder periods. T—Index of trophic state, H—Halobic index, δ^{18}O/−10—δ^{18}O GISP2 curve divided by −10

global climate change during the Late Pleistocene/Holocene transition. The original data (The Greenland Summit Ice Cores CD-ROM GISP2/GRIP) were divided by the coefficient of −10 to directly compare the climatic fluctuations with the indexes of trophic state and salinity (Fig. 1). The dating of periods was based on Stuvier et al. (1995) with the exception of the dating of Preboreal oscillations which followed Amman & Lotter (1989). This approach to chronology was adopted from Pokorný & Jankovská (2000).

Results

Diatom analysis and diatom-inferred variables

A total of 118 diatom taxa were identified in 55 samples analyzed. The profile was subdivided into four diatom accumulation zones and seven subzones, following the results of the cluster analysis (Fig. 2). The marginal segments (230–222 cm, 130–120 cm) only contained fragments of frustules or were devoid of diatoms. Benthic forms predominated during the entire existence of the lake. Many of the taxa found in the sediment samples (e.g., *Aulacoseira tethera* Haworth, *A.* aff. *fennoscandica* Cleve-Euler, *Encyonema hebridicum* Grunow ex Cleve) are typical components of recent northern-alpine diatom flora.

The index of trophic state and the halobic index (Fig. 1) were calculated for each layer of the profile using the identified diatom contents. The lower counts of frustules at 215 cm and in the layer between 140 and 130 cm were taken into account.

A transfer function reconstruction was applied on those samples that reached at least 700 valves per sample to prevent ambiguous results. Values from non-analogous diatom samples were marked with a cross in the Figs. 3 and 4. The best analogy from the available pH datasets fulfilled the Finnish dataset (Weckström et al., 1997). The generated values were close to the average value of the dataset ($pH_{finnish\ mean} = 7$); therefore, WA with inverse deshrinking was applied (RMSE = 0.30599; $r^2 = 0.68948$; Ave Bias = −0.021339; Max Bias = 0.70743) (see Fig. 3 for details).

The Northwest Europe dataset (Bennion et al., 1996) fitted best within TP datasets. Calculated values fell at the bottom boundary of the dataset (TPnwe min = 5 μm/l, TPnwe-max = 1,189 μm/l); therefore, LWWA with classical deshrinking was employed (RMSE = 0.26083; $r^2 = 0.74133$; Ave Bias = 0.011378; Max Bias = 0.55922) (see Fig. 4 for details).

DAZ-1 (220–216 cm; Bölling)

This zone typically contained small benthic, fragilarioid taxa [*Staurosira construens* (Ehrenberg) D. M. Williams et Round, *Staurosirella pinnata* (Ehrenberg) D. M. Williams et Round], reaching a relative abundance of 39%, the genus *Aulacoseira*, represented by *Aulacoseira valida* (Grunow) Krammer and *A. tethera* Haworth, with a relative abundance of 17%, and the genus *Encyonema* [*E. hebridicum* Grunow ex Cleve and *E. elginense* (Krammer) Mann] with a relative abundance of 16%.

The index of trophic state (T) suggests oligo-mesotrophic to mesotrophic conditions for this zone,

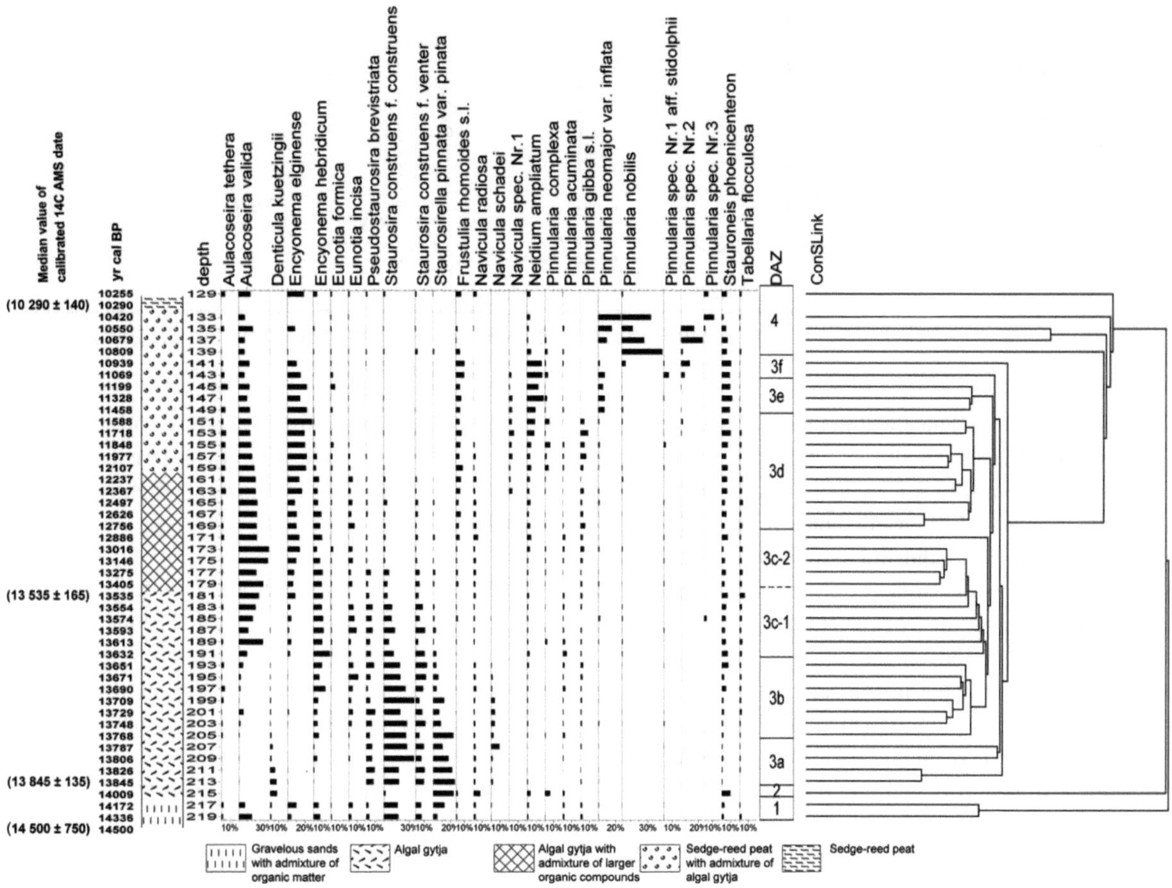

Fig. 2 Relative frequency diagram of sediments from the former lake Velanská Cesta, with diatom taxa reaching at least 5% abundance in any sample. Diatom Accumulation Zones (DAZ) are defined on the basis of ConSLink cluster analysis. Core chronology followed linear interpolation of calibrated radiocarbon dates

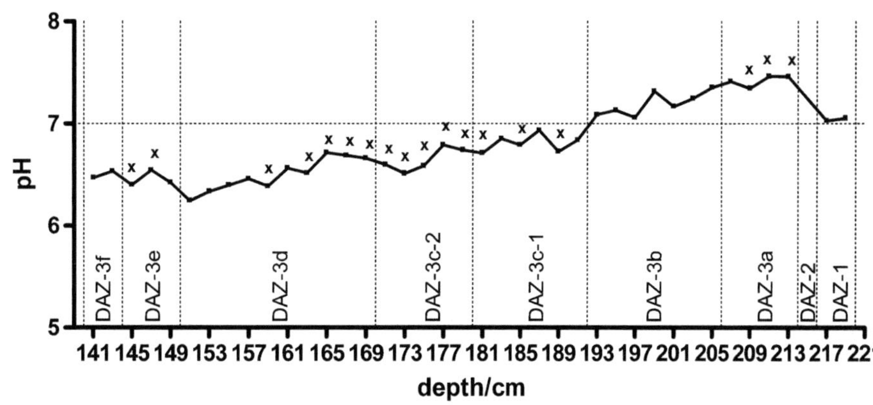

Fig. 3 Diatom-inferred pH within the core of the Velanská Cesta lake. Non-analogous samples are marked with a cross

while the values of inferred TP (~45 μg/l) were within the range of mesotrophy. The halobic index (H) indicates fresh to fresh-brackish conditions and the values of inferred pH (~7) correspond to circumneutral conditions for this zone.

DAZ-2 (216–214 cm; Older Dryas, Alleröd)

Species typical of fresh-brackish environments [*Staurosirella pinnata*, *Stauroneis phoenicenteron* (Nitzsch) Ehrenberg, *Navicula radiosa* Kützing, and *Denticula kuetzingii* var. *kuetzingii* Grunow] reached

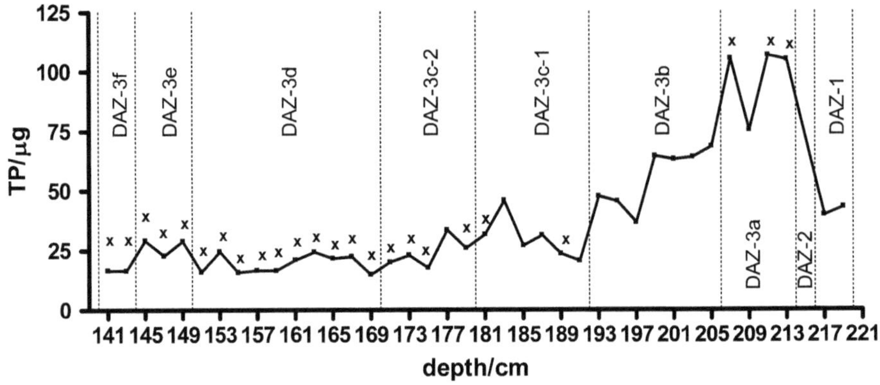

Fig. 4 Diatom-inferred total phosphorus (TP) within the core of the Velanská Cesta lake. Non-analogous samples are marked with a cross

their maximum abundance within this core, while taxa preferring freshwater conditions with lower nutrient contents (e.g., *Encyonema elginense*, *E. hebridicum*, and *Aulacoseira valida*) were not represented. The relative abundance of the low-temperature indicator *Staurosirella pinnata* (Bigler et al., 2000) attained its maximum within small fragilarioid species. The low concentration of diatoms in the sediment only allowed counting of 50 frustules in the sample. Low diatom abundance prevented the calculation of TP and pH. The T index and H index indicated mesotrophic, fresh-brackish conditions in this zone.

DAZ-3 (214–140 cm; Alleröd, Younger Dryas, and Preboreal)
This cluster is divided into seven subzones and described in detail in the following paragraphs.

DAZ-3a (214–206 cm; Alleröd interstadial)
Small fragilarioid taxa reached their maximum within this core (max 67%, mean 62%). The freshwater species *Staurosira construens* f. *construens* outnumbered the fresh-brackish *Staurosirella pinnata* in this subzone and was the most abundant taxon within small fragilaroid species. The relative frequency of the higher salinity level indicator *Denticula kuetzingii* var. *kuetzingii* declined gradually from 7% at the bottom to 3% at the top of the subzone. Abundance of some species frequently found in the zone DAZ 2 [e.g., *Frustulia rhomboides* s.l. (Ehrenberg) De Toni, *Stauroneis phoenicenteron*] decreased considerably, while some species abundant in DAZ-1 (e.g., *Encyonema hebridicum*, *Eunotia incisa* Gregory) reappeared.

The T index corresponded to meso-eutrophic conditions and TP (~75–100 µg/l) indicated eutrophic conditions for this zone. Both models reached their highest values within this core. H index was also at its highest here indicating fresh-brackish conditions. The pH was ~7.4 showing slightly alkaline conditions.

DAZ-3b (206–192 cm; Alleröd)
Small fragilarioid species prevailed (max 60%) in this zone, although the mean frequency (50%) was markedly lower compared with DAZ-3a (62%). This relative decline of the frequency was mainly caused by an increase in the abundance of *Encyonema hebridicum*, *Eunotia formica*, and *Stauroneis phoenicenteron*. *Staurosirella pinnata* continually decreased. The genus *Aulacoseira* reappeared in significant numbers.

This zone has a T index value that indicates mesotrophic conditions, while TP suggests eutrophic conditions, although it is significantly lower in this zone (~40–70 µg/l) than in DAZ-3a. The H index is lower than that in DAZ-3a, but still corresponds to fresh-brackish conditions. The pH decreased from ~7.3 to ~7.0 in this zone. An abrupt decline in the T index, TP, and H index values, and a less pronounced decrease in pH occurred in the middle part of the subzone, corresponding to the sample at a depth of 197 cm.

DAZ-3c-1 (192–180 cm; Alleröd)
This zone was characterized by the dominance of *Aulacoseira* (max 25%, mean 20%) and *Encyonema* (max 22%, 17%), whereas small fragilarioid taxa decreased (max 35%, mean 25%). *Encyonema elginense* (Krammer) Mann reappeared in significant numbers, while *Staurosirella pinnata* was not present in this zone.

The bottom of this zone was characterized by low values of TP (~20 µg/l) and T. Both variables rose progressively throughout the zone to a peak value of ~40 μg/l in the sample 183 cm. In general, the T index suggested oligo-mesotrophic and TP meso-eutrophic conditions. The H index indicated a freshwater environment and the pH was between ~6.8 and 7.0.

DAZ-3c-2 (180–170 cm; Alleröd interstadial including Intra Alleröd Cold Period)
This zone was not distinguished by the results of the cluster analysis, but on the basis of abrupt changes in lithostratigraphy, the structure of sediment (admixture of larger organic particles to algal gyttja), and the rate of sedimentation. The genus *Aulacoseira* (max 37%, mean 30%), mainly represented by *A. valida*, reached its maximum within this subzone. An increase in the abundance of the genus *Encyonema* (max 24%, mean 20%) was caused by a positive shift in the frequency of *E. elginense*. Small fragilarioid species were rare in this zone.

Diatom-inferred water parameters were slightly lower when compared with DAZ-3c-1, and indicated oligo-mesotrophic conditions by T index and meso-eutrophic by TP. The H index suggested freshwater conditions, and the pH was circumneutral to acidic. Peak values of T index, TP, H index, and pH occurred at 177 cm.

DAZ-3d (170–150 cm; Younger Dryas)
The frequency of the genus *Aulacoseira* was lower compared with the previous subzone (DAZ-3c-2; max 24%, mean 22%). The abundance of the genus *Encyonema* (max 29%, mean 22%) increased due to the presence of *E. elginense*, while *E. hebridicum* almost disappeared at the top of this subzone. Small fragilarioid taxa were rare. The abundance of the acidophilic species *Frustulia rhomboides* s.l. and *Neidium ampliatum* (Ehrenberg) Krammer was higher than that in the previous subzones. In general, the changes in the diatom community structure appeared to be less abrupt in this zone than that in the previous ones.

The inferred trophic state of this subzone is relatively complex. TP in the first sample of the subzone (169 cm) reached the lowest value in the core. T index also decreased compared with the previous zone. Both variables then increased until reaching the middle part of the subzone where they started to decline again. The last two samples marked a relatively steep increase of T index and a decrease of TP. The pH increased in the bottom part of the zone, and declined in the top part. The H index was relatively stable in this zone. The shifts in all inferred variables were less abrupt in comparison with previous DAZs.

DAZ-3e (150–144 cm; Preboreal)
This subzone was characterized by an increase in the abundance of acidophilic *Neidium ampliatum* (max 19%, mean 14%) and the first remarkable rise (max 9%, mean 8%) in the populations of larger *Pinnularia* taxa (*P. neomajor* Krammer var. *inflata* Krammer, *P. nobilis* (Ehrenberg) Ehrenberg var. *rectangularis* Krammer, *P.* sp. Nr.1, Nr.2 and Nr.3). *Stauroneis phoenicenteron* reached its maximum abundance within this core (max 12%, mean 7%).

All the inferred variables were higher in this zone than those in DAZ-3d. A peak for all the variables except TP was recorded at 147 cm.

DAZ-3f (144–140 cm; Preboreal)
This zone contained two independent layers, as shown by the cluster analysis. A rapid increase in the frequency of larger *Pinnularia* taxa (max and mean 18%) compared with previous zones was characteristic for these layers. *Frustulia rhomboides* s.l. (max and mean 10%) and *Neidium ampliatum* (max 18%, mean 17%) reached their maximum abundances within this subzone. The abundance of the genera *Encyonema* and *Aulacoseira* decreased, with very similar frequencies (max 15%, mean 13%).

The T index and H index (T and H indices) rose and reached their maxima at 145 cm. The pH was stable throughout this subzone, while TP decreased. The pH and TP were not inferred for the sample at 145 cm due to low valve counts.

DAZ-4 (140–128; Preboreal)
This zone mainly contained strongly silicified fragments of diatom frustules [central areas of larger *Pinnularia* taxa (max 72%, mean 47%), *Encyonema elginense*, *E. hebridicum*, *Frustulia rhomboides* s.l., *Stauroneis. phoenicenteron*, valve faces of *Aulacoseira valida*, and *A. tethera*]. A maximum of 400 valves per sample was counted due to the low abundance of valves.

The T index decreased in this zone, with a peak in the sample (139 cm). The H index indicated fresh-brackish conditions in the first two samples and then decreased.

Macrofossils

First aquatic macrofossils (oospores of charophytes) appeared at 223 cm, followed by seeds of *Potamogeton gramineus* L. from 221 cm. A continuous appearance of aquatic plants and cladoceran ephippia was recorded from 215 cm. Aquatic indicators vanished at 133 cm (see Figs. 5, 6).

Two low-temperature indicators were identified among the macrofossils found in the investigated sediment samples. First, *Betula nana* (Wahlenb.) P. Karst. ex Mussat, which presently inhabits Eurasian and North American tundra (Kolibáčová et al., 2001), and second, *Nuphar pumila* (Timm) DC, known from the Czech Republic and interpreted as a glacial relict. The presence of low-temperature indicators can be seen in Fig. 1.

Geochemistry

All the measured element concentrations (Ca, Si, Fe, and K) showed an increase in the samples 199, 189, 179 cm, and a peak was also recorded in the sample at 129 cm. (Fig. 7a–d). Results of LOI are shown in Fig. 8.

Discussion

Bölling

The median value of the calibrated ^{14}C date (14,500 ± 750 cal year BP; 221 cm, Table 1) from the sediment base sets the origin of the lake to Bölling interstadial. However, due to the possible error in dating, it should be extended to the period between Bölling and Alleröd. The first appearance of an aquatic environment, determined by aquatic indicators, is also unclear. Some indicators, such as cladoceran ephippia and vascular aquatic plants, appeared in the layers assigned to Alleröd interstadial (213 cm). By contrast, diatoms, oospores of charophytes, and a seed of *Potamogeton gramineus* were found in the samples 219 and 217 cm that were dated

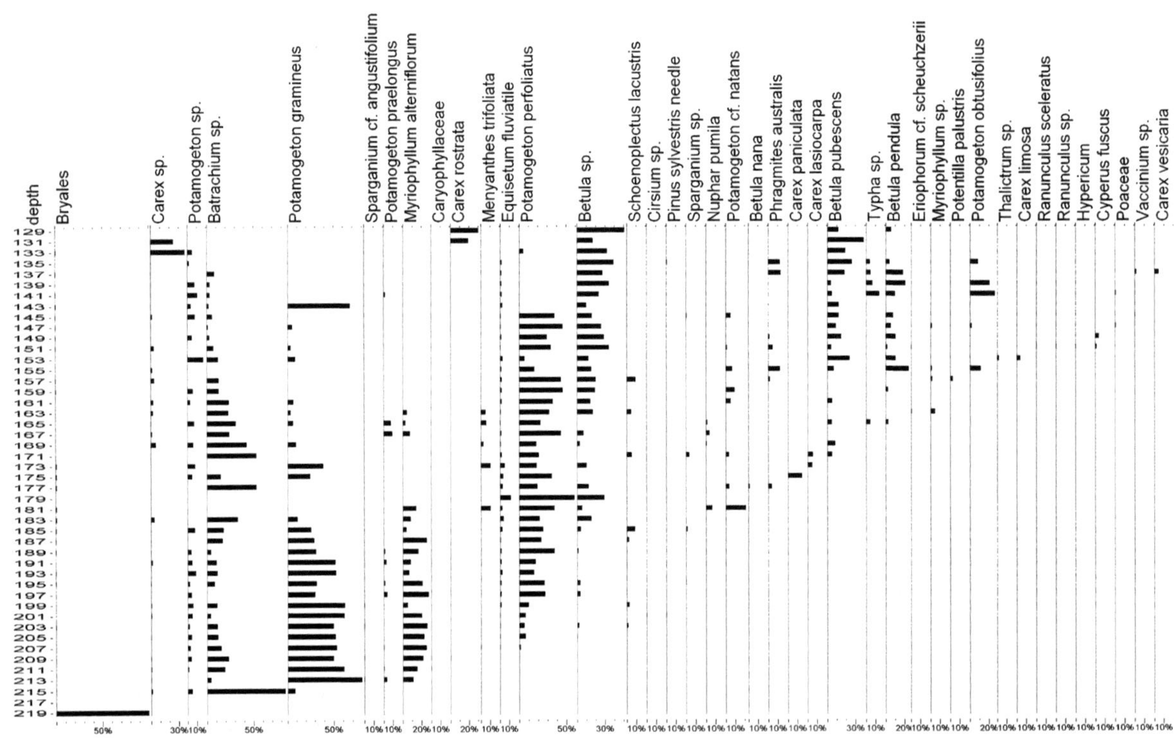

Fig. 5 Relative frequency diagram of macrofossils from sediments of the former lake Velanská Cesta showing the identified taxa except for charophytes

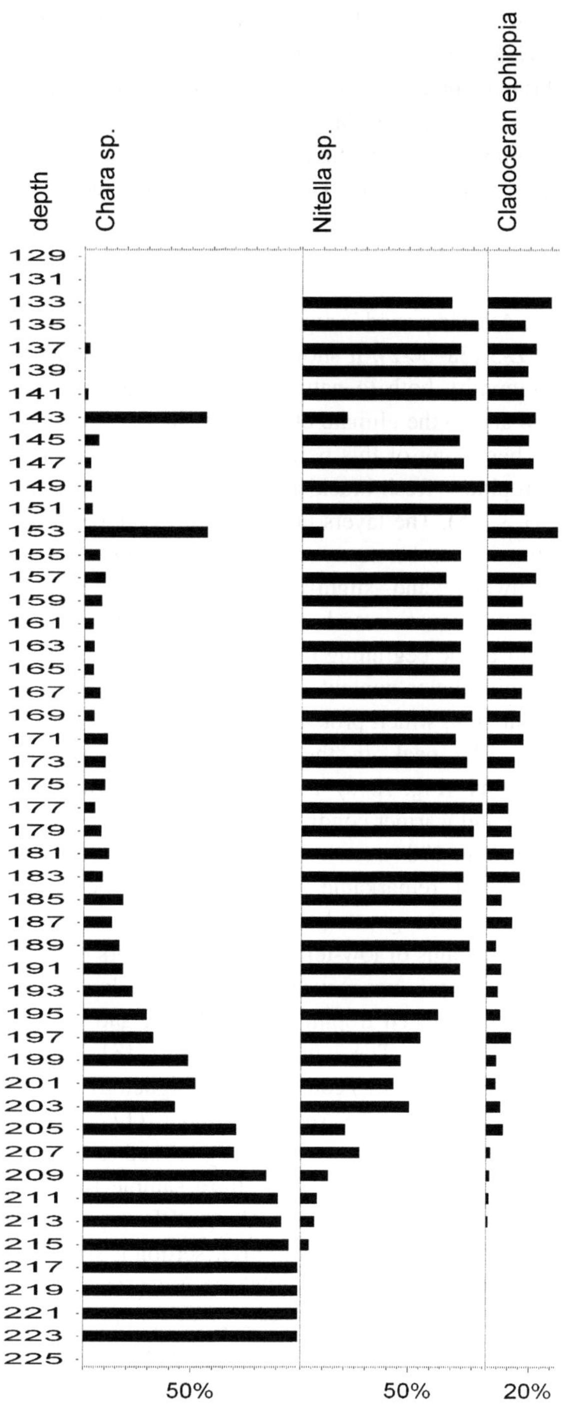

Fig. 6 Relative frequency diagram of charophytes and cladoceran ephippia found in the sediments of the former lake Velanská Cesta

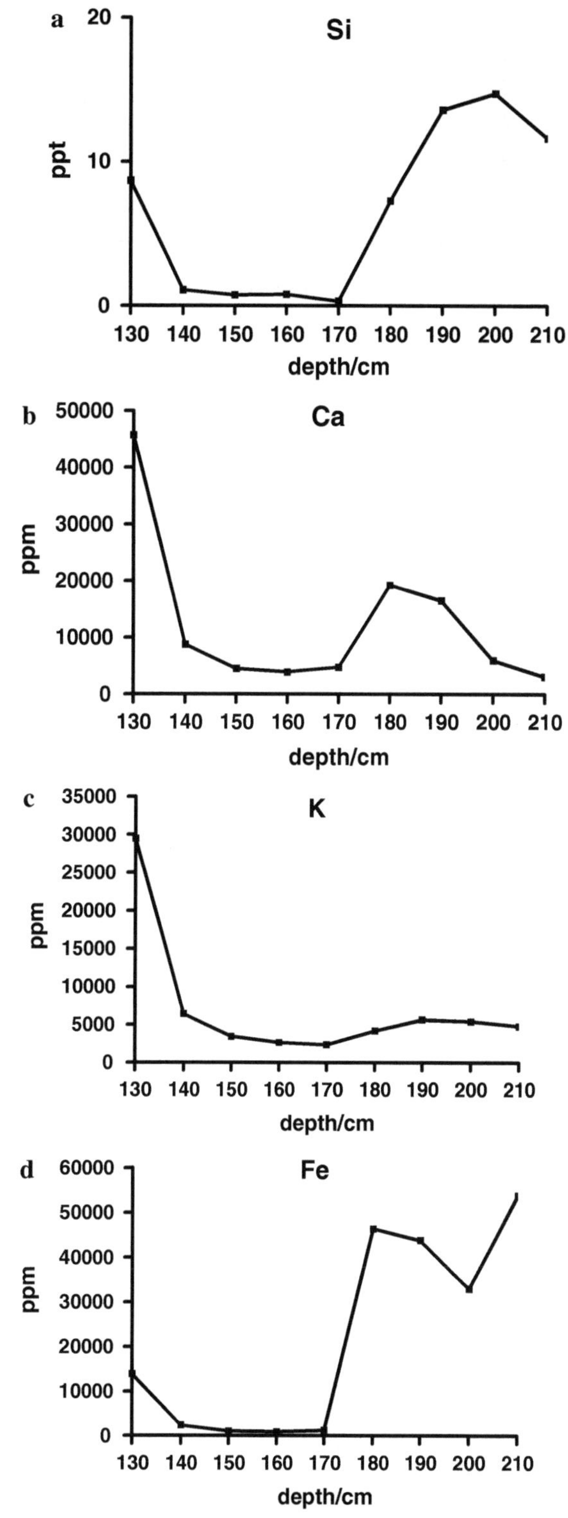

Fig. 7 Concentration of elements from the sediments of the former lake Velanská cesta expressed as ppm of minerogenic fraction

to Bölling. Contamination of sandy Bölling samples, which are characteristic in having a very low content of organic matter (Fig. 8), by remains of aquatic

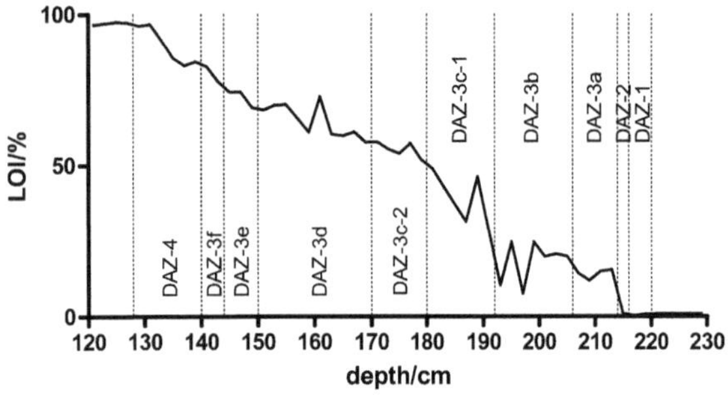

Fig. 8 Loss-on-ignition measured after igniting at 550°C for 4 h

organisms from upper layers is not likely due to a very different composition of the diatom community in Bölling sediments compared with later periods (Fig. 2).

The genesis of the lake was probably connected to an early rapid warming at the beginning of the Bölling interstadial. A shift to damper and warmer conditions may have enabled the formation of a shallow water body in the natural depression of the Novohradské Hory mountains, supplied by groundwater rising along a geological fault and/or from a small catchment area. Ecological range of diatom species suggests an alkaline to circumneutral, oligo-meso to mesotrophic (T was preferred in the estimation of general trophic conditions due to the high number of non-analogous TP samples), and fresh to fresh-brackish environment. Deposition of organic matter was the lowest in the lake history (Fig. 8).

Older Dryas

One sample (215 cm), forming DAZ-2, was assigned to the Older Dryas interval. Previous studies revealed that very low concentration of diatom frustules, and an increase in the trophic state (Rippey et al., 1997) and the pH (Bradbury, 1986; Webster et al., 1990; Carvalho & Moss, 1999) can indicate colder and dryer events in freshwater open basin lake ecosystems. As evident from Fig. 1, T index and H index abruptly increased, while valve counts decreased during this period documenting a strong response of the lake ecosystem to the changing conditions.

The Older Dryas period interval is sometimes disregarded as a period of minor importance in the climatological record (e.g.,Rasmussen et al., 2006), whereas in other studies, it is considered an important Late Glacial period (e.g., Stuvier et al., 1995; Pokorný & Jankovská, 2000). Our results, which show abrupt changes that are unique in the whole history of the lake, suggest that Older Dryas was the most dramatic event archived in the continuous late Pleistocene and early Holocene lacustrine record.

Alleröd

The Alleröd period is represented by four subzones of DAZ-3 (3a–3c-2). It shows a response of the diatom community both to natural oligotrophization of the lake and to the climate change within this period. At the beginning of this period, the lake appears meso-eutrophic, fresh-brackish, and slightly alkaline (pH ~7.5). The layers preceding the Younger Dryas interval can be summarized as oligo-mesotrophic, freshwater, and slightly acidic (pH ~6.5). Most vascular aquatics and cladoceran ephippia start to occur at the beginning of this period. The increasing organic matter content documents a higher primary production, which probably led to the infilling of the lake. The peak in the concentration of selected elements (Fig. 7a–d) also matches well with the moist and warmer conditions of the period (Engström & Wright, 1984).

Several remarkable fluctuations can be distinguished in the general trend outlined above. Macrofossil remains of low-temperature indicators (*Nuphar pumila*, *Betula nana*) were found in three layers with locally enhanced trophic state, salinity, and pH (see Figs. 1, 3). The first sample (181 cm—ca. 13,530 cal year BP) could not be assigned to a colder fluctuation, but the second sample (177 cm—ca. 13,270 cal year BP) was probably related to the intra Alleröd cold period, and the third sample from this zone (171 cm—ca. 12,890 cal year BP) signaled the upcoming Younger Dryas. This assertion is supported particularly by the shape of the T index (see Fig. 1) curve in context with GISP2 δ^{18}O, although some delay is apparent. This lag was probably caused by the nature of floating chronology and the coarse interval of sampling (2 cm).

During this period, *Chara* sp. was replaced by *Nitella* sp. (Fig. 6). This corresponds to a steep decrease in the TP content (Fig. 4) and to fluctuations in LOI (Fig. 8). A similar, but inverse, process driven by artificial eutrophication was observed in the Swiss Alps (Tobolski & Amman, 2000).

Younger Dryas

The Younger Dryas period is linked with the subzone DAZ-3d. At the beginning of this period, the trophic state and salinity increased. Later on (ca. 12,300 cal year BP), it started declining, and declined until the beginning of the Holocene. A similar pattern was also observed in the pH curve. Macrofossil low-temperature indicators (*Nuphar pumila* and *Betula nana*) were found in samples corresponding to these shifts (Fig. 1). A steep decrease in the concentrations of Si, Ca, K, and Fe is also apparent (Fig. 7a–d) at the beginning of the period. This suggests a response of the lake ecosystem to colder and dryer conditions of the Younger Dryas. The disappearance of low-temperature indicators beginning with the sample 163 cm (ca. 12,350 cal year BP), accompanied by a decrease in the nutrient level, salinity, and pH, suggests a shift toward a warmer climate. Warmer conditions during the later phase of Younger Dryas, by approximately 1–2°C when compared with the start of the period, were documented in studies based on macrofossils and pollen (e.g., Isarin & Bohncke, 1999) and beetle (Berglund et al., 1994) data across Europe. Results from stable isotope studies are in agreement with these suggestions (e.g., Stuvier et al., 1995).

Preboreal

The last three intervals (subzones DAZ-3e, DAZ-3f, and zone DAZ-4) were assigned to the Preboreal period. The start of the Holocene is traditionally described as an abrupt change to moister and warmer conditions (e.g., Stuvier et al., 1995) throughout Europe, which also affected our study area (e.g., Pokorný & Jankovská, 2000). A sharp decline in the abundance of diatoms indicates a change in the trophic state and salinity in the boundary sample (152 cm), which corresponds with the assumption of lower nutrient supply during warmer periods (Rippey et al., 1997). The climatic amelioration at the locality is characterized by the rising abundance of *Potamogeton obtusifolius* Mert. et W. D. J. Koch, and warmer conditions are demonstrated by the occurrence of *Typha latifolia* L., correlated with minimum July temperature of 14°C (Isarin & Bohncke, 1999).

The following colder Preboreal Oscillations have been described in numerous paleoclimatological studies (e.g., Rasmussen et al., 2006). However, data from the area of the Czech Republic are missing in these publications. The peaks found in the trophic state and salinity dated to these colder events probably document another response of the lake ecosystem to the climatic fluctuations.

Expansion of *Phragmites australis* (Cav.) Steud., *Typha* sp., *Schoenoplectus lacustris* (L.) Palla, and the genus *Carex*, represented by *C. lasiocarpa* Ehrh., *C. limosa* L., and *C. paniculata* L., illustrates the final terrestrialization. The natural infilling process terminated within the samples 139–133 cm (ca. 10,420–10,260 cal year BP). The last layers of the zone show a disappearance of indicators of an aquatic environment (cladoceran ephippia, charophytes, and hydrophytes). Most indicators disappeared at 133 cm and diatom valves appeared in countable concentrations to 129 cm where LOI reached a maximum, and a peak in selected elements indicated terrestrial conditions.

Conclusion

The documented evolution of a shallow lake at the locality of Velanská Cesta illustrates the response of a lacustrine ecosystem to climatic forcing during the late Pleistocene–early Holocene transition. Diatom record and inferred variables (Index of Trophic state and Halobic index, TP, pH), supported by results of macrofossil analysis and geochemistry data, show a remarkably quick response to climatic shifts at the background of the natural infilling process of the lake. Diatom accumulation zones (DAZ), distinguished on the basis of cluster analysis, correspond clearly with differing climatic periods and consequent different nutrient supply and salinity. The earliest evidence of an aquatic environment was recorded after the initial warming at the beginning of the Bölling interstadial. The most abrupt change, related to strong eutrophication and a rise in salinity and pH, appeared at the onset of the Older Dryas period. Diatom-inferred variables reverted to the Bölling and lower levels during Alleröd interstadial. Within this reversal, local peaks in trophic state and salinity were observed. Findings of macrofossil low-temperature indicators in the corresponding samples suggest their assignment to intra Alleröd short term colder events. Similarly, the initial colder period of Younger Dryas is indicated by the presence of low-temperature indicators and enhanced trophic state and salinity. This correlation, without the support of low-temperature macrofossil indicators, was repeated during the Preboreal oscillations for the last time. The natural infilling process of the lake finished around 10,300 cal year BP, and the lacustrine sediments were then covered by peat.

Acknowledgments The authors thank Jan Kaštovský, Václav Houk, Petr Marvan, and Jiří Vrba for advice, Eva Kaštovská for help with LOI measurement, Jan Chlumský for graphical advice, Marek Stibal for language corrections, the collective of EDDi for the free availability of transfer functions, and numerous persons for inspirative opinions. The research was funded by The Grant Agency of the Academy of Sciences of the Czech Republic grant KJB 6141405, and the Ministry of Education of the Czech Republic grant MSM 600 766 5801.

References

Amman, B. & A. F. Lotter, 1989. Late-glacial radiocarbon and palynostratigraphy of the Swiss Plateau. Boreas 22: 171–188.

Anderberg, A. L., 1981. Atlas of Seeds and Small Fruits of Northwestern-European Plant Species (Sweden, Norway, Denmark, East Fennoscandia and Iceland) with Morphological Descriptions, Part 4: Resedaceae–Umbelliferae. Swedish Museum of Natural History, Stockholm: 1–281.

Arzet, K., 1987. Diatomeen als pH-Indikatoren in subrezenten Sedimenten von Weichswasser-seen. Disertationen aus der Abteilung für Limnologie des Institut für Zoologie der Universität Innsbruck 24: 1–266.

Arzet, K., 1993. Diatoms. In Schmidt, R. (ed.), Acidification of Bohemian Lakes. Recent Trend and Historical Development. Závěrečná zpráva programu Ost-West [A Final Report from the Ost-West Program]: 24–32.

Battarbee, R. W., 1986. Diatom analysis. In Berglund, B. E. (ed.), Handbook of Holocene Paleoecology and Paleohydrology. John Wiley & Sons Ltd., Chichester: 527–570.

Bennion, H., S. Juggins & N. J. Anderson, 1996. Predicting epilimnetic phosphorus concentrations using an improved diatom-based transfer function and its application to lake eutrophication management. Environmental Science and Technology 30: 2004–2007.

Berglund, B. E., S. Björck, G. Lemdahl, H. Bergsten, K. Nordberg & E. Kolstrup, 1994. Late Weichselian environmental change in southern Sweden and Denmark. Journal of Quaternary Sciences 9: 127–132.

Bergren, G., 1969. Atlas of Seeds and Small Fruits of Northwestern-European Plant Species (Sweden, Norway, Denmark, East Fennoscandia and Iceland) with Morphological Descriptions, Part 2: Cyperaceae. Swedish Natural Science Research Council, Stockholm: 1–68.

Bergren, G., 1981. Atlas of Seeds and Small Fruits of Northwestern-European Plant Species (Sweden, Norway, Denmark, East Fennoscandia and Iceland) with Morphological Descriptions, Part 3: Salicaceae–Cruciferae. Swedish Museum of Natural History, Stockholm: 1–260.

Bigler, Ch., R. I. Hall & I. Rengberg, 2000. A diatom-training set for palaeoclimatic inferences from lakes in northern Sweden. In Williams, W. D. (ed.), Verhandlungen der Internationale Vereinigung für Theoretische und Angewandte Limnologie 27. Gantner Verlag KG, Ruggel: 1174–1182.

Bradbury, J. P., 1986. Late Pleistocene and Holocene paleolimnology of two mountain lakes in western Tasmania. Palaios 1: 381–388.

Břízová, E., 2002. Paleoekologický výzkum bývalých jihomoravských jezer [Paleoecological research of the former south Moravian lakes]. In Kirchner, K. & P. Roštínský (eds), Geomorfologický sborník 1, Masarykova univerzita, [Geomorfological Proceedings 1, Masaryk University] Brno.

Cappers, R. T. J., 1994. The identification of Potamogetonaceae fruits found in the Netherlands. In Cappers, R. T. J. (ed.), An Ecological Characterization of Plant Macro-Remains of Heveskesklooster (the Netherlands), A Methodical Approach. Rijksuniversiteit Gröningen: 75–161.

Carvalho, L. & M. Moss, 1999. Climate sensitivity of Oak Mere: a low altitude acid lake. Freshwater Biology 42: 585–591.

Dąmbska, I., 1964. Flora słodkowodna Polski, Charophyta–Ramienice [Freshwater Flora of Poland, Charophyta–Stoneworts]. Państwowe Wydawnictwo Naukowe [Polish Scientific Publishers PWN], Warszawa: 1–126.

EDDI home page. http://craticula.ncl.ac.uk/Eddi/jsp/index.jsp.

Engström, D. R. & H. E. Wright Jr., 1984. Chemical stratigraphy of lake sediments as a record of environmental change. In Haworth, E. Y. & J. W. G. Lund (eds), Lake sediments and environmental history. University of Minnesota Press, Minneapolis, MN: 11–68.

Fott, B., 1954. Pleurax, synthetická pryskyřice pro preparaci rozsivek [Pleurax, a synthetic resin for diatom preparation]. Preslia 26: 163–194.

Haas, J. N., 1994. First identification key for charophyte oospores from central Europe. European Journal of Phycology 29: 227–235.

Houk, V., 2003. Atlas of freshwater centric diatoms with a brief key and descriptions—Part 1 Melosiraceae, Orthoseiraceae, Paraliaceae and Aulacoseiraceae. Czech Phycology Supplement 1: 1–27.

Husák, S. & S. Hejný, 1978. General characteristic of the Trebon Basin and Lednice region. In Dykyjová, D., & J. Květ (eds), Pond Littoral Ecosystems, Ecological Studies 28. Springer-Verlag: 136–168.

Isarin, R. F. B. & S. J. P. Bohncke, 1999. Mean July temperatures during the Younger Dryas in Northern and Central Europe as inferred from Climate Indicator Plant Species. Quaternary Research 51: 158–173.

Jankovská, V., 1980. Paläogeobotanische Rekonstruktion der Vegetationsenwicklung im Becken Třeboňská pánev während des Spätglazials und Holozäns. Vegetace ČSSR A11, Academia, Praha: 1–149.

Jankovská, V., 2006. Late Glacial and Holocene history of Plešné Lake and its surrounding landscape based on pollen and palaeoalgological analyses. Biologia 61: 371–385.

Jankovská, V., 2007. Giant Mountains and pollen analytical research: new results and interesting palaeobotanical findings. In Štursa, J., & R. Knapik (eds), Geologické problémy Krkonoš, sborník mezinárodní vědecké konference 2006 [Geoecological Problems of the Giant Mountains, Proceedings of the International Conference 2006]. Svoboda n. Úpou. Opera Corcontica 44: 227–242.

Kac, J. N., C. V. Kac & M. G. Kipiani, 1965. Atlas i opredeliteľ plodov i semyan vstrechayushchikhsyza v chetvertichnykh otlozheniyakh SSSR [An Atlas and a Key to Fruits and Seeds Found in Quaternary Deposits of Soviet Union]. Nauka, Moskow.

Kolibáčová, S., P. Maděra & L. Úradníček, 2001. Dřeviny rašelinišť a reliktních borů ČR [Woody species of peat bogs and relictus pine woods in the area of the Czech Republic]. Živa [Life] 2: 57–59.

Körber-Grohne, U., 1991. Bestimmungsschlüssel für subfossile Gramineen-Früchte (Identification key for subfossil *Gramineae* fruits). Probleme der Küstenforschung im südlichen Nordseegebiet 18: 169–234.

Krammer, K., 1997a. Die Cymbelloiden Diatomeen. Eine Monagraphie der weltweit bekannten Taxa. Teil 1. Allgemeines und *Encyonema* part. Bibliotheca Diatomologica Band 36. J. Cramer, Stuttgart.

Krammer, K., 1997b. Die Cymbelloiden Diatomeen. Eine Monographie der weltweit bekannten Taxa. Teil 2. *Encyonema* part., *Encyonopsis* und *Cymbellopsis*. Bibliotheca Diatomologica Band 37. J. Cramer, Stuttgart: 1–469.

Krammer, K., 2000. Diatoms of Europe. Diatoms of European Inland Waters and Comparable Habitats. Vol. 1. The Genus *Pinnularia*. A.R.G. Gantner Verlag K.G., Ruggel.

Krammer, K., 2002. Diatoms of Europe. Diatoms of the European Inland Waters and Comparable Habitats. Vol. 3. *Cymbella*. A.R.G. Gantner Verlag K.G., Ruggell.

Krammer, K., 2003. Diatoms of Europe. Diatoms of the European Inland Waters and Comparable Habitats. Vol. 4. *Cymbopleura, Delicata, Navicymbulla, Gomphocymbellopsis, Afrocymbella*. A.R.G. Gantner Verlag K.G., Ruggell.

Krammer, K. & H. Lange-Bertalot, 1986. *Bacillariophyceae* (*Naviculaceae*). In Ettl, H., J. Gerloff, H. Heinig & D. Mollenhauer (eds), Süsswasserflora von Mitteleuropa, (2 (1)). Gustav Fisher Verlag, Stuttgart.

Krammer, K. & H. Lange-Bertalot, 1988. *Bacillariophyceae* (*Bacillariaceae, Epithemiaceae, Surirellaceae*). In Ettl, H., J. Gerloff, H. Heinig & D. Mollenhauer (eds), Süsswasserflora von Mitteleuropa, (2 (2)). Gustav Fisher Verlag, Stuttgart.

Krammer, K. & H. Lange-Bertalot, 1991a. *Bacillariophyceae* (*Centrales, Fragilariaceae, Eunotiaceae*). In Ettl, H., J. Gerloff, H. Heinig & D. Mollenhauer (eds), Süsswasserflora von Mitteleuropa, (2 (3)). Gustav Fisher Verlag, Stuttgart.

Krammer, K. & H. Lange-Bertalot, 1991b. *Bacillariophyceae* (*Achnanteaceae*, Kritische Ergänzungen zu *Navicula* (*Lineolateae*) und *Gomphonema* Gesamtliteraturverzeichnis). In Ettl, H., J. Gerloff, H. Heinig & D. Mollenhauer (eds), Süsswasserflora von Mitteleuropa, (2 (4)). Gustav Fisher Verlag, Stuttgart.

Lange-Bertalot, H., 1996. Iconographia Diatomologica. Annotated Diatom Micrographs. Vol. 2. Indicators of Oligotrophy. 800 Taxa Representative of Three Ecologically Distinct Lake Types. Koeltz Scientific Books, Konigstein.

Latałowa, M., 1999. Palaeoecological reconstruction of the environmental conditions and economy in Early medieval Wolin against a background of the Holocene history of the landscape. Acta Palaeobotanika 39: 183–271.

Mądalski, J., 1949. Cechy morfologiczne pestek europejskich gatunków Potamogeton (Tourn.) L. i klucz do oznaczenia ich szczątków dyluwialnych [Morphological classes of seeds of the genus *Potamogeton* (Tourn.) L. and a key to their diluvial macrofossil remains]. Prace wrocławskiego towarzystwa naukowego [Wrocław Scientific Society] 24: 11–23.

Marjatta, A., 1970. Potamogetonaceae fruits. I. Recent and subfossil endocarps of the fennoscandian species. Acta Botanica Fennica 88: 3–85.

Nalepka, D. & A. Walanus, 2003. PolPal—Program for counting pollen grains, diagrams plotting and numerical analysis. Acta Paleobotanica Supplementum 2: 659–661.

Petr, L. & P. Pokorný, 2008. Přirozená jezera na území české republiky. Jejich význam pro studium pravěkého osídlení a přírodního prostředí [Natural lakes in the Czech Republic. Their importance for the study of prehistoric occupation and environment]. In Beneš, J. & P. Pokorný (eds), Bioarcheologie v České republice, Jihočeská univerzita, České Budějovice a Archeologický ústav akademie věd České republiky, Praha [Bioarcheology in the Czech Republic, University of South Bohemia, České Budějovice & Institute of Archeology of the Academy of Sciences of the Czech Republic, Prague]: 73–98.

Pokorný, P. & V. Jankovská, 2000. Long-term vegetation dynamics and the infilling process of a former lake (Švarcenberk, Czech Republic). Folia Geobotanica 35: 433–457.

Ramsey, C., 2005. Oxford Radiocarbon Accelerator Unit. http://c14.arch.ox.ac.uk/embed.php?File=oxcal.html.

Rasmussen, S. O., K. K. Andersen, A. M. Svensson, J. P. Steffensen, B. M. Vinther, H. B. Clausen, M. L. Siggaard-Andersen, S. J. Johnsen, L. B. Larsen, D. Dahl-Jensen, M. Bigler, R. Rothlisberger, H. Fischer, K. Goto-Azuma, M. E. Hansson & U. Ruth, 2006. A new Greenland ice core chronology for the Last Glacial termination. Journal of Geophysical Research-Atmospheres 111: D06102.

Řeháková, Z., 1986. The Postglacial history of diatom-bearing sediments of the former Lake Komořany (North-West Bohemia). In Tyráček, J. (ed.), Journal of Geological Sciences. Anthropozoic 17: 87–134.

Řeháková, Z., 1988. Biostratigraphy and palaeoecology of diatom-bearing sediments within the Soos basin in western Bohemia, Czechoslovakia. 10th Diatom-Symposium 1988: 407–418.

Rippey, B., N. J. Anderson & R. H. Foy, 1997. Accuracy of diatom-inferred total phosphorus concentrations and the accelerated eutrophication of a lake due to reduced flushing. Canadian Journal of Fisheries and Aquatic Sciences 54: 2637–2646.

Soják, J., 1958. Klíč k určení plodů našich Cyperaceí (excl. Carex) [A key to fruits of Cyperacea of the Czech Republic (excl. Carex)]. Preslia 30: 43–58.

Stuvier, M., P. M. Grootes & T. Braziunas, 1995. The GISP δ^{18}O climate record of the past 16,500 years and the role of the sun, ocean and volcanoes. Quaternary Research 44: 341–354.

Tátosová, J., J. Veselý & E. Stuchlík, 2006. Holocene subfossil chironomid stratigraphy (Diptera: Chironomidae) in the sediment of Plešné Lake (the Bohemian Forest, Czech Republic). Biologia 61: 401–411.

The Greenland Summit Ice Cores CD-ROM, 1997. Available from the National Snow and Ice Data Center, University of Colorado at Boulder, and the World Data Center-A for Paleoclimatology, National Geophysical Data Center, Boulder, Colorado, USA.

Tobolski, K. & B. Amman, 2000. Macrofossils as records of plant responses to rapid Late Glacial climatic changes at three sites in the Swiss Alps. Palaeogeography, Palaeoclimatology, Palaeoecology 159: 251–259.

Van Dam, H., A. Martens & J. Sinkeldam, 1994. A coded checklist and ecological indicator values of freshwater diatoms from the Netherlands. Netherlands Journal of Aquatic Ecology 28: 117–133.

Wasylikowa, K., 1986. Analysis of fossil fruits and seeds. In Berglund, B. E. (ed.), Handbook of Holocene Palaeoecology and Palaeohydrology. John Wiley & Sons Ltd., Chichester: 571–590.

Watts, W. A., 1978. Plant macrofossils and Quaternary paleoecology. In Walker D. & J. C. Guppy (eds), Biology and Quaternary Environments. Australian Academy of Science: 53–67.

Webster, K. E., A. D. Newell, L. A. Barker & D. P. L. Brezonik, 1990. Climatically induced rapid acidification of a softwater seepage lake. Nature 347: 374–376.

Weckström, J., A. Korhola & T. Blom, 1997. Diatoms as quantitative indicators of pH and water temperature in subarctic Fennoscandian lakes. Hydrobiologia 347: 171–184.

Zelinka, M. & P. Marvan, 1963. Porovnani metod saprobialniho hodnoceni vody [Comparisson of methods for saprobial evaluation of water]. Vodní hospodářství [Aquatic Economy] 13: 291–293.

A multi-proxy Late-glacial palaeoenvironmental record from Lake Bled, Slovenia

Maja Andrič · Julieta Massaferro · Ueli Eicher · Brigitta Ammann · Markus Christian Leuenberger · Andrej Martinčič · Elena Marinova · Anton Brancelj

Originally published in the journal Hydrobiologia, Volume 631, No. 1, 121–141.
DOI: 10.1007/s10750-009-9806-9

Abstract This study investigates the palaeoecological record ($\delta^{18}O$, $\delta^{13}C$, pollen, plant macrofossils, chironomids and cladocera) at Lake Bled (Slovenia) sedimentary core to better understand the response of terrestrial and aquatic ecosystems to Late-glacial climatic fluctuations. The multi-proxy record suggests that in the Oldest Dryas, the landscape around Lake Bled was rather open, presumably because of the cold and dry climate, with a trend towards wetter conditions, as suggested by an increase in tree pollen as well as chironomid and cladocera faunas typical for well-oxygenated water. Climatic warming at the beginning of the Late-glacial Interstadial at ca. 14,800 cal yr BP is suggested by an increase in the $\delta^{18}O$ value, the appearance of *Betula* and *Larix* pollen and macrofossils, and a warmth-adapted chironomid fauna. With further warming at ca. 13,800 cal yr BP, broad-leaved tree taxa (*Quercus, Tilia, Ulmus*), *Artemisia*, and *Picea* increase, whereas chironomid data (*Cricotopus B*) suggest lowering of lake levels. After 12,800 cal yr BP (and throughout the Younger Dryas), the climate was colder and drier, as indicated by lower $\delta^{18}O$ values, decline of trees,

Electronic supplementary material The online version of this article (doi:10.1007/s10750-009-9806-9) contains supplementary material, which is available to authorised users.

Guest editors: K. Buczkó, J. Korponai, J. Padisák & S. W. Starratt
Palaeolimnological Proxies as Tools of Environmental Reconstruction in Fresh Water

M. Andrič (✉)
Institute of Archaeology, Scientific Research Centre of the Slovenian Academy of Sciences and Arts, Novi trg 2, P.B. 306, 1001 Ljubljana, Slovenia
e-mail: maja.andric@zrc-sazu.si

J. Massaferro
Lab de Biodiversidad Darwin INIBIOMA/CONICET, San Martin 24, 8400 Bariloche, RN, Argentina
e-mail: julimassaferro@crub.uncoma.edu.ar; julimassaferro@hotmail.com

U. Eicher · M. C. Leuenberger
Climate and Environmental Physics, Physics Institute, University of Bern, Sidlerstrasse 5, 3012 Bern, Switzerland

U. Eicher
e-mail: info@dialogpersonal.ch

B. Ammann
Institute of Plant Sciences, University of Bern, Altenbergrain 21, 3013 Bern, Switzerland
e-mail: brigitta.ammann@ips.unibe.ch

B. Ammann · M. C. Leuenberger
Oeschger Centre for Climate Change Research, University of Bern, Erlachstrasse 9a, 3012 Bern, Switzerland

M. C. Leuenberger
e-mail: leuenberger@climate.unibe.ch; info@oeschger.unibe.ch

increase of microscopic charcoal, xerophytes and littoral chironomids. A warmer climate, together with the spread of broad-leaved tree taxa and a deeper, more productive lake, mark the onset of the Late-glacial/Holocene transition. These results suggest that terrestrial and aquatic ecosystems at Lake Bled were very dynamic and sensitive to Late-glacial climatic fluctuations.

Keywords Late-glacial Lake Bled · Chironomids · Cladocera · Plant macrofossils · Stable isotopes · Pollen

Introduction

A new interest has developed in the rapid climatic shifts of the Late glacial of the last glaciation because these shifts were so rapid and of so large amplitude. Even if conditions were very different from the present in several respects, and if direct analogues to future climatic warming may not be correct in detail, we may learn about the behaviour of the systems from the past. This includes both the climatic system and various ecosystems. These data on past changes may also be useful to test models.

The study area is situated at a crossroad between south-central Europe, the southeastern Alps and the Balkan Peninsula. Several biogeographically important issues may therefore be tackled from a detailed study of the Late glacial, e.g. biogeographical questions about the position of refugia of plants and invertebrates and their migration routes. Palaeobotanical and genetic studies suggest that glacial refugia for coniferous and some deciduous trees were located in the Iberian, Italian and Balkan Peninsulas and in parts of central and eastern Europe (e.g. east of the Alps, Hungarian Plain and River Danube), with Slovenia being located in the vicinity or in central part of the suggested refuge areas (e.g. Willis et al., 2000; Petit et al., 2003; Willis & van Andel, 2004). For example, it has been suggested that eastern Alps was one of the most important refugia for *Fagus sylvatica* (Magri et al., 2006), and it is interesting that the study site lies between ranges of two distinctive haplotypes of *Pinus sylvestris* (Cheddadi et al., 2006).

Previous palynological research in Slovenia suggests that at the beginning of the Late glacial, the vegetation of Slovenia consisted of predominantly herbaceous steppe with few pine trees, but with the climatic warming in the Late glacial, Interstadial woodlands (*Pinus, Betula, Picea, Quercus, Corylus, Fagus, Tilia, and Ulmus*) were expanding (Culiberg, 1991). However, despite intensive palynological research in the area, studies including radiocarbon-dated multi-proxy analysis of the sediment are very rare. This article presents the results of such multidisciplinary palaeoecological research on Late-glacial sediment at Lake Bled (Slovenia) to better understand changes of Late-glacial vegetation, climate and hydrology.

Earlier research at Lake Bled focused on studies of the modern carbon budget (e.g. $\delta^{13}C$ and $\delta^{15}N$ analyses) and pollen records from short (ca. 45 cm) cores (Molnar et al., 1978; Dolenec et al., 1984; Ogrinc et al., 1997, 1998, 2002; Lojen et al., 1997, 1999). The results of pollen analysis for the last ca. 400 years indicate a forest dominated by *Fagus* and increasing evidence of human pressure (grazing and agriculture, suggested by *Fagus* decline and an increase of cereals, buckwheat and maize), followed by diminished human impact indicated by an increase of tree pollen, especially *Pinus* (Molnar et al., 1978).

In 1982, a 675-cm-long sediment core was taken from the deepest (eastern) part of the basin (Fig. 1), and a multi-proxy study was carried out. Results from analysis of Ostracoda, mosses, sedimentary pigments and pollen (Löffler, 1984; Schultze, 1984, 1988) suggested a Late-glacial age (from the bottom to 550 cm), with a pollen assemblage dominated by *Artemisia*, *Pinus* and *Betula*. Subsequently, the vegetation shifted to a mixed oak forest, with *Fagus* showing up at the beginning of the Holocene. The

A. Martinčič
Zaloška cesta 78A, 1000 Ljubljana, Slovenia
e-mail: andrej.martincic@siol.net

E. Marinova
Center for Archaeological Sciences, Katholike Universiteit Leuven, Celestijnenlaan 200E, 3001 Leuven, Belgium
e-mail: elena.marinova@bio.kuleuven.be

A. Brancelj
National Institute of Biology, Večna pot 111, P.B. 141, 1101 Ljubljana, Slovenia
e-mail: anton.brancelj@nib.si

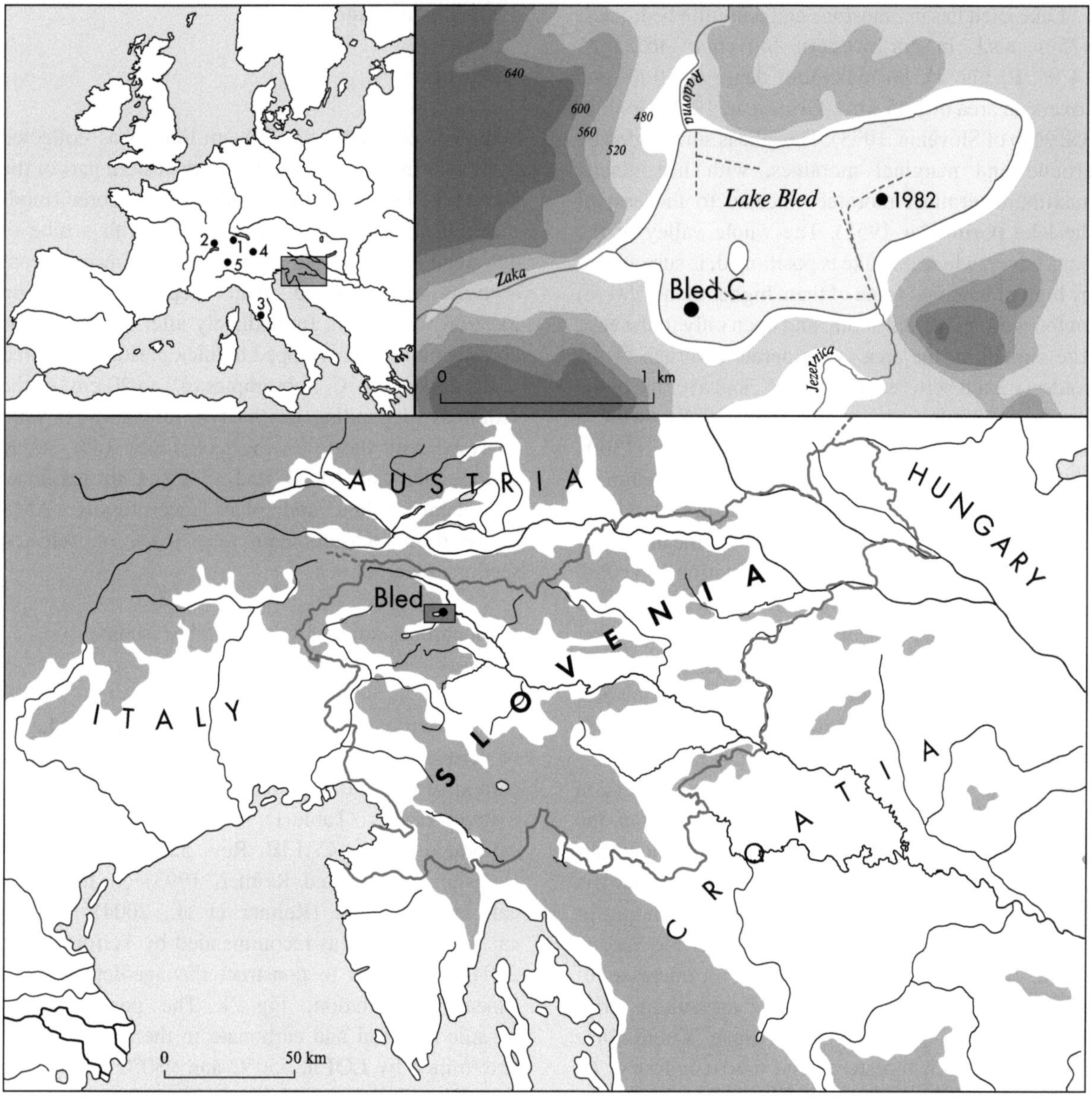

Fig. 1 The geographical position of Lake Bled (Slovenia) and the sediment core. The location of Olszewski tubes is marked by *dotted lines*. • 1982 marks the position of the sediment core collected by Schultze (1984). Study sites, mentioned in Table 3 are as follows: 1—Gerzensee, 2—Lake Lautrey, 3— Lake Accesa, 4—Lago di Origlio, 5—Lago Piccolo di Avigliano (study sites on map)

information about the vegetation during the Holocene at Lake Bled is vague, because sampling resolution in the upper section of the core was rather low (10-cm interval, no pollen was analysed above 470 cm), and the sequence was not radiocarbon dated. Despite the attempt to reconstruct Lake Bled environmental history using a multi-proxy approach, the lack of high sampling resolution and chronological control made it unfeasible. Hence in 2002, to complement the previous study, a new 521-cm sediment core was collected in the southwestern part of the lake and studied to determine the history of the vegetation, hydrology and climate. An age model was constructed using AMS radiocarbon dating and analysis of loss-on-ignition (LOI), stable isotopes ($\delta^{18}O$, $\delta^{13}C$), geochemistry ($CaCO_3$), plant macrofossils, mosses, pollen, chironomids and cladocerans. Results of this research are presented in this article.

Lake Bled lies in limestone and dolomite bedrock at 475 m a.s.l. in northwestern Slovenia (46°22′ N, 14°06′ E, Fig. 1). Its maximum depth is 30 m, and covers an area of 1.45 km^2 (Grimšičar, 1955; Geological Map of Slovenia, 1993). The lake is surrounded by ground and marginal moraines, with last glacial maximum terminal moraines located to the east of the lake (Grimšičar 1955). The whole valley of the Sava River, where the lake is positioned, is surrounded by high mountains (some of them higher than 2500 m) on the west, north and south, and open only to the east. The climate of the area is temperate (average of the coldest month between 0 and −3°C, and average of the warmest month between 15 and 20°C), with a submediterranean precipitation regime (1300–2800 mm annual precipitation with a maximum in autumn) (Ogrin, 1996). In Lesce, ca. 8 km east of Lake Bled, mean July temperature is 18°C, mean annual temperature 8.4°C and annual precipitation 1541 mm (Pak, 2001). Forest of predominant beech (*Fagus sylvatica*) grows to the west and north of the lake, whereas patchy woodlands of hornbeam (*Carpinus betulus*) and willow (*Salix* sp.) are located to the east (Čarni et al., 2003). Meadows and agricultural fields are also located in the vicinity of the lake.

Dimictic Lake Bled developed in a depression shaped by a glacier (Radinja et al., 1987). Today, the lake receives water from small natural tributaries (e.g. Zaka River) and artificial inflows (e.g. Radovna River) as well as precipitation. At the beginning of the twentieth century and especially after the Second World War, the lake experienced an increase of nutrient loading from sewage and agriculture, triggering blooms of blue-green algae *Oscillatoria rubescens*, which threatened the tourist industry. As a consequence, in 1964, clean water from the Radovna River was artificially diverted to the lake (2000 l s^{-1}), and later on in 1980 and 1982, the basin was drained by Olszewski tubes, and domestic sewage was rerouted from the lake (Fig. 1; Radinja et al., 1987). Pumping of hypolimnetic cold water from the bottom of the lake resulted in decrease of nutrients in the water column, and finally in the disappearance of *Oscillatoria rubescens* (Vrhovšek et al. 1982, 1984, 1985). After several years of anoxic conditions below a depth of 15–18 m during thermal stratification in late summer and late winter, oxygen conditions improved. Nowadays, only short-lasting anoxic events occurred from time to time.

Materials and methods

Fieldwork

Sediment core Bled C (521-cm-long) was collected from a coring platform in the southwestern part of the lake (Fig. 1) with a Livingstone piston corer (modification after Merkt & Streif, 1970) with a tube of 4.8 cm diameter. A single core with 1-m-long core sections was analysed. Water depth at the coring location was 9.6 m. Immediately after extrusion, the core sections were wrapped in thick plastic and stored in the dark at +4°C. Subsamples of 1 to 20 cm^3 of the sediment (depending on the type of analysis) were analysed with the following techniques: LOI, stable isotope analysis ($\delta^{18}O$ and $\delta^{13}C$), Chironomidae, Cladocera, pollen and plant macrofossils. AMS radiocarbon measurements were made on selected sections (see below).

Radiocarbon dating, LOI and stable isotope analyses

The age was determined by AMS radiocarbon dating of terrestrial plant macrofossils and bulk organic carbon, extracted from the sediment at 40, 60 and 210 cm depths (Table 1). Conventional ages were calibrated using CALIB Rev 5.0.1 (CALIB 5.0 Website; Stuiver and Reimer, 1993) on IntCal 04 calibration dataset (Reimer et al., 2004). Median cal yr BP values (as recommended by Telford et al., 2004) were used to construct the age-depth model (linear interpolation; Fig. 2). The percentage of organic material and carbonate in the sediment was determined by LOI at 550°C and 950°C (Bengtsson and Ennell, 1986). The isotopic composition of oxygen ($\delta^{18}O$) and carbon ($\delta^{13}C$) on bulk carbonate was measured following the conventional method described by Siegenthaler & Eicher (1986).

Plant macrofossil and moss analyses

For macrofossil analysis, 2.5-cm sediment slices with volume of approximately 20 cm^3 were taken with 10-cm sampling resolution. After brief soaking (5–10 min) in 10% KOH, the samples were washed through 400- and 160-µm sieves in two fractions. The material was sorted and identified under stereomicroscope with magnification up to 56×. For determination

Table 1 Radiocarbon dating in Lake Bled core

Sample no. (material dated, pre-treatment)	Depth (cm)	Conventional radiocarbon age (BP)	13C/12C ratio (‰)	2 Sigma calibration (Intcal 04) (BP)	Median (BP)
Beta-222472 (*Larix* leaves, part of cone scale and wings of coniferous seeds, acid/alkali/acid)	60	9340 ± 40	−26.9	10679–10426	10555
Beta-217805 (*Larix* twig and needles, acid/alkali/acid)	210	11930 ± 40	−28.2	13908–13699	13790
Beta-189953 (organic sediment, acid washes)	40	13860 ± 100	−33.8		
Poz-3123 (organic sediment, acid washes)	40	15430 ± 80	/		

The two dates based on terrestrial plant macrofossils (Beta-222472 and Beta-217805) are in good agreement with palynological events in Northern Italy (Vescovi et al., 2007)

The two dates based on organic sediments (Beta-189953 and Poz-3123) are too old and inconsistent with regional events and were not used for age-depth modelling (see text)

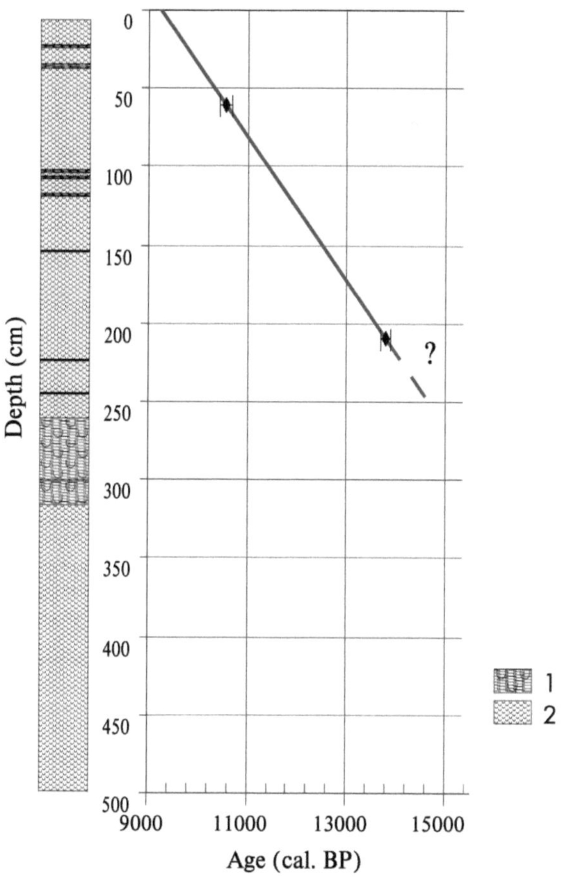

Fig. 2 Age-depth model (linear interpolation). The estimated age of Lake Bled sediment (cal yr BP) is plotted against depth (cm). Details of radiocarbon measurements are presented in Table 2. Lithology: 1—moss layers and 2—calcareous clay (lake marl)

of macrofossils, the reference collection from the Herbarium of Sofia University and specialised literature was used (Beijerinck 1947; Katz et al., 1977; Tobolski, 2000). Plant macrofossil concentrations (number of macrofossils in 20 cm^3 of the sediment) were integrated in the pollen diagram (Fig. 4). The sample volume was much smaller than that recommended for plant macrofossil analysis (Birks, 2002), and this could affect the results to some extent.

Samples for moss analysis were collected from distinct moss layers at 315, 299, 269, 244, 225, 153, 104, 36.5 and 24.5 cm and identified under light microscope at 600× magnification. Keys of Nyholm (1954–1969) and Smith (2004) were used for taxonomic identification.

Pollen analysis

For pollen analysis, 1 cm^3 of the sediment was used following standard laboratory procedures (Bennett & Willis, 2002). Two tablets with a known number of *Lycopodium* sp. spores were added prior to laboratory preparation to determine the pollen concentration (Stockmarr, 1971). A minimum of 500 pollen grains of terrestrial plants and spores was counted in each sample, with a sampling resolution of 8 cm. Pollen identification was carried out under a Nikon Eclipse E400 light microscope at (400× magnification) using a reference collection of the Institute of Archaeology from Ljubljana and ID keys by Moore et al., 1991 and

Reille, 1992, 1995. Microscopic charcoal particles and estimation of past fire activity in the area were counted according to the method of Clark (1982). Pollen data were analysed and plotted with the PSIMPOLL 3.00 program (Bennett, 1998; http://chrono.qub.ac.uk/psimpoll/psimpoll.html). The pollen diagram was divided into five statistically significant pollen zones by the method of optimal splitting by information content (Bennett, 1996, 1998).

Chironomid analysis

Subfossil chironomid analysis was carried out on 29 samples. Subsamples of 3–4 cm^3 wet weight (WW) were deflocculated in 10% KOH and heated to 70°C for 20 min. The sediment was subsequently passed through 212- and 95-μm mesh sieves. Chironomid larval head capsules were picked out from a Bogorov sorting tray with fine forceps under a stereo microscope at 25–40× magnification. Larval head capsules were mounted in Euparal, ventral side up, under a 6 mm diameter coverslip. Larvae were identified with reference to Cranston (1997) and Wiederholm (1983). Chrionomid data were processed, analysed and graphed with TILIA v. 1.12, TILIAGRAPH (Grimm 1991) and C2 software (Juggins, 2001, unpublished). The chironomid percentage diagram was created with TILIAGRAPH and C2 graphic tools. In order to detect zones of similar chironomid assemblages within the core, both stratigraphical-constrained cluster analysis with CONISS and optimal partitioning (Birks & Gordon, 1985) with ZONE v. 1.2 (Juggins, 1991, unpublished) were used.

Cladocera analysis

Cladoceran analysis was carried out on 1-g wet weight sediment on the same 29 depths as for chironomid analysis. Samples were washed under a 50-μm sieve to remove fine mineral and organic particles. The material left on the mesh was analysed for Cladocera remains (for details see Korhola & Rautio, 2001). Most of the samples were screened under stereomicroscope at magnification of 40 or 100×. Due to the very low concentration of Cladocera remains, the whole sample was analysed, and only qualitative results can be presented.

Results

Radiocarbon dates, LOI and stable isotopes

The results of radiocarbon measurements are presented in Table 1. Radiocarbon dates of organic sediment at 40 cm (Beta-189953 and Poz-3123; Table 1) were not used in the age-depth model due to the errors presumably caused by the reservoir effect (i.e. hard-water effect). Linear interpolation between median values of radiocarbon dates at 210 cm and 60 cm was used for age-depth modelling (Fig. 2), suggesting that most of the Holocene sediment is missing. Conversely, Late-glacial and early-Holocene sediments deposited between 13,790 ca. and 10,555 cal yr BP show an average sedimentation rate of 0.046 cm yr^{-1}. The Holocene sediment is often absent on subaquatic (littoral) terraces and is usually explained by either non-sedimentation or erosion due to wave action (Digerfeldt, 1986), lower lake levels (Scholz, 2001) and movement of sediment from littoral to profundal zone following breakdown of thermocline (Davis, 1976, after Dearing & Foster, 1986) during the early Holocene and/or after the basin was drained by Olszewski tubes.

The amount of organic material, carbonate and the remaining (after burning at 950°C) inorganic residue was obtained by LOI and was calculated as a percentage of the sediment dry weight (Fig. 3). LOI shows that below 310 cm, there is a high percentage of inorganic material (65–75%) remaining. The percentage of organic material stays low throughout the core, whereas carbonate increases to 40% (and the remaining inorganic material decreases to 55%) above 270 cm.

The samples below 400 cm show positive $\delta^{13}C$ values (+1‰), whereas $\delta^{18}O$ values are ca. −5‰ and $CaCO_3$ 40–45% (Fig. 3). At about 300 cm, carbonates increase to ca. 80%, $\delta^{13}C$ values are negative and a decrease of $\delta^{18}O$ to ca. −10.50‰ is observed between 300 and 262 cm. The $\delta^{18}O$ values decrease sharply between depths of 168- and 152-cm, and the values remain at these low values until 102 cm but with a slightly increasing trend towards shallower depth. The upper part of the profile from 102 cm up to 50 cm (some sediment is missing due to coring problems) exhibits a step-like increase by 0.8‰. Towards the core surface, the values remain relatively high at −8.1‰.

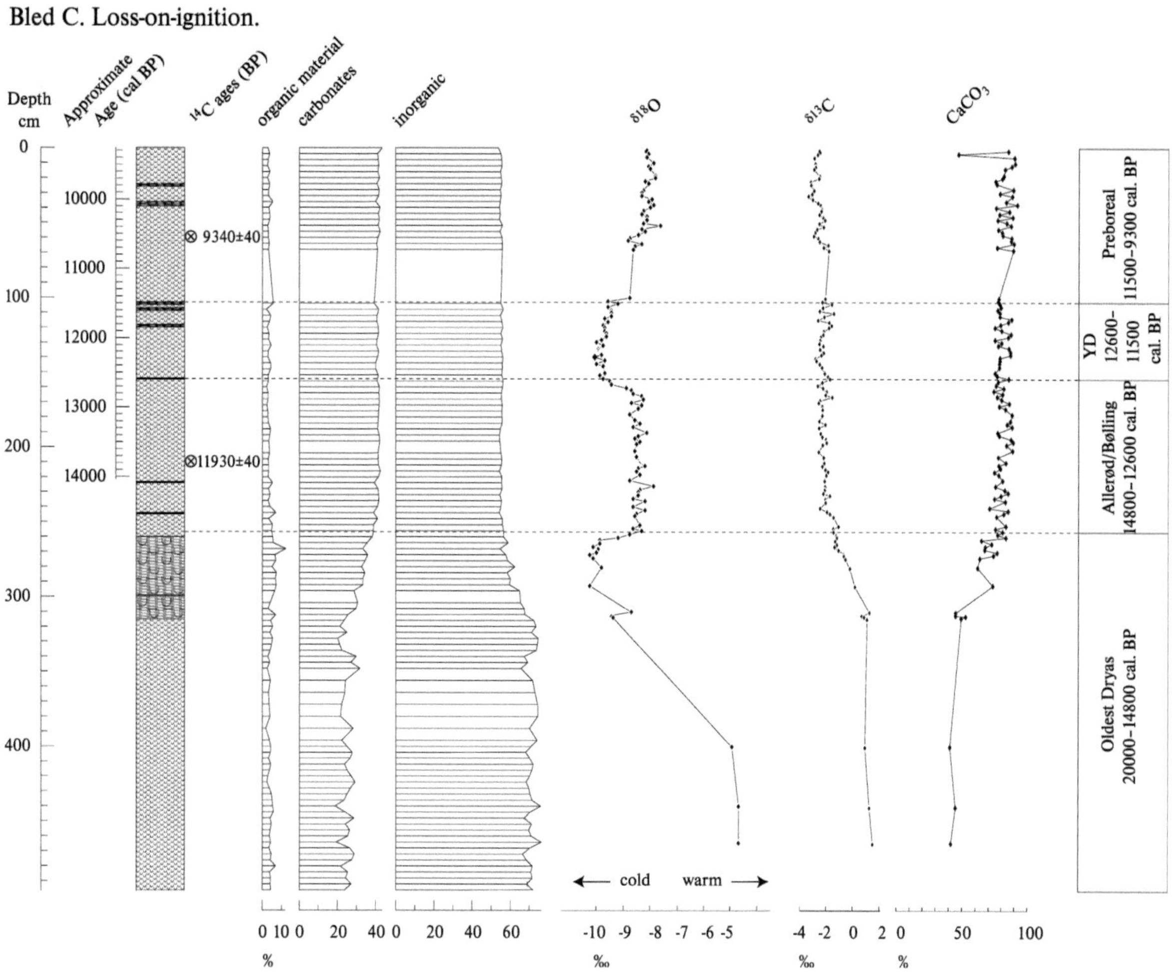

Fig. 3 Chronology and physical and chemical characteristics (loss-on-ignition: organic material, carbonates, inorganic residue; stable isotopes: $\delta^{18}O$, $\delta^{13}C$; $CaCO_3$) of a sediment core Bled C

Plant macrofossils and moss

Moss layers were found between 315 and 223 cm, 155 and 107 cm, and 29 and 24 cm, with a thick moss layer between 315 and 260 cm, (indicated as dark horizontal lines in sedimentary column of Figs. 3, 4a, b, 5, 6). Moss species recovered belong to the genera *Scorpidium* (*S. scorpioides*), *Pseudocalliergon* (*P. trifarium*) and *Calliergon* (*Calliergon* cf. *C. cordifolium—giganteum*). Plant macrofossils appear only in the upper part of the sequence (from ca. 320 cm upward), and the dominant assemblages are a combination of needles of *Larix* and *Betula* catkins. *Picea* (and *Populus* in B-5) macrofossils appear in the last two pollen zones (Fig. 4a).

Pollen

Due to the poor preservation and low concentration of pollen in pollen zones B-1 and B-2 (496–390, 390–315 cm) (ca. 500 pollen grains per 1 cm^3 of wet sediment, with more than 20% of degraded pollen grains, Fig. 4b), the pollen sum in most of the samples was below the statistically required minimum (300 grains per count). The main taxa present are *Pinus* and herbs. In pollen zone B-3 (315–200 cm), pollen concentration increases to ca. 2000–30,000 pollen grains cm^{-3}, with high percentage of *Pinus* (ca. 75–95%) and much lower percentage of herbs than in the previous two zones. At the beginning of pollen zone B-4 (200–115 cm), the percentage of *Pinus* starts to

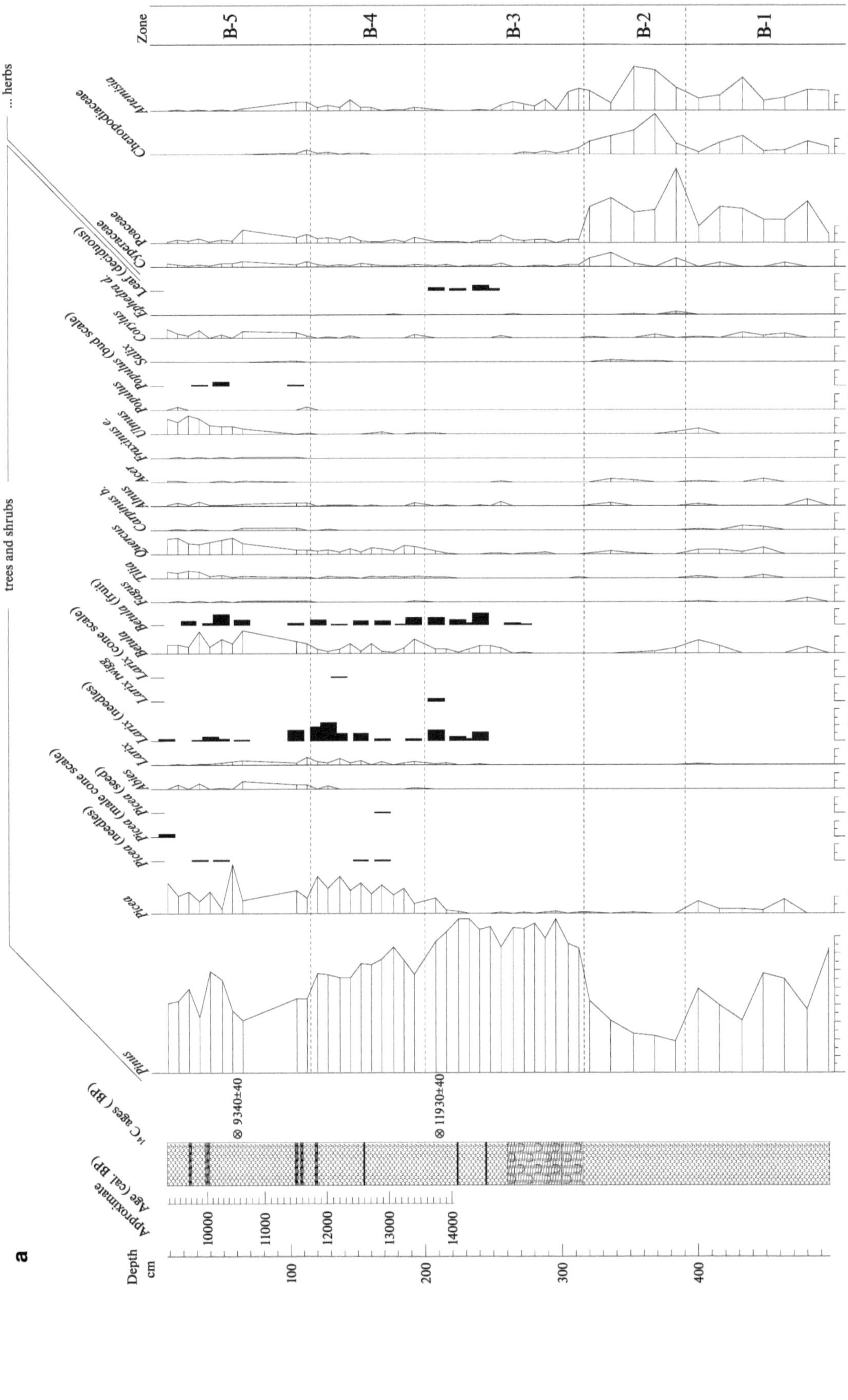

Fig. 4 Pollen and plant-macrofossil diagram of selected taxa. Pollen is presented as percentage data. Macrofossils are presented as number of macrofossils per 20 cm^3 of sediment (*black columns*)

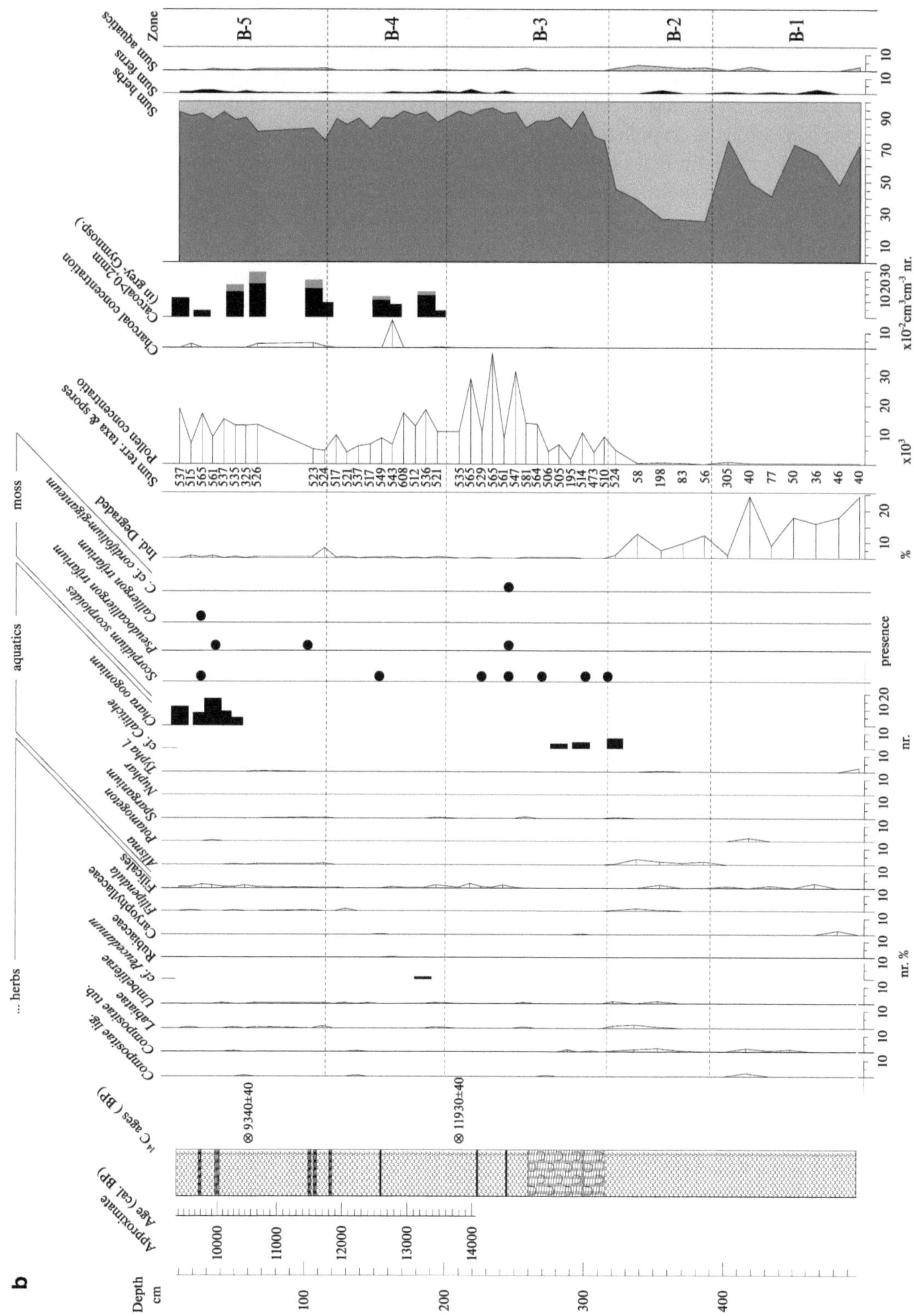

Fig. 4 continued

decline, whereas other tree taxa (*Picea, Larix, Tilia, Ulmus, Quercus, Corylus*) increase with short tree taxa declining in the upper half of the zone (160–115 cm). Finally, at the beginning of zone B-5 (115–0 cm), tree taxa start to increase again. In pollen zones B-4 (192 and 160 cm) and especially B-5 (104, 64, and 16 cm), charcoal peaks occur more frequently than in the lower part of the core.

Chironomids

A total of 28 chironomid taxa were identified in the Lake Bled sediment sequence (for taxonomic notes, see Electronic supplementary material), and the distribution of selected taxa is shown in Fig. 5. Optimal partitioning and CONISS analysis allowed us to recognise six zones along the core: Zone Ch-1 (496–312 cm) was devoid of chironomids below 336 cm. At the top of this zone, head-capsule concentrations were very low, and in most samples, not more than 20 head capsules were counted. Cold stenohermic taxa such as *Paracladius* and *Heteotrissocladius* were present. At the top of the zone, warm-adapted *Chironomus antracinus* and *Tanytarsus* B appear. Zone Ch-2 (312–264 cm) is characterised by the increase of *Micropsectra radialis* and the appearance of *Paratanytarsus penicillatus* and several species of the cold stenothermic *Heterotrisocladius*. At the beginning of the zone Ch-3 (264–200 cm), a decline in *M. radialis* coincides with an increase in *P. penicillatus*. Appearing for the first time are *Chironomus plumosus* and *Arctopelopia*. Zone Ch-4 (200–168 cm) is characterised by the decrease of *M. radialis* and *Ch. antracinus,* and the increase of *Cricotopus* B, and specially *Arctopelopia* and *Parakiefferiella*. In zone Ch-5 (168–120 cm), *Cricotopus* B reaches its maximum value. In Zone Ch-6 (120–90 cm), there is a decrease in the number of head capsules. The zone is characterised by the increase of *M. radialis* and the decease of *P. penicillatus, T. chineyensis, Arctopelopia* and *Cricotopus* B. In zone Ch-7 (90–0 cm), there is a marked drop in *M. radialis* mirrored by the increase of *P. penicillatus*.

Cladocera

Very few remains of Cladocera (always >10 remains g^{-1}) were found in the sediment, and only head shields and valves were used to identify them to the species level. In total, only four taxa of Cladocera (all species are benthic) were detected along the core: *Acroperus harpae*, *Alona affinis*, *Alonella nana* and *Chydorus sphaericus* (Table 2). No Cladocera remains were present below the depth of 272 cm, and neither Ostracoda nor Mollusca were found below 320 cm. The most common species recovered in most of the samples up to the depth of 272 cm was *Alona affinis,* found in 10 samples (out of 13). The section between 48 cm and the top of the core is characterised by very poor Cladocera assemblage, with no remains or very low concentrations of *A. affinis* and *A. harpae*.

Discussion

Oldest Dryas, undated, presumably > ca. 14,800 cal yr BP (496–260 cm)

The isotope values from the deepest part of the core below ca. 300 cm are influenced by micritic, allochtonous carbonates, transported into the lake from the surrounding watershed; therefore, they cannot be used to determine temperature (Lister 1988; Finsinger et al., 2008). Below 336 cm, no biological remains were found, except for pollen in very low concentrations. This absence of fauna could be related to the low percentage of organic matter and extremely cold waters, making insufficient food for animals to survive. Soon after, at 325 cm, macrofossil remains of *Callitriche* indicate very shallow water at the coring point. A rather high percentage of herbaceous pollen (e.g. Poaceae) and xerophytes ($\sim$10–50%, e.g. Chenopodiaceae and *Artemisia, Pinus* stays around 50%, *Picea* goes up to almost 10%, Fig. 4a and 6) suggests that the landscape was rather open, presumably due to cold and dry climate. Sporadic pollen (but no macrofossils) of broad-leaved tree taxa such as *Quercus, Ulmus, Tilia, Carpinus, Acer* and *Fagus* probably does not reflect local vegetation and can be linked to two possible sources: either long-distance transport or reworking from older sediment. If we compare Lake Bled palaeoecological record with other study sites in northern Italy and southern Switzerland (e.g. Lago di Origlio, Lago Piccolo di Avigliano and Lago della Costa with local presence of Larix and Betula in LGM, Kaltenrieder et al., in press a), a similar silty clay is visible above glacial

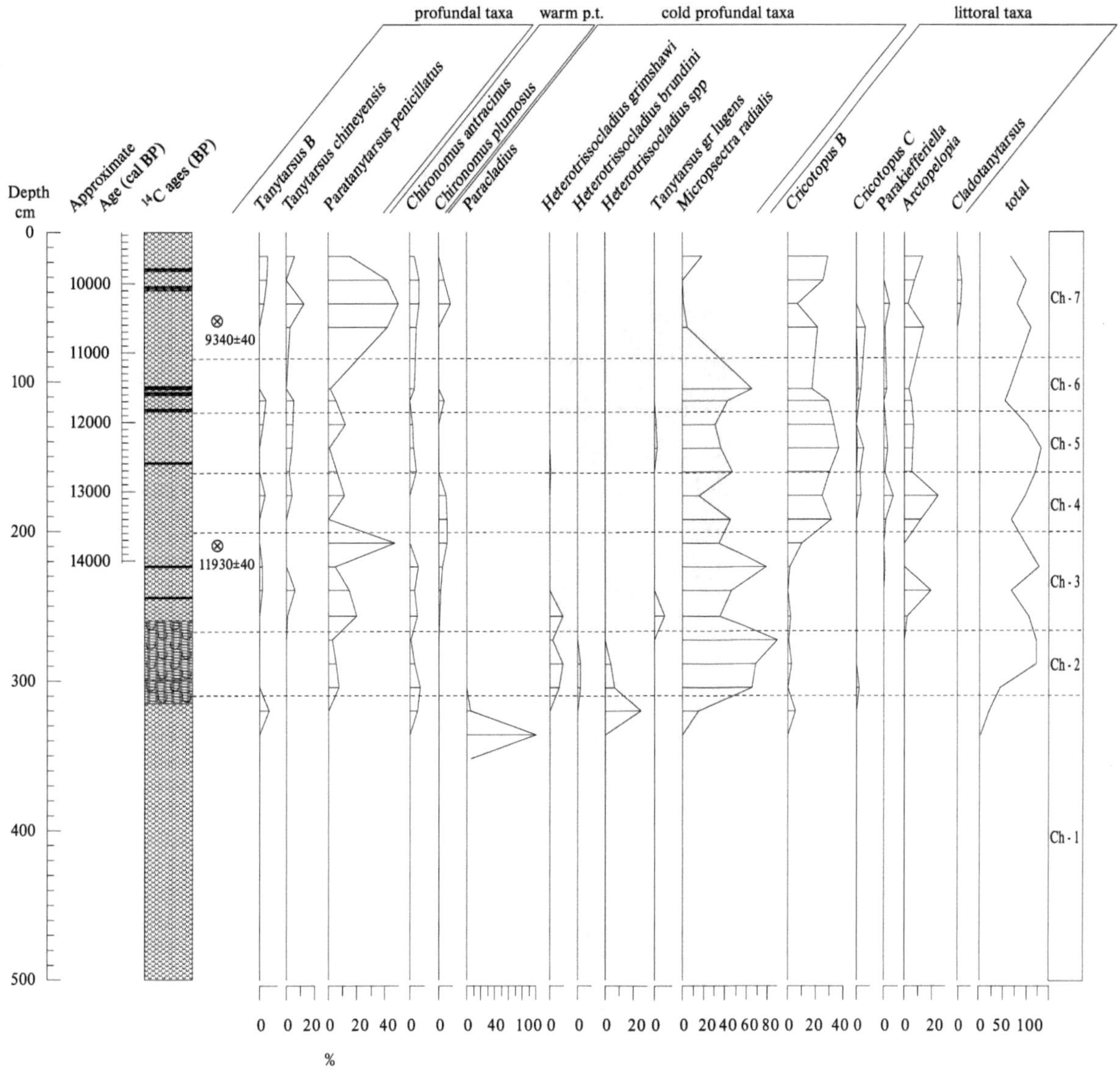

Fig. 5 Chironomid stratigraphy. Data are presented as percentage of relative abundance plotted against depth (cm). Only selected taxa are shown

till (at ca. 18,000 cal yr BP), after glacial retreat. Pollen, stomata and plant-macrofossil data from those study sites suggested that before 16,000 cal yr BP, the lowlands of northern Italy and southern Switzerland were not yet covered by dense forest vegetation (Vescovi et al., 2007, Table 3). A similar vegetation, namely an herb-dominated steppe tundra, probably with some shrubs of *Salix* and *Betula*, was growing also around Lake Bled. The presence of *Pinus* and *Picea* trees cannot be completely ruled out, although plant macrofossils, proving that they were growing in the vicinity of the lake, were not found.

Results from the Lake Bled core during the Oldest Dryas are in accordance with palaeoclimatic research based on a simple glacier-flow model and statistical glacier-climate models of the Gschnitz glacier (Ivy-Ochs et al., 2006b) suggesting that during the Gschnitz cold period (from 19,000 to 18,000 cal yr BP, coinciding with Heinrich 1 ice rafting event), the precipitation was about one-third of modern-day

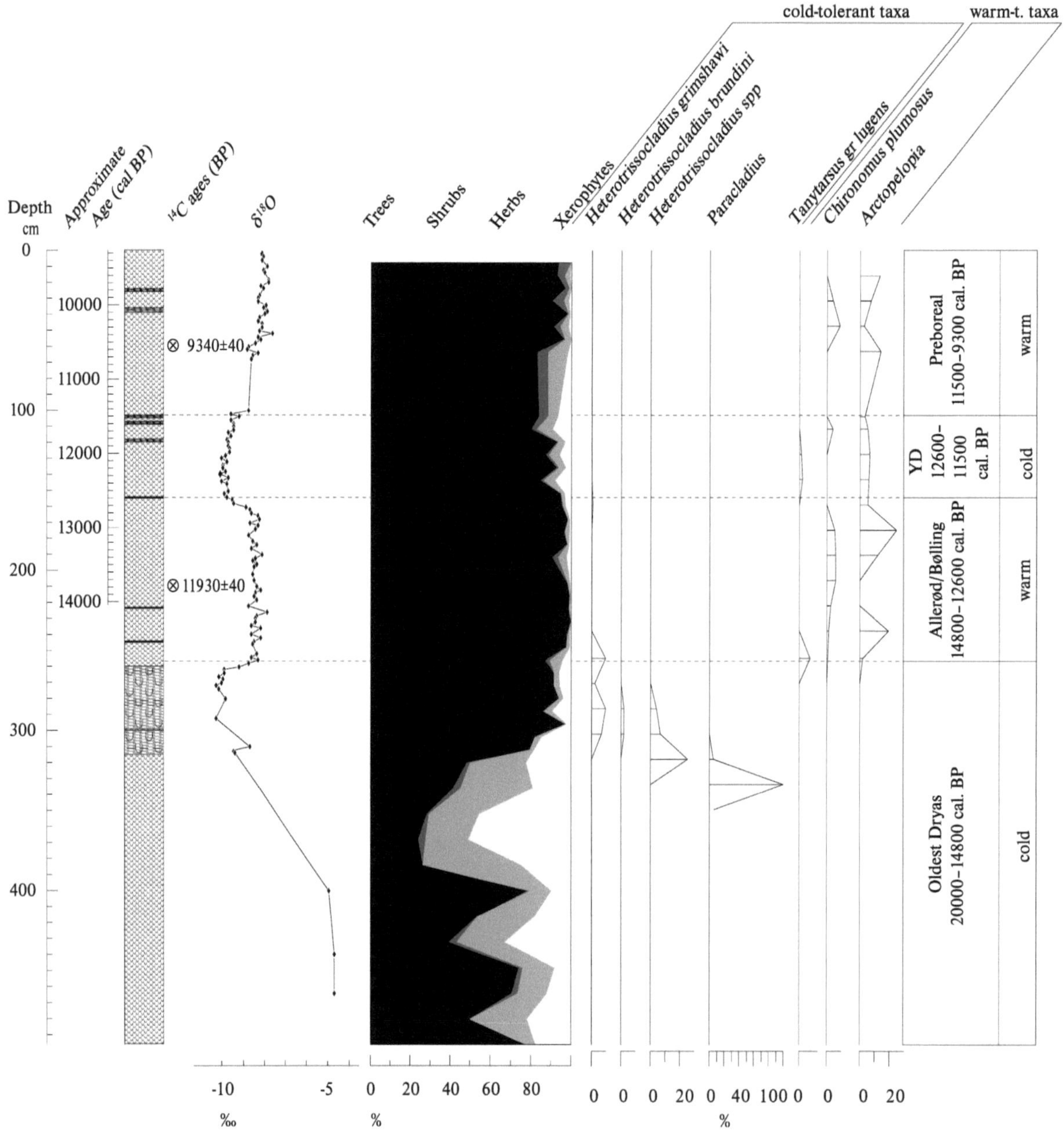

Fig. 6 Comparison of selected multi-proxy data (stable isotopes-$\delta^{18}O$, pollen and chironomid—in percentages) plotted against depth and estimated age

values, and summer temperatures were around 10°C lower than today (Ivy-Ochs et al., 2006a; Kerschner & Ivy-Ochs, 2008). Also, pollen- and chironomid-inferred reconstructions of Oldest Dryas episode for Lake Lautrey (E France) suggest a 3.0–5.5°C lower temperature of the warmest month than today and a drier climate (Peyron et al., 2005, Table 3). Similarly, drier Late-glacial climate was suggested for eastern-central Europe by quantitative, pollen-based methods (Feurdean et al., 2008).

Later on (at 315 cm depth), the palaeoecological record of Lake Bled changes significantly with negative $\delta^{13}C$ values, and $CaCO_3$ (most probably of biogenic origin, as suggested by the appearance of

Table 2 Cladocera remains (presence/absence analysis) and some other groups of organisms (mollusca, ostracoda) from Lake Bled

No. of sample	Depth (cm)	*Acroperus harpae*	*Alona affinis*	*Alonella nana*	*Chydorus spahericua*	Ostracoda	Mollusca	Mosses
1	16	X	X			X		
2	32					X		
3	48					X	X	
4	64	X	X	X	X	X	X	
5	112	X	X			X	X	
6	128	X	X			X		
7	144		X			X		
8	160	X	X	X		X		
9	176	X	X			X		
10	192		X			X		
11	208		X			X		
12	224					X		X
13	240	X	X			X		
14	256		X			X		
15	272		X			X		X
16	288							X
17	304					X		X
18	320					X		
19	336							
20	352							
21	368							
22	384							
23	400							
24	416							
25	432							
26	448							
27	464							
28	480							
29	496							

ostracoda and *Callitriche* macrofossils) increases to 80%. Low $\delta^{18}O$ values (−10.5‰) indicate low temperatures. However, climate conditions were rather favourable (increased availability of moisture) and supported the growth of macrophytes, as suggested by the presence of a 55-cm thick *Scorpidium scorpidioides* moss layer at a depth of 315–260 cm, which was most probably growing 'in situ'. In Slovenia today, this moss grows in marshy areas on wet soil or in shallow (<1 m depth) permanent wet pools (Martinčič, 1996). However, from the literature, it is also known that *Scorpidium* spp. can grow in shallow lakes up to 20-m depth (Light & Smith 1976). All moss taxa discovered in Lake Bled tolerate various climatic conditions, and can be found from subarctic to southern Europe today (with the exception of the Mediterranean), at 250–1200 m a.s.l. The chironomid assemblage consists of a typical cold-profundal fauna represented by *Micropsectra radialis* and several species of genera *Heterotrissocladius* and *Paracladius*, typical for well-oxygenated waters. All Cladocera remains are present in a much lower concentration than in present-day samples from other Slovenian lakes (Brancelj et al., 2002). All are benthic and very eurytopic (ubiquitous), tolerant of high oscillations of physical and chemical

Table 3 Comparison of environmental change at Lake Bled with some other European lakes

	Lake Bled (Slovenia) this study 475 m a.s.l.	Lake Gerzensee (Switzerland) Schwander et al. (2000), Wick (2000), von Grafenstein et al. (2000) and in prep., Lotter et al. (2000) and in prep. 603 m a.s.l.	Lake Lautrey (E France) Heiri and Millet (2005), Peyron et al. (2005) 877 m a.s.l.	N Italy, S Switzerland Vescovi et al. (2007)	Lake Accesa (Italy) Magny et al. (2006)
Preboreal After 11,500 cal yr BP (105–0 cm)	Warmer and wetter Mixed pine-broad-leaved forest High water levels with abundant vegetation, mesotrophic/eutrophic lake and lower oxygen concentrations	Rapid warming, summer temperature 11.5–15°C Immigration of *Corylus, Ulmus, Quercus*	Warmer and wetter, summer temperatures: 14–17.5°C, annual precipitation: >450 mm, *Betula-Pinus* forest	Thermophilous tree taxa expanded	Warmer, deciduous *Quercus-Corylus*-NAP zone, first cold and low lake levels, later highstand
Younger Dryas 12,600–11,500 cal yr BP (155–105 cm)	Cold and dry Slightly more open landscape (Chenopodiaceae and *Artemisia* increase) Low water levels, but still well-oxygenated water	Depression in the oxygen isotopes of bulk sediments and of ostracods, as well as in tree pollen. Summer temperatures 9–10°C	Onset colder, later warmer, drier, summer temperatures: 12–14°C, annual precipitation: 300–450 mm, increase of NAP (*Artemisia*, Poaceae)	Cooling, forest cover diminished, herbs, xerophytes increase	Colder, Poaceae-*Artemisia*-Chenopodiaceae zone, after 12,150 cal yr BP drier, increased microcharcoal
Late-glacial interstadial (Allerød, Bølling) 14,800–12,600 cal yr BP (260–155 cm)	Warmer Mixed woodland (*Betula, Larix, Picea, Quercus*) warm, well-oxygenated water, rich in vegetation Towards the end of this time period, climate gets colder and dryer	Very rapid warming at 14670 cal yr BP, then minor fluctuations Summer temperatures 12–16°C Open woodlands of juniper, sea-buckthorn and then tree birches Immigration of pine 14,000 cal yr BP. Minor lake-level fluctuations	Warmer, summer temperatures: 16°C, annual precipitation: 800 mm, increase of *Juniperus* and *Betula*, later also *Pinus*	Ca. 14,800–14,400 cal yr BP change of forest structure and density due to warming Palughetto: expansion of *Pinus, Betula, Larix, Picea, Quercus*, mixed oak forest at 13,000–12,600 cal yr BP	Warmer, deciduous oak zone, colder and lower lake levels at 14,300–14,200 and 13,900–13,700 cal yr BP, cold and high lake level at 13,400–13,100 cal yr BP
Oldest Dryas 20,000–14,800 cal yr BP (496–260 cm)	Cold and dry, becoming wetter after 16,000 cal yr BP Open, predominantly herbaceous vegetation. Well-oxygenated lake water, rich in vegetation and increase of trees after 16,000 cal yr BP	Cold and dry, shrub tundra: summer temperatures $\leq$11°C.	Cold and dry, summer temperatures: 10–12.5°C, annual precipitation: 200–300 mm, open, herbaceous landscape (*Artemisia*, Poaceae)	Herb-dominated steppe tundra, open woods and shrublands (*Juniperus, Betula, Larix, Pinus cembra* from 17,500 cal yr BP, 16,000 cal yr BP: afforestation (*Pinus, Betula*) in lowlands, treeline ascended to 800–1000 m a.s.l.	Cold, *Artemisia*-Poaceae-*Juniperus* zone

parameters. *Alona affinis* is common in different types of stagnant water, but it prefers an environment rich in vegetation, either algae or macrophytes (e.g. dense moss stands growing in the lake). This suggests that although climatic conditions were still cold, lake water was well oxygenated and warm enough to support the growth of macrophytes.

Better pollen preservation, as well as, an increase of *Pinus* and decline of herbaceous taxa indicate moist conditions. Arboreal pollen (mostly *Pinus*) starts to increase above 315 cm, but the first tree macrofossils (*Betula* catkins, demonstrating that *Betula* was present close to the lake margin) occur only at 270 cm. Although the percentage values for *Pinus* are rather high (75–95%, >14,000 cal yr BP), surprisingly no macrofossils were discovered in the core. It is possible that either only very few pine trees were growing in the vicinity of the lake, or pine pollen came from long-distance transport. Local absence of plants cannot be proved by the absence of macrofossils for taphonomic reasons, as they have much more limited dispersal than pollen, and macrofossils of terrestrial taxa tend to be under-represented in fossil assemblages (Birks, 2003; Jackson & Booth, 2007). *Pinus* pollen accumulation rates (PAR) in dated section of the core (with exception of 13,300–12,800 cal yr BP interval) are, namely, below the threshold value for the presence of local pine (= 500 pollen grains cm^{-2} yr^{-1}) according to the modern PAR studies in the Alps and in Scandinavia (van der Knaap et al., 2001; Seppä & Hicks, 2006). Taphonomic reasons for the absence of *Pinus* macrofossils are not excluded, especially in the depth between 315 and 260 cm characterised by a moss layer. Such moss vegetation, if growing in very shallow water, can also play a role of 'filter' for the macrofossils. No pine wood or charcoal was found in the vicinity of the lake, but *Pinus* wood, discovered in Late-glacial sediment of the Soča River valley ca. 50 km to the west, was radiocarbon dated to 12,450 ± 70 BP (14,205–14,959 cal yr BP) (Culiberg, 1991). A similar change in vegetation composition at 16,000 cal yr BP is noticeable in northern Italy, with pollen, stomata,and plant-macrofossil data indicating afforestation by *Pinus* and *Betula* (and termophilous taxa at Lago della Costa, Kaltenrieder et al., in press b) between 16,000 and 15,800 cal yr BP, and the treeline shifted to 800–1000 m a.s.l. due to an increase in temperature (Vescovi et al., 2007).

Late-glacial interstadial, undated before 13,790 cal yr BP, presumably ca. 14,800–12,600 cal yr BP (260–155 cm)

Climatic warming at the beginning of the Late-glacial interstadial is suggested by $\delta^{18}O$ increase by 1.5–2‰ to ca. −8.5‰ (Fig. 3). Warmer condition is also suggested by the pollen and plant-macrofossil record (Figs. 4a, b). Pollens of tree taxa (e.g. *Betula*) and *Larix* increase, whereas plant-macrofossil records suggest that *Larix* and *Betula* (after 13,000 cal yr BP also *Picea*) were growing on the lakeshore. The cold-adapted chironomid *Micropsectra radialis* and *Heteroterissocladius* spp. disappear and warm-adapted fauna, such as the representatives from the Pentaneurini tribe, appear. A mixed warm-adapted chironomid fauna including *Arctopelopia, Dicrotendipes, Chironomus pulmosus* and *Tanytarus chineyensis* appears, indicating that although the lake became warmer and probably more productive, it was still well oxygenated. Also, the presence of littoral-dweller cladoceran *Acroperus harpae*, together with *Alona affinis*, which is a common inhabitant in modern littoral vegetation, suggests warm water. Similar environmental change was also detected in northern Italy and southern Switzerland between 14,800 and 14,400 cal yr BP, when abrupt changes in forest composition and density were associated with climatic warming (von Grafenstein et al., 2000; Lowe et al., 2001; Heiri & Millet, 2005; Vescovi et al., 2007), which is also in accordance with northern-hemispheric reconstructions (Björck et al., 1998; Lowe et al., 2008). Besides, pollen- and chironomid-inferred temperatures for the Lake Lautrey suggest strong climatic warming and increase of precipitation by that time (Peyron et al., 2005). Also pollen-based climate estimations for northwestern Romania suggest summer temperatures close to modern values (Feurdean et al., 2008). Further in south east, the increasing moisture availability is recorded by the palynological and plant macrofossil assemblages of the Rila, Pirin Mountains (Bozilova et al., 1996; Stefanova & Ammann, 2003) and Thracian Plain (Magyari et al., 2008). At Lake Bled, the local presence of *Picea* in the late Alleröd, as indicated by macroremains, demonstrates the more easterly position of refugia in the southeastern Alps and southwestern Carpathians, as indicated by van der Knaap et al. (2005), Latałowa & van der Knaap (2006), Ravazzi et al. (2006), Vescovi et al. (2007) and Feurdean et al. (2007).

A change in vegetation composition also occurs at ca. 13,800 cal yr BP. The decrease of *Pinus* and the increase of *Quercus, Tilia, Ulmus*, as well as *Artemisia* (drier) and *Picea* could be linked to warmer conditions, although stable isotope record at Lake Bled does not show significant increase of $\delta^{18}O$. A sharp increase of chironomid larvae of *Cricotopus* B at the same time suggests that the lake levels decreased. Again, similar vegetation change and lowering of lake levels occurred over a wider area on the southern side of the Alps (Vannière et al., 2004; Magny et al., 2006; Vescovi et al., 2007).

After 12,800 cal yr BP, $\delta^{18}O$ started to decrease, suggesting climatic cooling. Climate was possibly also getting drier, as suggested by the increase of microcharcoal concentrations, the decline of pollen of tree taxa, the increase of xerophytes and the presence of littoral chironomids *Parakiefferiella, Arctopelopia* and *Cricotopus*.

Younger Dryas, 12,600–11,500 cal yr BP (155–105 cm)

Oxygen-isotope records indicate that both the onset and termination of Younger Dryas as recorded in the presented Lake Bled record are remarkably sharp (Fig. 3). Climatic conditions during the Younger Dryas were cold ($\delta^{18}O$ −9.7‰) and dry, as suggested by the increase of Chenopodiaceae and *Artemisia*, and the recurrent presence of moss layers. Chironomid taxa are mostly littoral, suggesting that throughout the YD the lake was probably shallow. The presence of *Alonella nana*, which is common in the environment rich in organic debris and well oxygenated water, is also an indication of cold and dry conditions. Among the trees, *Larix* (cold and drought adapted species, also present at other sites in central and eastern Europe, e.g. Willis et al., 2000; Feurdean et al., 2007) is more abundant than *Quercus*, as in northern Italian pollen records (Vescovi et al., 2007). The palaeoecological records at Lake Lautrey (Peyron et al., 2005) and Gerzensee (von Grafenstein et al., 2000; Lotter et al., in prep.) also suggest colder and drier climate, although lake level at Gerzensee was 1.7 m higher than today (0.4 m higher than during the Alleröd). This is probably due to less vegetation and longer season of frozen soils, causing reduced percolation of water to groundwater and increased direct runoff into the lake. The palaeoecological record of Lake Kremensko-5, Pirin Mountains covering YD show also increasing aridity and remarkable retreat of the *Pinus, Picea* and *Betula* curves related to colder climatic conditions (Atanassova & Stefanova, 2003; Stefanova et al., 2006). Also, palaeoclimatic modelling based on simple glacier-flow model and statistical glacier-climate models of Egesen maximum advance (ca. 12,400–12,300 cal yr BP) suggest that after 12,700–12,600 cal yr BP, summer temperature was ca. 3.5°C lower, with 20–30% less precipitation in the interior of the Alps. Winters were cold and dry, but summers were presumably only moderately drier or even wetter than today (Kerschner & Ivy-Ochs, 2008).

Preboreal, 11,500–9300 cal yr BP (105–0 cm)

Climatic warming at the Late-glacial–Holocene transition is inferred from a sharp increase of $\delta^{18}O$ to ca. −8‰. Tree taxa including *Betula, Fagus, Tilia, Quercus, Carpinus betulus, Carpinus orientalis/Ostrya, Alnus, Acer, Fraxinus excelsior* type, *Ulmus, Salix, Corylus* and *Abies* started to increase, whereas Chenopodiaceae and *Artemisia* declined. The increase of profundal chironomids *Micropsectra radialis* and the decrease of littoral *Cricotopus* at the beginning of the zone may indicate an increase of water level. However, at 11,200 cal yr BP, *Micropsectra radialis* decreases sharply, and all the fauna typical for well-oxygenated water disappears. Both species of *Chironomus* reach their maximum values. This is the typical situation in a mesotrophic/eutrophic lake that only support *Chironomus*, and other species adapted to survive under low oxygen concentrations (Hofmann, 1986; Walker, 2001). At ca. 10,400 cal yr BP, the conditions must have been favourable for cladocera, as it is the only interval when all the four taxa (*Allona affinis, Acroperus harpae, Allonella nana* and *Chydorus spahericus*) coexist. This assemblage is quite different from present-day cladoceran assemblages at Lake Bled, which are more planktonic (*Daphnia hyalina, D. hyalina × galeata, Bosmina longispina, Diaphanosoma brachyurum* and *Scapholeberis kingi*) (Brancelj, 1991), whereas benthic Cladocera are reduced due to the eutrophic condition of the lake. *Allonella nana*, which is present in the core, has not yet been found in the present-day fauna of Lake Bled (Brancelj, unpublished). The species is quite common in the littoral zone of oligotrophic–mesotrophic cold-water lakes. This conditions are no more existing in

Lake Bled, which is mesotrophic–eutrophic warm water. Palaeoecological research at Lakes Gerzensee and Lautrey suggest a temperature rise of ca. 3°C at the Holocene transition in less than 50 years (Lotter et al., 2000; Lotter et al., in prep.; Schwander et al., 2000; von Grafenstein et al., 2000; Peyron et al., 2005).

In summary, at the beginning of Preboreal, the climate became warmer and wetter, with a mixed pine-broad-leaved forest around the lake. The water level increased, and the lake became deep again, holding favourable conditions for Cladocera and profundal chironomid assemblages.

Conclusions

Late-glacial terrestrial and aquatic ecosystems at Lake Bled were very dynamic, and several proxies responded to climatic change simultaneously, enabling more detailed reconstruction of environmental changes. Both ecosystems, i.e. terrestrial and aquatic, responded to changes of temperature, precipitation and hydrological conditions; therefore, our assumptions about past climate (e.g. temperature) are affected also by local (e.g. lake level) conditions.

Whereas in the Oldest Dryas the climate was cold and dry, later a trend towards wetter and warmer climate occurred, with the beginning of the precipitation of biogenic carbonates, suggested by the appearance of ostracods and an increase of aquatic plant macrofossils (e.g. *Callitriche*). By that time, Lake Bled water was well oxygenated and rich in vegetation (e.g. *Callitriche, Scorpidium scorpidioides*), and the lake was probably surrounded by predominantly herbaceous vegetation, with only very small populations of shrub and tree taxa. The climate became warmer, with an increase of tree cover (*Betula, Larix*) and warmer chironomid fauna, suggesting a well-oxygenated, more productive lake. Further climatic warming at 13,800 cal yr BP led to increase of broad-leaved tree taxa (*Quercus, Tilia, Ulmus*) and *Picea* and lower lake levels, as indicated by Chironomid record.. Towards the end of this interstadial, the climate became colder and drier, as suggested by the $\delta^{18}O$ record, increase of xerophytes and microcharcoal and lower lake levels (presence of littoral chironomids) after 12,800 cal yr BP. A warmer and wetter Preboreal climate after 11,500 cal yr BP enabled the spread of broad-leaved tree taxa, whereas the lake became deeper (increase of profundal Chironomids) and mesotrophic/eutrophic, with lower oxygen content.

The local presence of a wide variety of tree taxa is not a surprise, given the proximity of Slovenia to full-glacial refugia (e.g. Willis & van Andel, 2004; Cheddadi et al., 2006). Plant macrofossils suggest the local presence of *Betula* (probably from ca. 15,000 cal yr BP), *Larix* (probably ca. 14,500 cal yr BP), *Picea* (ca. 13,000 cal yr BP), *Populus* (ca. 12,000 cal yr BP) and *Ephedra* (probably ca. 14,700 cal yr BP), but surprisingly, no *Pinus* macrofossils were found. Although general vegetation development at Lake Bled is comparable to that of the neighbouring regions (e.g. northern Italy; Vescovi et al., 2007, Hungary; Willis et al., 1995, 2000), the stable-isotope record, in contrast to many lakes south of the Alps (Eicher, 1987), shows a distinct negative Younger Dryas $\delta^{18}O$ signal (colder climate). A similar stable-isotope record is characteristic only for Lago Piccolo di Avigliana (Finsinger et al., 2008).

Future research should focus also on multi-proxy studies of Late-glacial and Holocene environmental change and human versus climatic impact on the environment at Lake Bled and other study sites in the region to better understand the present and future environmental changes.

Acknowledgments This research was partly funded by the Slovenian Research Agency (project Z6-4074-0618-03). We would like to thank Timotej Knific for his support, Willy Tanner and Mike Tanner for coring Lake Bled, and Mateja Belak for preparing figures. We thank Steve Brooks from the Natural History Museum of London for lab facilities during Julieta Massaferro postdoc and for his help with the chironomid identifications. We are very grateful to Herb Wright for critical comment on the manuscript and checking the English language. Comments from two anonymous referees are gratefully acknowledged.

References

Atanassova, J. & I. Stefanova, 2003. Late-glacial vegetational history of Lake Kremensko-5 in the northern Pirin Mountains, southwestern Bulgaria. Vegetation History and Archaeobotany 12: 1–6.

Beijerinck, W., 1947. Zadenatlas der nederlandsche Flora. H. Veeman & Zonnen, Wageningen.

Bengtsson, L. & M. Ennell, 1986. Chemical analysis. In Berglund, B. E. (ed.), Handbook of Holocene Palaeoecology and Palaeohydrology. Wiley, New York: 423–451.

Bennett, K. D., 1996. Determination of the number of zones in a biostratigraphical sequence. New Phytologist 132: 155–170.

Bennett, K. D., 1998. Documentation for PSIMPOLL 3.00 and PSCOMB 1.03: C programs for plotting pollen diagrams and analysing pollen data. http://chrono.qub.ac.uk/psimpoll/psimpoll.html.

Bennett, K. D. & K. J. Willis, 2002. Pollen. In Smol, J. P., H. J. Birks & W. M. Last (eds), Tracking Environmental Changes Using Lake Sediments. Volume 3. Terrestrial, Algal and Siliceous Indicators. Kluwer Academic Publishers, Dordrecht: 5–32.

Birks, H., 2002. Plant macrofossils. In Smol, J. P., H. J. Birks & W. M. Last (eds), Tracking Environmental Changes Using Lake Sediments. Volume 3. Terrestrial, Algal and Siliceous Indicators. Kluwer Academic Publishers, Dordrecht: 49–74.

Birks, H., 2003. The importance of plant macrofossils in the reconstruction of Late-glacial vegetation and climate: examples from Scotland, western Norway, and Minnesota, USA. Quaternary Science Reviews 22: 453–473.

Birks, H. J. B. & A. D. Gordon, 1985. Numerical Methods in Quaternary Pollen Analysis. Academic Press, London.

Björck, S., M. J. C. Walker, L. C. Cwynar, S. Johnsen, K. L. Knudsen, J. J. Lowe, B. Wolfarth & INTIMATE members, 1998. An event stratigraphy for the Last Termination in the North Atlantic region based on Greenland ice-core record: a proposal by the INTIMATE group. Journal of Quaternary Science 13: 283–292.

Bozilova, E., M. Filipova, L. Filipovich & S. Tonkov, 1996. Bulgaria. In Berglund, B. E., H. J. B. Birks, M. Ralska-Jasiewiczowa & H. E. Wright (eds), Palaeoecological Events During the Last 15,000 Years: Regional Syntheses of Palaeoecological Studies of Lakes and Mires in Europe. Wiley, Chichester: 701–728.

Brancelj, A., 1991. A model of the zooplankton production assessed in *Daphnia hyalina* (Crustacea: Cladocera) in the lake Blejsko jezero. Ph.D. Dissertation. University of Ljubljana, Department of Biology: 144 pp (in Slovene).

Brancelj, A., M. Šiško, G. Muri, P. Appleby, A. Lami, E. Shilland, N. L. Rose, C. Kamenik, S. J. Brooks & J. A. Dearing, 2002. Lake Jezero v Ledvici (NW Slovenia)—changes in sediment records over the last two centuries. Journal of Paleolimnology 28: 47–58.

CALIB 5.0 Website, 2006. http://calib.qub.ac.uk/.

Cheddadi, R., G. G. Vendarmin, T. Litt, L. François, M. Kageyama, S. Lorentz, J.-M. Laurent, J.-L. de Beaulieu, L. Sadori, A. Jost & D. Lunt, 2006. Imprints of glacial refugia in the modern genetic diversity of Pinus sylvestris. Global Ecology and Biogeography 15: 271–282.

Clark, R. L., 1982. Point count estimation of charcoal in pollen preparations and thin sections of sediments. Pollen et Spores 24: 523–532.

Cranston, P. S., 1997. Identification Guide to the Chironomidae of New South Wales. Australian Water Technologies Pty Ltd., West Ryde.

Culiberg, M., 1991. Late Glacial Vegetation in Slovenia [Kasnoglacialna vegetacija v Sloveniji]. Dela SAZU IV/29, SAZU, Ljubljana.

Čarni, A., A. Marinček, A. Seliškar & M. Zupančič, 2003. Vegetation map of Slovenian forest plant associations, Institute of Biology, Scientific Research Centre of the Slovenian Academy of Sciences and Arts. http://bijh.zrc-sazu.si/bio/si/zbirke/400/400.asp.

Davis, M. B., 1976. Erosion rates and land-use history in Southern Michigan. Environmental Conservation 3: 139–148.

Dearing, A. J. & I. D. L. Foster, 1986. Lake sediments and palaeohydrological studies. In Berglund, B. E. (ed.), Handbook of Holocene Palaeoecology and Palaeohydrology. Wiley, New York: 67–90.

Digerfeldt, G., 1986. Studies on past lake-level fluctuations. In Berglund, B. E. (ed.), Handbook of Holocene Palaeoecology and Palaeohydrology. Wiley, New York: 127–143.

Dolenec, T., J. Pezdič, B. Ogorelec & M. Mišič, 1984. Izotopska sestava kisika in ogljika v recentnem sedimentu iz Blejskega jezera in v pleistocenski jezerski kredi Julijskih Alp [The isotopic composition of oxygen and carbon of the recent sediment from the Bled Lake and of the Pleistocene lacustrine chalk from the Julian Alps]. Geologija 27: 161–170.

Eicher, U., 1987. Die spätglazialen sowie die frühpostglazialen Klimaverhältnisse im Bereiche der Alpen: Sauerstoffisotopenkurven kalkhaltiger Sedimente. Geographica Helvetica 42: 99–104.

Feurdean, A., B. Wohlfarth, L. Björkman, I. Tantau, O. Bennike, K. J. Willis, S. Farcas & A. M. Robertsson, 2007. The influence of refugial population on Lateglacial and early Holocene vegetational changes in Romania. Review of Palaeobotany and Palynology 145: 305–320.

Feurdean, A., S. Klotz, S. Brewer, V. Mosbrugger, T. Tămas & B. Wohlfarth, 2008. Lateglacial climate development in NW Romania—Comparative results from three quantitative pollen-based methods. Palaeogeography, Palaeoclimatology, Palaeoecology 265: 121–133.

Finsinger, W., C. Bellis, S. E. P. Blockley, U. Eicher, M. Leuenberger, A. F. Lotter & B. Ammann, 2008. Temporal patterns in lacustrine stable isotopes as evidence for climate change during the late glacial in the Southern European Alps. Journal of Palaeolimnology, Open Access. doi 10.1007/s10933-008-9205-7.

Geological Map of Slovenia, 1993. 1: 500 000, Ljubljana, Geodetski zavod Slovenije.

Grimm, E., 1991 of Tilia and Tilia Graph. Illinois State Museum.

Grimšičar, A., 1955. Zapiski o geologiji Bleda. Notes about the geology of Bled area. Geologija 3: 220–225.

Heiri, O. & L. Millet, 2005. Reconstruction of Late Glacial summer temperatures from chironomid assemblages in Lac Lautrey (Jura, France). Journal of Quaternary Science 20: 33–44.

Hofmann, W., 1986. Chironomid analysis. In Berglund, B. E. (ed.), Handbook of Holocene Palaeoecology and Palaeohydrology. Wiley, New York: 715–723.

Ivy-Ochs, S., H. Kerschner, P. W. Kubik & C. Schlüchter, 2006a. Glacier response in the European Alps to Heinrich Event 1 cooling: the Gschnitz stadial. Journal of Quaternary Science 21: 115–130.

Ivy-Ochs, S., H. Kerschner, A. Reuther, M. Maisch, R. Sailer, J. Schaefer, P. W. Kubik, H.-A. Synal & C. Schlüchter, 2006b. The timing of glacier advances in the northern European Alps based on surface exposure dating with cosmogenic ^{10}Be, ^{26}Al, ^{36}Cl, and ^{21}Ne. Geological Society of America Special Paper 415: 43–60.

Jackson, S. T. & R. K. Booth, 2007. Using plant macrofossils to refine and validate pollen studies. In Elias, S. A. (ed.), Encyclopedia of Quaternary Sciences. Elsevier, New York: 2413–2442.

Kaltenrieder, P., C. A. Belis, S. Hofstetter, B. Ammann, C. Ravazzi & W. Tinner, in press a. Environmental and climatic conditions at a potential Glacial refugial site of tree species near the southern Alpine glaciers. New insights from multiproxy sedimentary studies at Lago della Costa (Eugaean Hills, northeastern Italy), Quaternary Science Reviews.

Kaltenrieder, P., G. Procacci, B. Vanniere & W. Tinner, in press b. Postglacial vegetation and fire history of the Euganean Hills (Colli Euganei) as recorded by sedimentary pollen and charcoal series from Lago della Costa (northeastern Italy). The Holocene.

Katz, N., S. Katz & M. Kipiani, 1977. Atlas and keys of fruits and seeds occurring in the quaternary deposits of the USSR, Moscow (in Russian).

Kerschner, H. & S. Ivy-Ochs, 2008. Palaeoclimate from glacier: examples from the Eastern Alps during the Alpine Lateglacial and early Holocene. Global and Planetary Change 60: 58–71.

Korhola, A. & M. Rautio, 2001. Cladocera and other branchiopod crustaceans. In Smol, J. P., H. J. B. Birks & W. M. Last (eds), Tracking Environmental Change Using Lake Sediments. Volume 4: Zoological indicators. Kluwer Academic Publishers, Dordrecht, The Netherlands: 5–41.

Latałowa, M. & W. O. van der Knaap, 2006. Late Quaternary expansion of Norway spruce *Picea abies* (L.) Karst. in Europe according to pollen data. Quaternary Science Reviews 25: 2780–2805.

Light, J. J. & R. I. Lewis Smith, 1976. Deep-water bryophytes from the highest Scottish lochs. Journal of Bryology 9: 55–62.

Lister, G. S., 1988. A 15 000-year isotopic record from Lake Zürich of deglaciation and climatic change in Switzerland. Quaternary Research 29(2): 129–141.

Löffler, H., 1984. The palaeolimnology of Lake Bled (Blejsko jezero). Verhandlungen der Internationalen Vereingung für theoretische und angewandte Limnologie 22: 1409–1413.

Lojen, S., N. Ogrinc & T. Dolenec, 1997. Carbon and nitrogen stable isotope fractionation in the sediment of Lake Bled (Slovenia). Water, Air and Soil Pollution 99: 315–323.

Lojen, S., N. Ogrinc & T. Dolenec, 1999. Decomposition of sedimentary organic matter and methane formation in the recent sediment of Lake Bled (Slovenia). Chemical Geology 159: 223–240.

Lotter, A. F., H. J. B. Birks, U. Eicher, W. Hoffmann, J. Schwander & L. Wick, 2000. Younger Dryas and Allerød summer temperatures at Gerzensee (Switzerland) inferred from fossil pollen and cladoceran assemblages. Palaeogeography, Palaeoclimatology, Palaeoecology 159(3–4), Special Issue: Ammann, B. (ed.), Biotic Responses to Rapid Climatic Changes Around the Younger Dryas: 349–361.

Lotter, A. F., O. Heiri, J. N. F. van Leeuwen, U. Eicher, J. Schwander & B. Ammann, in prep. Rapid summer temperature changes during Termination 1: multi-proxy reconstructions from Gerzensee (Switzerland), Poster at the EGU 2008.

Lowe, J. J., W. Z. Hoek & INTIMATE Group, 2001. Inter-regional correlation of palaeoclimatic records for the Last Glacial-Interglacial Transition: a protocol for improved precision recommended by the INTIMATE project group. Quaternary Science Reviews 20: 1175–1187.

Lowe, J. J., S. O. Rasmussen, S. Björck, W. Z. Hoek, J. P. Steffenson, M. J. C. Walker, Z. C. Yu & INTIMATE group, 2008. Synchronisation of palaeoenvironmental events in the North Atlantic region during the Last Termination: a revised protocol recommended by the INTIMATE group. Quaternary Science Reviews 27: 6–17.

Magny, M., J.-L. De Beaulieu, R. Drescher-Schneider, B. Vannière, A.-V. Walter-Simonnet, L. Millet, G. Bousset & O. Peyron, 2006. Climatic oscillations in central Italy during the Last Glacial-Holocene transition: the record from Lake Accesa. Journal of Quaternary Science 21: 311–320.

Magri, D., G. G. Vendramin, B. Comps, I. Dupanloup, T. Geburek, D. Gömöry, M. Latałowa, T. Litt, L. Paule, J. M. Roure, I. Tantau, W. O. van der Knaap, R. J. Petit & J.-L. de Beaulieu, 2006. A new scenario for the Quaternary history of European beech populations: palaeobotanical evidence and genetic consequences. New Phytologist 171: 199–221.

Magyari, E., J. Chapman, B. Gaydarska, E. Marinova, T. Deli, J. Huntley, J. Allen & B. Huntley, 2008. The 'oriental' component of the Balkan flora: evidence of presence on the Thracian Plain during the Weichselian late glacial. Journal of Biogeography 35: 865–883.

Martinčič, A., 1996. Barja. wetlands. In Gregori, J., A. Martinčič, K. Tarman, O. Urbanc-Brečič, D. Tome & M. Zupančič (eds), Narava Slovenije, stanje in perspektive [Nature of Slovenia, Current State and Prospects for the Future]. Društvo ekologov Slovenije [Ecological Society of Slovenia] Ljubljana: 122–132.

Merkt, J. & H. Streif, 1970. Stechrohr-Bohrgeräte für limnische und marine Lockersedimente. Geologische Jahrbuch 88: 137–148.

Molnar, F. M., P. Rothe, U. Förstner, J. Štern, B. Ogorelec, A. Šercelj & M. Culiberg, 1978. Lakes Bled and Bohinj. Origin, composition, and pollution of recent sediments. Geologija 21: 93–164.

Moore, P. D., J. A. Webb & M. E. Collinson, 1991. Pollen Analysis, 2nd ed. Blackwell Science, Oxford.

Nyholm, E., 1954–1969. Illustrated moss flora of Fennoscandia II. Musci (fasc. 1–6). Botanical Society of Lund, Stockholm: 799 pp.

Ogrin, D., 1996. Podnebni tipi v Sloveniji [The climate types in Slovenia]. Geografski vestnik 68: 39–56.

Ogrinc, N., S. Lojen & J. Faganeli, 1997. The sources of dissolved inorganic carbon in pore waters of lacustrine sediment. Water, Air and Soil Pollution 99: 333–341.

Ogrinc, N., S. Lojen, J. Faganeli, B. Čermelj, T. Dolenec & J. Pezdič, 1998. Carbon cycling in a lacustrine environment (Lake Bled). RMZ—Materials and Geoenvironment 45: 136–140.

Ogrinc, N., S. Lojen & J. Faganeli, 2002. A mass balance of carbon stable isotopes in an organic-rich methane-producing lacustrine sediment (Lake Bled, Slovenia). Global and Planetary Change 33: 57–72.

Pak, M., 2001. Savska raven. In Perko, D. & O. M. Adamič (eds), Slovenija. Pokrajine in ljudje. Založba Mladinska knjiga, Ljubljana: 84–95.

Petit, R., I. Aguinagalde, J.-L. de Beaulieu, C. Bittkau, S. Brewer, R. Cheddadi, R. Ennos, S. Fineschi, D. Grivet, M. Lascoux, A. Mohanty, G. Müller-Starck, B. Demesure-Musch, A. Palmé, J. P. Martin, S. Rendell & G. G. Vendramin, 2003. Glacial refugia: hotspots but not melting pots of genetic diversity. Science 300: 1563–1565.

Peyron, O., C. Bégeot, S. Brewer, O. Heiri, M. Magny, L. Millet, P. Ruffaldi, E. Van Campo & G. Yu, 2005. Late-Glacial climatic changes in Eastern France (Lake Lautrey) from pollen, lake-levels, and chironomids. Quaternary Research 64: 197–211.

Radinja, D., S. Buser & D. Vrhovšek, 1987. Blejsko jezero. Lake Bled. In Adamič, F., V. Bračič, M. Čepič, R. Godec, B. Grafenauer, M. Kališnik, J. Kastelic, A. Kornhauser, G. Počkar, M. Rožman, F. Šali & I. Tavčar (eds), Enciklopedija Slovenije 1 [Encyclopedia of Slovenia 1]. Mladinska knjiga, Ljubljana: 289–291.

Ravazzi, C., M. Donegana, E. Vescovi, E. Arpenti, M. Caccianiga, P. Kaltenrieder, L. Londeix, S. Marabini, S. Mariani, R. Pini, G. Battista Vai & L. Wick, 2006. A new Late-glacial site with Picea abies in the northern Apennine foothills: an exception to the model of glacial refugia of trees. Vegetation History and Archaeobotany 15: 357–371.

Reille, M., 1992. Pollen et Spores d'Europe et d'Afrique Du Nord. Laboratoire de Botanique Historique et Palynologie, URA, CNRS, Marseille.

Reille, M., 1995. Pollen et Spores d'Europe et d'Afrique Du Nord (Supplement). Laboratoire de Botanique Historique et Palynologie, URA, CNRS, Marseille.

Reimer, P. J., M. G. L. Baillie, et al., 2004. INTCAL04 terrestrial radiocarbon age calibration, 0–26 cal kyr BP. Radiocarbon 46: 1029–1058.

Scholz, C. A., 2001. Applications of seismic sequence stratigraphy in lacustrine basins. In Last, W. M. & J. P. Smol (eds), Tracking environmental change using lake sediments. Volume 1: Basin analysis, coring and chronological techniques. Kluwer Academic Publishers, Dordrecht: 7–22.

Schultze, E., 1984. Beiträge zur Paläolimnologie des Bled-Sees in Slowenien (Jugoslawien) und zur Vegetations- und Klimaentwicklung der Umgebung. Geologija 27: 97–106.

Schultze, E., 1988. Fallstudien zur Paläolimnologie [Case studies on paleolimnology]. Geologija 31(32): 437–516.

Schwander, J., U. Eicher & B. Ammann, 2000. Oxygen isotopes of lake marl at Gerzensee and Leysin (Switzerland), covering the Younger Dryas and two minor oscillations, and their correlation to the GRIP ice core. Palaeogeography, Palaeoclimatology. Palaeoecology 159: 203–214.

Seppä, H. & S. Hicks, 2006. Integration of modern and past pollen accumulation rate (PAR) records across the arctic tree-line: a method for more precise vegetation reconstructions. Quaternary Science Reviews 25: 1501–1516.

Siegenthaler, U. & U. Eicher, 1986. Stable oxygen and carbon isotope analyses. In Berglund, B. E. (ed.), Handbook of Holocene Palaeoecology and Palaeohydrology. Wiley, Chichester: 407–422.

Smith, A. J. E., 2004. The Moss Flora of Britain and Ireland, 2nd ed. Cambridge University Press, Cambridge: 1012 pp.

Stefanova, I. & B. Ammann, 2003. Lateglacial and Holocene vegetation belts in the Pirin Mountains (southwestern Bulgaria). The Holocene 13: 97–107.

Stefanova, I., J. Atanassova, M. Delcheva & H. E. Wright, 2006. Chronological framework for the Lateglacial pollen and macrofossil sequence in the Pirin Mountains, Bulgaria: Lake Besbog and Lake Kremensko-5. The Holocene 16: 877–892.

Stockmarr, J., 1971. Tablets with spores used in absolute pollen analysis. Pollen et Spores 13: 615–621.

Stuiver, M. & P. Reimer, 1993. Extended 14C database and revised CALIB radiocarbon calibration program. Radiocarbon 35: 215–230.

Telford, R. J., E. Heegaard & H. J. B. Birks, 2004. The intercept is a poor estimate of a calibrated radiocarbon age. The Holocene 14(2): 296–298.

Tobolski, K., 2000. A Guide to Peat and Lake Deposits. Vademecum Geobotanicum. Naukowe PWN, Warszawa: 508 pp (in Polish with English summary).

Van der Knaap, W. O., J. F. N. van Leeuwen & B. Ammann, 2001. Seven years of annual pollen influx at the forest limit in the Swiss Alps studied by pollen traps: relations to vegetation and climate. Review of Palaeobotany and Palynology 117: 31–52.

Van der Knaap, W. O., J. F. N. van Leeuwen, W. Finsinger, E. Gobet, R. Pini, A. Schweizer, V. Valsecchi & B. Ammann, 2005. Migration and population expansion of *Abies, Fagus, Picea* and *Quercus* since 15000 years in and across the Alps, based on pollen-percentage threshold values. Quaternary Science Reviews 24: 645–680.

Vannière, B., G. Bossuet, A.-V. Walter-Simonnet, P. Ruffaldi, T. Adatte, M. Rossy & M. Magny, 2004. High-resolution record of environmental changes and tephrachronological markers of the Last Glacial-Holocene transition at Lake Lautrey (Jura, France). Journal of Quaternary Science 19: 797–808.

Vescovi, E., C. Ravazzi, E. Arpenti, W. Finsinger, R. Pini, V. Valsecchi, L. Wick, B. Ammann & W. Tinner, 2007. Interactions between climate and vegetation during the Late glacial period as recorded by lake and mire sediment archives in Northern Italy and Southern Switzerland. Quaternary Science Reviews 26: 1650–1669.

Von Grafenstein, U., U. Eicher, H. Erlenkeuser, P. Ruch, J. Schwander & B. Ammann, 2000. Isotope signature of the Younger Dryas and two minor oscillations at Gerzensee (Switzerland): palaeoclimatic and palaeolimnologic interpretation based on bulk and biogenic carbonates, Palaeogeography, Palaeoclimatology, Palaeoecology 159(3–4), Special Issue: Ammann, B. (ed.), Biotic Responses to Rapid Climatic Changes Around the Younger Dryas: 215–229.

Vrhovšek, D., G. Kosi & M. Zupan, 1982. The effect on water chemistry and phytoplankton of artificial inflow of the River Radovna into Lake Bled (Yugoslavia). Hydrobiologia 96: 225–242.

Vrhovšek, D., G. Kosi, M. Kralj & M. Bricelj, 1984. The effect of three consequent sanation measures in Lake Bled on physico-chemical and biological variables. Ekologija 19: 31–52.

Vrhovšek, D., G. Kosi, M. Kralj, M. Bricelj & M. Zupan, 1985. The effect of lake restoration measures on the physical, chemical and phytoplankton variables of Lake Bled. Hydrobiologia 127: 219–228.

Walker, I. R., 2001. Midges: chironomidae and related Diptera. In Smol, J. P., H. J. B. Birks & W. M. Last (eds), Tracking Environmental Change Using Lake Sediments, Vol. 4. Zoological Indicators. Kluwer Academic Publishers, Dordrecht: 43–66.

Wiederholm, T., 1983. Chironomidae of the Holarctic region. Keys and diagnoses. Part 1. Larvae. Entomologica Scandinavica Supplement 19: 1–457.

Willis, K. J. & T. H. van Andel, 2004. Trees or no trees? The environments of central and eastern Europe during the Last Glaciation. Quaternary Science Reviews 23: 2369–2387.

Willis, K. J., P. Sümegi, M. Braun & A. Tóth, 1995. The late Quaternary environmental history of Bátorliget, N. E. Hungary. Palaeogeography, Palaeoclimatology, Palaeoecology 118: 25–47.

Willis, K. J., E. Rudner & P. Sümegi, 2000. The Full-Glacial forests of central and southeastern Europe. Quaternary Research 53: 203–213.

Lake–peat bog transformation recorded in the sediments of the Stare Biele mire (Northeastern Poland)

Michał Gąsiorowski · Mirosława Kupryjanowicz

Originally published in the journal Hydrobiologia, Volume 631, No. 1, 143–154.
DOI: 10.1007/s10750-009-9807-8

Abstract The history of Stare Biele paleolake (northeast Poland) has been reconstructed using subfossil Cladocera remains and pollen and spores of aquatic and mire plants from a sediment core. Sediment accumulation began approximately 12,000 years ago during the Older Dryas chronozone. Throughout the entire Late Glacial period, the basin was a small, low-trophic state lake with a developed open-water zone. A well-recorded increase in the trophic state started at the beginning of Holocene. The lake reached its highest trophic level during the early and middle Atlantic chronozone. The first human activity in the lake catchment area occurred at this time, as recorded by fern spores and numerous charcoal grains. Repeated rises in lake water level are documented at the beginning and throughout the early part of the Younger Dryas. Two clear events of decreasing lake water levels are recorded, first during the middle part of the Younger Dryas and second in the Preboreal. Terrestrialization processes first intensified at the end of Atlantic period, which appears to correspond to a decrease in pH.

Keywords Pollen · Vegetation history · Cladocera · Terrestrialization · Late Vistulian · Holocene

Guest editors: K. Buczkó, J. Korponai, J. Padisák & S. W. Starratt
Palaeolimnological Proxies as Tools of Environmental Reconstruction in Fresh Water

M. Gąsiorowski (✉)
Institute of Geological Sciences, Polish Academy of Sciences, Twarda 51/55, 00-818 Warszawa, Poland
e-mail: mgasior@twarda.pan.pl

M. Kupryjanowicz
Institute of Biology, Department of Botany, University of Białystok, Świerkowa 20B, 15-950 Bialystok, Poland

Introduction

A terrestrialization process is identified by the infilling of a lake basin, and produces specific and sharply modified living conditions. Progressive shallowing leads to the migration of submergent and emergent vegetation as a lake fills and becomes a mire. Though the primary factor controlling this process is climate change, human activity is also very important, especially during the later Holocene.

The disappearance of an open-water zone and expansion of a littoral zone promote cladoceran species that are associated and/or largely restricted to plants. Therefore, the manner, timing, and rate of terrestrialization can be easily identified by cladoceran fossils from the sediments.

Cladoceran community changes in response to terrestrialization have been studied by Korhola (1990, 1992), Szeroczyńska & Gąsiorowski (2002) and Mirosław-Grabowska & Niska (2007). Nilssen & Sandoy (1990) investigated cladoceran changes primarily caused by pH variations. This study confirms very strong relationships between terrestrialization process, changes in submerged vegetation, changes in pH, and cladoceran community diversity.

This article presents the results of pollen and Cladocera analyses of Holocene sediments deposited in the Stare Biele fossil lake. This lake is an ideal location for studying past water level fluctuations; it is located in an area sensitive to precipitation, the sediment sequence is relatively thick, and fossil plant and cladoceran communities are relatively diverse. Therefore, analysis of the core provided data on many aspects of water environment evolution under changing hydrological conditions.

Site description

The Stare Biele mire (53°13′55″–53°15′35″ N and 23°30′20″–23°32′40″ E) is located in the Knyszyn Forest, 25 km to the northeast from Białystok (northeast Poland) at an altitude 144 m a.s.l. The Stare Biele mire corresponds generally to the Derazina stream catchment, which has an area of 256 ha (Fig. 1). It is an isolated complex of moor and moss with alder and pine forests, surrounded by moraine and kame hills. A basin developed in sands and gravels from the Riss glaciation, and was then filled by organic material during the late Würm glaciation and Holocene. The morphology of the basin bedrock varies and contains many deep, well-like pits.

Methods

The sediment core for palaeoecological studies was collected in 1995 with a Russian peatcorer from the site where the thickest organic deposits were found (Fig. 1), but it did not reach mineral levels. The 10-m-long profile was sub-sampled every 5 cm.

Pollen and spore analysis

Sediment samples for pollen analysis were treated with standard methods (Berglund & Ralska-Jasiewiczowa, 1986). The material was heated in 10% potassium hydroxide (KOH) in a water bath and then treated by acetolysis process (Faegri & Iversen, 1989). Mineral matter was removed with cold hydrofluroic acid (HF). At least 1000 arboreal pollen (AP) grains were counted for each sample. Non-arboreal pollen (NAP) and aquatic and mire taxa were also counted. The percentages of terrestrial pollen taxa were calculated from their total sum (AP + NAP = 100%). Percentage calculations for aquatic and mire taxa were calculated using the

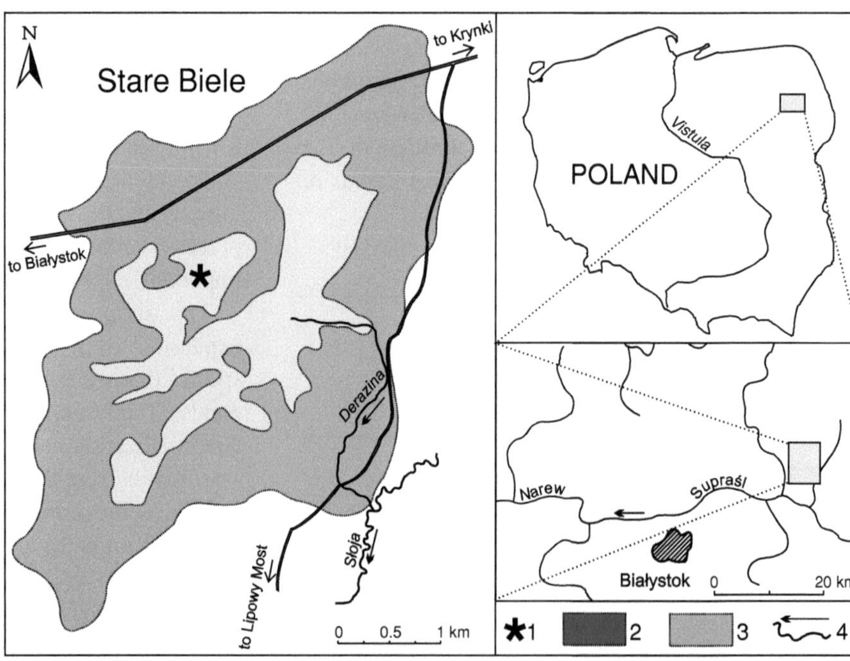

Fig. 1 Location of the Stare Biele range in northeastern Poland. 1, location of the core; 2, Stare Biele range; 3, Stare Biele peat bog; 4, Streams and rivers and its outflow directions

formula AP + NAP + taxon = 100%. Pollen diagrams were prepared and drawn with the POLPAL (Walanus & Nalepka, 1999).

Cladoceran analysis

Samples for cladoceran analysis were stored in the dark at 4°C. Each 1 cm^3 sample was prepared by the standard procedure (Frey, 1986); the sediment was treated with 10% hydrochloric acid (HCl) to remove calcareous material, and then heated in 10% potassium hydroxide (KOH) for 20–30 min using a magnetic stirrer to deflocculate the organic material. Samples were sieved through a 33-μm screen and transferred to distilled water. Before counting, samples were colored with a safranin dye. A minimum of 400 Cladocera remains (usually three or four slides, each 0.1 ml) were examined for each level, and total concentration per cm^3 was calculated. Results are presented in a total concentration diagram drawn using POLPAL software (Walanus & Nalepka, 1999), and cluster analysis was done using MSVP version 3.12 software (Kovach, 1998). The cladoceran zones were distinguished on the basis of cluster analysis results. The unweighted minimum variance method (Ward's method or sum-of-squares) was used along with squared Euclidean distance for cluster analysis (Birks, 1986). In order to enable easier dendrogram reading, $\log_{10}$ data transformation was applied prior to applying Ward's method. Only species with at least 5% abundance in each level were included in the statistical calculations.

Results

Pollen stratigraphy and changes in terrestrial vegetation

Seventy-six pollen samples were counted and 98 pollen and spore types were identified. A simplified pollen percentage diagram for terrestrial plants is shown in Fig. 2. The analyzed sequence was divided into eight local pollen assemblage zones (LPAZ). The profile of the Stare Biele sediment spans a period from the Older Dryas to the present. The ^{14}C data from the bottom of the sequence indicates an age of 13,900 ± 310 BP (17,650–15,650 cal years BP), which is older than that would be expected, perhaps due to the basin effect (Czerwiński et al., 2000). The sample from the upper portion of the profile gave a date of 4,000 ± 120 BP (4,850–4,150 cal years BP). As our study was limited to only two radiocarbon dates, a time scale was constructed on the basis of correlating local pollen zones from the Stare Biele profile with regional pollen zonation for north Poland (Noryśkiewicz & Ralska-Jasiewiczowa, 1989) and northeastern Poland (Ralska-Jasiewiczowa, 1989). The vegetation history was reconstructed primarily based on palynological results, but also draws on plant macrofossil investigations by Marek (Czerwiński et al., 2000).

Within the SB-1 *Juniperus–Hippophaë*—NAP LPAZ portion of the core (10.00–9.80 m), which represents the Older Dryas chronozone, the proportion of NAP is very high. The pollen curve of *Juniperus* culminates within this segment of the core. Values of *Salix* and *Hippophaë* are relatively high. Poaceae and Cyperaceae communities with *Ranunculus acris*-type and *Betula nana*-type shrubs dominated in humid places, while open vegetation with prominent *Artemisia* and Chenopodiaceae colonized drier sites. Shrubs of *Juniperus, Salix, Hippophaë,* and *Ephedra* were scattered across the landscape. A relatively high proportion of *Betula alba*-type pollen is found, as supported by *Betula pubescens* and *B. verrucosa* nutlets (Czerwiński et al., 2000), which indicate the existence of local birch stands. The role of pine was insignificant as suggested by its low pollen values, but the macrofossils confirm its local presence. Spruce might be occurring in this time period in more fertile and moist habitats, as is documented by the presence of *Picea abies* needles in the sediment (Czerwiński et al., 2000). The vegetation types were clearly the result of more severe, dry, and continental climatic conditions.

Within the SB-2 *Pinus–Betula* LPAZ core slice, (9.75–8.85 m), which represents the Alleröd, *Pinus sylvestris*-type and *Betula alba*-type pollens are dominant. Curves of NAP, *Juniperus,* and *Hippophaë* decline in this period compared to the prior period.

The subzone SB-2a *Betula–Salix* (9.75–9.35 m) is characterized by a peak of *Betula alba* type and a rise in the portion of *Pinus sylvestris*-type pollen. The pollen spectra reflect the development of birch–pine tree stands that were typical during the earlier part of the Alleröd chronozone, as reported from numerous regions of Poland and Europe as a whole

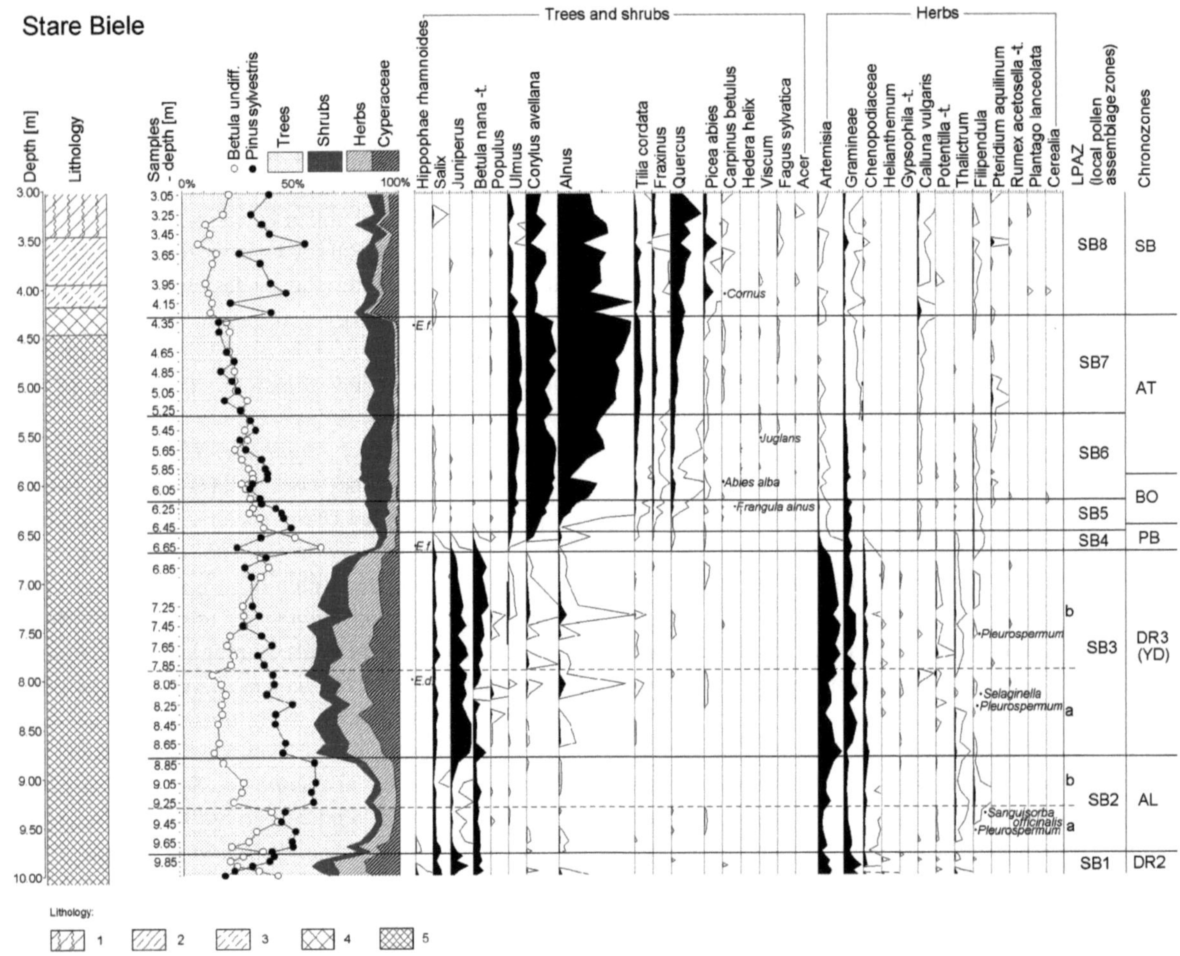

Fig. 2 Simplified pollen percentage diagram from Stare Biele—selected pollen curve of the terrestrial plants. The hollow silhouette curve denotes the 10× exaggeration of the percentages. Lithology: 1, sedge peat with *Phragmites*; 2, sedge-moss peat; 3, moss peat; 4, course detritus gyttja; 5, fine detritus gyttja. Pollen: *E. f.*—*Ephedra fragilis*, *E. d.*—*Ephedra dystachia*

(Noryśkiewicz & Ralska-Jasiewiczowa, 1989; Ralska-Jasiewiczowa, 1989; Ralska-Jasiewiczowa et al., 1998). The shrubs are still present, though gradually decreasing in abundance. The heliophyte herb vegetation was still widespread.

Subzone SB-2b *Pinus–Filipendula* (9.25–8.85 m) is characterized by a peak in the *Pinus sylvestris*-type pollen curve and lower values of *Filipendula*. The pollen values of *Betula alba* type decline during this period. The subzone corresponds to the later part of the Alleröd chronozone, which was a time of dominant open pine–birch forest with an understory of juniper and birch shrubs. The pine expanded in the catchment, partially replacing the birch. Damper places were occupied by tall herbs, such as *Filipendula*, *Thalictrum*, and *Salix* shrubs. Their development indicates a warmer and more oceanic type climate than in the SB-2a subzone. The dry open vegetation is still rich in taxa, but more limited in area.

The SB-3 *Juniperus*–Cyperaceae–*Artemisia* LPAZ zone (8.75–6.75 m) represents the Younger Dryas. Herein, the proportion of NAP is very high, and *Juniperus*, *Betula nana*-type, Cyperaceae, *Artemisia*, and Chenopodiaceae attain their maximum values. The vegetation of that time was probably of the parkland type, with well-developed shrub communities of *Juniperus*, *Ephedra* in drier places and *Betula nana*-type and *Salix* in moister soils, along with scattered groups of trees—mainly birch and rare pine. The dry habitats during this time period supported xeric grassland with abundant *Artemisia*, Chenopodiaceae, and a variety other plant taxa.

The onset of the SB-3a *Juniperus* LPASZ subzone (8.75–7.95 m) is characterized by a sudden decrease in *Pinus sylvestris*-type pollen, and by a distinct increase in *Juniperus*. Such a *Juniperus* peak in the later phase of Younger Dryas is typical of many pollen diagrams from northern (Ralska-Jasiewiczowa et al., 1998) and northeastern Poland (Noryśkiewicz & Ralska-Jasiewiczowa, 1989; Ralska-Jasiewiczowa, 1989); during this time, the climate was probably the coldest and driest of all the time periods represented in the core.

In the SB-3b *Betula–Betula nana* LPASZ subzone (7.85–6.75 m), tree pollen values are still low whereas shrub percentages are relatively high. The *Juniperus* percentage pollen curve decreases. Within the NAP, *Artemisia*, Cyperaceae, Poaceae, and Chenopodiaceae are the dominant pollen types. At the top of this zone, some tall herbs have reappeared. These changes suggest climatic change toward more humid conditions during that time.

During the SB-4 *Betula–Corylus–Ulmus* LPAZ zone (6.65–6.55 m), *Betula alba*-type pollen is the dominant taxon. *Pinus sylvestris*-type pollen also remains. The continuous curves of *Ulmus* and *Corylus avellana* begin here, rising slightly near top, or the youngest part, of the zone. The *Juniperus*, *Betula nana*-type, and *Salix* pollen curves gradually decline. A fall in NAP values is seen caused mostly by a reduction in Cyperaceae and *Artemisia* percentages. This zone presumably corresponds with the initial phase of Holocene—Early Preboreal subchronozone. The pollen record reflects the rapid spread of birch woodland, which replaced the *Juniperus* and *Betula nana*-type shrub communities. Grasslands with abundant *Artemisia* were still common at the beginning of the zone, but the variety of herbs contributing to this community is reduced. The developing tall herbs (*Filipendula* and *Thalictrum*) may have been associated with increase of climate humidity. Therefore, the development of birch forest suggests a maritime climate with relatively warm summers and mild, moist winters (Berglund et al., 1996). The proportion of non-arboreal pollen declines during this zone, indicating a closing of the forest canopy. Toward the end of the period, *Corylus avellana* and *Ulmus* appeared in the Stare Biele area.

The SB-5 *Pinus–Corylus–Ulmus* LPAZ zone (6.45–6.20 m) was characterized mainly by the pollen types *Pinus sylvestris* and *Betula alba*. *Corylus avellana* pollen values rise, while the *Ulmus* pollen curve declines. *Quercus*, *Fraxinus*, *Tilia cordata*, and *Alnus* pollen are continuously present from the base of the zone. *Hedera helix* pollen appears for the first time. This zone presumably corresponds with Late Preboreal and Early Boreal subchronozones. At the beginning of the zone, *Pinus sylvestris*-type taxa expanded within the birch woodland area. The pine forests were initially rather open, but with hazel expanding in the understory. On more fertile, humid soils, *Ulmus* expanded slowly. *Alnus* began its rapid colonization of damp lakeshores. *Tilia cordata*, *Fraxinus*, and *Quercus* arrived in the area. Tall herbs were still common on damper soils, and grasslands were widespread but with a reduced number of species. A continuous low-percentage curve of *Pteridium aquilinum* begins in this zone, and the appearance of numerous charcoal particles and fragments of charred plant parts may reflect of the presence of Mesolithic man.

The SB-6 *Alnus–Pinus–Corylus–Ulmus* LPAZ zone (6.15–5.35 m) begins with a rise in the share of *Alnus* pollen. *Pinus sylvestris*-type percentages drop, reflecting the gradual decline of the pine forest. *Betula alba*-type pollen shows a slightly declining trend. *Corylus avellana* values remain at 15%, while percentages of *Ulmus*, *Tilia cordata*, *Fraxinus*, and *Quercus* pollen rise slowly. *Viscum* pollen appears for the first time in this zone, which corresponds approximately with the Late Boreal and Early Middle Atlantic subchronozones. This was the time of dominant pine–birch forests, with deciduous trees slowly expanding onto more fertile soils. Fern (*Pteridium aquilinum*) spores and charcoal grains were found at both the bottom and top of the zone.

During the SB-7 *Alnus–Ulmus–Tilia–Fraxinus* LPAZ zone (5.25–4.35 m), *Ulmus*, *Alnus*, *Tilia cordata*, and *Fraxinus* attained their maximum pollen values. The pollen curve of *Corylus avellana* oscillates. The NAP values are the lowest here relative to anywhere in the profile. This zone corresponds with Middle and Late Atlantic subchronozone; at that time, the mixed deciduous forests reached their maximum Holocene development in the investigated area. The pine forests were restricted to the sandy soils, and they might have been encroached upon by oak. At the bottom of this zone, the spores of bracken (*Pteridium aquilinum*) attain their maximum values, and charcoals and charred plant parts are relatively abundant.

During the SB-8 *Pinus–Quercus* LPAZ zone (4.25–3.05 m), pollen values of the *Pinus sylvestris* type remain at 20–50% and those of *Alnus* at 10–30%. The *Ulmus* pollen curve oscillates, falling to 3% near the beginning of the zone and then rising temporarily in the middle of the zone. *Tilia cordata* and *Fraxinus* percentages decline. The *Quercus* curve rises gradually. This zone corresponds with the Early and Middle Subboreal subchronozones. At the beginning, *Picea abies* percentages start to increase, which reflects the immigration of spruce to northeastern Poland. The low middle Holocene pollen values of *Picea alba* in Stare Biele are due to the early- and middle Holocene existence of isolated spruce stands in Central Europe and possibly elsewhere in northeastern Poland; the simultaneous rise of *Picea* is caused by the expansion of this type of forest. The NAP values increase in relation to the abundance of Cyperaceae, which reflects only changes of local environment. For the first time, clear anthropogenic indicators such as Cerealia type and *Plantago lanceolata* (Behre, 1981) appear in this period. In the upper section of the core (above 3.05 m), pollen frequency is very low and sporomorphs are poorly preserved.

A pollen analysis of terrestrial plants does not always reflect lake conditions. Therefore, a reconstruction of lake history should be based mainly on the analysis of water plants and aquatic fauna development.

Development of water and mire vegetation

Changes in the pollen curves measured from local mire and water plants enabled specific zonation related exclusively to these taxa (Fig. 3). These pollen assemblage zones are marked with "bis" for distinction from the original zonation in the diagram (Fig. 2). Cyperaceae pollen was omitted during description of sediments below 4.30 m. The absence of its macrofossils (Czerwiński et al., 2000) suggests that these plants are present in the lake surroundings but absent in the water.

SB-1bis *Pediastrum* LPAZ (10.00–9.80 m). In this zone, coenobia of *Pediastrum* (*P. kawraiskyi*, *P. integrum* are noted) dominate among local plants. Single spores of *Equisetum* and pollen grains of *Potamogeton* and *Cladium mariscus* were present.

SB-2bis *Botryococcus* LPAZ (9.75–8.75 m). Measured values of *Botryococcus* are relatively high. Percentages of *Pediastrum* are significantly lower than in the previous zone. A peak for *Equisetum* was noted. Single pollen grains of *Typha latifolia*, *Menyanthes trifoliata*, *Potamogeton*, *Stratiotes aloides*, and spores of Filicales monolete and *Dryopteris thelypteris* are present.

SB-3bis *Sphagnum–Pediastrum* LPAZ (8.65–8.35 m). The quantity of water vascular plants pollen and algae was diminished. A continuous curve of *Sphagnum* starts and increases slightly. *Equisetum* spores are still frequent.

SB-4bis *Pediastrum–Sphagnum–Botryococcus* LPAZ (8.25–7.45 m). Coenobia of *Pediastrum* are in high abundance. Values of *Botryococcus braunii* rise a little. *Sphagnum* curve attains its maximum. *Equisetum* forms a continuous curve.

SB-5bis *Botryococcus–Pediastrum–Myriophyllum* LPAZ (7.35–6.75 m). Peak of *Myriophyllum spicatum/verticillatum* is the most characteristic feature of this zone. Measured values of *Botryococcus braunii* increase. Percentages of *Pediastrum* are still fairly high. The curves of *Equisetum* and *Sphagnum* decline.

SB-6bis *Nymphaea–Equisetum* LPAZ (6.65–6.45 m). This zone is characterized by peaks of *Nymphaea alba*, *Strartiotes aloides*, and *Equisetum*. The values of *Botryococcus braunii* and *Pediastrum* are low. The continuous curves of Filicales monolete and those of *Sphagnum* slightly increase.

SB-7bis *Botryococcus*–Filicales monolete–*Sphagnum* LPAZ (6.35–5.65 m). *Botryococcus braunii* is dominant. *Pediastrum duplex*, *P. boryanum*, and *Tetraedron* are present. Pollen of *Nuphar*, *Nymphaea alba*, *N. candida*, and *Potamogeton* were present.

SB-8bis *Botryococcus* LPAZ (5.55–5.25 m). Maximum values of *Botryococcus braunii* are characteristic feature of this zone. Percentages of Filicales monolete, *Equisetum*, and *Pediastrum* are similar to the previous zone. The *Sphagnum* curve declines.

SB-9bis *Nymphaea–Botryococcus–Pediastrum* LPAZ (5.15–4.35 m). The zone is characterized by maximum values of *Nymphaea alba* and *Alnus*. The proportion of *Botryococcus braunii* is lower than in the previous zone. The *Pediastrum* curve increased slightly.

SB-10bis Cyperaceae–Filicales monolete–*Menyanthes* LPAZ (4.25–3.05 m). Percentages of Cyperaceae are very high and for *Botryococcus braunii*, *Pediastrum*, and *Sphagnum* very low. Values of Filicales monolete fluctuate from 0.3% to 50%.

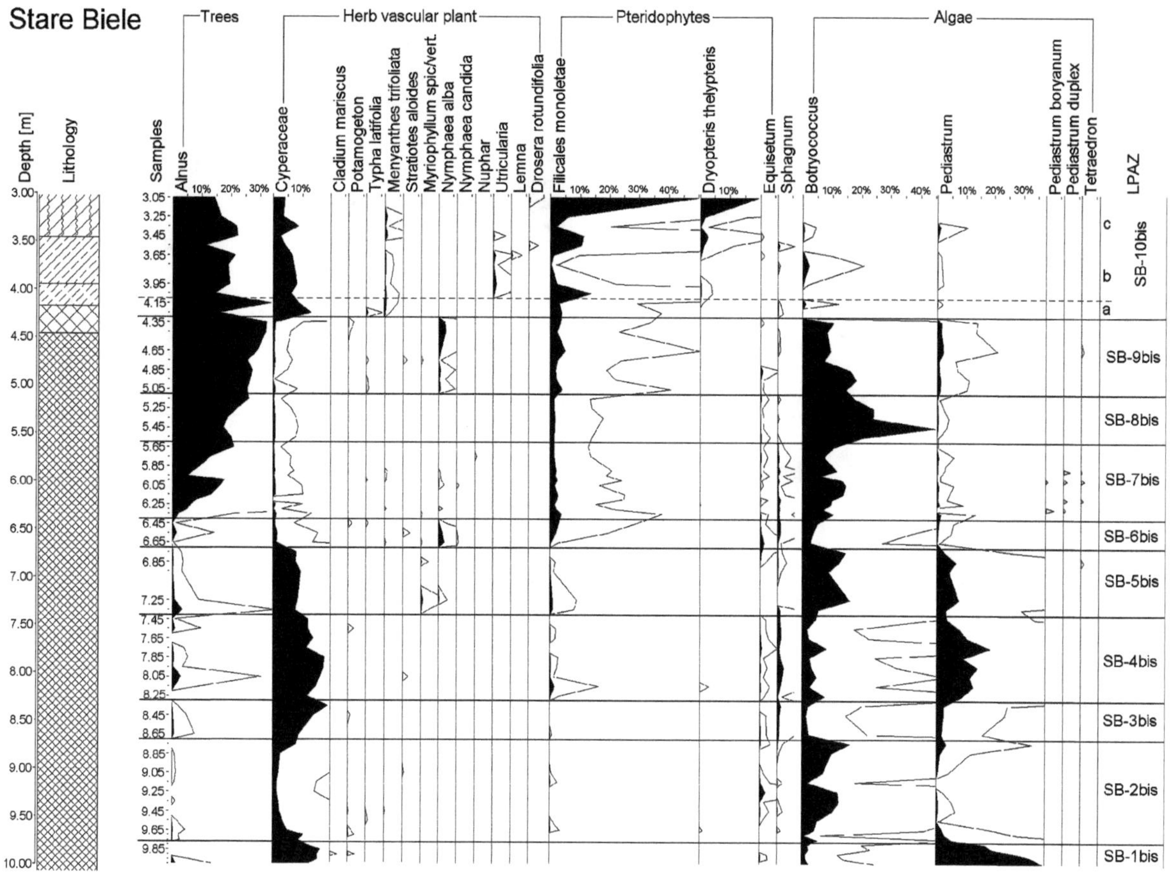

Fig. 3 Simplified pollen percentage diagram from Stare Biele—selected pollen curve of the mire and water plants. The hollow silhouette curve denotes the 10× exaggeration of the percentages. Lithological key is shown in Fig. 2

Continuous pollen curve of *Menyanthes trifoliata* starts.

SB-10abis *Typha*-Cyperaceae LPASZ (4.25–4.15 m). The pollen of *Typha latifolia* attains its maximum.

SB-10bbis *Utricularia* LPASZ (4.05–3.65 m). There is culmination of *Utricularia*. The pollen of *Lemna* is present.

SB-10cbis *Dryopteris thelypteris–Menyanthes* LPASZ (3.55–3.05 m). The *Dryopteris thelypteris* curve rapidly increases. *Menyanthes trifoliata* attains its maximum. The pollen of *Drosera rotundifolia* occurs for the first time.

Cladoceran zonation—CAZ (cladoceran assemblage zones)

The remains of 31 species of Cladocera were found: 26 species belong to Family Chydoridae, three to Bosminidae, one to Sididae, and one to the *Daphnia* group (Fig. 4). The identification of the species was based mainly on work by Frey (1958) and Szeroczyńska & Sarmaja-Korjonen (2007). The taxa were grouped in planktonic or littoral communities according to Müller (1964) and Alhonen (1970), and in ecological groups according to Whiteside (1970). Remains of the planktonic taxa were present mainly in the lower part of the profile and littoral species were dominant everywhere.

The presence of only two species and a low concentration of their remains characterize zone A (bottom to 9.80 m). *Alona affinis* was dominant and a few carapace shells of *Chydorus sphaericus* were found.

The beginning of zone B (9.80–8.30 m) is marked by the first occurrence of the planktonic species *Bosmina* and some Chydoridae species. The curves of *Pleuroxus uncinatus* and *Alona guttata* start to

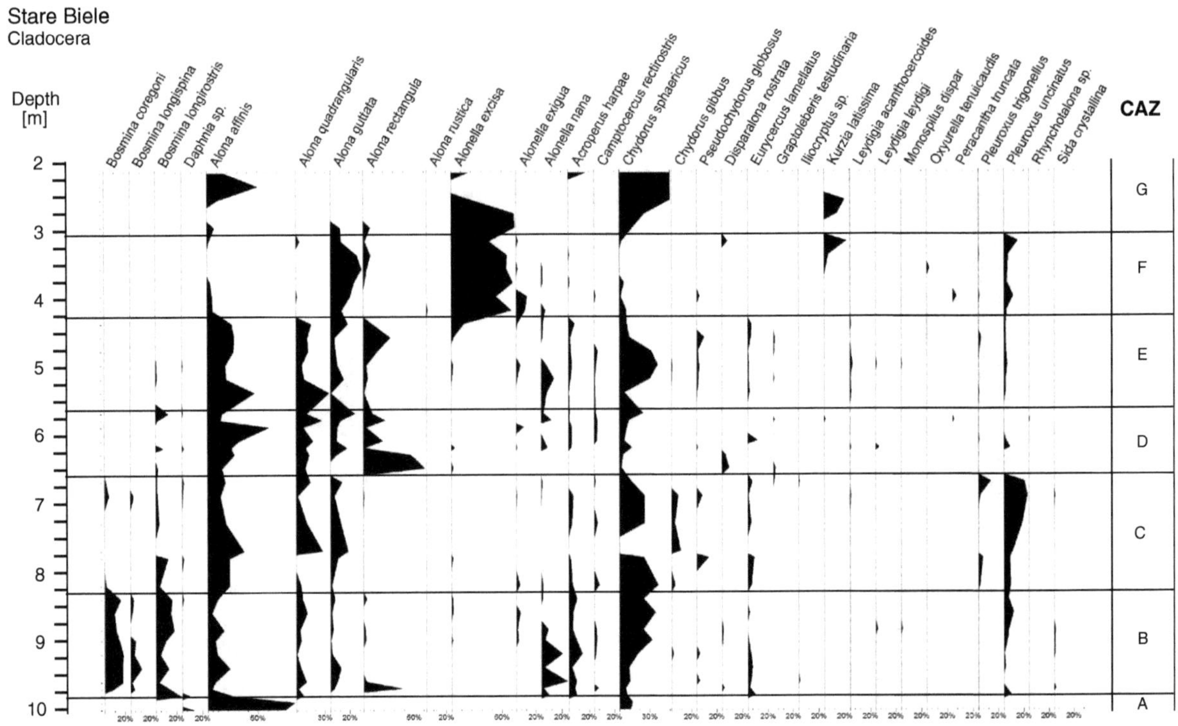

Fig. 4 Cladocera percentage diagram. CAZ—cladoceran assemblage zones

increase. *Alonella nana* and *Acroperus harpae* attained their maximums in this zone. The increase of the *Chydorus sphaericus* value is also significant. The concentration of *Alona affinis* decreases. The Cladocera assemblage showed maximum diversity expressed by an increasing value of the Shannon–Wiener Index (Fig. 5).

At the beginning of zone C (8.30–6.60 m) Bosminidae disappear and a sharp decrease in Chydoridae is observed. In the top part of this zone, *Bosmina coregoni* and *Bosmina longispina* are noted for the last time in Stare Biele history, and *Pluroxus uncinatus* attained its maximum.

Zone D (6.60–5.60 m) starts with decreasing *Pleuroxus* species. *Alona rectangular* dominates and attained its maximum percentage.

The zone E (5.60–4.30 m) is characterized by the maximum concentration of Cladocera (almost 20,000 individuals per 1 cm^3 of wet sediments), and the domination of *Chydorus sphaericus* and *Alona* species. In the lower part of this zone, *Alonella nana* is also common.

The concentration of most species decline at the beginning of zone F (4.30–3.10 m), and a complete change of cladoceran assemblage structure is observed. *Alonella excisa* dominates this zone and *Alona guttata* is abundant. This zone contains evidence of the transition from *A. guttata* to *A. guttata* var. *tuberculata.* The occurrence of usually common species like *Alona affinis* and *Chydorus sphaericus* is rare.

In zone G (3.10–2.20 m) only a few cladoceran remains were found. They belong only to few species: *Alona affinis, A. guttata, A. rectangular, Alonella excisa, Chydorus sphaericus*, and *Kurzia latissima.* Over 2.20 m, no cladoceran remains were observed.

Discussion

Cladocera and pollen comparison diagrams show good correlation (Fig. 6). Differences in Cladocera and general pollen division are characterized in these analyses. Cladocera mirror most changes in the water, while the pollen analysis indicates changes that took place in entire region. However, the boundaries of cladoceran zones correspond to the pollen zones boundaries as defined by aquatic and mire plants.

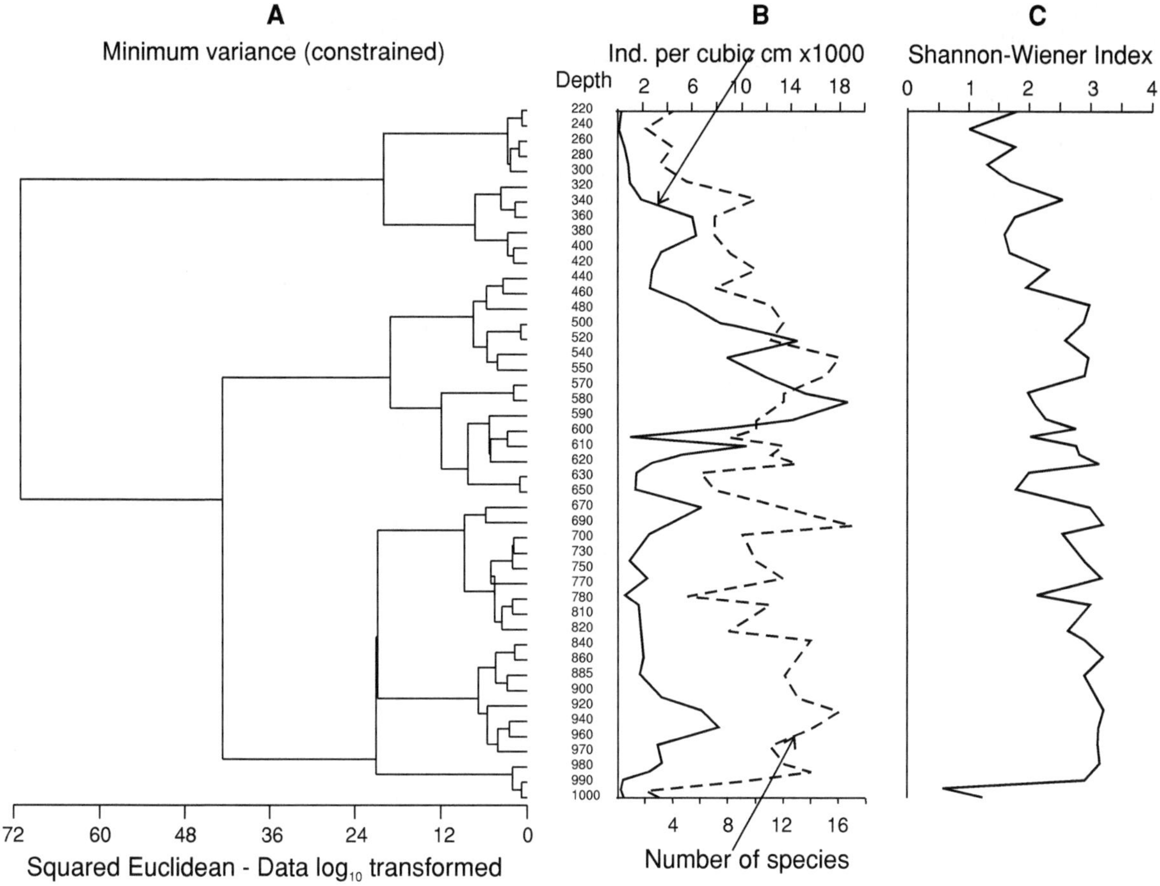

Fig. 5 Numerical analyses of cladoceran data: A, cluster analysis diagram; B, total number of specimens and species; C, Shannon–Wiener diversity index

Sediment accumulation began at the end of the last glacial period—approximately 12,000 years ago, during the Older Dryas chronozone. Stare Biele was a small, mid-forest water body, which evolved from a relatively oligotrophic lake with an open-water zone (with *Bosmina coregoni* and *Bosmina longispina*) to become eutrophic, and finally dystrophic, with only a few species of Cladocera.

Masses of *Pediastrum kawraiskyi* and *P. integrum*, during the first phase of the lake development (SB-1bis LPAZ), indicate there was oligotrophic, clean, and cool water (Jankovská & Komárek, 2000). Climatic conditions were too severe for most Cladocera. This is indicated by the presence of only 'arctic species'.

The profuse occurrence of *Botryococcus* with the sporadic occurrence of other coccal green algae (SB-2bis LPAZ) suggests specific lake conditions at Alleröd. It is likely that *Botryococcus* dominates in relatively extreme environments, which prevents the occurrence of others, such as *Pediastrum*. The water still was cold, clear, and oligotrophic (Jankovská & Komárek, 2000). This is also confirmed by the presence planktonic *Bosmina longispina* and *Bosmina coregoni*. The higher diversity and richness of the cladoceran community indicate that mild climatic conditions prevailed.

The cooling at the beginning of Younger Dryas is documented by the rapid reduction of the green alga *Botryococcus* (SB-3bis LPAZ) and Cladocera frequencies (the highest part of the zone B) in the lake. The gradual warming of climate in the later part of the Younger Dryas is confirmed by the rapid development of *Pediastrum* (SB-5bis LPAZ) and *Botryococcus* (SB-5 LPAZ), and a decrease of the Cladocera frequency (the zone C) without a significant drop in diversity. The peak of *Myriophyllum spicatum* (SB-5

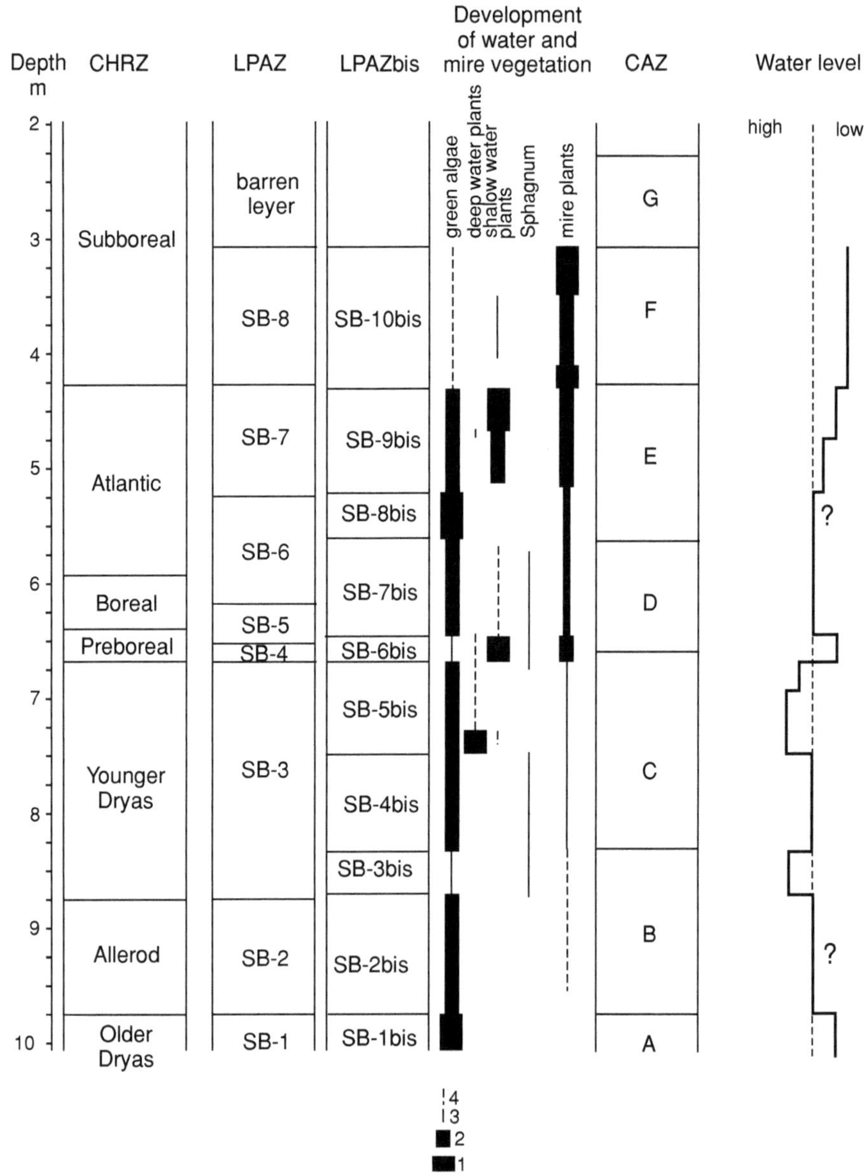

Fig. 6 Comparison between pollen and cladoceran zonation, reconstruction of aquatic and mire vegetation, changes of water level. CHRZ—chronozones, LPAZ—local pollen zones, LPAZbis—local pollen zones of water and mire vegetation, CAZ—Cladocera zones; 1, very abundant occurrence; 2, abundant occurrence; 3, regular presence; 4, sporadic presence

LPAZ) might suggest a rise in the lake level. This change is also indicated by the appearance of *Potamogeton praelongus* stones in the sediment (Marek, 2000). This species grows in water to the depth of ca. 3 m (Podbielkowski & Tomaszewicz, 1982).

The middle stages of Stare Biele evolution, beginning in the early Holocene, were also mirrored in the Cladocera grouping. Increases in *Alona affinis, A. quadrangularis, A. guttata, A. rectangula*, and *Chydorus sphaericus* clearly indicate an increase in nutrient load starting at the beginning of zone D. The pollen record (SB-6bis LPAZ) shows the presence of numerous macrofossils of *Nymphaea alba*, *Nuphar luteum*, and *Batrachium aquatile*, and document a decrease of water level during the Preboreal. However, the presence of *Bosmina longirostris* shows that an open-water zone still existed at this time. In central and northern Poland, a general decrease of water level occurs as a consequence of the climate warming and deforestation (Ralska-Jasiewiczowa & Starkel, 1988). The decline of green algae was attributed to the dense cover of water plants.

The assemblage of alga species, characteristic of the Early Atlantic (SB-7bis LPAZ), is typical for the eutrophic lakes and for warmer periods of the Holocene. This community of coccal green algae, qualitatively highly variable and quantitatively abundant, can be found in many central Europe localities since the Holocene climatic optimum. It is also common in lowland eutrophic lakes (Jankovská & Komárek, 2000). The Middle Atlantic was the time of the highest eutrophication of the lake. This community is documented by the pollen record (maximum values of *Botryococcus*—SB-8 LPAZ), and a very high number of Cladocera is still present (the zone E). The most frequent *Alona affinis*, *Alona quadrangularis*, and *Chydorus sphaericus* had maximums at that time. The appearance of macrofossils of *Najas minor* (Czerwiński et al., 2000) also documents this community. At present, this species is characteristic of shallow and eutrophic lakes (Podbielkowski & Tomaszewicz, 1982). The process of eutrophication might be connected to human activity at this time. This is suggested by the continuous curve of *Pteridium aquilinum* and numerous charcoal particles (Fig. 2). The pollen (SB-9bis LPAZ) and Cladocera records of the Late Atlantic show a significant decrease in the water level.

The end of gyttja deposition and beginning of peat development is evidence of the continued drop in the water table at the beginning of the Subboreal (Fig. 6). The dominance by *Alonella excisa* in zone F, also showed the intensification of terrestrialization processes, because this species is strongly related to macrophytes (Boronat et al., 2001) and lower lake productivity (Bos & Cumming, 2003). This process seems to be connected to decreasing pH, which is mirrored by the presence of *A. guttata* var. *tuberculata* (Korhola, 1990). The presence of these two species and *Kurzia latissima* is very strongly correlated. The terrestrialization of some lakes at 5,500–4,500 BP was characterized throughout Central Europe (Ralska-Jasiewiczowa & Starkel, 1988, Szeroczyńska & Gąsiorowski, 2002).

At the end of the Early Subboreal, some patches similar to the modern *Thelypteridi–Pragmitetum* Kuiper 1957 may have developed at Stare Biele (Podbielkowski & Tomaszewicz, 1982). This is suggested by the rapid rise of *Drypteris thelypteris* and Filicales monolete spores (SB-10c LPASZ). Only a few species of Cladocera survived, which suggests the continued terrestrialization and acidification of the lake (Korhola 1992).

Conclusions

Stare Biele has evolved from a lake to bog and all stages of its evolution were recorded in sediment. The process of transition from lake to peat bog was relatively rapid and clearly reflected by both the water plant and zooplankton communities. This change was correlated with the beginning of Subboreal chronozone and was probably related to changes in the climate conditions. This transition is also confirmed by changes in the forest composition. The terrestrialization process was slow. Specific conditions (i.e., low pH, low nutrient content) caused the long-term existence of an extraordinary cladoceran community dominated by *Alonella excisa*, and *Alona gutatta*, and the presence of *Kurzia latissima*. These species usually occur as less important or rare elements of zooplankton. The terrestrialization process was completed at the end of the Middle Subboreal chronozone. Many other lakes also disappeared at this time in northern Poland (Ralska-Jasiewiczowa & Starkel, 1988). Therefore, this change was likely controlled by drier climate conditions in the entire region. The activity of Mesolithic and Neolithic settlements is not reflected in the sediments. Clear human indicators were found only in the Subboreal sediments. However, as opposed to many other studies in central and northern Poland (Szeroczyńska, 1998a, b, c), they are not correlated with the zooplanktonic indicator of increased nutrient load. This is because the human influence on the lake and its surroundings was not significant.

Although the terrestrial plant record is generally well correlated with the zooplankton record, there are some differences. The terrestrial plant community usually mirrors regional conditions, and zooplankton reflects the state of a lake only. Zonation based only on water plant frequencies was better correlated to cladoceran zonation, because both reflected changes in the lake ecosystem.

References

Alhonen, P., 1970. On the significance of the planktonic/littoral ratio in the cladoceran stratigraphy of lake sediments. Societas Scientiarum Fennica, Commentationes Biologicae 35: 1–9.

Behre, K. E., 1981. The interpretation of anthropogenic indicators in pollen diagrams. Pollen and Spores 23: 225–245.

Berglund, B. E. & M. Ralska-Jasiewiczowa, 1986. Pollen analysis and pollen diagrams. In Berglund, B. E. (ed.), Handbook of Holocene Palaeoecology and Palaeohydrology. John Wiley & Sons, Chichester: 455–484.

Berglund, B. E., L. Barnekow, D. Hammarlund, P. Sandgren & I. F. Snowball, 1996. Holocene forest dynamics and climate changes in the Abisko area, northern Sweden: the Sonesson model of vegetation history reconsidered and confirmed. Ecological Bulletins 45: 15–30.

Birks, H. J. B., 1986. Numerical zonation, comparision and correlation of Quaternary pollen-stratigraphical data. In Berglund, B. E. (ed.), Handbook of Holocene Palaeoecology and Palaeohydrology. John Wiley & Sons, Chichester: 743–774.

Boronat, L., M. R. Miracle & X. Armengol, 2001. Cladoceran assemblages in a mineralization gradient. Hydrobiologia 442: 75–88.

Bos, D. G. & B. F. Cumming, 2003. Sedimentary Cladoceran remains and their relationship to nutrients and other limnological variables in 53 lakes from British Columbia, Canada. Canadian Journal of Fisheries and Aquatic Sciences 60: 1177–1189.

Czerwiński, A., A. Kołos & B. Matowicka (eds), 2000. Przemiany siedlisk i roślinności torfowisk uroczyska Stare Biele w Puszczy Knyszyńskiej. [Habitat and Peatbog Vegetation Changes in the Stare Biele Range in Knyszyn Forest. In Polish with English summary]. Białystok: 225 pp.

Faegri, K. & J. Iversen, 1989. Textbook of Pollen Analysis, 4th ed. John Wiley & Sons, Chichester: 328.

Frey, D. G., 1958. The late glacial cladoceran fauna of a small lake. Archiv für Hydrobiologie 54: 209–275.

Frey, D. G., 1986. Cladocera analysis. In Berglund, B. E. (ed.), Handbook of Holocene Palaeoecology and Palaeohydrology. John Wiley & Sons, Chichester: 667–692.

Jankovská, V. & J. Komárek, 2000. Indicative value of *Pediastrum* and other coccal green algae in palaeoecology. Folia Geobotanica 35: 59–82.

Korhola, A., 1990. Paleolimnology and hydroseral development of the Kotasuo Bog, Southern Finland, with special references to the Cladocera. Annales Academiae Scientiarum Fennicae 155: 1–40.

Korhola, A., 1992. The Early Holocene hydrosere in a small acid hill-top basin studied using crustacean sedimentary remains. Journal of Paleolimnology 7: 1–22.

Kovach, W. L., 1998. MVSP—A Multivariate Statistical Package for Windows, ver. 3.0. Kovach Computing Services, Pentraeth, Wales, UK.

Marek, S., 2000. Analiza biostratygraficzna profili torfowych. [Biostratigraphic analysis of peat cores] (in Polish). In Czerwiński, A., A. Kołos & B. Matowicka (eds), Przemiany siedlisk i roślinności torfowisk uroczyska Stare Biele w Puszczy Knyszyńskiej [Habitat and Peatbog Vegetation Changes in the Stare Biele Range in Knyszyn Forest]: 70–77.

Mirosław-Grabowska, J. & M. Niska, 2007. Isotope and Cladocera data and interpretation from the Eemian optimum and postoptimum deposits, Kaliska palaeolake (Central Poland). Quaternary International 175: 155–167.

Müller, W. P., 1964. The distribution of cladoceran remains in surfical sediments from three northern Indiana lakes. Investigation Indiana Lakes & Streams 6: 1–63.

Nilssen, J. P. & S. Sandoy, 1990. Recent lake acidification and cladoceran dynamics: surface sediment and core analysis form lakes in Norway, Scotland and Sweden. Philosophical Transactions of the Royal Society of London, Series B 327: 299–310.

Noryśkiewicz, B. & M. Ralska-Jasiewiczowa, 1989. Type region P-w: Dobrzyń-Olsztyn Lake Districts. In Ralska-Jasiewiczowa, M. (ed.), Environmental Changes Recorded in Lakes and Mires of Poland During the Last 13,000 years. Part Three. Acta Palaeobotanica 29: 85–93.

Podbielkowski, Z. & H. Tomaszewicz, 1982. Zarys hydrobotaniki. [The Epitome of Hydrobotany. In Polish]. Państwowe Wydawnictwo Naukowe, Warszawa: 531.

Ralska-Jasiewiczowa, M., 1989. Type region P-x: Masurian Great Lakes District. In Ralska-Jasiewiczowa, M. (ed.), Environmental Changes Recorded in Lakes and Mires of Poland During the Last 13,000 years, Part Three. Acta Palaeobotanica 29: 95–100.

Ralska-Jasiewiczowa, M. & L. Starkel, 1988. Record of the hydrological changes during the Holocene in the lake, mire and fluvial deposits of Poland. Folia Quaternaria 57: 91–127.

Ralska-Jasiewiczowa, M., D. Demske & B. van Geel, 1998. Late-Glacial vegetation history recorded in the lake Gościąż sediments. In Ralska-Jasiewiczowa, M., T. Goslar, T. Madeyska & L. Starkel (eds), Lake Gościąż, Central Poland. A Monographic Study. W. Szafer Institute of Botany, Polish Academy of Sciences, Kraków: 128–142.

Szeroczyńska, K., 1998a. Palaeolimnological investigations in Poland based on Cladocera (Crustacea). Palaeogeography Palaeoclimatology Palaeoecology 140: 335–345.

Szeroczyńska, K., 1998b. Anthropogenic transformation of nine lakes in Central Poland from Mesolithic to modern times in the light of Cladocera analysis. Studia Geologica Polonica 112: 123–165.

Szeroczyńska, K., 1998c. Correlation between human activity and trophic stages in the Lake Gościąż development based on cladoceran analysis. In Ralska-Jasiewiczowa, M., T. Goslar, T. Madeyska & L. Starkel (eds), Lake Gościąż, Central Poland. A Monographic Study. W. Szafer Institute of Botany, Polish Academy of Sciences, Kraków: 294–297.

Szeroczyńska, K. & M. Gąsiorowski, 2002. Palaeohydrological aspect of transformation of lakes into peat bogs during Middle Holocene on the basis of Cladocera analysis in the northern Poland. Studia Quaternaria 19: 55–60.

Szeroczyńska, K. & K. Sarmaja-Korjonen, 2007. Atlas of subfossil Cladocera from central and northern Europe. Friends of the Lower Vistula Society, Poland.

Walanus, A. & D. Nalepka, 1999. Polpal Program for counting pollen grains, diagrams plotting and numerical analysis. Acta Palaeobotanica, Supplementum 2: 659–661.

Whiteside, M. C., 1970. Danish Chydorid Cladocera: modern ecology and core studies. Ecological Monographs 40: 79–118.

Diatoms as a proxy in reconstructing the Holocene environmental changes in the south-western Baltic Sea: the lower Rega River Valley sedimentary record

Andrzej Witkowski · Bernard Cedro · Agnieszka Kierzek · Damian Baranowski

Originally published in the journal Hydrobiologia, Volume 631, No. 1, 155–172.
DOI: 10.1007/s10750-009-9808-7

Abstract This study focuses on diatom assemblages occurring in cores of Late-glacial and Holocene deposits retrieved from the mouth of the lower Rega River valley, of the southern coast of the Baltic Sea. Sediment samples from four cores were the subject of the present study. Diatom-inferred environmental characteristics, e.g., water level; water salinity (conductivity), trophic status and pH, within each core are presented. Diatom assemblage zones (DAZ) were distinguished, based on differences in the distribution of particular ecological groups. Each DAZ appears to be related to environmental changes during the deposition of a given sediment interval. The Late-glacial (Allerød) sediments originated in a shallow lake with increasing concentrations of solutes and nutrients. The Holocene record begins in the early Atlantic Chronozone and the diatoms point to weakly brackish-water sediments deposited in a shallow water environment. During the period of 8,500–5,800 cal year BP sedimentation took place in a shallow embayment and/or lagoon. From ca. 5,800 cal year BP onwards sedimentation took place in a peat bog environment alternating with Aeolian deposition. Changes in diatom community structure imply a close relationship with the climate-controlled eustatic rise of the ocean level and its consequence Littorina transgression. As with other southern Baltic Sea localities, brackish-water diatoms appear in the sediments, signaling the onset of marine transgression somewhat earlier than previously accepted. Differences and similarities in diatom assemblages and the palaeogeographic development of nearby regions within the Baltic Sea basin and lagoons (coastal areas) from different geographic regions, are also discussed.

Keywords Diatom assemblages · Palaeoenvironmental reconstruction · Late-glacial and Holocene sediments · Sea level change · Southern Baltic Sea · Lower Rega valley

Guest editors: K. Buczkó, J. Korponai, J. Padisák & S. W. Starratt
Palaeolimnological Proxies as Tools of Environmental Reconstruction in Fresh Water

A. Witkowski (✉) · A. Kierzek · D. Baranowski
Department of Palaeoceanology, Institute of Marine Sciences, University of Szczecin, Mickiewicza 18, 70-383 Szczecin, Poland
e-mail: witkowsk@univ.szczecin.pl

B. Cedro
Department of Geology and Palaeogeography, Institute of Marine Sciences, University of Szczecin, Mickiewicza 18, 70-383 Szczecin, Poland

Introduction

Diatoms (Bacillariophyceae) are among the most useful tools for monitoring the environmental status of recent inland (e.g. Stevenson & Pan, 1999; Smol & Cumming, 2000) and marine (Witkowski, 1994; Snoeijs, 1999) water bodies. This is due to the fact that the diatom flora species composition responds to shifts in environmental variables. Studies of recent environments provide calibration sets for the inclusion of diatoms in palaeoceanographic, palaeogeographic and palaeoecological reconstructions (e.g. Gasse, 1994; Zielinski & Gersonde, 1997; Fritz et al., 1999; Weckström & Juggins, 2005). Inland water bodies, lakes in particular, are sensitive to climatic changes, and diatom assemblages can be used as indicators of climate variability (e.g. Fritz et al., 1999). Changes in the species composition of diatom assemblages and proportions of particular ecologic groups are frequently applied to reconstructing past environmental conditions. Results of diatom analysis allow reconstruction of a variety of environmental variables, e.g. palaeotemperature of surface waters, pH, salinity (conductivity), variation in fresh water inflow in estuarine environment, water depth, and productivity (e.g. Fritz et al., 1991; Zielinski & Gersonde, 1997; Starratt, 2007).

Diatom studies have long been used in palaeoenvironmental reconstructions of the Baltic Sea, in time spans from deglaciation in the Weichselian Lateglacial up to recent times. Reconstruction of the palaeogeographic development and in stratigraphy of the Baltic Sea basins with diatoms have been performed by, e.g. Witkowski (1994), Sohlenius et al. (1996), and Andrén et al. (2000a, b). Diatom analysis has also been applied to studies of shallow water and lagoon deposits, e.g. Przybyłowska-Lange (1974, 1979), Jensen et al. (1997), Witkowski et al. (2004, 2005), Yu et al. (2005), Berglund et al. (2005), Janczak-Kostecka & Kostecki (2006). The major aim of these diatom analyses was to document changes in environmental variables. Earlier studies performed in 1970s and 1980s, e.g. Przybyłowska-Lange (1974, 1979), were usually limited to reconstructing palaeosalinity and its relationship to the sea level rise and habitat changes as a result of water level rise. Beginning from 1990s, the diatom application was expanded and reconstructions of changes in salinity were supplemented with reconstructions of trophy, pH and of human impact (e.g. Denys, 1990; Witkowski and Pempkowiak, 1995; Andrén, 1999; Clarke et al., 2003; Yu et al., 2005). Reliable inferences of palaeoenvironmental conditions were possible due to progress in calibration studies on diatom assemblages inhabiting lagoons and coastal areas. Most of these species have well-documented autecological characteristics in terms of salinity, pH and nutrients requirement (Weckström et al., 2002; Weckström & Juggins, 2005). Diatoms were also recognized as useful biostratigraphic markers of particular stages of the Baltic Sea development (e.g. Witkowski, 1994; Snoeijs, 1999; Andrén et al., 2000a, b; Emeis et al., 2003).

Over the past several decades, application of diatom analysis in the study of lagoon and coastal deposits has received more attention worldwide. The primary aims of these studies have been climate change impact and palaeogeographic developments of coastal areas. Dated cores of lagoon and coastal sediments were studied from Belgian coast (Denys, 1990), Iberian Peninsula (Bao et al., 2007), Australia (Fluin et al., 2007; Saunders et al., 2008), Argentina (Hassan et al., 2006), Uruguay (Inda et al., 2006), to name a few.

In general, studies in the Baltic Sea basin, other lagoons referred to above, and in saline lakes point out that diatom assemblage composition is regulated by climatic (precipitation) factors. Precipitation impacts the mineralization (salinity) of water and further pH, trophy (e.g. Fritz et al., 1999) and freshwater inflow into the coastal (estuarine/lagoonal) systems (Starratt, 2007). A somewhat different situation is found in newly deglaciated areas in which leaching of the solutes can cause a distinct increase in water mineralization which can be reconstructed from diatom assemblages (Jensen et al., 1997). The latter situation was quite common in the Baltic Sea basin after the Vistulian ice sheet retreat and as a result carbonate rich lacustrine deposits originated (Marciniak, 1981; Jensen et al., 1997).

Interestingly diatom flora of the Baltic Sea lagoons and of lagoons from other geographic regions show high degree of similarity. Frequently dominant are the small fragilarioid species (e.g. *Fragilaria brevistriata* Grunow, *F. lapponica* Grunow, *Fragilaria pinnata* Ehrenberg, *Staurosira construens* Ehrenberg, *S. venter* (Ehrenberg) Cleve & Möller and *S. subsalina* (Grunow) Lange-Bertalot). Some species of broad geographic distribution in lagoons and coastal sediments have been recently described in the case of the Baltic Sea. Included in this group are, e.g. *Fragilaria geocollegarum* Witkowski et Lange-Bertalot, *F. sopotensis* Witkowski et Lange-Bertalot and *Staurosira punctiformis* Witkowski, Lange-Bertalot & Metzeltin (Witkowski et al., 2000). At least a few of them, i.e. *F. geocollegarum* and *S. punctiformis* are worldwide in their distributions. These fragilarioid species mentioned above occur, sometimes in high relative abundance, at such distant sites as Baltic Sea coastal waters (e.g. Busse & Snoeijs, 2002; Weckström & Juggins, 2005), coastal plains of the North Sea (Denys, 1990), western coast of Iberian Peninsula (Bao et al., 2007), Brazil (Sylvestre et al., 2005), Uruguay (Bracco et al., 2005), Argentina (Hassan et al., 2006) or South-Eastern Australia lagoons and saline lakes (Fluin et al., 2007; Saunders et al., 2008). The reasons for dominance of fragilarioid taxa in unstable conditions (such as contact zones between marine and freshwater environments) were discussed by, e.g. Stabell (1985), Denys (1990) and Weckström & Juggins (2005). These taxa are referred to as opportunistic, fast reproducing with a broad tolerance, making them highly competitive (e.g. Weckström & Juggins, 2005) in these types of habitats. However, as shown by Weckström & Juggins (2005), they find optimum living conditions and reach the highest abundances, exceeding 20–30%, in brackish-water, i.e. with 1–6 psu. Here, they may also be abundant in salinities exceeding the above-mentioned optimum range (e.g. Hassan et al., 2006; Fluin et al., 2007).

Palaeolimnological development of coastal lagoons is complex and regulated by a combination of climatic, eustatic, isostatic and tectonic factors (e.g. Rotnicki et al., 1999; Rotnicki, 2001; Berglund et al., 2005; Harff et al., 2007; Uścinowicz et al., 2007). At open oceanic coasts with their regional to local variation, eustatic sea level rise is the dominant factor responsible for the development of lagoons and coastal wetlands (e.g. Garcia-Rodriguez & Witkowski, 2003; Bao et al., 2007). In the Baltic Sea basin, in general, coastal lagoon development depends on two factors, i.e. isostatic uplift/sinking related to the deglaciation of the Scandinavian Peninsula, and eustatic sea level rise (e.g. Yu et al., 2005; Witkowski et al., 2005; Uścinowicz et al., 2007; Harff et al., 2007). Isostatic uplift impacts the development of lagoons along the coasts of the Scandinavian Peninsula. This results in isolation of the brackish water embayments which terminate as freshwater lakes (Miller, 1986; Westman and Hedenström, 2002; Yu et al., 2005).

Lagoons located along the southern Baltic Sea coasts are primarily developed under the impact of eustatic sea level rise. However, as already postulated by Rotnicki (2001) in the development of these lagoons, eustatic sea level rise is combined with tectonics, which acts as a local factor. Hence, in quite closely located lagoons, e.g. Leba Lake, Rega Valley and Szczecin Lagoon, the beginning of marine transgression can differ by ca. 1,500 cal year (Rotnicki, 2001; Witkowski et al., 2004; Borówka et al., 2005).

In this article, we describe the development of the area near the mouth of the Rega river during the Lateglacial and Holocene. The extent of changes in environmental conditions is shown by means of diatoms deposited in the limnic, lagoonal and littoral marine sediments. The results of our study enable reconstruction of changes in the relative sea level variation and in salinity (conductivity), trophic and pH conditions in the study area during the Lateglacial and Holocene.

Materials and methods

Diatom analysis was performed on sediment cores retrieved from the mouth of the lower Rega River valley in the vicinity of Mrzeżyno between the Baltic Sea coast and the oxbow of the so-called "Old Rega" (Fig. 1). In total, 21 cores were drilled in this area with depths ranging from 7 to 13 m. Drilling was completed by means of Meres® device, which allows sub-sampling without contamination or mixing (Rotnicki et al., 1999).

Analyses of the diatom flora were performed on core Trz-15 (43 samples), Trz-20 (65 samples), Trz-21 (19 samples) and Trz-14 (5 samples). Microscopic slides were prepared according to the method of Battarbee (1986). For examination by scanning electron microscope, selected samples were prepared by means of the standard method.

Diatom analysis was performed on a Leica DMLB microscope equipped with 100× PlanApo objective (n.a. 1.4). SEM examination was performed by means of Hitachi S-4500 field emission apparatus. The diatom slides are stored in the Diatom Collection of Andrzej Witkowski in the Institute of Marine Sciences at the University of Szczecin (SZCZ).

Approximately, 300 valves were counted from each sample, but the counts ranged from 293 to 383. Identification of diatoms was aided with Krammer & Lange-Bertalot (1986, 1988, 1991a, b); Lange-Bertalot (2001); Witkowski et al. (2000) and specific taxonomic publications. The percentages of particular ecological

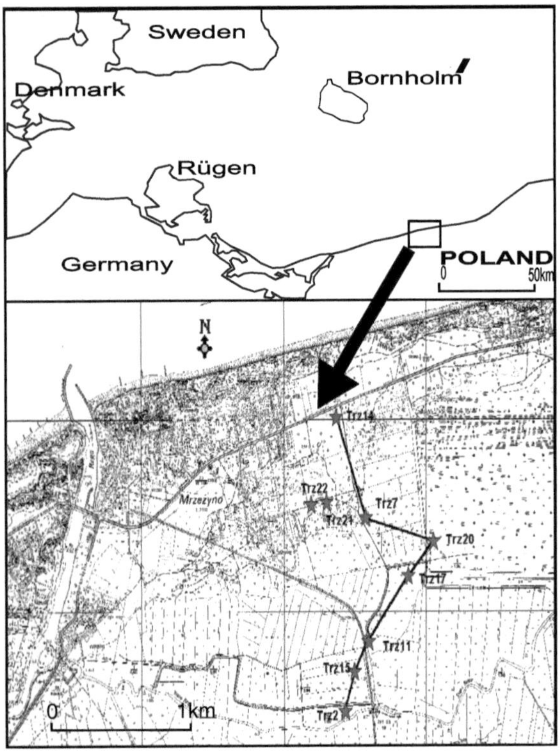

Fig. 1 Location of the study area

groups were computed by means of the OMNIDIA ver. 3 software, which has a database (Omnis7) with information on more than 11,000 species. Most of the taxa have fairly well-defined ecological characteristics, though for a significant number of them, such information is missing. Depending on the reconstructed variable, the percentage of taxa with unknown autecologies was nearly 40%. In cases where taxa are absent in the OMNIDIA data set, information available in the diatom literature and our own observations were included. Diagrams showing the percentages of dominant taxa and ecological groups were constructed. This was done with the help of Tilia® software. The diatom assemblage zones (DAZ) were distinguished by means of Tilia dendrograms which involve sum of least squares after being transformed with square root.

Sediment dating

Radiocarbon measurements were performed at Radiocarbon Laboratory, Silesian University of Technology and at the Poznan Radiocarbon Laboratory. Most of the dates were obtained from samples of bulk sediment, but some others from AMS analyses of bivalve shells. Altogether, 15 samples have been dated by conventional ^{14}C methods and two shells, one *Cardium* sp. and the second *Mytilus* sp. were dated by means of ^{14}C AMS method. Among the sediment cores included in this article, eight samples were dated (cf. Table 1). Radiocarbon dates of conventional age were calibrated with the INTCAL98 table (Stuiver et al., 1998).

Environmental setting

The Rega River valley constitutes a part of the coastal lowlands of the southern Baltic Sea incised by rivers discharging from the Pomeranian Lakeland Plateau. The palaeogeographic development of the mouth of the Rega river valley began during the Late-glacial. The oldest morphologic unit of the area studied is ground morainic plateau composed of boulder clay and glacial sands (Dobracka, 1992). The dominant form in the surface morphology of the study area is peat lowland which originated from the inundation and after marginal infill of the valley melt water. Significant in terms of the surface area is aeolian sands plain. On this plain, dunes varying in size and relative altitude (2–14 m) developed (Fig. 1). All the cores involved in this study were retrieved from lowland areas. The altitude of particular coring sites ranged between 0.5 and 2.0 m a.s.l. (Table 1).

Results

The geological N–S profile (Figs. 1, 2) extends from the road connecting Mrzeżyno and Rogowo (North) to the Old Rega River bridge towards the village Roby (South).

Core Trz-15

In core Trz-15, five local DAZ Trz-15-I to Trz-15-V were distinguished (Figs. 3, 4). A total of 294 diatom taxa representing 53 genera were identified (18 centric taxa in 6 genera and 276 pennate taxa in 47 genera). However, only 12 taxa attained a relative abundance exceeding 10% in at least one sample, and these are illustrated in the diagram. The following taxa were included in this group: *Amphora pediculus* (Kützing) Grunow, *Aulacoseira ambigua* (Grunow in Van

Table 1 Information on ^{14}C dates of the material studied

Sample code	Type of material dated	Depth (m below surface or bottom)	Altitude (m a.s.l.)	Latitude	Longitude	Radiocarbon age ^{14}C (BP)	Calibrated age range 68% cal year BP	Lab code
Trz2/43	Peat	5.85–5.90	−5.1	54°7′59″	15°18′55″	7680 ± 140	8640–8330	Gd-30067
Trz7/46	Peat	7.80–7.85	−6.8	54°8′38″	15°19′8″ E	7680 ± 40	8480–8400	Gd-12641
Trz7/74	Peat	10.95–11.00	−10.2	54°8′38″	15°19′8″ E	10390 ± 220	12850–11750	Gd-15670
Trz11/69	Gyttja	7.00–7.05	−6.2	54°8′10″	15°19′3″ E	7560 ± 140	8460–8180	Gd-18298
Trz14/39	Organic-rich sand	5.3–5.5	−4.5	54°8′51″	15°18′57″	4130 ± 80	4730–4530	Gd-15894
Trz15/12	Peat	2.5–2.7	−2.2	54°8′7″	15°18′5″	4835 ± 75	5660–5570	Gd-30117
Trz15/38	Peat	5.8–5.9	−5.4	54°8′7″	15°18′55″	6990 ± 80	7880–7730	Gd-12838
Trz15/47	Peat	7.0–7.1	−6.6	54°8′7″	15°18′55″	7500 ± 80	8390–8290	Gd-12832
Trz15/50	Peat	7.5–7.6	−7.1	54°8′7″	15°18′55″	7530 ± 220	8600–8050	Gd-17390
Trz15/90	Peat	10.7–10.8	−10.3	54°8′7″	15°18′55″	12160 ± 140	14200–13820	Gd-12839
Trz17/58	Peat	6.3–6.35	−5.8	54°8′20″	15°19′18″	7230 ± 140	8200–7930	Gd-15900
Trz20/31	Organic-rich sand	3.5–3.55	−3.0	54°8′24″	15°19′30″	5900 ± 70	6800–6630	Gd-15910
Trz20/54	Peat	5.75–5.85	−5.3	54°8′24″	15°19′30″	7170 ± 75	8050–7930	Gd-12835
Trz21/34	Shell	5.7–5.8	−4.8	54°8′36″	15°18′49″	6540 ± 40	7250–7110	Poz 20538
Trz22/32	Shell	6.8–6.9	−5.9	54°8′37″	15°18′53″	6740 ± 40	7420–7300	Poz 20539
Trz29/10	Peat	1.05–1.10	−0.05	54°8′38″	15°19′5″	1700 ± 50	1680–1600	GdS-766
Trz29/19	Peat	1.8–1.9	−0.9	54°8′38″	15°19′5″	5630 ± 50	6520–6440	Gd-11985
Trz29/57	Peat	5.7–5.8	−4.8	54°8′38″	15°19′5″	7320 ± 140	8265–8060	GdS-767

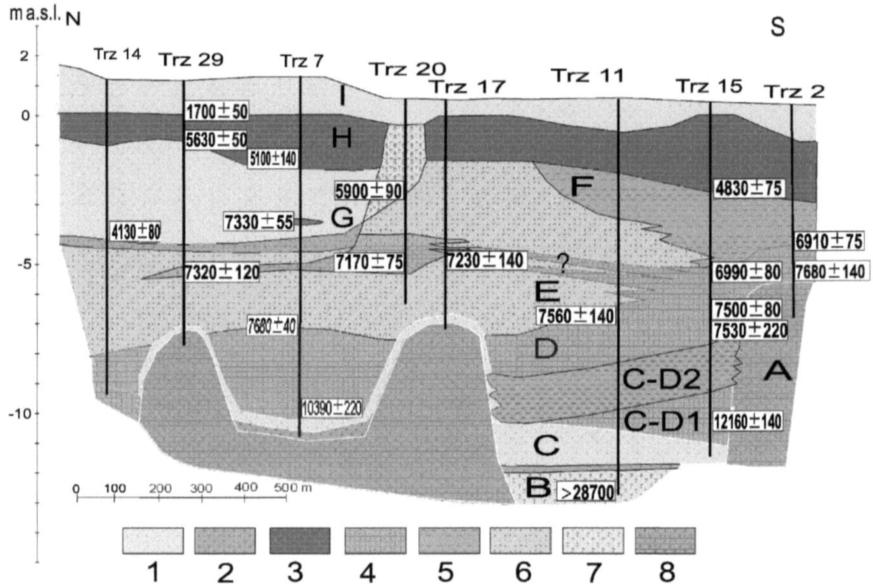

Fig. 2 Geological cross section of the sedimentary record studied. Ryc. 1. Geological cross section: Mrzeżyno-Rogowo—Stara Rega. 1—sand, 2—sand and gravel, 3—peat, 4—lagoon deposits (peat, gyttja, clay and mud), 5—gyttja and mud, 6—marine sand, 7—organic matter-rich sand, 8—boulder clay (unit symbols explained in the text)

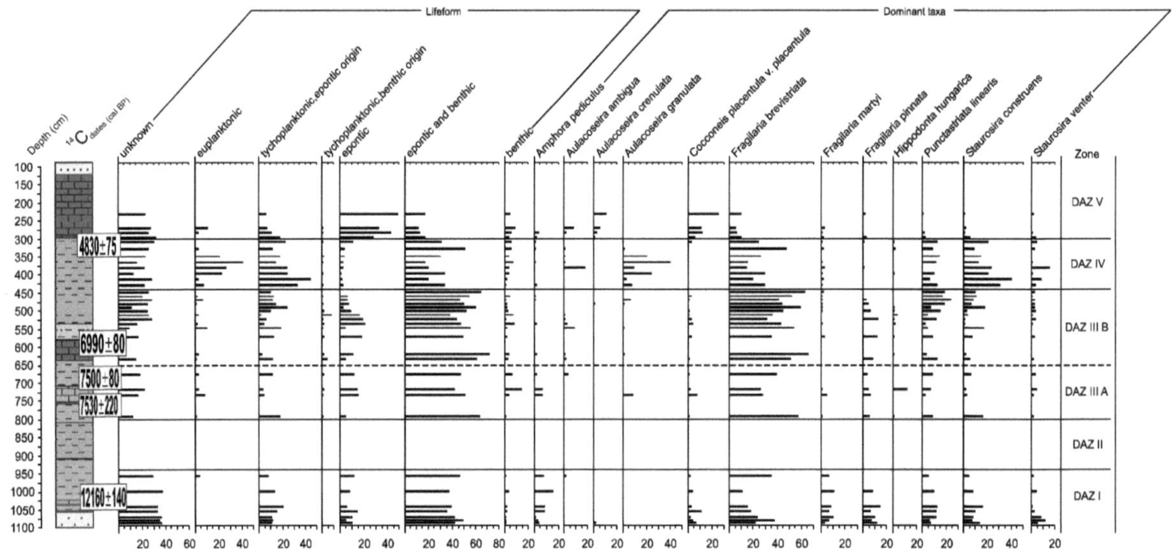

Fig. 3 Distribution of diatom ecological groups (habitat) and of the diatom species with relative abundance exceeding 10% (in at least one sample) in core Trz-15

Heurck) Simonsen, *Aulacoseira crenulata* (Ehrenberg) Thwaites, *Aulacoseira granulata* (Ehrenberg) Simonsen, *Cocconeis placentula* Ehrenberg var. *placentula*, *Fragilaria brevistriata*, *Fragilaria martyi* (Héribaud) Lange-Bertalot, *Fragilaria pinnata*, *Hippodonta hungarica* (Grunow) Lange-Bertalot, Metzeltin & Witkowski, *Punctastriata linearis* Williams & Round, *Staurosira construens*, and *Staurosira venter* (Fig. 3). The core can be divided into two parts: The first part includes DAZ Trz-15-I in the lowermost part of the core, and the upper part includes the stretch from DAZ Trz-15-III to DAZ Trz-15-V. They are separated by diatom-poor DAZ Trz-15-II. Although the diatom assemblages differ in terms of species composition, it is possible to characterize both parts in a more general section, and herebelow, we will concentrate on these factors which are subject to changes in particular DAZ. In terms of salinity, the sediments of core Trz-15 show very strong domination of fresh-brackish forms. This results from relative

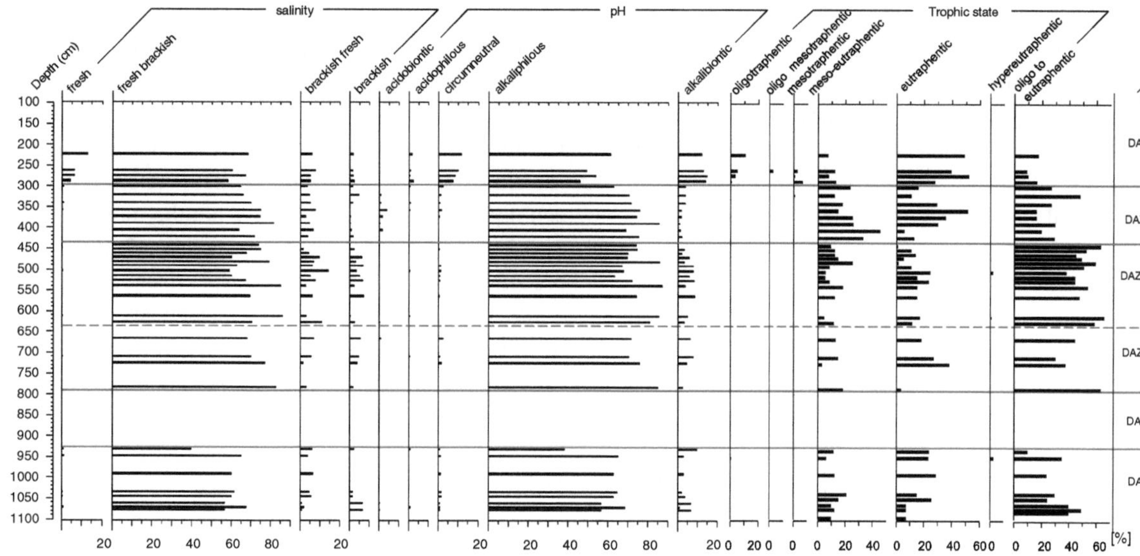

Fig. 4 Distribution of diatom ecological groups (salinity, pH and trophy) in core Trz-15

abundance of *Fragilaria brevistriata* which attained a relative abundance of up to ca. 60%. Similar situation with an exception of DAZ V is observed regarding pH. Dominant throughout are the alkaliphilous forms which reach relative abundance up to 80%. More variable is the situation in respect to trophic status. Although dominant are taxa of broad trophic tolerance (oligo- to eutraphentic), their content is not that high as in the case of other ecological groups. A distinct change in trophic preferences is observed beginning from DAZ IV. The hitherto dominant forms decrease, and an increase in eutraphentic taxa is observed. In the uppermost DAZ V, the latter forms become more dominant (Fig. 4). As the distribution of habitat categories and of dominant species showed significant variation (Fig. 3), these features will be characterized below.

DAZ Trz-15-I includes the core interval of 1,037–890 cm b.s.l. and encompasses lithological units CD1 and D. The lowermost part (1,020–1,030 cm b.s.l.) of this unit is dated at 14,200 cal year BP (Table 1). *Fragilaria brevistriata*, *F. pinnata*, *Punctastriata linearis* and *Staurosira construens* are the most abundant species. Dominant in terms of habitat are periphytic/benthic taxa. Up to the depth of 993 cm b.s.l., an increase in tychoplanktonic forms which originate from the periphyton is observed (e.g. *Staurosira construens* and *S. subsalina*. At the top of the zone, there euplanktonic forms appear (e.g. *Stephanodiscus hantzschii* Grunow and *Aulacoseira ambigua*), but their abundance remains low. The abundance of fresh-brackish forms remains nearly stable (ca. 60%); this is due to strong dominance of *Fragilaria brevistriata*. Taxa also occur, which in terms of halobous tolerance are considered to be either brackish-fresh or typical brackish-water. Included in these groups are *Staurosira subsalina* (exceeding 5%) and *Anomoeoneis sphaerophora* (Ehrenberg) Pfitzer; however, their total abundance is rather low.

The DAZ Trz-15-II encompasses the sediment interval 890–742 cm b.s.l. and the sediment unit CD2. The diatom concentration was too low to allow counting. Single valves of *Staurosira construens*, *Punctastriata linearis*, *Fragilaria martyi*, *Rhoicosphenia abbreviata* and *Amphora pediculus* were identified.

In DAZ Trz-15-III, two sub-zones were distinguished. Sub-zone DAZ Trz-15-IIIA includes the depth interval of 742–550 cm b.s.l. and the sediment unit D. Radiocarbon ages of the sediment from the depth interval of 700–710 and 650–660 cm b.s.l. was 8,600 cal year BP and 8,390 cal year BP, respectively (Table 1). In this sub-zone, taxa representing periphyton/benthos strongly dominate, although their abundance decreases up to the depth of 718 cm, and they are replaced to a certain extent by an increasing number of benthic forms, e.g. *Fragilaria brevistriata* and *Hippodonta hungarica* and of periphytic taxa. Noticeable is also the increased content of euplanktonic taxa, e.g. *Aulacoseira ambigua* and *A. granulata*. They occur in the middle part of the subzone, but have small relative abundances. Dominant are fresh-brackish forms, but brackish-fresh and brackish-water taxa also occur. They attain rather low relative abundance (maximum 5–10%) and are represented by, e.g. *Cyclotella meneghiniana* Kützing, *Planothidium delicatulum*, *Rhoicospenia abbreviata* and *Staurosira subsalina*.

Sediment of the sub-zone DAZ Trz-15-IIIB occurred at the depth interval of 550–390 cm b.s.l. and included middle and upper part of lithological unit D with a thin (a few cm) intercalation of unit E. The sediment from the depth interval of 530–540 cm b.s.l. was dated to 7,880 cal year BP (cf. Table 1). With respect to habitat in this sub-zone, periphytic/benthic forms dominate. They show the highest relative abundance in the lowermost part of the sub-zone. Euplanktonic forms reach low abundances. The latter group is represented by *Aulacoseira ambigua*. Taking into consideration halobous preferences, fresh-brackish forms are still dominant, although increased abundances of brackish-freshwater and brackish-water forms were observed. *Cocconeis scutellum* Ehrenberg, *Fragilaria sopotensis* and *Fragilaria schulzii* Brockmann appear in the samples.

The DAZ Trz-15-IV encompasses the depth interval of 390–255 cm b.s.l. and the lithological unit F. In this zone, a distinct change in respect to habitat groups is observed. The hitherto strongly dominant periphytic/benthic forms show significant decrease and are replaced by euplanktonic ones. In this zone, euplanktonic forms attain the highest relative abundance and are represented by *Aulacoseira granulata*, and to a lesser extent by *A. ambigua*. They are accompanied by the tychoplanktonic taxa of periphytic origin; however, the latter forms attain highest abundance at the lower limit of the zone and show upward decrease. The most abundant representative of this group is *Staurosira construens*, which attained a relative abundance up to 40%. However, taking into consideration the whole zone, the relative abundance of the latter group decreases from 50% in the lower part of the zone to ca. 20% in the uppermost one. Characteristic for DAZ IV is a slight decrease in brackish-freshwater and strong decrease in brackish-water forms. This zone is also characterized by changes in relationships of trophic diatom groups. Dominate eutraphentic and meso-eutraphentic taxa while oligo- to eutraphentic forms show distinct decrease when compared to the preceding zone.

Sediment interval 255–182 cm b.s.l. constituted the DAZ Trz-15-V. This included the lithological unit F and H. In this zone, diatoms occur only in the lowermost part. Changes in the distribution of diatom ecological groups in the DAZ V are the most distinct. Species which were the most abundant in the preceding zones, i.e. *Fragilaria brevistriata* and *Staurosira construens* almost disappear. Dominant in this zone are *Cocconeis placentula*, *Aulacoseira crenulata* and *A. ambigua*. Most abundant in this zone in terms of habitat are periphytic forms, and they show an upward increasing trend up to ca. 50%. With respect to salinity preferences, the fresh-brackish taxa still dominate. Increase is observed in freshwater taxa and is connected with the occurrence of *A. crenulata*. A change is also observed in the distribution of pH groups. Alkaliphilous taxa decrease and are replaced by upwardly increasing circumneutral ones. The next group in terms of abundance, up to 15%, were alkalibiontic forms including, e.g. *Rhopalodia gibba* (Ehrenberg) O. Müller. Analysis of trophic preferences showed a dominance of eutraphentic forms and *Fragilaria brevistriata* has significantly contributed to this. Their relative abundance increases from 15.9% at the depth of 260 cm b.s.l. to 52% at 133 cm b.s.l. From this depth onwrds it is lower, though still high.

Core Trz-20

The core profile was divided into six DAZ: Trz-20-I to Trz-20-VI, Figs. 4 and 5. A total of 308 diatom taxa were identified (18 centric representing 7 genera and 290 pennate representing 64 genera). However, only eight of them attained a relative abundance exceeding 10% in at least one sample. The following taxa were included in this group: *Catenula adhaerens* Mereschkowsky, *Fragilaria brevistriata*, *F. heidenii*, *F. martyi*, *Navicula vaneei*, *Planothidium delicatulum*, *Staurosira binodis* and *S. construens*. As there were quite few species with increased abundances in this core, we illustrate in a diagram also a few with a relative abundance exceeding 5% in at least one of the samples and these are: *Fragilaria neoelliptica* Witkowski & Lange-Bertalot, *F. pinnata*, *Punctastriata ovalis* Williams & Round, *Staurosira subsalina* and *S. venter* (Fig. 4).

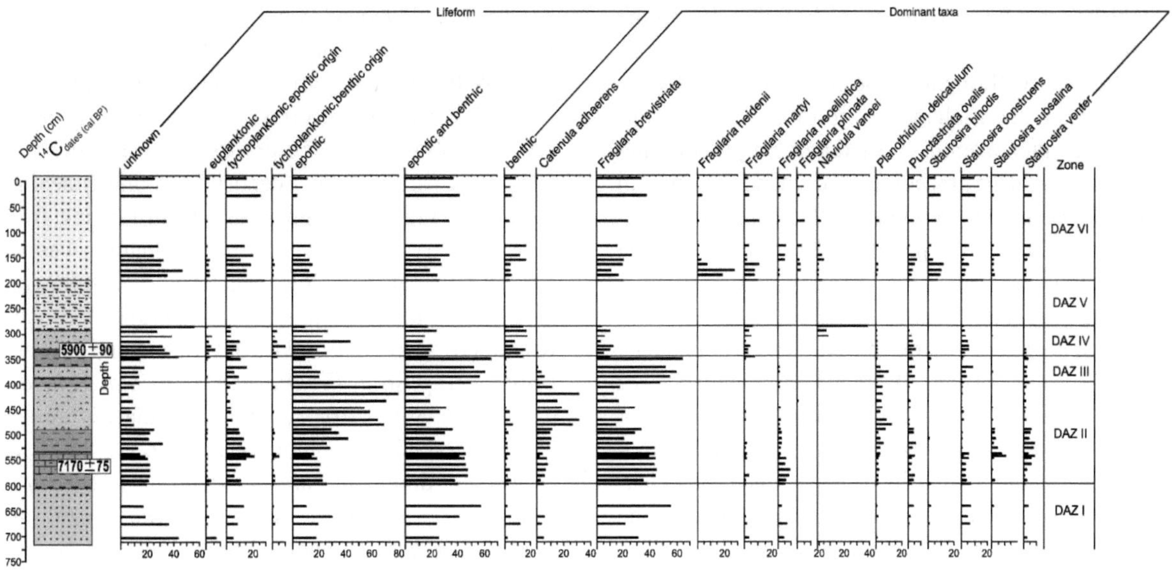

Fig. 5 Distribution of diatom ecological groups (habitat) and of the diatom species with relative abundance exceeding 5% (in at least one sample) in core Trz-20

DAZ Trz 20-I includes the depth range of 670–550 cm b.s.l. These are sands and in the uppermost part gyttja with sand and clay admixture and represent unit E. In uppermost part of the zone, a few samples were barren. Dominant in the sediments of this zone are *Fragilaria brevistriata*, *Catenula adhaerens*, and *Staurosira construens*. With respect to the habitat, dominant in this zone are periphyton/benthos forms including *Fragilaria brevistriata*. Their abundance showed an upward increase from 24.8% at 670 cm to 56.8% at 593.3 cm b.s.l. The abundance of periphytic and of tychoplanktonic forms of periphytic origin, e.g. *Staurosira venter* was also noticeable. Analysis with respect to pH revealed a very strong domination of alkaliphilous forms. Their abundance shows a significant increase from 41.7% at the lower limit of the zone to ca. 78.4% at the upper limit of the zone. With respect to salinity, fresh-brackish forms are dominant, and this is related to high contents of *Fragilaria brevistriata*. Their abundance shows upward increase from 38.1% to 71.6%. Brackish-fresh forms attain ca. 15%. True brackish-water forms occur only sporadically, e.g. *Melosira lineata* (Dillwyn) C. Agardh.

DAZ Trz-20-II represents the depth interval of 550–350 cm b.s.l. and is composed of gyttja, and mud and at the depth of ca. 445 cm b.s.l. changes into sand. The sediment from the depth interval of 520–530 cm b.s.l. was dated to 8,050 cal year BP. The change in lithology is very well correlated with change in diatom flora. In the lower part of the zone with respect to habitat, periphytic and benthic taxa dominate. However, they show upward decrease in abundance and when the sediment changes into sand, they are replaced by periphytic taxa. The most interesting in this zone are changes in proportions of particular halobous groups which also are related to lithological change. In the lower part, strongly dominant are the fresh-brackish forms (e.g. *F. brevistriata*). However, they show strong decrease from ca. 60% in the lower part of the zone to ca. 20%. This is best shown in the decrease of *F. brevistriata*. In the upper limit of the zone, the former group is replaced by brackish-fresh forms (up to 50.3%) represented by, e.g. *Staurosira subsalina*. The highest abundance, for the entire core, attained in this zone (in the lowest part of sandy sediments) were brackish-water forms, and their abundance exceeds 10%. Included in this group are, e.g. *Catenula adhaerens* Mereschkowsky, *Cocconeis peltoides* Hustedt, *Fragilaria atomus* Hustedt, *Planothidium delicatulum*, *Diploneis didyma* Ehrenberg, *Diploneis smithii* (Brèbisson) Cleve, species abundant in the marine littoral. The minimum abundance in this DAZ is attained by freshwater forms. Although the abundance of *C. adhaerens* thereafter shows strong decrease, the content of brackish-water forms remains fairly stable due to the contribution of *P. delicatulum*.

DAZ Trz-20-III includes the depth interval of 350–300 cm b.s.l. and is characterized by changing lithology. Here, intercalations of sand and gyttja with mud occur. Unlike in the preceding zone, changes in diatom assemblages are not correlated with changes in lithology. Strong increase of periphytic/benthic taxa is best expressed in domination of *Fragilaria brevistriata*. The latter species contributes also to the strong domination of fresh-brackish taxa. Their abundance reached up to 79.6%. Decrease in abundance of *Staurosira subsalina* is reflected in strong decrease of brackish-freshwater forms. Despite low abundance of *C. adhaerens* and finally its disappearance, brackish-water forms reach up to 10% until the middle of the DAZ, and then show a dramatic decrease.

DAZ Trz20-IV includes depth interval of 300–240 cm b.s.l. and lithological unit E. The sediment from the depth interval of 290–300 cm b.s.l. was dated to 6,800 cal year BP (Table 1). With respect to habitat, an increase in abundance of periphytic taxa was found. Their abundance attained ca. 25%, with a maximum (at 320 cm) of 43.7%. Periphytic/benthic taxa decrease in abundance to ca. 15% while planktonic forms, e.g. *Aulacoseira ambigua* (max. 6% at 286.6 cm b.s.l.) show the opposite trend. With respect to pH, alkaliphilous taxa, such as *Amphora pediculus* were dominant. A slight peak in circumneutral taxa, e.g. *Achnanthes conspicua* A. Mayer, is also observed. Analysis of halobous preferences revealed a strong dominance of fresh-brackish forms and an upward increase in freshwater taxa. They are represented by, e.g. *Ellerbeckia arenaria* (Moore) Crawford which appears for the first time in this zone. Brackish-water forms attain a few percent of relative abundance, though *C. adhaerens* is absent and *P. delicatulum* almost disappears from the record. DAZ IV is the first zone in core Trz-20 with strong dominance of eutraphentic forms. Their relative abundance attains ca. 30% on average in the whole zone (max. 37% at 270 cm b.s.l.). In DAZ Trz-20-IV, oligotraphentic taxa appear.

DAZ Trz-20-V includes sediment interval of 240–150 cm b.s.l. and represents organic rich sands which originated in an oxbow with the abundance of diatom valves too low to allow any statistical analysis. Solitary valves of *Amphora copulata*, *A. inariensis*, *A. pediculus*, *Cavinula scutelloides*, *Fragilaria brevistriata*, *F. martyi*, *Staurosira construens*, *S. subsalina*, *Navicula vaneei* and *Punctastriata ovalis* D.M. Williams & Round were recorded in this zone.

DAZ Trz-20-VI encompasses the sediment interval of 150 b.s.l. to 50 cm a.s.l. It includes the uppermost part of oxbow deposits and unit I (aeolian sediments, Fig. 2). Again at the depth ranging from 66 b.s.l. to 0.0 cm, the abundance of diatoms was too low to allow diatomologic analysis. The following taxa *Amphora copulata*, *Caloneis molaris* (Grunow) Krammer, *Cocconeis disculus* (Schumann) Cleve, *Cosmioneis pusilla*, *Fragilaria brevistriata*, *F. heidenii* Østrup, *Staurosira construens*, *Navicula vaneei* were observed in low numbers. In DAZ Trz-20-VI, as a whole, dominant are *Fragilaria brevistriata*, *F. heidenii*, *Staurosira binodis*, *S. construens* and *S. venter*. Regarding habitat preferences, this zone is dominated by periphytic/benthic taxa, e.g. *Amphora copulata*, *F. brevistriata*, and their relative abundance amounted to ca. 40%. Tychoplanktonic forms of periphytic origin reached increased abundance (max. 23.7%), whereas periphytic taxa occur with constant abundance of ca. 15%. With respect to pH, the most abundant taxa were alkaliphilous forms. Their abundance reached ca. 60% and increased by ca. 30% in comparison to DAZ Trz-20-V. Less abundant were the alkalibiontic forms. Regarding halobous preferences, fresh-brackish taxa were dominant. Their abundance rose from 46% at 83.3 cm b.s.l. to 71.3% at 17.3 cm a.s.l. Beginning from this depth, relative abundance of the latter forms decreases to 57%. Brackish-fresh forms (e.g. *Cyclotella meneghiniana*, *Navicula vaneei*) maintained a relative abundance of 7%. The abundance of brackish-water forms increases from 1.4% at the lower limit of the zone to 11.3% at 83.3 cm b.s.l., and again decreases to ca. 2% in the uppermost part of the zone (e.g. *Mastologia smithii* Thwaites, *Cosmioneis pusilla*, *Diploneis interrupta* (Kützing) Cleve, *Planothidium delicatulum*). Analysis of trophic preferences showed a distinct decrease of eutraphentic taxa with a maximum abundance of 20.6% recorded at 83.3 cm b.s.l. and then decreasing to 6.6% in the uppermost part of the core (Fig. 6).

Core Trz-21

Along the whole core Trz-21 profile, randomly distributed sub-samples were collected. Nineteen

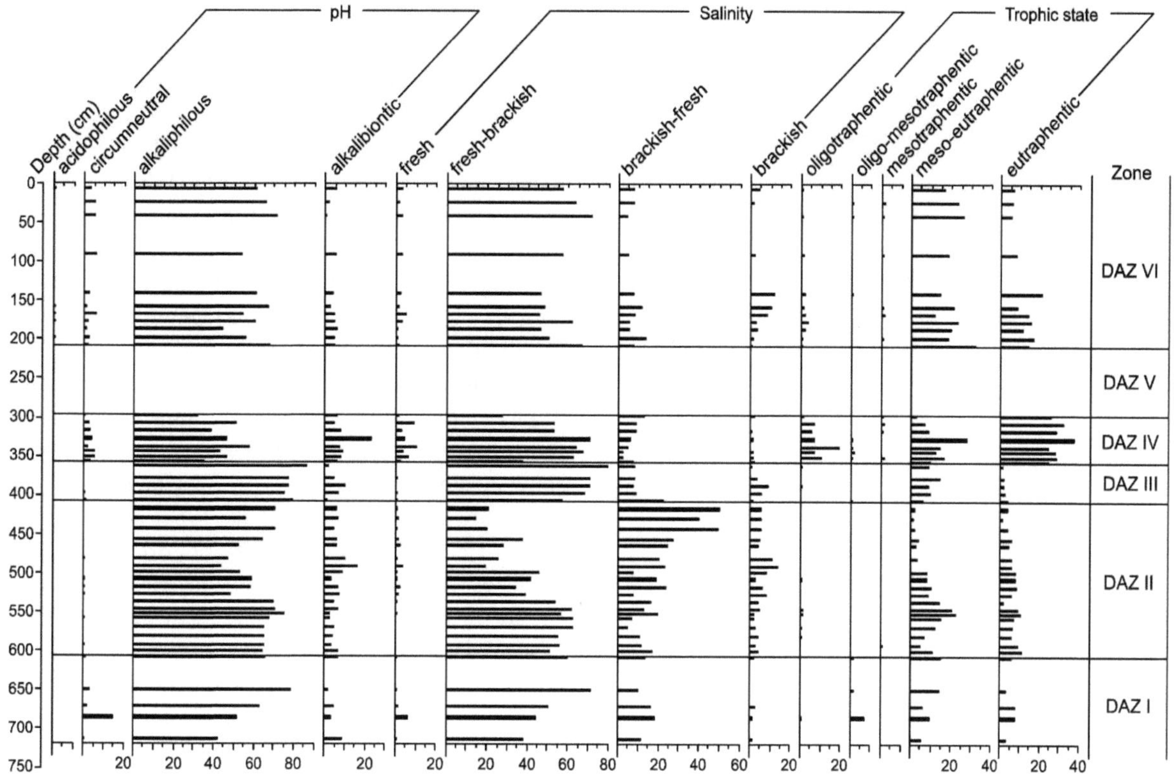

Fig. 6 Distribution of diatom ecological groups (salinity, pH and trophy) in core Trz-20

samples originated from the depth interval of 240–950 cm. Almost all of these samples were completely barren. Only samples from the depth of 550–560, 710–720, 720–725 and 730–735 cm contained fragmented diatom valves and diatom detritus. The species composition in all the enumerated samples was very similar. Robust taxa, e.g. *Anomoeoneis sphaerophora*, *Diploneis didyma* and *Epithemia turgida* were dominant. In addition, fairly abundant, but always fragmented were *Campylodiscus clypeus* Ehrenberg, *C. echeneis* Ehrenberg and *Ellerbeckia arenaria*. This diatom assemblage when developed in the sediments is named "clypeus flora" and is meant to represent the lagoon stage and the isolation onset (Miller, 1986). Altogether a few dozen taxa were identified, which are considered to be typical of lagoon or the shallow littoral zone of weakly saline environments. Included are, e.g. *A. sphaerophora*, *Brebissonia boeckii* (Ehrenberg) Grunow, *C. clypeus*, *C. echeneis*, *Ellerbeckia arenaria*, *Epithemia turgida* and *Diploneis didyma*.

Discussion

This study is the first palaeogeographic contribution to the reconstruction of the development of the Rega River mouth area on the southwestern coast of the Baltic Sea. As shown in Fig. 7, the sediments occurring in the Rega valley cover the time span of 14,200 cal years (12,160 ± 140 year BP). In Fig. 7, the sediments studied are plotted against the Baltic Sea development stages. The Late-glacial through the Holocene history of the Baltic Sea basin is divided into several stages which are named after leit fossil species (Björck, 2008). Beginning of the sedimentary record of the Rega corresponds with the part of the Baltic Ice Lake period (sediment unit C and CD1). In the material so far obtained from the cores, we can not subdivide further stages, i.e. Yoldia Sea and Ancylus Lake. Our sediment unit CD2 corresponds with these stages. Our sediment units D-I correspond with the Littorina and Postlittorina stages (Björck, 2008).

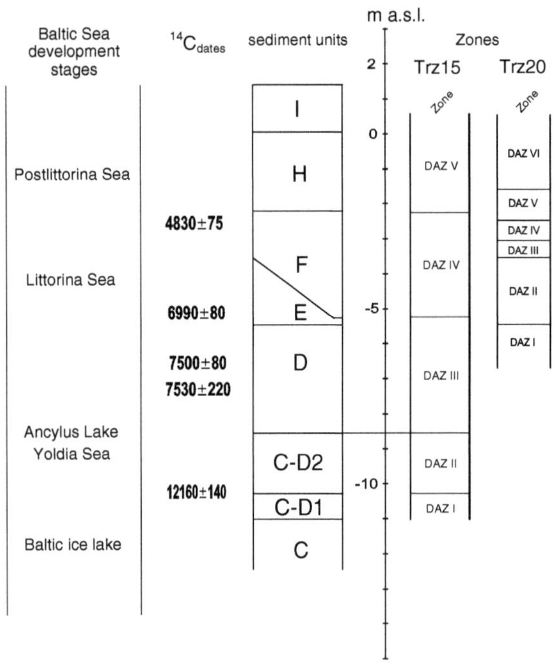

Fig. 7 Plot of the diatom assemblage zones distinguished in our study and of the lithological units versus the developmental stages of the Baltic Sea

Changes in the relative abundances of diatom taxa, and of diatom ecological groups, allowed us to characterize the depositional environments in which particular lithological units were deposited (Fig. 7). The oldest chronozone found in core Trz-15 was Allerød in age (sediment unit C-D1). The radiocarbon date was confirmed by the pollen stratigraphy (Cedro et al. in preparation). Allerød represents one of the warm oscillations of the Late-glacial. An amelioration of climatic conditions is also reflected in the species composition of diatom floras in temperate climate zone. Of special importance for interpretation of our results is the diatom analysis supported by pollen stratigraphy of Late-glacial fossil lake sediments from nearby Niechorze (Marciniak, 1981). The sediments of Oldest Dryas, Bølling and Older Dryas were characterized either by species poor (*Fragilaria* spp. dominated) with flora or were barren. A distinct change in the diatom flora in terms of species richness, preservation state and abundance is first observed at the transition from Older Dryas to Allerød. The diatom flora of Allerød in Niechorze was dominated by *Fragilaria* spp. (as in our study). In addition, taxa appear which prefer increased conductivity (i.e. halophilous or fresh-brackish, Marciniak, 1981; our study). Included in this group are, e.g. *Anomoeoneis sphaerophora* and *Staurosira subsalina*.

The diatom assemblages preserved in lacustrine sediments of Late-glacial and Early Holocene age are consistently similar in the Polish Lowland lakes (e.g. Marciniak, 1981), in lagoon deposits (e.g. Przybyłowska-Lange, 1974), and in shallow basins like Mecklenburg Bay and Arkona Basin (e.g. Jensen et al., 1997). Distribution of the diatom ecological groups in these sediments points to shallow waters with increased trophy and increased conductivity. In general, diatom floras of the Late-glacial in the localities mentioned show a strong dominance of *Fragilaria* spp. and *Staurosira* spp. (e.g. Marciniak, 1981; Jensen et al., 1997, 1999). The dominance of *Fragilaria* spp. indicates alkaline waters with increased nutrients (meso- to eutrophic conditions) content. Occurrence of several taxa, e.g. *Actinocyclus normanii* (Witkowski et al., 2004) and *Anomoeoneis sphaerophora* reflect relatively high water conductivity.

The earliest Holocene diatom flora occurs in sediments dated to the Atlantic chronozone. The diatom flora of sediment units D, E and F dated to ca. 8,600–5,800 cal year BP is in both cores strongly dominated by *Fragilaria brevistriata*, but diatom assemblages include taxa which inhabit slightly brackish to brackish-water habitats. Diatom studies from extant (Busse & Snoeijs, 2002; Weckström & Juggins, 2005) and fossil (Jensen et al., 1997; Hassan et al., 2006) coastal and lagoonary settings show that fragilarioid taxa are dominant components. In this respect, Rega valley sediments resemble other lagoons. The diatom assemblages indicate higher salinities in waters of the open marine littoral in which sandy sediments were deposited (unit E). Diatom assemblages dominated by fragilarioid taxa with low content of brackish-water species were characteristic for lagoons in which gyttja and muds (unit D and F) were deposited. Sediment units D and F with diatom assemblages typical for slightly saline lagoon waters occurred in core Trz-15, while unit E with brackish-water flora occurred in cores Trz-20 and Trz-21 located seaward of core Trz-15. The major difference between both sites is the occurrence of species representative for open marine littoral, e.g. *Catenula adhaerens* and *Planothidium delicatulum* in unit E. They are either rare (*P. delicatulum*) or absent

(*C. adhaerens*) in units D and F (Trz-15). Characteristics of diatom assemblages in units D and F (core Trz-15) point to a lagoon environment with a weak connection to the sea. Absence of *C. adhaerens* indicates that the salinity in this lagoon did not exceed 5 psu (Sundbäck & Medlin, 1986).

It seems that the major factors controlling species composition were salinity and the sediment type. In our study, these factors are correlated. Brackish-water and brackish-fresh diatom assemblages constitute a distinct proportion in sediments of core Trz-20 (DAZ I-II). Their vertical distribution is related to the lithology and relative abundances are the highest in lower part of unit E (sands, uppermost part of DAZ II). It seems that this part of the core represents the maximum salinity level and is dated to ca. 7,000 cal year BP. The diatom assemblage points out to values exceeding 5 psu, and similar to the recent salinities in the southern Baltic Sea (7–8 psu, Elken & Matthäus, 2008). As the lithology changes to gyttja (DAZ III), a drastic change in diatom halobic preferences is observed. This is evidenced by disappearance of *C. adhaerens* and strong decrease in *Planothidium delicatulum*. Such change is not recorded in core Trz-15 as in this locality lagoonary deposits with weakly saline flora occur throughout. Therefore, change in lithology accompanied by disappearance of brackish-water diatoms is interpreted as a signal of isolation of the study area from the sea and the sea waters retreat. It seems that the isolation episode was rather short, and the sandy sediments of unit E appear again in DAZ Trz-20-IV, but brackish-water diatom assemblage was not restored in the area. *Catenula adhaerens* occurs only in low quantity in the lowermost part of DAZ-IV, while *P. delicatulum* attains low abundances.

Decrease in *F. brevistriata* in DAZ-IV confirms change in the habitat and increasing salinity. Berglund et al. (2005) have illustrated such relationship in the coastal sediments from Blekinge in southeastern Sweden. Similarly, in our study the abundance of fragilarioid diatoms was high and showed a strong decrease with an increase in brackish-water diatoms and with the appearance of marine forms (Berglund et al., 2005). Weckström & Juggins (2005) have quantified relationship between fragilarioid taxa and some of the environmental variables including salinity and trophy. It is obvious that *Fragilaria brevistriata* and *F. pinnata* can not thrive salinities higher than 6–7 psu. This would mean that as described in recent publications from different geographic regions referring to these species as dominant in lagoonary deposits with high salinity, e.g. from Australia (Saunders et al., 2008) and from a hypersaline lagoon in Brazil (Sylvestre et al., 2005) *Fragilaria brevistriata* and *F. pinnata* are rather unlikely to occur. *Fragilaria brevistriata* can be misconstrued to be as *F. geocollegarum*, a species occurring in marine littoral habitats (Witkowski et al., 2000).

In addition to climatic issues, our study also contributes to an understanding of the age of the Atlantic transgression, the so-called Littorina transgression. The most recent results show that North Sea water incursions into the Baltic Sea must have started significantly earlier than previously believed (e.g. Berglund et al., 2005; Björck et al., 2008). Reports raising this problem were also published earlier, but were less convincing and hence received less attention. Such reports were published from Southern Baltic Sea coasts (e.g. Rotnicki et al., 1999; Rotnicki, 2001) and basins, e.g. Bornholm Basin (Zachowicz, 1995; Andrén et al., 2000b). The relative Holocene sea level rise for the southern Baltic Sea was either regional, (e.g. Rotnicki, 2001 for the middle part of the Polish coast) or for the whole coast (e.g. Uścinowicz, 2003). It was not our aim to construct the sea level rise curve, but results of our study allow us to distinguish particular stages of either sea level rise (transgression) or drop (regression). Our observations indicate that sea level rise did occur in the area of Rega valley in a few stages. This supports observations of Rotnicki (2001) and Berglund et al. (2005) and differs from the curve proposed by Uścinowicz (2003). We assume that DAZ Trz-20-I represents the time when the brackish-water conditions were already established. Sea level rise has resulted in the beginning of the sedimentation on the fossil boulder clay plateau dated to ca. 8,640 cal year BP. This matches very well with Berglund et al.'s (2005) first transgressive stage of the Littorina sensu stricto (L1) dated to 8,500–8,200 cal year BP. Similarly, as in Blekinge, diatom assemblages in our material point to rather low salinities of brackish-water conditions. Further Rega valley brackish-water sediments of unit E (core Trz-20 and Trz-29, Trz-17, Trz-14, Fig. 2) gyttja and peat intercalations dated to 8,355–7,930 cal year BP occur. Change in the sediment is associated with retreat of brackish-water diatom assemblages. Instead, fragilariod taxa dominated in this intercalation. This episode is also distinguished in the Blekinge sediments, as regression dated to 8,200–8,000 cal year BP (Berglund et al., 2005).

The data we posses so far and the poor diatom record (DAZ IV) which are probably attributed to riverine erosion (Figs. 2, 5, 6) show that in the Rega valley conditions similar to those in the preceding zones were not restored. Although the sandy sediments we distinguish as unit E were deposited again, the diatom flora differs significantly from that in the lower part of the core. The upper part of unit E is supposedly related to the second (L2, 7,800–7300 cal year BP) and third (L3 6,800–6,200 cal year BP) stages of Littorina transgression (cf. Berglund et al., 2005), but we have no dates to make this subdivision. Finally, the sedimentations of both littoral (unit E) and lagoon sediments (units D and F) ceased around 5,800 cal year BP. We assume that this decline could have had its beginning from the second stage of the Littorina transgression (ca. 7,800 cal year BP) because of the shallow water conditions and the neighborhood of the land as well as the local factors (e.g. morphology, water dynamics, riverine valley migration, water exchange between the sea and lagoon and ground water level, development of peat bog and of aeolian plain).

In DAZ Trz-20-II, the diatom flora has a quite stable composition with *C. adhaerens* being one of the most abundant species. We interpret this as an indication of well developed and stable environmental conditions with salinity exceeding 5 psu. The habitat preferred by *C. adhaerens* suggests an open marine embayment (e.g. Sundbäck & Medlin, 1986; Witkowski et al., 2000). This environmental setting most likely disappeared ca. 5,800 cal year BP and was replaced by peat bog and finally aeolian environment. The major indication of such drastic change is lithology. The hitherto lacustrine (biogenic) and sandy marine littoral sediments were followed by peat deposits. Swampy deposits of lithological unit H covered the whole study area. The final stage of the development involved aeolian sands (unit I) with some organic-rich intercalations which contain representatives of salt marsh diatom flora.

Change of the habitat into swampy environment and deposition of peat is accompanied by a drastic change in diatom assemblage. In core Trz-15 which is located further inland, dominant fragilarioid taxa, with the exception of *Fragilaria brevistriata*, almost disappear. Habitat change resulted in noticeable decrease in salinity and pH. This is evidenced by appearance of freshwater taxa and increase in circumneutral forms. Seemingly the swampy and aeolian habitats in core Trz-15 were so unfavorable that diatoms occur only in the lowermost part of unit H (peat) and then completely disappear from the record. Unlike in the latter core in Trz-20 species rich diatom assemblages occur. This core is closer to the sea shore and the deposition processes were affected by the saline waters. Hence, the species composition in the uppermost sediments of unit I is nearly the same as in the preceding units and differs significantly from core Trz-15. It appears that both the peat bog and the aeolian plain were impacted by the seawaters. Distribution of dominant species and diatom ecological groups (habitat, salinity, pH and trophy) indicates that environmental conditions during deposition of DAZ VI sediments were fairly stable.

Diatom assemblages show that the major factor controlling species composition was climatic amelioration in Late-glacial (Allerød) deposits. This situation is common in Europe due to the ice sheet retreat. The diatom assemblages indicate that in the study area, a shallow lake existed. This lake was rich in solutes and nutrients. The diatom assemblage was dominated by opportunistic fragilarioid species (Marciniak, 1981; Stabell, 1985; Denys, 1990). In the Holocene, the diatom assemblages are controlled by the sea level rise which was triggered by the climatic change and followed the deglaciation of high latitudes of both hemispheres. These large scale (global) factors impacted the area, the sediment deposition processes and the diatom assemblages until ca. 7,800 cal year BP. From this time onwards, local factors have probably taken over priority in the palaeogeographic development of the study area.

In the area of this study, the diatom species composition shows, for the whole deposition period dated to ca. last 14,800 cal B.P., shallow to very shallow water conditions. These were observed during the lacustrine and during the lagoonary stages of development of the mouth of the Rega River. Sometimes, we have even observed periods of aerophilous diatom development, which can be interpreted as reflecting an exposed habitat. Representative of these kinds of conditions are *Diadesmis* spp., *Cosmioneis pusilla*, *Diploneis interrupta*, *Caloneis molaris* and *Navicula vaneei* (Krammer & Lange-Bertalot, 1986; Witkowski et al., 2000; Lange-Bertalot, 2001). Increased abundances of *Diadesmis* spp. occurred in core Trz-20 at the depths of 614 cm b.s.l. and 3-50 cm a.s.l., whereas the remaining species occurred in the upper part of the profile and encompassed unit H and I. These units mark the sea level close to the recent one and deposition in a very shallow water to exposed environment resembling recent salt marshes. Species composition of such diatom flora was reported by Witkowski (1994) for the area of Hel Peninsula and the Puck Bay coasts. Uścinowicz et al. (2007) in their multiproxy study of sea level oscillations during the middle and late Holocene in the area of Puck Bay (Poland's eastern coast) made use of salt marsh diatom assemblages in reconstructing the start of sea level rise episodes.

The radiocarbon dates of the sediment interval with brackish-water diatoms, which we studied seem to approach 8,600 cal year BP and also confirm the Early Atlantic as a beginning of the marine transgression at the mouth of the Rega River. The sediments containing brackish-water flora in the Rega mouth (unit E) are underlined by sediment unit D rich in fresh-brackish and brackish-fresh and brackish-water diatoms. Przybyłowska-Lange (1979) published results of her study on diatom flora of Lake Jamno sediments, which is located between Rega mouth (our study) and Leba-Gardo Lowland (Rotnicki et al., 1999). In Lake Jamno, brackish-water and brackish-fresh (halophilous) diatoms appear in palinologically dated Boreal sediments. Przybyłowska-Lange (1979) concluded that the flora must be redeposited, but in connection with our findings and the results published by Rotnicki et al. (1999) and Janczak-Kostecka & Kostecki (2006), the presence of halophilous and brackish-water diatoms must be interpreted as early signal of marine transgression.

Another problem of interest is the heterochronous appearance of the Littorina transgression in some parts of the coast which are located close to each other. Such a situation is recognized when the age of the oldest marine sediments which we studied is compared to the Szczecin lagoon located westward from the Rega valley. The results of a multiproxy-based palaeogeographic analysis of the Szczecin Lagoon were published by, e.g. Borówka et al. (2005) and Witkowski et al. (2004). Results of diatom, malacological and macroremnant analyses show clearly that the onset of brackish-water sedimentation in the Szczecin Lagoon is dated to ca. 6,250 year BP (7,200 cal year BP). Results of this study and those of Rotnicki (2001), cf. also Janczak-Kostecka & Kostecki, 2006) reflect an earlier transgressive episode of the marine transgression. Rotnicki has indicated that the first marine water incursion in the Gardno-Leba Lowland to be older than 8,000 year BP (9,500 cal year BP, i.e. Early Atlantic). The study area of Rotnicki (2001) is located ca. 100 km east from the mouth of the Rega. The different age of Littorina transgression along the Polish coast can be explained in terms of tectonic activity in some of its parts. As summarized by Zuchiewicz et al. (2007), the Polish coast of the Baltic Sea belongs to three neotectonic structural units. These are East Baltic Graben System, Pomeranian Depression and West Baltic Step. The Rega valley is located within Pomeranian Depression. The Pomeranian Depression is located eastward from West Baltic Step containing Szczecin Lagoon. These two units also differ significantly in terms of amplitudes of Quaternary vertical movement. The area of Szczecin Lagoon has the movement amplitude of -50 to -100 m, whereas the Rega River valley has an amplitude of 0 to -50 m. Values similar those calculated for Rega valley were given further eastward up to the Vistula mouth (Zuchiewicz et al., op. cit.).

We estimate that at ca. 5,800 cal year BP, the sedimentation of the gyttja and muddy deposits ceased. This is consistent with the postulated duration of the Littorina transgression, which according to Uścinowicz et al. (2007) terminated at ca. 5,000 BP (5,800 cal year BP). However, diatom flora species composition shows an impact of marine waters, either as direct inflow or during, e.g. extreme storm surges or as indirect inflow by, e.g. ground waters seepage. The result of both phenomena is the development of diatom assemblages typical for salt meadows (marshes). Typical for such habitat are brackish water forms tolerant to rapid changes in osmotic pressure, high content of organic matter and decreased pH values (e.g. Witkowski, 1994; Trobajo-Pujadas, 2007). Representatives of such assemblages are, e.g. *Cosmioneis pusilla*, *Diploneis interrupta*, *Staurosira subsalina*, *Navicula vaneei*.

It seems that floristic composition is regulated by climatic (precipitation) factor, whereas palaeolimnological development by a combination of climatic, isostatic and tectonic factors. Diatom species composition appears to be controlled by large scale (global) factors, hence the similarity of lagoon floras on a global scale. The Baltic Sea lagoons have similar flora in general and roughly the same dominant taxa as those of Atlantic Ocean coasts, and Australia (e.g. Hassan et al., 2006; Bao et al., 2007; Fluin et al., 2007; Saunders et al., 2008). Palaeolimnologic development is more complex and the factors of crucial importance involve isostatic, eustatic sea level rise, tectonics and recently human activity (hydro-engineering and industrialization). The spatial extent of these factors can change on a local-to-regional scale, hence the differences occurring in the development of lagoons even of those located close to each other as we have been able to show it in the case of the southern Baltic Sea coast.

Conclusions

The diatom record of Rega River valley is of variable quality. The best and most informative in terms of palaeoinferences are diatom assemblages of Allerød and of Early to Middle Atlantic chronozones. This is due to the fact that the global factors, i.e. climate change and sea level rise, impacted the diatom assemblages. Beginning from Middle Atlantic, the global factors are overwhelmed by local factors, such as morphology, water dynamics, and sedimentation.

The diatom record of the sediments deposited in the area of the Rega River reflects several changes in the environmental conditions during which deposition took place through the Late-glacial and Holocene.

The Late-glacial sediments dated to Allerød were deposited in a shallow lacustrine environment. The lake contained water with increased conductivity and nutrient concentrations.

The Holocene record in the cores studied begins in early Atlantic and the diatoms indicate brackish-water sediments. During the period of 8,600–5,800 cal year BP, sedimentation took place in a shallow lagoon/littoral embayment. The sediment is further subdivided into

- Lagoon—units D and F (core Trz-15: DAZ III and DAZ IV; core Trz-20: DAZ III, IV).
- Littoral embayment—unit E (core Trz-20, DAZ I and DAZ II)

The lower part of unit D (Trz-15, DAZ III) and unit E (Trz-20, DAZ I-II) correspond to the Littorina transgression (L1 sensu Berglund et al., 2005) and were deposited in a brackish water embayment/ lagoon. Transgressive sediments are covered by a rather short episode of sea level decrease—regression (unit D, core Trz-15, DAZ IV; core Trz-20 DAZ III) which corresponds to the 8,200 cal year BP event. The upper part of units D and E and unit F were deposited in shallow water embayment/lagoon, but with limited water exchange with the nearby located marine littoral, and hence, the diatom assemblages show decrease in salinity.

Beginning ca. 5,800 cal year BP, the sedimentation has been taking place in a swampy and, finally in an aeolian environment, which we distinguish as

- Peat bog—unit H (Trz-20, DAZ VI)
- Aeolian plain—unit I (Trz-15, DAZ V)

Acknowldegements The authors would like to express their gratitude to Manfred Ruppel (Frankfurt am Main, Germany) for operating SEM apparatus, and Dr. Anna Pędziszewska for pollen analysis. Valuable suggestions of two anonymous reviewers are appreciated. Correction of English language and critical reading of manuscript by Dr. John P. Kociolek is appreciated.

References

Andrén, E., 1999. Changes in the composition of the diatom flora during the last century indicate increased eutrophication of the Oder Estuary, South-western Baltic Sea. Estuarine, Coastal and Shelf Science 48: 665–676.

Andrén, E., T. Andrén & H. Kunzendorf, 2000a. Holocene history of the Baltic Sea as a background for assessing records of human impact in the sediments of the Gotland Basin. The Holocene 10: 687–702.

Andrén, E., T. Andrén & G. Sohlenius, 2000b. The Holocene history of the southwestern Baltic Sea as reflected in a sediment core from the Bornholm Basin. Boreas 29: 233–250.

Bao, R., A. Alonso, C. Delgado & J. L. Pages, 2007. Identification of the main driving mechanisms in the evolution of a small coastal wetland (Traba, Galicia, NW Spain) since its origin 5700 cal yr BP. Palaeogeography, Palaeoclimatology, Palaeoecology 247: 296–312.

Battarbee, R. W., 1986. Diatom analysis. In Berglund, B. E. (ed.), Handbook of Holocene Palaeoecology and Palaeohydrology. Wiley, Chichester, New York, Brisbane, Toronto, Singapore: 527–570.

Berglund, B. E., P. Sandgren, L. Barnekow, G. Hannon, H. Jiang, G. Skog & Y. Shi-Yong, 2005. Early Holocene history of the Baltic Sea, as reflected in coastal sediments in Blekinge, southeastern Sweden. Quaternary International 130: 111–139.

Björck, S., 2008. The late quaternary development of the Baltic Sea. In Bolle, H.-J., M. Meneti & I. Rasool (eds), Regional climate studies, assessment of climate change for the Baltic Sea Basin. Annex A.2. Springer, Berlin, Heidelberg: 398–407.

Björck, S., T. Andrén & J.-B. Jensen, 2008. An attempt to resolve the partly conflicting data and ideas on the Ancylus-Littorina transition. Polish Geological Institute Special Papers 23: 21–26.

Borówka, R. K., A. Osadczuk, A. Witkowski, B. Wawrzyniak-Wydrowska & T. Duda, 2005. Late Glacial and Holocene depositional history in the eastern part of the Szczecin Lagoon (Great Lagoon) basin—NW Poland. Quaternary International 130: 87–96.

Bracco, R., H. Inda, L. del Punto, C. Castinera, P. Sprechmann & F. Garcia-Rodriguez, 2005. Relationships between Holocene sea-level variations, trophic development and climatic change in Negra Lagoon, Southern Uruguay. Journal of Paleolimnology 33: 253–263.

Busse, S. & P. Snoeijs, 2002. Gradient responses of diatom communities in the Bothnian Bay, northern Baltic Sea. Nova Hedwigia 74: 501–525.

Clarke, A., S. Juggins & D. Conley, 2003. A-150 year reconstruction of the history of coastal eutrophication in Roskilde Fjord, Denmark. Marine Pollution Bulletin 46: 1615–1618.

Denys, L., 1990. Fragilaria blooms in the Holocene of western coastal plain of Belgia. In Simola, H. (ed.), Proceedings of the 10th International Diatom Symposium. Koeltz Sci, Koenigstein: 397–406.

Dobracka, E., 1992. Geological map of Poland 1:50000. Trzebiatów. Wydawnictwa Geologiczne, Warszawa.

Elken, J. & W. Matthäus, 2008. Baltic Sea oceanography. In Bolle, H.-J., M. Meneti & I. Rasool (eds), Regional Climate Studies, Assessment of climate change for the Baltic Sea Basin. Annex A.1.1. Springer, Berlin, Heidelberg: 379–385.

Emeis, K. C., U. Struck, T. Blanz, A. Kohly & M. Voß, 2003. Salinity changes in the central Baltic Sea (NW Europe) over the last 10000 years. The Holocene 13: 411–421.

Fluin, J., P. Gell, D. Haynes & J. Tibby, 2007. Paleolimnological evidence for the independent evolution of neighboring terminal lakes, the Murray Darling Basin, Australia. Hydrobiologia 591: 117–134.

Fritz, S. C., S. Juggins, R. W. Battarbee & D. R. Engstrom, 1991. Reconstruction of past changes in salinity and climate using a diatom-based transfer function. Nature 352: 706–708.

Fritz, S. C., B. F. Cumming, F. Gasse & K. R. Laird, 1999. Diatoms as indicators of hydrologic and climatic change in Saline Lakes. In Stoermer, E. F. & J. P. Smol (eds), The Diatoms: Applications for the environmental and earth sciences. Cambridge University Press, Cambridge: 41–72.

Garcia-Rodriguez, F. & A. Witkowski, 2003. Inferring sea level variation from relative percentages of *Pseudopodosira kosugii* in Rocha Lagoon, SE Uruguay. Diatom Research 18: 49–59.

Gasse, F., 1994. Lacustrine Diatoms for reconstructing past hydrology and climate. In Duplessy, J. C. & M. T. Spyridakis (eds), Long-Term Climatic Variations: Data and Modelling. NATO ASI Series. Springer-Verlag, Berlin, Heidelberg: 335–369.

Harff, J., W. Lemke, R. Lampe, F. Lüth, H. Lübke, M. Meyer, F. Tauber & U. Schmölcke, 2007. The Baltic Sea coast-A model of interrelations among geosphere, climate and anthroposphere. The Geological Society of America Special Paper 426: 133–142.

Hassan, G. S., M. A. Espinosa & F. I. Isla, 2006. Modern diatom assemblages in surface sediments from estuarine systems in the southeastern Buenos Aires Province, Argentina. Journal of Paleolimnology 35: 39–53.

Inda, H., F. Garcia-Rodriguez, L. Del Puerto, V. Acevedo, D. Metzeltin, C. Castineira, R. Bracco & J. B. Adams, 2006. Relationships between trophic state, paleosalinity and climatic changes during the first Holocene marine transgression in Rocha Lagoon, southern Uruguay. Journal of Paleolimnology 35: 699–713.

Janczak-Kostecka, B. & R. Kostecki, 2006. Diatomological and geochemical investigations of lagoon environmental on the Gardno-Łeba barier. Quaestiones Geographicae 25A: 23–26.

Jensen, J. B., O. Bennike, A. Witkowski, W. Lemke & A. Kuijpers, 1997. The Baltic Ice Lake in the southwestern Baltic: sequence-, chrono- and biostratigraphy. Boreas 26: 217–236.

Jensen, J. B., O. Bennike, A. Witkowski, W. Lemke & A. Kuijpers, 1999. Early Holecene history of the southwestern Baltic Sea: the Ancylus Lake stage. Boreas 28: 437–453.

Krammer, K. & H. Lange-Bertalot, 1986. Bacillariophyceae 1, Naviculaceae. In Ettl, H. J. Gerloff, H. Heynig & D. Mollenhauer (eds), Süsswasserflora von Mitteleuropa. Fisher, Stuttgart 2(1): 876.

Krammer, K. & H. Lange-Bertalot, 1988. Bacillariophyceae 2, Epithemiaceae, Bacillariaceae, Surirellaceae. In Ettl, H., Gerloff, J., Heynig, H. & D. Mollenhauer (eds), Süsswasserflora von Mitteleuropa. Fisher, Stuttgart 2 (2): 596.

Krammer, K. & H. Lange-Bertalot, 1991a. Bacillariophyceae 3, Centrales, Fragilariaceae, Eunotiaceae. In Ettl, H., J. Gerloff, H. Heynig & D. Mollenhauer (eds), Süsswasserflora von Mitteleuropa. Fisher, Stuttgart 2 (3): 576.

Krammer, K. & H. Lange-Bertalot, 1991b. Bacillariophyceae 4, Achnanthaceae. In Ettl, H., G. Gärtner, J. Gerloff, H. Heynig & D. Mollenhauer (eds), Süsswasserflora von Mitteleuropa. Fisher, Stuttgart 2 (4): 437.

Lange-Bertalot, H., 2001. *Navicula* s. str. 10 genera separated from *Navicula* s. lato. *Frustulia*. In Lange-Bertalot, H. (ed.), Diatoms of Europe, Vol. 2. A. R. G. Gantner Verlag K. G, Germany: 526.

Marciniak, B., 1981. Late-Glacial diatom phases in Western Pomerania. Acta Geologica Polonica 31: 127–137.

Miller, U. 1986. Ecology and palaeoecology of brackish water diatoms with special reference to the Baltic Sea Basin. In Ricard, M. (ed.), Proceedings of the 8th International Diatom Symposium, Koeltz Sci., Koenigstein: 601–611.

Przybyłowska-Lange, W., 1974. Development of the Vistula Firth in the light of diatom analysis. Prace Instytutu Meteorologii i Gospodarki Wodnej 9: 129–164 (in Polish).

Przybyłowska-Lange, W., 1979. Diatoms of lake deposits from the Polish Baltic coast. II. Lake Jamno. Acta Palaeobot 20: 227–244.

Rotnicki, K., 2001. Relative Southern Balic Sea level along the middle part of the polish coast in the Holocene in the light of studies of the Gardno-Łeba lowland. In Rotnicki, K. (ed.), Vistulian and Holocene Transformations of Geographic Environment of the Southern Baltic Sea Coastal Lowlands. Bogucki, Poznań: 63–80 (in Polish).

Rotnicki, K., R. K. Borówka, A. Pazdur, S. Hałas, J. Krzymińska & A. Witkowski, 1999. Chronology of the Balic Sea Holocene transgression in the region of Leba spit bar. In Pazdur, A., A. Bluszcz, W. Stankowski & L. Starkel (eds), Geochronology of the Upper Quaternanry in Poland in the Light of Radiocarbon and Luminescence Datings. WIND-J, Wojewoda, Wrocław: 229–242 (in Polish).

Saunders, K. M., D. A. Hodgson, J. Harrison & A. McMinn, 2008. Palaeoecological tools for improving the management of coastal ecosystems: a case study from Lake King (Gippsland Lakes) Australia. Journal of Paleolimnology 40: 33–47.

Smol, J. P. & B. F. Cumming, 2000. Tracking long-term changes in climate using algal indicators in lake sediments. Journal of Phycology 36: 986–1011.

Snoeijs, P., 1999. Diatoms and environmental change in brackish waters. In Stoermer, E. F. & J. P. Smol (eds), The diatoms. Application for the Environmental and Earth Sciences. Cambridge University Press, Cambridge: 298–333.

Sohlenius, G., J. Sternbeck, E. Andrén & P. Westman, 1996. Postglacial development in the Baltic Proper - as recorded in a sediment core from the Gotland Deep. Marine Geology 134: 183–201.

Stabell, B., 1985. The development and succession of taxa within the diatom genus *Fragilaria* Lyngbye as a response to basin isolation from the sea. Boreas 14: 273–386.

Starratt, S. W., 2007. Diatoms in estuaries and tidal marshes. In Starratt S. W. (ed.), Pond Scum to Carbon Sink, Geological and Environmental Applications of the Diatoms. The Paleontological Society Papers 13: 85–109.

Stevenson, R. J. & Y. Pan, 1999. Assessing ecological conditions in rivers and streams with diatoms. In Stoermer, E. F. & J. P. Smol (eds), The Diatoms: Applications for the Environmental and Earth Sciences. Cambridge University Press, Cambridge: 11–41.

Stuiver, M., P. J. Reimer, E. Bard, J. W. Beck, G. S. Burr, K. A. Hughen, B. Kromer, F. G. McCormac, J. V. D. Plicht & M. Spurk, 1998. INTCAL98 radiocarbon age calibration, 24, 000–0 cal BP. Radiocarbon 40: 1041–1083.

Sundbäck, K. & L. K. Medlin, 1986. A light and electron microscopic study of the epipsammic diatom *Catenula adhaerens* Mereschkowsky. Diatom Res 1: 283–290.

Sylvestre, F., A. Sifeddine, B. Turcq, E. Lallier-Verges & J. Abrao, 2005. Hydrological changes related to the variability of tropical South American climate from the Cabo Fri lagoonal system (Brazil) during the last 5000 years. The Holocene 15: 625–630.

Trobajo-Pujadas, R., 2007. Ecological analysis of periphytic diatoms in Mediterranean coastal wetlands (Emporda wetlands, NE Spain). In Witkowski, A. (ed.), Diatom Monographs 7. A. R. G. Gantner Verlag K. G, Germany: 1–210.

Uścinowicz, S., 2003. Relative sea level changes, glacio-isostatic rebound and shoreline displacement in the Suthern Balic. Polish Geological Institute Special Papers 10: 1–80.

Uścinowicz, S., J. Zachowicz, G. Miotk-Szpiganowicz & A. Witkowski, 2007. Southern Baltic sea-level oscillations: New radiocarbon, pollen and diatom proof of the Puck Lagoon (Poland). The Geological Society of America, Special Paper 426: 143–258.

Weckström, K. & S. Juggins, 2005. Coastal diatom-environment relationships from the Gulf of Finland, Baltic Sea. Journal of Phycology 42: 21–35.

Weckström, K., A. Korhola & P. Shemeikka, 2002. Physical and chemical characteristics of shallow embayments on the southern coast of Finland. Hydrobiologia 477: 115–127.

Westman, P. & A. Hedenström, 2002. Environmental changes during the isolation from the Littorina Sea as reflected by diatoms and geochemical parameters–a case study. The Holocene 12: 531–540.

Witkowski, A., 1994. Recent and fossil diatom flora of the Gulf of Gdańsk, Southern Baltic Sea. Bibliotheca Diatomologica, 28. J. Cramer, Berlin, Stuttgart.

Witkowski, A. & J. Pempkowiak, 1995. Reconstructing the development of human impact from diatoms and Pb210 sediment dating (the Gulf of Gdansk—southern Baltic Sea). Geographia Polonica 65: 63–77.

Witkowski, A., H. Lange-Bertalot & D. Metzeltin, 2000. Diatom Flora of Marine Coasts I. Iconographia Diatomologica 7: 1–925.

Witkowski, A., M. Latałowa, R. K. Borówka, P. Gregorowicz, M. Bąk, A. Osadczuk, J. Święta, M. Lutyńska, B. Wawrzyniak-Wydrowska & R. Woziński, 2004. Paleoenvironmental changes in the area of the Szczecin Lagoon (the south western Baltic Sea) as recorded from diatoms. Studia Quaternaria 21: 153–165.

Witkowski, A., A. Broszinski, O. Bennike, B. Janczak-Kostecka, J. B. Jensen, W. Lemke, R. Endler & A. Kuijpers, 2005. Darss Sill as a biological border in the fossil record of the Baltic Sea: evidence from diatoms. Quaternary International 130: 97–109.

Yu, S. Y., B. E. Berglund, P. Sandgren & S. C. Fritz, 2005. Holocene palaeoecology and shoreline displacement along the Blekinge coast, SE Sweden, and ist implications for climate and sea-level changes. The Holocene 15: 278–292.

Zachowicz, J., 1995. Quaternary geological profiles. In Dadlez, R., J. Mojski & J. Zachowicz (eds), Geological Atlas of the southern Balic. Polish Geological Institute, Sopot, Warszawa.

Zielinski, U. & R. Gersonde, 1997. Diatom distribution in Southern Ocean surface sediments (Atlantic sector): Implications for paleoenvironmental reconstructions. Palaeogeography, Palaeoclimatology, Palaeoecology 129: 213–250.

Zuchiewicz, W., J. Badura & M. Jarosiński, 2007. Noetectonics of Poland: An overview of active faulting. Studia Quaternaria 24: 5–20.

Reconstruction of human influence during the last two centuries on two small oxbow lakes near Warsaw (Poland)

Luiza Galbarczyk-Gąsiorowska ·
Michał Gąsiorowski · Krystyna Szeroczyńska

Originally published in the journal Hydrobiologia, Volume 631, No. 1, 173–183.
DOI: 10.1007/s10750-009-9809-6 © Springer Science+Business Media B.V. 2009

Abstract Short sediment cores from two small oxbow lakes located in the Vistula valley near Warsaw (Poland) were analysed for chemical composition and zooplankton (Cladocera) remains. The results of these analyses were evaluated using ordination techniques and the results were used to reconstruct the history of human impact of the lakes. The cores were dated using the ^{210}Pb method. These analyses show that the collected sediments (60 cm- and 67 cm-long-cores) were deposited during the last 150–200 years. Cladocera assemblages were different in the two lakes; however, *Bosmina longirostris* was dominant among subfossil Cladocera species in both lakes. The presence of this species indicates the high trophic state of the lakes. Changes in Cladocera species composition suggest that a significant increase in the lakes' trophic levels has occurred since the 1920s. The change could be related to the construction of flood dikes along the river Vistula, the isolation of the studied lakes from the river, changes in land use, or some combination of these factors. The chemical composition of sediments, particularly the increase in phosphorus and sulphur concentrations, confirms that the construction of dikes against flood was the major factor inducing eutrophication of the oxbow lakes. This study shows that the Warsaw small water bodies are especially sensitive to local disturbances caused by human activity, mainly water management, land use changes and agriculture.

Keywords Oxbow lakes · Eutrophication · Flood dikes · Land use changes · Cladocera · ^{210}Pb dating

Guest editors: K. Buczkó, J. Korponai, J. Padisák & S. W. Starratt
Palaeolimnological Proxies as Tools of Environmental Reconstruction in Fresh Water

L. Galbarczyk-Gąsiorowska
Institute of Geochemistry, Mineralogy and Petrology, Faculty of Geology, Warsaw University, Al. Zwirki i Wigury 93, 02-089 Warszawa, Poland

M. Gąsiorowski (✉) · K. Szeroczyńska
Institute of Geological Sciences, Polish Academy of Sciences, Twarda 51/55, 00-818 Warszawa, Poland
e-mail: mgasior@twarda.pan.pl

Introduction

Natural lakes face changes to their environment as a result of imposed anthropogenic activities (Villena & Romo, 2003). Lake sediments contain a record of climatic and ecological changes taking place during the time of their deposition. These changes are recorded as vertical variations of physical and chemical properties (Eskola & Peuraniemi, 2008) and changes in species composition of plant and animal remains (Milecka & Szeroczyńska, 2005; Sienkiewicz et al., 2006; Szeroczyńska et al., 2007). Most changes are connected with natural processes, such as aging of the lake and climatic changes; however, these natural processes are often interrupted by human-induced changes, in particular, overloading of the system by biogenic substances (Kozerski et al., 1999; Reed et al., 2008).

Changes in the aqueous environment in recent millennia have often been driven by human activity (e.g. Lamentowicz et al., 2008; Magny et al., 2008). Anthropogenic influences have been identified since the Neolithic and have become increasingly prominent (Niinemets & Saarse, 2007; Vanniere et al., 2008). The effects of human activity on a given lake depend on many factors, including lake morphometry, water circulation, catchment characteristics, land use and the types of local and regional human activities. In general, small and shallow lakes such as oxbows that are located in urban and agricultural landscapes are the most impacted by human activity. The oxbow lakes are affected not only by changes in local land use (Chalupova & Jansky, 2007), but also by changes in the watershed (Schneider et al., 2007; Babek et al., 2008). Changes in land use, primarily deforestation, in the drainage areas of many European rivers have resulted in the transformation of fluvial systems, e.g. increased overbank floodplain sedimentation or lateral migration of a river (de Moor et al., 2008). Additionally, the construction of flood dikes and other structures have isolated many oxbows from rivers (Hudson et al., 2008).

The Vistula is one of the last great rivers of Europe that has retained its natural appearance. It has source in southern Poland in the western part of Carpathian Mountains. It is 1,092 km long and drains an area of 194,424 km^2, mainly lowlands (Starkel, 2001). It empties with several branches of its delta into the Vistula Lagoon or directly to the Gdansk Bay of the Baltic Sea. Despite natural appearance of its channel and floodplain, many segments of the Vistula are regulated and navigable, especially those close to towns and cities.

Cladocerans are the most representative aquatic fauna whose remains are preserved in lake sediments. Subfossil specimens are usually well-preserved and can be identified to the sub-species level. For hundreds of thousands years, the cladoceran species have not changed (Korhola & Rautio, 2001). Some species live in open water while others exist in shallow parts of the lake. Cladocerans are divided into eurybionthic species that can survive in various environmental conditions and stenobionthic species that are not tolerant of environmental changes (Szeroczyńska, 1998). Understanding the ecological requirements of particular species of Cladocera allow us to form conclusions about their living environment. Cladocera are used as bioindicators to determine and reconstruct physical and chemical changes in lake environments (Bos & Cumming, 2003; Schneider et al., 2007). They are used as proxies for temperature, lake water level and trophic changes (Sarmaja-Korjonen & Alhonen, 1999; Duigan & Birks, 2000; Lotter et al., 2000). In particular, trophic level reconstructions have enabled many studies regarding the impact of natural and anthropogenic activities on lakes (Kamenik et al., 2007; Szeroczyńska, 2002).

The geochemistry of lake sediments has been used to effectively evaluate catchment weathering, paleoclimatic variations and environmental changes (Last & Smol, 2001; Rose et al., 2004). The chemical record of lake sediments is affected by a variety of factors, including source rock characteristics, catchment weathering, sorting and processes that occur during transport, deposition, post-depositional reactions and anthropogenic inputs (Abraham, 1998; Jin et al., 2006; Schmidt et al., 2008). Additionally, changes in element concentration in deep parts of lake sediments control ion exchange between the bottom water and sediments, reduction-oxidation conditions and sediment compaction (Boyle, 2001; Haggard et al., 2005; Wilson et al., 2008).

In this study, we investigate two small and shallow oxbow lakes in the Vistula basin that are located close to the Warsaw metropolitan area of Poland. The Warsaw oxbow lakes have been subject to human impacts for many centuries; therefore, a history of human activity in the Warsaw region can be derived from analyses of sediments in these lakes. Sediment geochemistry and Caldocera from the oxbow lakes were analysed, with special regard to tracing the influence of human economic activity during the last two centuries.

Study sites

Two oxbow lakes located in the Vistula river valley were selected to analyse human influences on the ecosystems (Fig. 1). The lakes are located in a rural area and have been significantly affected by agriculture and settlements for several decades. Lake Kazuńskie Dolne (52°24′41,7″ N, 20°40′56,6″ E) is located 30 km to the northwest of Warsaw and Lake

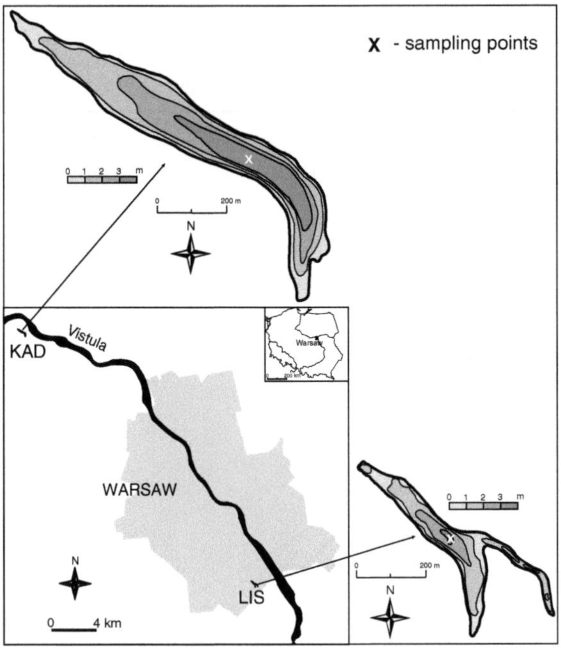

Fig. 1 Location and bathymetry of studied lakes: KAD—Lake Kazuńskie Dolne, LIS—Lake Lisowskie. X-marks indicate location of coring points

Lisowskie (52°07′43″ N, 21°07′26,1″ E) is located 10 km to the south of Warsaw. Both lakes are located on the floodplain of the Vistula River and are elevated 2–3 m above the mean level of the river. The lakes are shallow, narrow and elongated (Table 1). Small streams and ditches flow into and out of the lakes. The lake water has a pH between 6.9 and 7.2 and an oxygen concentration between 2.6 and 5.7 mg dm^{-3}. *Phragmites australis* dominates the shores and floating-leaved *Nuphar lutea* and *Nymphaea alba* dominate the littoral region. The dominant fish species are naturally-occurring tench (*Tinca tinca*), crucian carp (*Carassius carassius*), pike (*Esox lucius*), roach (*Rutilus rutilus*) and introduced grass carp (*Ctenopharyngodon idella*), silver crucian carp (*Carassius auratus gibelio*), carp (*Carpinus carpio*), wels (*Silurus glanis*), and sturgeon (*Acipenser* sp.).

Methods

Sampling

Sediment cores were collected from the central part of both lakes (Fig. 1) using a Kajak-type gravity corer. The cores were cut into 1 cm-thick samples and stored in plastic bags. The sediments were subsampled in the laboratory for radiometric dating, chemical composition and Cladocera analyses.

^{210}Pb dating

The ^{210}Pb activity of sediments was determined indirectly using alpha-spectrometry measurements of ^{210}Po ($E_\alpha = 5.31$ MeV, $T_{1/2} = 138$ days) activity (Flynn, 1968) in the Laboratory of the Institute of Geological Sciences of the Polish Academy of Sciences in Warsaw, Poland. A model assuming a constant rate supply of unsupported ^{210}Pb (CRS model) was used to calculate the sediment age (Appleby, 2001). An age-depth function (Fig. 2) was calculated using the randomisation method and was fitted using the LOESS procedure (Cleveland & Devlin, 1988).

Cladocera analysis

The samples were prepared for Cladocera analysis according to the methods described in Szeroczyńska & Sarmaja-Korjonen (2007). For each sample, 1 cm^3 of wet sediment was heated and stirred in 10% KOH on a hot plate for 20 min, sieved through a 33 μm mesh, dissolved in 10 cm^3 of distilled water and coloured with safranine dye. A slide was prepared from 0.1 ml of each sample and examined under a light microscope (magnification 100× and 200×) to count the number of cladoceran remains. Two to five slides were scanned from each sample. The total of all samples was more than 400 cladoceran remains (headshields, shells, ephippia, postabdomen). The most abundant body part was chosen for each species to represent the number of individuals and the percentages were calculated from the sum of individuals. Cladoceran remains were identified based on keys by Flössner (2000) and Szeroczyńska & Sarmaja-Korjonen (2007). Nomenclature follows that proposed in Szeroczyńska & Sarmaja-Korjonen (2007). Cladocera Local Zones (CLZ) were defined based on ConsLink analysis (Gordon & Birks, 1972). Ecological preferences of cladoceran taxa were determined on the basis of Whiteside (1970) and Szeroczyńska (1998). The results were plotted as percentage diagrams (Figs. 3, 4).

Table 1 Characteristics of studied oxbow lakes

	Latitude	Longitude	Altitude (m a.s.l.)	Length (m)	Wide (m)	Elongation index	Area (ha)	Maximal depth (m)	Mean Depth (m)	Volume (× 1,000 m^3)
Lake Kazuńskie Dolne	52°25′ N	20°41′ E	71.1	1540	200	7.70	18.82	3.5	1.8	343.0
Lake Lisowskie	52°08′ N	21°08′ E	83.8	765	110	6.95	6.40	3.1	1.2	78.6

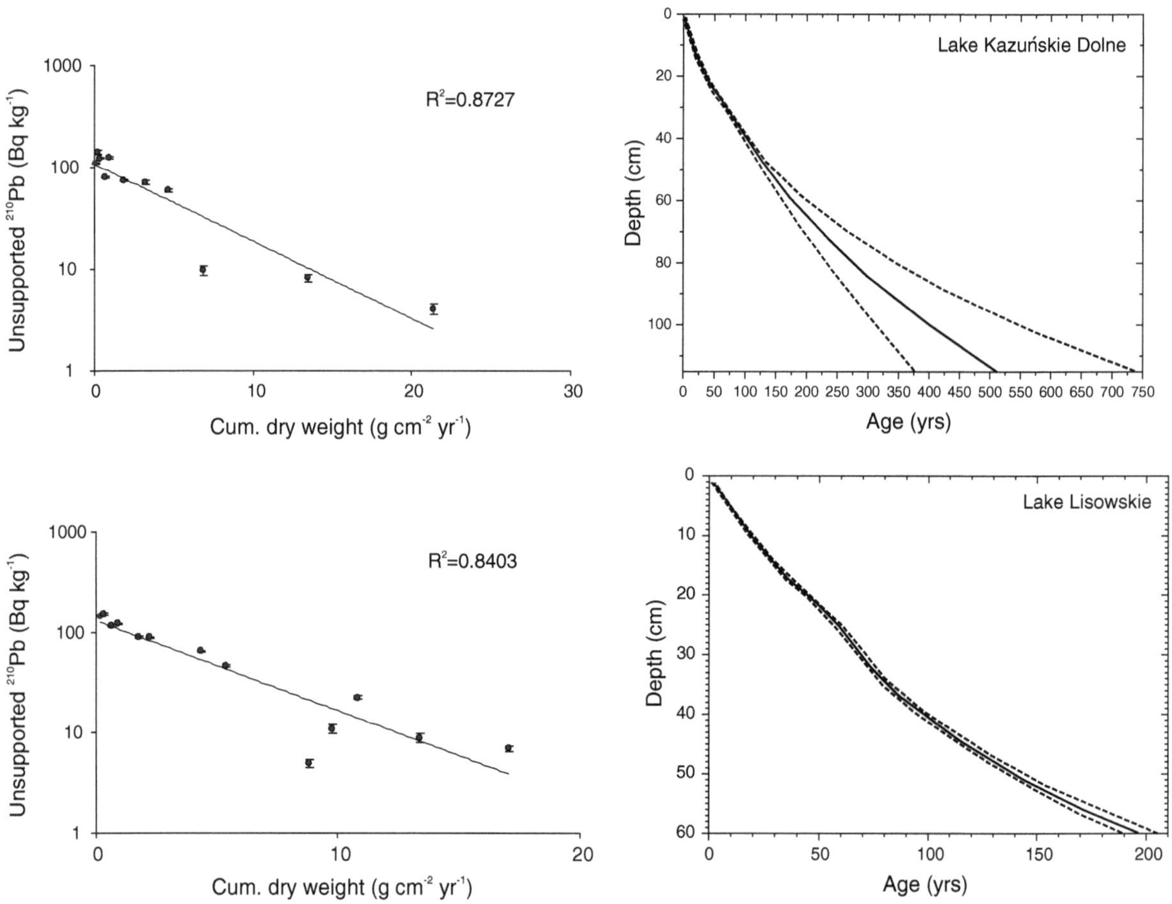

Fig. 2 Unsupported ^{210}Pb activity profiles and age-depth models for sediment sequences of Lake Kazuńskie Dolne and Lake Lisowskie

Chemical composition

The samples analysed for chemical composition (Fig. 5) were dried at 550°C overnight. The reagent used for demineralization was aqua regia (6 ml HCl + 2 ml HNO_3 0.5 g^{-1} of sample) plus 1 ml H_2O_2. The resulting solutions were diluted to 50 ml with deionised water. In all samples, the major elements (such as Ca, Mg, K, Na, Al, Ti, Cu, Zn, Mn, Fe, S and P) were determined by Inductively Coupled Plasma-Atomic Emission Spectrometry (ICP-AES) at the Institute of Geochemistry, Mineralogy and Petrology, Faculty of Geology, Warsaw University. The loss-on-ignition (LOI) was determined by igniting the oven-dried samples at 550°C for 1 h. The LOI was used as a measure of organic matter in the sediments.

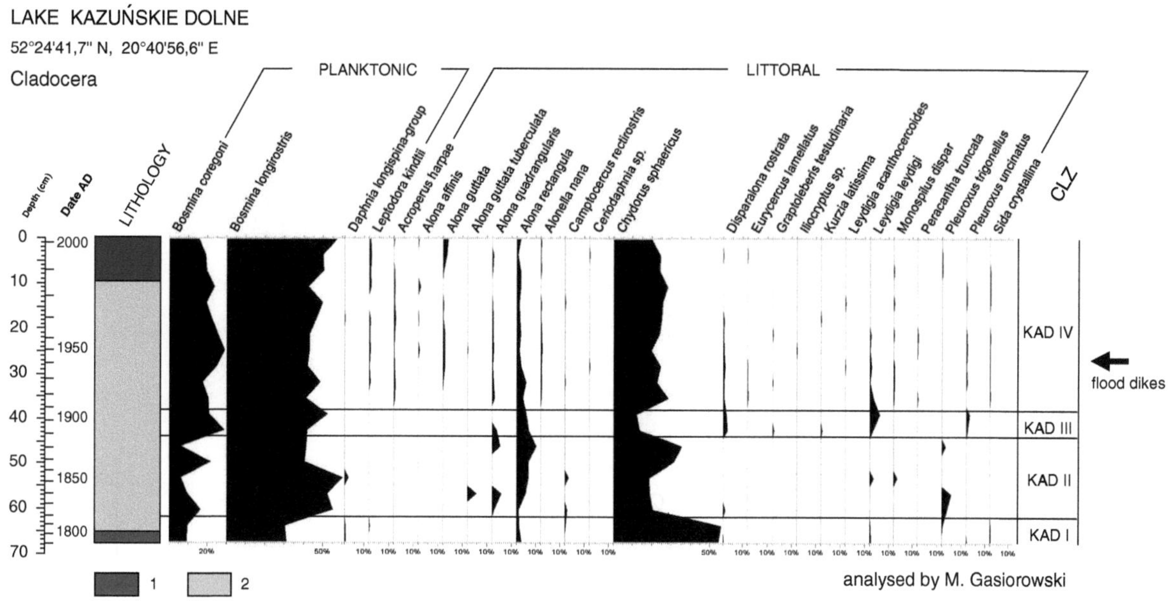

Fig. 3 Percentage diagram of Cladocera assemblages from the sediment of Lake Kazuńskie Dolne. Key to lithology: 1—gyttja with course detritus, 2—gyttja with fine detritus

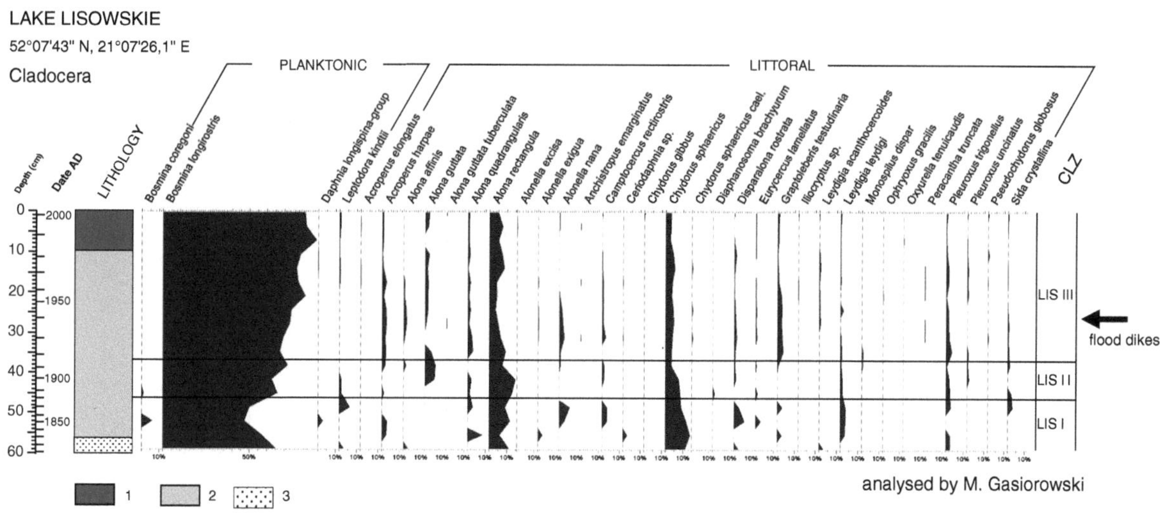

Fig. 4 Percentage diagram of Cladocera assemblages from the sediment of Lake Lisowskie. Key to lithology: 1—gyttja with course detritus, 2—gyttja with fine detritus, 3—fluvial sand

Statistics

Detrended correspondence analysis (detrending by segments) was used to test homogeneity of the cladoceran datasets. Since all datasets present the length of gradients <2 standard deviation unit, redundancy analysis (RDA) was applied to detect a relationship between Cladocera (species data) and the concentration of selected elements ('environmental' data). Environmental variables were ranked by forward selection and the statistical significance of each variable was tested by a Monte Carlo permutation test (299 permutations under a reduced model). Only statistically significant variables ($P < 0.05$) and species with more than 1% frequency in any sample and at least three occurrences were chosen for the

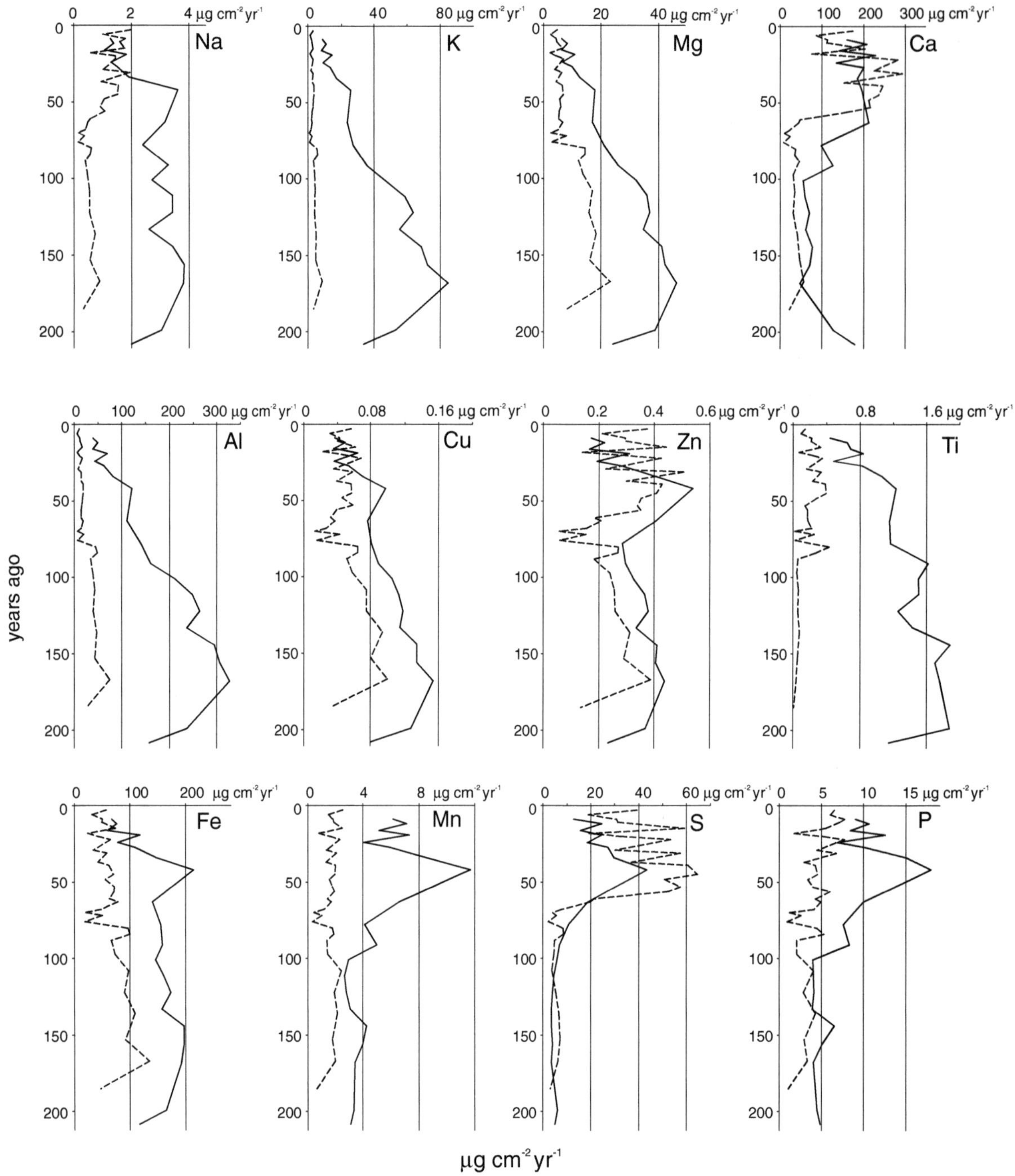

Fig. 5 Sedimentation rate (µg cm^{-2} yr^{-1}) of selected elements in sediment sequence from Lake Kazuńskie Dolne (*solid lines*) and Lake Lisowskie (*dashed lines*)

analysis. The species data for RDA were log-transformed. The DCA and RDA (Fig. 6) were performed with the CANOCO program, version 4.02 (ter Braak & Smilauer, 1998).

Results

Cores from Lake Kazuńskie Dolne and Lake Lisowskie were 67 cm and 60 cm long, respectively. The

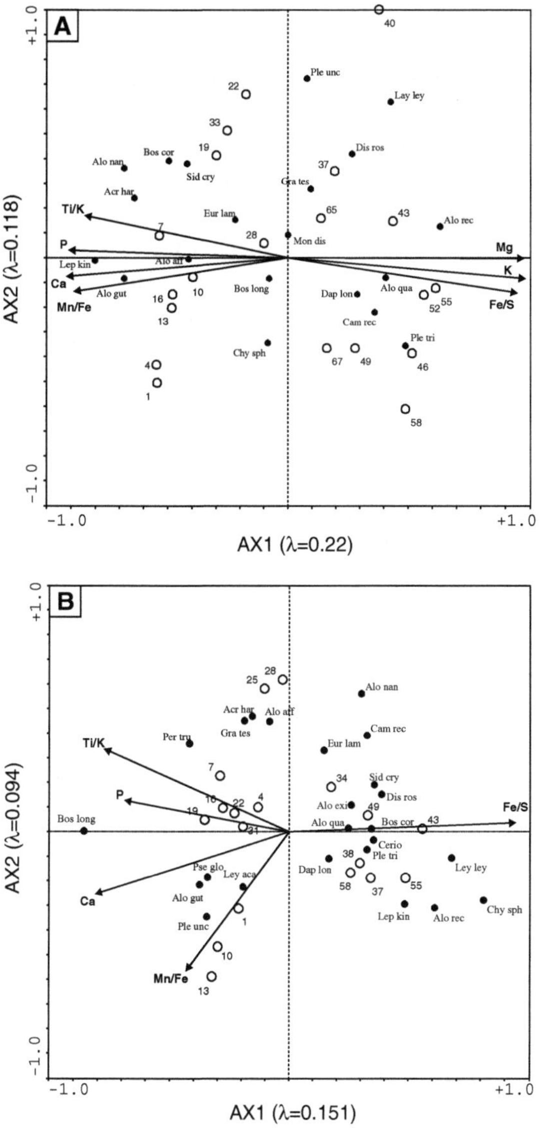

Fig. 6 Redundancy analysis of subfossil cladoceran assemblages from Lake Kazuńskie Dolne (A) and Lake Lisowskie (B). Samples were given as open circles, geochemical parameters (explanatory data) are arrows and cladoceran species (species data) are solid circles. Abbreviation: Acr har *Acroperus harpae*, Alo aff *Alona affinis*, Alo gut *Alona guttata*, Alo qua *Alona quadrangularis*, Alo rec *Alona rectangula*, Alo exi *Alonella excisa*, Alo nan *Alonella nana*, Bos cor *Bosmina coregoni*, Bos long *Bosmina longirostris*, Cam rec *Camptocercus rectirostris*, Cerio *Ceriodaphnia* spp., Chy sph *Chydorus sphaericus* s.l., Dap lon *Daphnia longispina* group, Dis ros *Disparalona rostrata*, Eur lam *Eurycercus lamellatus*, Gra tes *Graptoleberis testudinaria*, Ley aca *Leydigia acanthocercoides*, Ley ley *Leydigia leydigi*, Lep kin *Leptodora kindti*, Mon dis *Monospilus dispar*, Per tru *Pleuroxus truncatus*, Ple tri *Pleuroxus trigonellus*, Ple unc *Pleuroxus uncinatus*, Pse glo *Pseudochydorus globosus*, Sid cry *Sida crystallina*

collected sediments were fluvial sand, limnic clay and gyttja. Generally, unsupported ^{210}Pb declined with depth, and some irregularities were only present in the uppermost portions of the cores (Fig. 2). The age-depth models indicate that the analysed sediments were deposited during the last two centuries. The lithology and age of sediment sequences were discussed in detail by Gąsiorowski & Hercman, (2005).

Lake Kazuńskie Dolne

Twenty-six taxa of Cladocera were identified from the sediments of Lake Kazuńskie Dolne (Fig. 3) and four cladoceran zones were described. Zone KAD I (67–62 cm) is dominated by *Chydorus sphaericus* sensu lato (O.F. Müller) and *Bosmina* spp. Only a few benthic taxa (*Alona rectangula* Sars, *Leydigia leydigi* (Schoedler) were found. Zone KAD II (62–44 cm) was characterised by a decrease in *Chydorus sphaericus* sensu lato. *Bosmina longirostris* (O.F. Müller) and *Alona rectangula* increased until they reached a maximum percentage for the core. Benthic Cladocera were represented by *Alona quadrangularis* (O.F. Müller) and plant-associated *Pleuroxus trigonellus* (O.F. Müller). Zone KAD III (44–37 cm) was delineated by a decline in *Chydorus sphaericus* s.l. and an increase in *Bosmina* spp., *Leydigia leydigi* and *Disparalona rostrata* (Koch). These species inhabited open parts of the lake bottom, while *Pleuroxus uncinatus* Baird and *Graptoleberis testudinaria* (Fischer) lived between submerged macrophytes. Zone KAD IV (37–0 cm) was the zone with the highest species diversity. Plant-associated taxa were frequent. *Bosmina* spp. and predatory *Leptodora kindtii* (Focke) were present in the open-water zone.

The concentration of K, Mg, Al, Cu, Ti and Fe were relatively high in the sediments deposited between ca. 210 and 70 years ago (Fig. 5). These elements decreased significantly in abundance after that period and concentrations of Ca, Mn, P and S increased.

The K, Mg, Ca, P, Ti/K, Mn/Fe and Fe/S concentrations were used as explanatory data in redundancy analysis (RDA, Fig. 6A). The two first axes of ordination diagram explain 33.8% of variance in species data. The samples were split by axis 2 into two clusters. The first cluster (samples 67–37 cm) has positive values of axis 1 and is characterised by high values of magnesium, potassium and iron/sulphur ratio. The second cluster (samples 33–1 cm) has negative values of axis 1 and is characterised by higher values of calcium, phosphorus and titanium/potassium and manganese/iron ratios. The borderline between 37 and 33 cm occurred during the 1920s (Fig. 2).

Lake Lisowskie

Thirty-five Cladocera taxa were identified in the Lake Lisowskie sediment sequence. The most significant feature of the subfossil Cladoceran assemblage was the clear dominance of *Bosmina longirostris* (47–88%). Three Cladoceran assemblage zones were described (Fig. 4). Zone LIS I (58–45 cm) was characterised by a relatively high diversity of Cladocera assemblages. Despite the dominance of *B. longirostris* and the presence of other planktonic species (i.e. *B. coregoni* Baird, *Daphnia longispina* O.F. Müller group, *Leptodora kindtii*), benthic species were also relatively numerous. Zone LIS II (45–36 cm) was characterised by lower percentages of benthic and plant-associated taxa. Zone LIS III (36–0 cm) was delineated by an increase in plant-associated taxa. *Bosmina longirostris* reached a maximum percentage (88%, sample 7 cm) in this zone.

Major (K, Ca, Mg, Na, Al) and rare (Cu, Zn, Ti, Mn, Fe) elements and S and P concentrations were almost constant in the sediments deposited in the initial stage of lake development The sedimentation rate for all elements was also constant during this time (from ca. 0.08 $\mu g\ cm^{-1}\ yr^{-1}$ for copper to more than 40 $\mu g\ cm^{-1}\ yr^{-1}$ for calcium, Fig. 5). After a period of stable conditions, the accumulation rate of almost all elements decreased (e.g. to 2.5 $\mu g\ cm^{-1}\ yr^{-1}$ for Mg, <1 $\mu g\ cm^{-1}\ yr^{-1}$ for K, 19 $\mu g\ cm^{-1}\ yr^{-1}$ for Fe and >0.03 $\mu g\ cm^{-1}\ yr^{-1}$ for Ti) and the lowest concentrations were in the samples deposited 70–80 years ago (Fig. 5). After this episode, a strong increase in P and S concentrations and variable concentration of Ca were the most characteristic features of the chemical composition of the sediments.

Statistically significant geochemical data (i.e. Ca, P, Ti/K, Mn/Fe, Fe/S) explain 24.5% variation in Cladocera species composition (Fig. 6B). The plot was similar to that of Lake Kazuńskie Dolne (Fig. 6A). The samples were split by axis 2 into two clusters (58–34 cm and 31–1 cm) and 'boundary' was dated at 1920.

Discussion

The analysed core of Lake Kazuńskie Dolne does not contain the entire sequence of sediments deposited in this lake; the core is 67 cm long while the total sediment sequence is 115 cm long. Therefore, the lowest analysed sample does not represent an initial stage in the lake's development. This is evident from cladoceran species composition. The Cladocera assemblage of KAD I zone is relatively rich and much more diverse than typical pioneer assemblages. The dominance of *Chydorus sphaericus* s.l. indicates the relatively high trophic state of the lake during this period. The sample from 67 cm depth was dated to be from 1804 ± 15 yr, which was a period of extensive Dutch and German settlement near the lake (Szałygin, 2004). The settlers, unlike Polish farmers, significantly changed the natural floodplain environment through deforestation, irrigation and agriculture. All of these processes, directly or indirectly, may have accelerated an increase in the lake's trophic state. A similar situation was recorded in the Lake Lisowskie sediments. There, the first cladoceran zone (LIS I), which was dated to begin around 1810, also recorded a period of high percentages of *Chydorus sphaericus* s.l. A Dutch village existed at this time near the shore of Lake Lisowskie, and agriculture may have been responsible for the increase in the lake's trophic state. In this lake, however, the increase was not very strong and was not recorded in phosphorus or sulphur profiles (Figs. 5, 6). Human influence could also have caused decreased pressure from planktonivorous fishes, either directly by intense fishing or indirectly by increasing water turbidity (Jeppesen et al., 2003). As a result, a relatively high proportion of large body cladoceran taxa (i.e. *Leptodora kindti*, *Daphnia longispina* group, *Eurycercus lamellatus*) were noted at the end of the LIS I zone. The period of higher trophic state (and/or turbidity) ended with a decline in *Chydorus sphaericus* s.l. percentages and an increase in planktonic *Bosmina*. The lakes' trophic state decreased (KAD II and LIS II) as a result of the decline of Dutch settlement in the 1860s (Szałygin, 2004).

Changes in the abundance of *Bosmina coregoni* (only in Lake Kazuńskie Dolne), an absence of plant-associated Cladocera, and the presence of benthic taxa may also indicate frequent flood events. Several great floods were recorded in the Vistula basin during the 19th century (1813, 1844 and 1888/89) and the first half of 20th century (1924, 1934 and 1947) (Plit, 1992; Starkel, 2002). The decrease in some benthic species and the increase in soft-bottom dwelling species (e.g. *Leydigia leydigi*) in KAD III and LIS II may be related to the flood of 1844. This flood event was also recorded in the chemical composition of the Lake Kazuńskie Dolne and Lake Lisowskie sediments (Fig. 5) as increasing concentrations of elements related to mineral fraction (e.g. Al, Fe, K and Cu) and low concentrations of P and S. On the other hand, exceptionally heavy flood of 1924 was not clearly recorded by cladoceran species composition, but only by decreasing concentrations of almost all elements (Fig. 5).

The final zone (KAD IV and LIS III) was a period of more stable sedimentological conditions and a continuous increase in the trophic level of the lakes. The planktonic cladoceran species were dominant, but littoral and benthic taxa were also present. This diversity suggests coexistence of open-water zone and littoral macrophyte communities. According to historic data, abundant flood events occurred during the deposition of KAD IV and LIS III (Plit, 1992; Starkel, 2002); however, only the great flood episode of 1924 was recorded in the sediments as evidenced by an increase in the sedimentation rate (Gasiorowski & Hercman, 2005) and a decrease in the concentrations of almost all elements (Fig. 5). The sedimentation rate in both oxbow lakes stabilised after this flood, which is related to the artificial isolation of the lakes from the Vistula River after the construction of flood-control dikes against flood. The shift from the oxbows connected to the river to isolation from fluvial influences is clearly visible on ordination plots (Fig. 6). The flood dikes in this section of the Vistula Valley were completed during World War II. They affected water circulation, sedimentation rate and grain size, land use and induced intensive agriculture and settlement. As a result, they changed the ecological state of the lakes and induced eutrophication processes, as seen in changes of the Cladocera species composition (Figs. 3, 4) and the chemical composition of sediments (Fig. 5). The peaks of phosphorous and sulphur are the most prominent. The lakes' environments were also affected by direct manipulation, e.g. significant amounts of calcium carbonates (lacustrine chalk) were added into Lake Lisowskie in 1960s, 1970s and 1980s due to control lake pH and submerge macrophytes density. That is reflected by strong variation in Ca and S concentration (Fig. 5).

Conclusions

Large flood events and extensive agriculture contributed to the state of the oxbow lakes studied in this work. Changes in the size of nearby settlements were clearly recorded in lake sediments as changes in eutrophication. Floods were recorded as peaks in sedimentation rate and in relative increases in the number of benthic Cladocera species dwelling on the soft lake bottom. The construction of flood dikes significantly changed the oxbows' ecosystems. Isolation of the lakes from the river accelerated the eutrophication process. Today both oxbows are eutrophic lakes.

A precise reconstruction of artificial changes in sediment chemistry (especially Cu, Zn, S and P) is needed to calculate the geochemical background conditions. Also, sequential extraction and mineralogical studies might help to estimate the possible migration of specific elements and the sorption capability of lake sediments. Such studies could help to forecast the future response of the oxbow lakes to the hydrological and hydrochemical parameters (e.g. pH, redox, water depth and oxygen depletions) and could be used to develop an appropriate conservation programme.

Acknowledgments We thank two anonymous reviewers for detail and very helpful comments on an early version of the article. This study was supported by Polish Committee for Scientific Research grant no P04D07121 and the grant of Institute of Geochemistry, Mineralogy and Petrology, University of Warsaw no 1797/13.

References

Abraham, J., 1998. Spatial distribution of major and trace elements in shallow reservoir sediments: an example from Lake Waco, Texas. Environmental Geology 36: 349–363.

Appleby, P. G., 2001. Chronostratigraphic techniques in recent sediments. In Last, W. M. & J. P. Smol (eds), Tracking environmental change using lake sediments. Kluwer Academic Publishers, Dordrecht: 171–203.

Babek, O., K. Hilscherova, S. Nehyba, J. Zeman, M. Famera, J. Francu, I. Holoubek, J. Machat & J. Klanova, 2008. Contamination history of suspended river sediments accumulated in oxbow lakes over the last 25 years. Journal of Soils and Sediments 8: 165–176.

Bos, D. G. & B. F. Cumming, 2003. Sedimentary cladoceran remains their relationship to nutrients and other limnological variables in 53 lakes from British Columbia, Canada. Canadian Journal of Fisheries and Aquatic Sciences 60: 1177–1189.

Boyle, J., 2001. Redox remobilization and the heavy metal record in lake sediments: a modelling approach. Journal of Paleolimnology 26: 423–431.

Chalupova, D. & B. Jansky, 2007. Anthropogenic impact on selected oxbow lakes in the Elbe River floodplain. Journal of Hydrology and Hydromechanics 55: 86–97.

Cleveland, W. S. & S. J. Devlin, 1988. Locally weighted regression: an approach to regression analysis by local fitting. Journal of the American. Statistical Association 83: 596–610.

de Moor, J. J. W., C. Kasse, R. van Balen, J. Vandenberghe & J. Wallinga, 2008. Human and climate impact on catchment development during the Holocene —Geul River, the Netherlands. Geomorphology 98: 316–339.

Duigan, C. A. & H. H. Birks, 2000. The late-glacial and early-Holocene palaeoecology of cladoceran microfossil assemblage at Kråkenes, western Norway, with a quantitative reconstruction of temperature changes. Journal of Paleolimnology 23: 67–76.

Eskola, T. & V. Peuraniemi, 2008. Natural and anthropogenic impacts on the composition of lake sediments. Mineralogical Magazine 72: 411–413.

Flössner, D., 2000. Die Haplopoda und Cladocera (ohne Bosminidae) Mitteleuropas. Backhuys Publishers, Leiden: 428.

Flynn, W. W., 1968. The termination of low-levels of polonium-210 in environmental materials. Analytica Chemica Acta 43: 221–227.

Gąsiorowski, M. & H. Hercman, 2005. Recent changes of sedimentation rate in three Vistula oxbow lakes determined by ^{210}Pb dating. Geochronometria 24: 33–39.

Gordon, A. D. & H. J. B. Birks, 1972. Numerical methods in palaeoecology. I. Zonation of pollen diagrams. New Phytologist 71: 961–979.

Haggard, B. E., P. A. Moore & P. B. DeLaune, 2005. Phosphorus flux from bottom sediments in Lake Eucha, Oklahoma. Journal of Environmental Quality 34: 724–728.

Hudson, P. F., H. Middelkoop & E. Stouthamer, 2008. Flood management along the lower Mississippi and Rhine Rivers (The Netherlands) and the continuum of geomorphic adjustment. Geomorphology 101: 209–236.

Jeppesen, E., J. P. Jensen, T. L. Lauridsen, S. L. Amsinck, K. Christoffersen, M. Sondergaard & S. F. Mitchell, 2003. Sub-fossils of cladocerans in the surface sediment of 135 lakes as proxies for community structure of zooplankton, fish abundance and lake temperature. Hydrobiologia 491: 321–330.

Jin, Z., F. Li, J. Cao, S. Wang & J. Yu, 2006. Geochemistry of Daihai lake sediments, inner Mongolia, north China: implication for provenance, sedimentary sorting, and catchment weathering. Geomorphology 80: 147–163.

Kamenik, Ch., K. Szeroczyńska & R. Schmidt, 2007. Relationships among recent Alpine Cladocera remains and their environment: implications for climate-change studies. Hydrobiologia 594: 33–46.

Korhola, A. & M. Rautio, 2001. Cladocera and other branchiopod crustaceans. In Last, W. M. & J. P. Smol (eds), Tracking environmental change using lake sediments volume 4: zoological indicators. Kluwer Academic Publishers, Dordrecht: 5–41.

Kozerski, H. P., H. Behrendt & J. Kohler, 1999. The N and P budget of the shallow, flushed lake Muggelsee: retention, external and internal load. Hydrobiologia 408: 159–166.

Lamentowicz, M., M. Obremska & E. A. D. Mitchell, 2008. Autogenic succession, land-use change, and climatic influences on the Holocene development of a kettle-hole mire in Northern Poland. Review of Palaeobotany and Palynology 151: 21–40.

Last, W. M. & J. P. Smol, 2001. Tracking environmental change using lake sediments. Kluwer Academic Publishers, Dordrecht, The Netherland.

Lotter, A. F., H. J. B. Birks, U. Eicher, W. Hofmann, J. Schwander & L. Wick, 2000. Younger Dryas and Allerod summer temperatures at Gerzensee (Switzerland) inferred from fossil pollen and cladoceran assemblages. Palaeogeography Palaeoclimatology Palaeoecology 159: 349–361.

Magny, M., E. Gauthier, B. Vanniere & O. Peyron, 2008. Palaeohydrological changes and human-impact history over the last millennium recorded at lake Joux in the Jura mountains, Switzerland. Holocene 18: 255–265.

Milecka, K. & K. Szeroczyńska, 2005. Changes in macrophytic flora and planktonic organisms in lake Ostrowite, Poland, as a response to climatic and trophic fluctuations. Holocene 15: 74–84.

Niinemets, E. & L. Saarse, 2007. Mid- and late-Holocene land-use changes inferred from pollen records, in a south-eastern Estonian upland area. Review of Palaeobotany and Palynology 146: 51.

Plit, J., 1992. Historical changes of chosen elements of geographic environment of the Łomianki commune. Polish Ecological Studies 18: 167–184.

Reed, J. M., M. J. Leng, S. Ryan, S. Black, S. Altinsacli & H. I. Griffiths, 2008. Recent habitat degradation in karstic lake Uluabat, western Turkey: a coupled limnological-palaeolimnological approach. Biological Conservation 141: 2765–2783.

Rose, N. L., J. F. Boyle, Y. Du, C. Yi, X. Dai, P. G. Appleby, H. Bennion, S. Cai & L. Yu, 2004. Sedimentary evidence for changes in the pollution status of Taihu in the Jiangsu region of eastern China. Journal of Paleolimnology 32: 41–51.

Sarmaja-Korjonen, K. & P. Alhonen, 1999. Cladoceran and diatom evidence of lake-level fluctuations from a Finnish lake and the effect of aquatic-moss layer in microfossil assemblages. Journal of Paleolimnology 22: 277–290.

Schmidt, R., M. Roth, R. Tessadri & K. Weckstrom, 2008. Disentangling late-Holocene climate and land use impacts on an Austrian alpine lake using seasonal temperature anomalies, ice-cover, sedimentology, and pollen tracers. Journal of Paleolimnology 40: 453–469.

Schneider, H., D. Hoefer, R. Mausbacher & M. Gude, 2007. Past flood events reflected in Holocene floodplain records of East-Germany. Geomorphology 92: 208–219.

Sienkiewicz, E., M. Gąsiorowski & H. Hercman, 2006. Is acid rain impacting Sudetic lakes? Sciences of the Total Environment 369: 139–149.

Starkel, L., 2001. Historia Doliny Wisły od ostatniego zlodowacenia do dziś (Evolution of the Vistula River valley since the last glaciation till present). Polish Academy of Sciences, Institute of Geography and Spatial Organization, Monographies 2, Warszawa: 263 pp (in Polish).

Starkel, L., 2002. Change in the frequency of extreme events as the indicator of climatic change in the Holocene (in fluvial systems). Quaternary International 91: 25–32.

Szałygin, J., 2004. Katalog zabytków osadnictwa holenderskiego na Mazowszu (The catalogue of monuments of the Dutch settlement in the Mazovian region), Wydawnictwo DiG, Warszawa: 285 pp.

Szeroczyńska, K., 1998. Cladocera (Crustacea) jako źródło informacji w badaniach osadów jeziornych (Cladocera (Crustacea) as information source in studies of lake sediments). Studia Geologica Polonica 112: 9–28 (in Polish).

Szeroczyńska, K., 2002. Human impact on lakes in the remains of cladocera (Crustacea). Quaternary International 95–96: 165–174.

Szeroczyńska, K. & K. Sarmaja-Korjonen, 2007. Atlas of subfossil cladocera from central and northern Europe. Friends of the Lower Vistula Society, Poland.

Szeroczyńska, K., A. Tatur, J. Weckstrom, M. Gąsiorowski, A. Noryskiewicz & E. Sienkiewicz, 2007. Holocene environmental history in northwest Finnish lapland reflected in the multi-proxy record of a small subarctic lake. Journal of Paleolimnology 38: 25–47.

ter Braak, C. J. F. & P. Smilauer, 1998. CANOCO reference manual. Centre of Biometry, Wageningen: 351 pp.

Vanniere, B., D. Colombaroli, E. Chapron, A. Leroux, W. Tinner & M. Magny, 2008. Climate versus human-driven fire regimes in Mediterranean landscapes: the Holocene record of Lago dell'Accesa (Tuscany, Italy). Quaternary Science Reviews 27: 1181–1196.

Villena, M. J. & S. Romo, 2003. Phytoplankton changes in a shallow Mediterranean lake (Albufera of Valencia, Spain) after sewage diversion. Hydrobiologia 506: 281–287.

Whiteside, M. C., 1970. Danish chydorid cladocera: modern ecology and core studies. Ecological Monographs 40: 79–118.

Wilson, T. A., S. A. Norton, B. A. Lake & A. Amirbahman, 2008. Sediment geochemistry of Al, Fe, and P for two historically acidic, oligotrophic Maine lakes. Science of the Total Environment 404: 269–275.

Larval chaoborid mandibles in surface sediments of small shallow lakes in Finland: implications for palaeolimnology

Tomi P. Luoto · Liisa Nevalainen

Originally published in the journal Hydrobiologia, Volume 631, No. 1, 185–195.
DOI: 10.1007/s10750-009-9810-0 © Springer Science+Business Media B.V. 2009

Abstract The aim of this study was to assess the value of phantom midge *Chaoborus* (Diptera: Chaoboridae) larvae as a palaeoenvironmental proxy in northern Europe. The presence of *Chaoborus* was examined in 80 small shallow lakes across Finland (60–70°N), based on their subfossil mandibles preserved in lake sediments. *Chaoborus flavicans* was present in 33 lakes, but was the most abundant midge taxon in only one of the lakes. *Chaoborus crystallinus/obscurripes* mandibles were present in two lakes. Statistical analyses showed that the distribution of *Chaoborus* was significantly correlated with lake depth and mean July air temperature. *Chaoborus* was absent in the northern study lakes, whereas the results indicated that *C. flavicans* was frequently found in small shallow lakes in southern Finland. *Chaoborus flavicans* was particularly abundant in fish-free, macrophyte-rich, dystrophic lakes, but absent in oligotrophic clear-water lakes having fish. The results of this study indicate that the subfossil mandibles of *Chaoborus* can provide valuable information in multiproxy palaeolimnological studies, especially when investigating past changes in water level and temperature in small shallow lakes in northern Europe.

Keywords *Chaoborus* · Subfossils · Lake depth · Air temperature · Lake sediments

Guest editors: K. Buczkó, J. Korponai, J. Padisák & S. W. Starratt
Palaeolimnological Proxies as Tools of Environmental Reconstruction in Fresh Water

T. P. Luoto (✉) · L. Nevalainen
Department of Geology, University of Helsinki,
P.O. Box 64, 00014 Helsinki, Finland
e-mail: tomi.luoto@helsinki.fi

Introduction

The most common chaoborid (Diptera: Chaoboridae) species in Scandinavia are *Chaoborus flavicans* Meigen and *C. obscurripes* Wulp (Hongve, 1975; Stenson, 1981). Chaoborids are holometabolous aquatic insects that have four larval instars. The larvae of *Chaoborus* Lichtenstein, commonly referred to as phantom midges, are transparent planktonic predators. They have hydrostatic organs that regulate their buoyancy and use their prehensile antennae to sweep up prey, mainly water fleas (Cladocera). After overwintering as third or fourth larval instars, *C. flavicans* and *C. obscurripes* pupate in the following year and after transforming into flying adults in July–August, the number of larvae in lakes becomes low (Hongve, 1975; Paasivirta, 1989). Chaoborid larvae are known to avoid predators by diurnal vertical migration, although *C. americanus* Johannsen, *C. crystallinus* De Geer, and *C. obscurripes* do not migrate, and are restricted to fish-free lakes most likely for this reason (Uutala, 1990). Migrating chaoborids, e.g.,

C. flavicans, are able to migrate vertically to the hypolimnion or into the sediments during the day to avoid visual predation by planktivorous fish (Dawidowicz et al., 1990).

Chaoborus larvae are able to tolerate a variety of environmental conditions, e.g., different pH levels (Bendell & McNicol, 1987) and high trace metal contents (Hare & Tessier, 1998). The indirect influence of nutrient levels, which determines the size of zooplankton communities, may be the most important limnological factor affecting *Chaoborus* abundance (e.g., Liljendahl-Nurminen et al., 2002). However, the main factor limiting the presence and abundance of *Chaoborus* in lakes is very often the presence or absence of fishes (Stenson, 1978; Uutala, 1990), especially in the case of species that do not migrate the same way as *C. flavicans* (e.g., *C. obscurripes*) (Stenson, 1981). Liljendahl-Nurminen (2006) showed that clay turbidity disturbs visual predation by fish, and therefore, *C. flavicans* can be particularly successful in deep clay-turbid lakes where they can remain near their epilimnetic prey without long-distance diurnal vertical migration. *Chaoborus* is also very successful in humic brown-water lakes in Scandinavia (e.g., Paasivirta, 1982a), where light, temperature, and oxygen profiles reduce fish predation (Wissel et al., 2003). However, Yan et al. (1985) concluded that the presence of fish is not the prime determinant of *Chaoborus* abundance in central Canada, and in contrast to the suggestion by Eriksson et al. (1980) that the abundance should increase in acidified lakes due to the elimination of fish, *Chaoborus* larvae were not more abundant in acidified lakes. Brundin (1949) and Hofmann (1978) showed that *C. flavicans* is usually abundant only in deep lakes exhibiting low hypolimnetic oxygen content during summer stratification.

The chitinous mandibles of chaoborids are well preserved and commonly found in lake sediments (Frey, 1964). Subfossils from lake sediments can be used not only in palaeoenvironmental research but also in contemporary ecological studies, considering that the topmost sediment layer, if not disturbed, represents the present fauna. The greatest advantage in 'modern' subfossil analysis of macroinvertebrates, compared with traditional contemporary sampling, is that surface sediment sampling is comparatively easy and the samples are usually representative (cf. Frey, 1988). Single-point surface sediment sampling can also be used to characterize the lake environment for palaeoenvironmental calibration studies (Bunbury & Gajewski, 2008). Unfortunately, the identification of species from subfossil specimens is sometimes difficult; thus, species-type or genus-level identification is occasionally required (cf. Brooks et al., 2007). Some animal remains, however, are preserved exceptionally well and can be identified to species level (cf. Szeroczyńska & Sarmaja-Korjonen, 2007). Subfossil *Chaoborus* remains have been used in palaeolimnological research as indicators of predator–prey interactions, e.g., changes in fish (Johnson et al., 1990; Uutala, 1990; Uutala & Smol, 1996; Sweetman & Smol, 2006) or zooplankton (Sarmaja-Korjonen, 2002; Nevalainen, 2004) populations. They have also been used as indicators of climate change (Ilyashuk et al., 2005) and eutrophication (Luoto et al., 2008) due to their ability to tolerate reduced hypolimnetic oxygen conditions (Hongve, 1975; Scholz & Zerbst-Boroffka, 1998; Jager & Walz, 2002; Hynynen et al., 2004). The examination of *Chaoborus* mandibles has often been included in subfossil chironomid and cladoceran studies (e.g., Sarmaja-Korjonen, 2002; Heiri & Lotter, 2003; Heiri, 2004; Ilyashuk et al., 2005; Sarmaja-Korjonen et al., 2006) because they can be identified and counted using both analyses. Although considered as potential palaeoindicators, chaoborids have been underexploited in palaeolimnological studies (Smol, 2002).

In order to increase future paleolimnological exploitation of subfossil *Chaoborus* mandibles, more data are needed on the ecology, distribution patterns, and limiting factors of *Chaoborus*, especially in small shallow lakes (commonly used study sites in palaeolimnology) in the Eurasian region, where the number of studies is currently limited. Thus, the purpose here was to determine the occurrence of *Chaoborus* species in surface sediments of 80 small shallow lakes across a latitudinal transect in Finland and to identify the influence of selected environmental variables (surface area, depth, elevation, mean July air temperature (TJul), pH, conductivity, and hypolimnetic oxygen) on their distribution patterns, using univariate methods (linear regression). In addition, the influence of certain other environmental factors, e.g., fish predation and lake-type preferences, was considered for the assessment of the usability of subfossil *Chaoborus* as a palaeoenvironmental proxy in Northern Europe.

Materials and methods

Study lakes

The 80 study lakes are located in Finland (60°13′–69°53′ N and 22°00′–30°13′ E) (Fig. 1). The sites span a 1080 km transect from boreal coniferous forests to tundra vegetation. All the 80 lakes are small (<1.0 km^2) and shallow (<7.0 m) and represent different environmental conditions. The range and mean of the measured limnological and environmental data are shown in Table 1. Most of the study lakes are situated in woodlands, but some of the lakes are in urban areas (#9, 22, 37, and 62) and thus subject to anthropogenic influence, e.g., increased nutrient conditions.

Sediment sampling and data collection

The surface sediment samples (topmost 0–1 cm) were obtained with a Limnos gravity corer (Kansanen et al., 1991) between February and April in 2005, and the samples were stored in plastic bags in a coldroom at 4°C. The limnological data were acquired during collection of the sediment samples and measured on-site as single measurements before the sediment sampling. The measurements were taken from the hypolimnion (app. 0.5 m from the bottom) for oxygen

Table 1 Comparison of the environmental data for the 80 study lakes in Finland

	N	Minimum	Maximum	Mean
July T (TJul; °C)	80	11.3	17.1	14.9
Annual T (°C)	80	−2.0	5.8	1.8
Elevation (m a.s.l.)	80	11.3	404.0	147.7
Surface area (km^2)	80	0.003	0.988	0.104
Depth (m)	80	0.5	7.0	2.7
Oxygen (mg l^{-1})	31	0.5	11.8	5.5
pH (units)	59	3.8	9.3	6.2
Conductivity (μS cm^{-1})	58	8	312	58.1

Fig. 1 Location of the 80 study sites in Finland, and occurrence of *Chaoborus* in the lakes examined

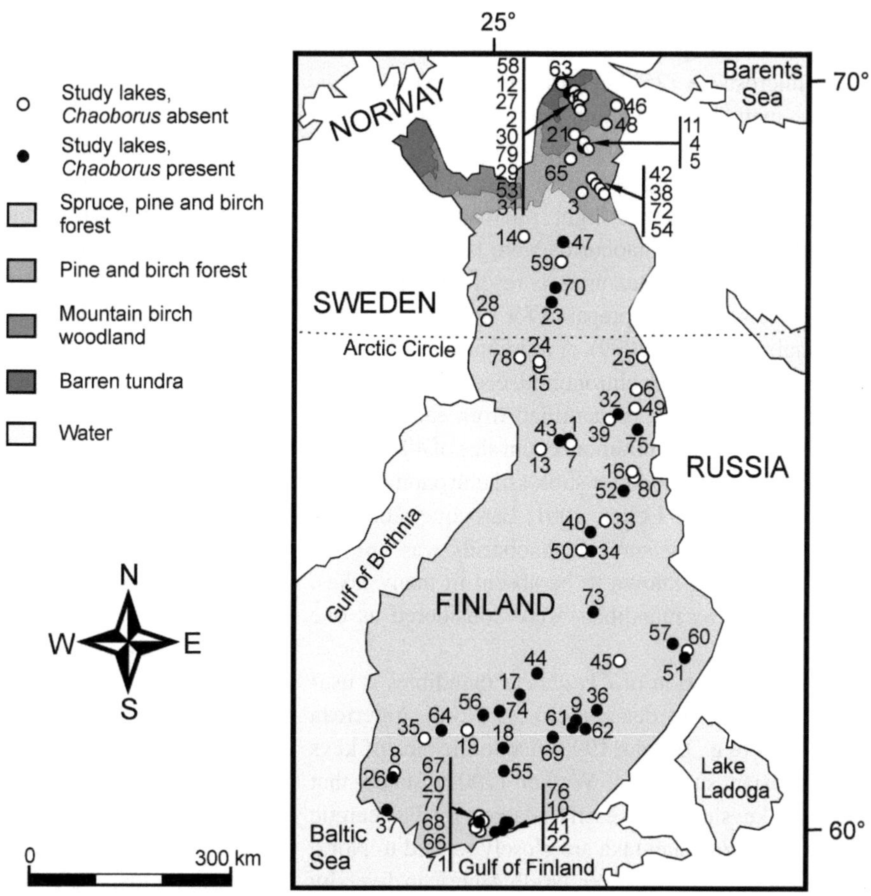

and from the epilimnion for conductivity and pH. The mean TJul were estimated, using a geographic information system (GIS)-based method (data provided by the Finnish Meteorological Institute) that is based on the climate norms for 1971–2000 and that takes into account all Finnish meteorological data. The temperature data were interpolated from the data of the meteorological stations to a 50 × 50 grid, which takes into account elevations and lake percentages, for example. If the elevation of the sample point was markedly different from the grid in question, then the elevation was reconsidered (Seija Kultti, pers. comm.). The lake depth was measured from the sampling point and was in fact the actual sampling depth.

Methods for subfossil *Chaoborus*

The sediment samples used in the examination of subfossil *Chaoborus* remains were prepared, using the standard methods in chironomid analysis described in Brooks et al. (2007). The wet sediment was treated with warm 10% KOH for 10–20 min and sieved through a 100-μm sieve. The residue was examined using a Bogorov counting chamber or a Petri dish under a stereomicroscope (25-times magnification) for *Chaoborus* mandible extraction with fine forceps. The mandibles were mounted permanently in Euparal® on microscope slides together with other midge remains and identified under a light microscope at 400× magnification. The chaoborids were later reidentified with more detailed taxonomic resolution from the original sample slides prepared for subfossil chironomid analysis (Luoto, 2009). A minimum of 100 midge individuals, including chironomid, ceratopogonid, and chaoborid remains was identified from each sample, which exceeded the minimum count size of 40–50 head capsules recommended for subfossil chironomid analysis (e.g., Heiri & Lotter, 2001; Larocque, 2001). A minimum counting sum of chaoborids was not set, because they are known to be absent in many lakes. Two *Chaoborus* mandibles were considered as one individual.

The identification of *Chaoborus* mandibles is usually based on the descriptions of North American specimens (e.g., Uutala, 1990) due to the lack of keys for Eurasian subfossils. Walker (2001) stated that Nearctic keys could be easily adapted for Palaearctic work, since Eurasian taxa are closely related to North American species. However, North American diversity and abundance are much greater than that in Eurasia, and the taxa differ in their morphology and ecology.

Larval *Chaoborus* mandibles have three major teeth (anterior, medial, and posterior), a subordinate tooth (between the medial and posterior teeth), and a row of minute teeth on which identification is based (Fig. 2). The most important characteristics used to identify the mandibles are the size and location of the subordinate tooth (Uutala, 1990), which is short with *C. flavicans* and is placed at the base of the medial tooth or midway between the medial and posterior teeth (Fig. 2). *Chaoborus crystallinus* and *C. obscurripes* are ecologically and morphologically closely related taxa and difficult to separate; therefore, they are considered here as a taxon group *C. crystallinus/obscurripes*. The subordinate tooth of the *C. crystallinus/obscurripes* mandible is larger than that in *C. flavicans*, and is placed above the base of the medial tooth. *Chaoborus flavicans* often has dark tooth tips, but in the *C. crystallinus/obscurripes* type, the pigmentation extends to larger parts of the mandible (see Fig. 2).

Statistical analysis

Univariate methods (linear regression) were used to identify the influence of selected environmental variables (surface area, depth, elevation, mean TJul, pH, conductivity, and hypolimnetic oxygen) on *Chaoborus flavicans* distribution patterns. Variables were considered significant if the test value for the significance level of variation found (P) was ≤ 0.05. The relationship between the environmental variables and *C. flavicans* was tested with Pearson's correlation coefficient (r). Since the pH, oxygen, and conductivity values were not available from all the lakes, only those lakes with measured values were included in these specific analyses. Since *Chaoborus* remains were not found in many of the lakes and there is a possibility that the reason for this may be methods used, the r was also applied by deleting the lakes where *Chaoborus* was absent, for comparison.

Results

From the dataset of 80 lakes in Finland, 9751 midge individuals were enumerated, of which 9252 were chironomids, 408 were chaoborids, and 91 were

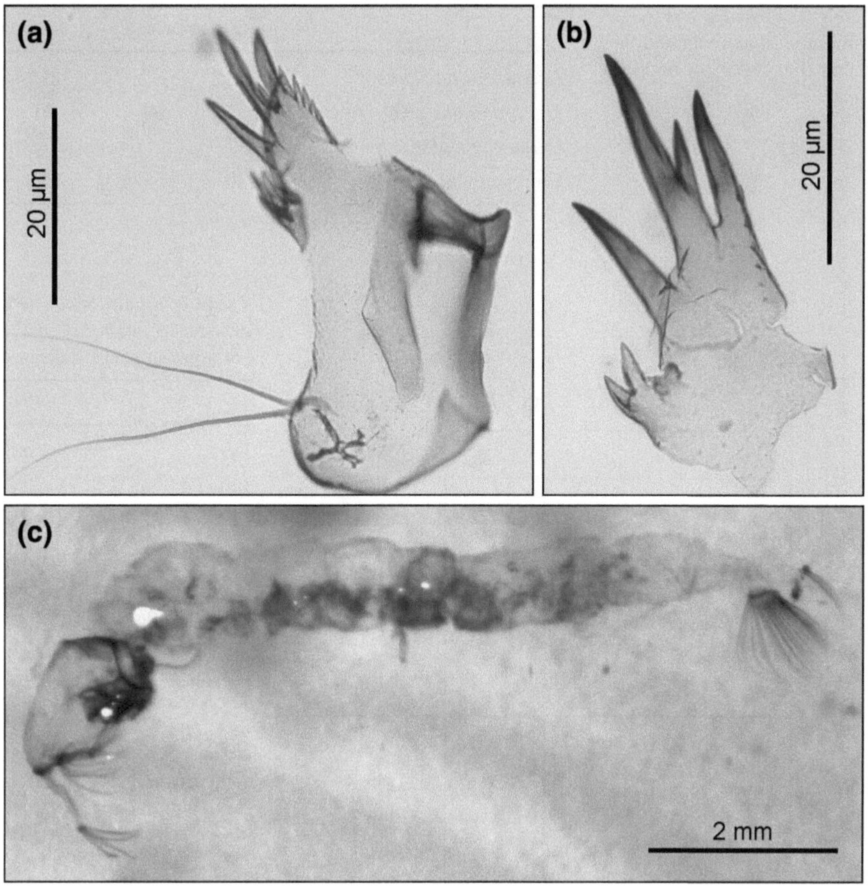

Fig. 2 Subfossil mandibles of (**a**) *Chaoborus flavicans* and (**b**) *C. crystallinus/obscurripes* from surface sediment samples, and (**c**) an intact transparent phantom midge larva sampled with a plankton net. Specimens (**a**) and (**c**) are from Lake Jousjärv (#76) and specimen (**b**) from Lake Huoneseljänlampi (#70)

ceratopogonids. The distribution of *Chaoborus flavicans* was widespread in southern Finland, unlike in northernmost Finland where it was rare (Fig. 1). *Chaoborus flavicans* was present in 33 lakes, with a mean abundance of 3.4% and a maximum abundance in one lake of 51.4% of the total midge fauna (Table 2, Fig. 3). *Chaoborus crystallinus/obscurripes* was found in only two lakes (#23, 70) with abundances of 0.9% of total midges.

The statistical analyses showed that lake depth ($P = 0.008$) and mean TJul ($P = 0.024$) were significantly related ($P \leq 0.05$) to the abundance of *Chaoborus* in the lakes (Table 3). The r between *Chaoborus* and environmental variables showed medium correlation for depth (0.31) and mean TJul (0.40). When lakes where *Chaoborus* was absent were deleted, the r for depth showed large correlation (0.53). Lake surface area, elevation, pH, conductivity, and hypolimnetic dissolved oxygen showed no statistically significant influence on *Chaoborus* distribution (Table 3). However, elevation and surface area showed small negative correlations (r), oxygen medium negative and conductivity small positive correlation with *Chaoborus* distribution and abundance.

Discussion

The distribution of *Chaoborus* mandibles in small shallow lakes in southern Finland correlated positively with lake depth and mean TJul (Table 3), corresponding well with results by Barley et al. (2006) from northwestern North America. The development time of the larval phases of chaoborids is strongly influenced by temperature and food availability (Ouimet, 2001), and chaoborids are restricted only to the warmest lakes in northwestern Finnish Lapland (Korhola et al., 2002). The present results also showed that the number of *C. flavicans*

Table 2 Presence and minimum, maximum and mean abundances of midge taxa

	Occurrences	Minimum	Maximum	Mean
Chaoborus flavicans	33	0	51.4	3.4
C. crystallinus/obscurripes	2	0	0.9	0.0
Ceratopogonidae	52	0	7.1	1.6
Chironomidae	80	47.0	100.0	95.0

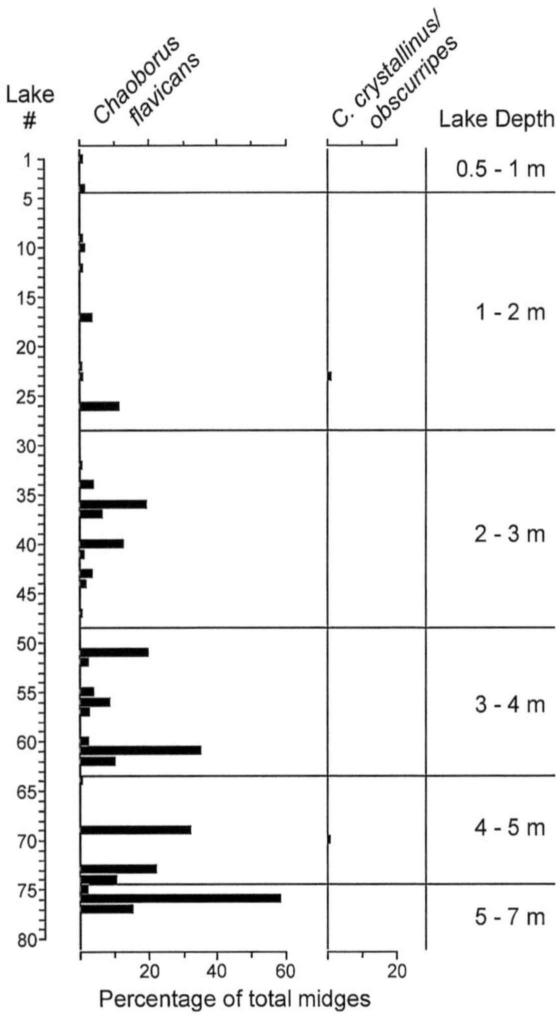

Fig. 3 Abundance of *Chaoborus* taxa in relation to other midges in the 80 lakes examined. The lakes are numbered in ascending order according to increasing lake depth

remains in northernmost Finland was very low, most likely because a colder climate prevails and zooplankton is scarce (Rautio, 1998) due to cool water temperatures and short open-water seasons. *Chaoborus flavicans* was present in only two lakes (#4, 12)

Table 3 Significance (*P*) and Pearson's correlation (*r*) between *Chaoborus flavicans* and different environmental variables in small shallow lakes in Finland

	N	*P*	*r*
Depth	80	0.008	0.31 (0.53)
July T (TJul; °C)	80	0.024	0.40 (0.42)
Elevation	80	0.186	−0.29 (−2.2)
Area	80	0.432	−0.19 (−0.11)
Oxygen	31	0.708	−0.38 (−0.21)
Conductivity	58	0.790	0.21 (−0.10)
pH	59	0.892	0.07 (−0.03)

The values in brackets indicate Pearson's correlations after the deletion of the lakes where *C. flavicans* remains were not encountered. Some of the measurements are missing from some of the lakes; therefore, data analyses consist only of the lakes with measured values (*N*)

located above the boundary of spruce (*Picea* A. Dietr.)–pine (*Pinus* L.)–birch (*Betula* L.) forests and pine–spruce forests (Fig. 1); furthermore, the abundance of the species was very low (Fig. 3). The presence of *C. flavicans* in lake #12, located in mountain birch (*Betula pubescens* ssp. *czeropanovii* (Orlova) Hämet-Ahti) woodland, is probably explained by the fact that the lake is situated in a river valley, where the climatic conditions are favorable (indicated also by the occasional presence of pine). The presence of *C. flavicans* in lake #4 is also intriguing; the lake is located in the pine–spruce forest zone and is known to have a very special macroinvertebrate fauna, dominated by an uncommon cladoceran species and warm associated chironomids (Kaarina Sarmaja-Korjonen, unpubl.; Luoto, 2009). The lakes #4 and #12 are commonly very small (<0.01 km^2) and shallow (<1.3 m). *C. flavicans* may survive in these lakes due to the lack of predation pressure, because the lakes are too small to maintain fish populations that could survive the winter. During winters in northern Finland, ice cover extends to the bottom of very shallow lakes, except for possibly the deepest part, where, however, oxygen depletion and thus winter kills would likely occur. However, the current results indicate that *C. flavicans* is not particularly successful in lakes # 4 and 12, as it occurs in very low abundance (Fig. 3).

Chaoborus is absent in oligotrophic, usually present in eutrophic, and especially common in dystrophic lakes (Thienemann, 1925). Chaoborid abundance is often controlled by the degree of hypolimnetic oxygen (Liljendahl-Nurminen et al., 2002) and poorly oxygenated lakes with high chaoborid abundance are often eutrophic or dystrophic (Sæther, 1997). The current results suggest that *C. flavicans* is more common in lakes with poor hypolimnetic oxygen conditions (Table 3). However, the relation was not statistically significant (Table 3), perhaps because chaoborids are likely to be more sensitive to changes in summer conditions (Brundin, 1949; Hofmann, 1978). In addition, there was a weak relationship between *C. flavicans* and conductivity, although statistically insignificant (Table 3). Elevated conductivity is related to eutrophic conditions in Finland (e.g., Kauppila et al., 2002); thus, the conductivity may provide an indication of trophic status of the lakes. Therefore, according to the present results, it seems that *C. flavicans* is slightly more common in small shallow lakes with enhanced nutrient conditions.

Chaoborus flavicans is also usually absent in oligotrophic clear-water forest lakes in Finland (Paasivirta, 1982b). In the Nuuksio Upland area in southern Finland, abundant *Chaoborus* fauna is present in dystrophic lake #77 (Lake Kalatoin), while in the same region *Chaoborus* is absent (Nevalainen, pers. obs.) in the oligotrophic clear-water lakes (#20, 66–68, 71) (Fig. 1). These observations on living fauna are in agreement with the present subfossil results (Figs. 1, 3); *Chaoborus* was abundant in lake #77 but absent in all the other lakes in the Nuuksio area. In addition, lake #77 is known to be fish-free (Nevalainen, 2008), while the other lakes experienced recent fish introductions (Nevalainen & Sarmaja-Korjonen, 2008). *Chaoborus* is absent in the oligotrophic lakes probably because it was easily consumed by the introduced predatory fishes due to the clear water and resulting high visibility in the lakes. However, the palaeolimnological results of Nevalainen et al. (2008) suggest that the lake #68 has not been occupied by *Chaoborus* at least during the last 2000 years. Unfortunately, palaeolimnological evidence for the presence of fish populations is lacking because fish remains are rarely preserved; therefore, it is difficult to know whether former populations of chaoborids are missing due to fish predation. In addition, the oligotrophic status of the lakes profoundly influences *Chaoborus* via food availability, since the biomass of zooplankton is lower in nutrient-poor lakes. However, in acidified fish-free lakes, *Chaoborus* is known to be a common midge genus and can regulate the community structure of zooplankton (Yan et al., 1991). The current results showed that *Chaoborus flavicans* was most abundant in lake #76 (Lake Jousjärv), constituting 51.6% of the midge taxa (Fig. 3). Lake Jousjärv is a small, dystrophic brown-water lake in southern Finland (Fig. 1), which is currently fish-free (Nevalainen & Sarmaja-Korjonen, 2008) and therefore dominated by *C. flavicans*.

Chaoborus crystallinus/obscurripes occurred in only two lakes (#23, 70), and had an abundance of 0.9% in both lakes. *Chaoborus flavicans* is known to co-occur with fishes, but *C. crystallinus/obscurripes* is found exclusively in small fish-free lakes and even bogs in Europe (Paasivirta et al., 1988; Sæther, 1997, Berendonk & Bonsall, 2002). Lakes #23 and #70 are situated relatively near each other above the Arctic Circle in the spruce–pine–birch forest zone (Fig. 1). Both lakes are small and probably fish-free, fitting well with the known habitat requirements of *C. crystallinus/obscurripes*. Lakes in Finland are commonly occupied by fish due to natural spreading or introductions, and it is therefore not surprising that the distribution and abundance of *C. flavicans* is far greater than that of *C. crystallinus/obscurripes*. Berendonk & Bonsall (2002) showed that the extinction risk for *C. crystallinus* is higher than that for *C. flavicans*; *C. flavicans* also occurs often in large populations, while *C. crystallinus/obscurripes* lives only in small groups (Borkent, 1981).

The statistically significant positive correlation between *Chaoborus flavicans* remains and lake depth (Table 3) is interesting because the dataset consists only of shallow lakes. For comparison, when lakes not having *Chaoborus* were deleted, the r increased from 0.31 (medium correlation, 0.30–0.50) to 0.53 (large correlation, 0.50–1.00). Since counting sums for *Chaoborus* remain relatively low when using the current method (percentage of total midges), it is possible that if the counting sum were to be increased, then it would reveal individuals in the lakes currently identified as *Chaoborus* free. However, this would require more sediment material that is often limited.

Borkent (1981) and Sæther (1997) showed that *C. flavicans* larvae are found in both large and small lakes, which is consistent with our present results, but deeper lakes provide more refuge for *C. flavicans* against fish predation (cf. Liljendahl-Nurminen, 2006). Liljendahl-Nurminen et al. (2002) also found that *C. flavicans* was lacking from the shallow parts and was most abundant in the deepest part of large eutrophic Lake Hiidenvesi in southern Finland. In addition, Horppila & Liljendahl-Nurminen (2005) suggested that the importance of *C. flavicans* in relation to vertebrate predators may change along turbidity and depth gradients. Berendonk & Bonsall (2002) showed that *C. flavicans* cannot exist in shallow ponds because the anoxic layer is too small for the larvae to hide from fish predation. However, Korhola et al. (2000) showed that chaoborids were indicative of low water levels in northwestern Finnish Lapland, but provided no information on the fish status. Therefore, it may very well be that the importance of the water depth varies whether there are fishes in a lake or not. In this study, we found that *C. flavicans* was clearly more abundant in deeper lakes (Fig. 3, Table 3), but the fish status remained mostly unknown due to the remoteness of many of the study sites. Therefore, in future studies, it would be important to analyze the fish status for quantifying the influence of fish predation.

Implications for palaeolimnology

The results of this study are consistent with the previous study on chaoborid distribution (Borkent, 1981; Sæther, 1997). Subfossil chaoborid remains are indicative of chaoborid-living environments; thus, they can be used in palaeolimnological studies in northern Europe, as they were previously in North America (review in Sweetman & Smol, 2006). However, since the present results indicate that *Chaoborus* is absent or rare in many lakes and usually represented by only one species (predominantly *C. flavicans*) in northern Europe, studies based only on *Chaoborus* mandibles cannot give a true indication of the status of aquatic environment. Luckily, analysis of subfossil *Chaoborus* remains can be combined with analyses of other related invertebrates (chironomids and cladocerans), thus contributing as a valuable supplementary proxy, e.g., changes in lake depth, air temperature, productivity, and fish status. However, more research is needed to reliably analyze the chaoborid environmental optima and tolerances and the influence of fish.

It should be recognized that even though *Chaoborus* is related to chironomids, since both have aquatic larval stages, their ecology and behavior differ considerably. The most prominent distinction between these taxa is in their living environment, since chaoborids are planktonic while the few partially planktonic species of aquatic chironomids dwell on sediment or on aquatic macrophytes. In addition, unlike chaoborids, few chironomid species are predators. In combined analysis with subfossil cladocerans, chaoborids may provide valuable information on food-web structure. However, aquatic ecosystems are multidimensional and very complicated structures; thus, the changes in cladoceran communities and chaoborid abundances cannot be explained solely by changes in predation conditions (cf. Frey, 1986). Therefore, all the above-mentioned taxa should be considered separately in subfossil analyses, and it would be preferable not to calculate their percentage abundances from the total sum of chaoborids and chironomids/cladocerans, but to calculate independent concentrations using volumetric or dry weighted samples.

Conclusions

Chaoborus flavicans is a common midge in small shallow lakes of southern Finland, but is rare in northern Finland. Statistical analysis showed that the distribution of *Chaoborus* is significantly and positively correlated with lake depth and mean TJul. In southern Finland, *C. flavicans* was absent in oligotrophic lakes, but abundant in dystrophic lakes. In addition, fish-free lakes were more commonly occupied by chaoborids in southern Finland than lakes with fish populations, although co-occurrence is not exceptional in larger lakes. Although the presence of *C. crystallinus/obscurripes* was infrequent, its distribution may be considered to be restricted to small, fish-free lakes.

The results of this study suggest that the subfossil mandibles of chaoborids have good potential to be more extensively used in multiproxy palaeolimnological studies to interpret past changes in water level and temperature in small shallow lakes in Finland. In addition, there were indications that reconstructions of lake productivity and fish populations may also be possible. However, to fully utilize the potential of chaoborids in palaeolimnological studies in the Eurasian region, substantially more information on their responses to these potential environmental forcing factors is required. This could be accomplished by sampling additional lakes, taking more precise limnological measurements, and by analysing the fish status.

Acknowledgements We are grateful to Seija Kultti, Susanna Kihlman, Anu Hakala, and the Kevo Research Station staff for assistance with the fieldwork. Thanks are also due to Mikko Haaramo for helping with the photographs and Kaarina Sarmaja-Korjonen for fruitful discussions. The two anonymous reviewers are thanked for their constructive reviews of our manuscript. This study was funded by the Finnish Graduate School in Geology, Finnish Entomological Society, and Nordenskiöld Foundation for Tomi P. Luoto and by the Ephippium Project (Academy of Finland, Grant No. 1107062) for Liisa Nevalainen.

References

Barley, E. M., I. R. Walker, J. Kurek, L. C. Cwynar, R. W. Mathewes, K. Gajewski & B. P. Finney, 2006. A northwest North American training set: distribution of freshwater midges in relation to air temperature and lake depth. Journal of Paleolimnology 36: 295–314.

Bendell, B. E. & D. K. McNicol, 1987. Fish predation, lake acidity and the composition of aquatic insect assemblages. Hydrobiologia 150: 193–202.

Berendonk, T. & M. B. Bonsall, 2002. The phantom midge and a comparison of metapopulation structures. Ecology 83: 116–128.

Borkent, A., 1981. The distribution and habitat preferences of the Chaoboridae (Culicomorpha: Diptera) of the Holarctic region. Canadian Journal of Zoology 59: 122–133.

Brooks, S. J., P. G. Langdon & O. Heiri, 2007. The identification and use of Palaearctic Chironomidae larvae in palaeoecology. QRA Technical Guide No. 10. Quaternary Research Association, London: 276 pp.

Brundin, L., 1949. Chironomiden und andere Bodentiere der Südschwedischen Urgebirgsseen. Report of the Institute for Freshwater Research, Drottningholm 30: 1–914.

Bunbury, J. & K. Gajewski, 2008. Does a one point sample adequately characterize the lake environment for paleoenvironmental calibration studies? Journal of Paleolimnology 39: 511–531.

Dawidowicz, P., J. Pijanowska & K. Ciechomski, 1990. Vertical migration of *Chaoborus* larvae is induced by the presence of fish. Limnology and Oceanography 35: 1631–1637.

Eriksson, M. O. G., L. Henrikson, B. I. Nilsson, G. Nyman, H. G. Oscarson, A. E. Stenson & K. Larsson, 1980. Predator–prey relations important for the biotic changes in acidified lakes. Ambio 9: 248–249.

Frey, D. G., 1964. Remains of animals in Quaternary lake and bog sediments and their interpretation. Ergebnisse der Limnologie 2: 1–114.

Frey, D. G., 1986. Cladocera analysis. In Berglund, B. E. (ed.), Handbook of Holocene Palaeoecology and Palaeohydrology. John Wiley and Sons, New York: 667–692.

Frey, D. G., 1988. Littoral and offshore communities of diatoms, cladocerans and dipterous larvae, and their interpretation in paleolimnology. Journal of Paleolimnology 1: 179–191.

Hare, L. & A. Tessier, 1998. The aquatic insect *Chaoborus* as a biomonitor of trace metals in lakes. Limnology and Oceanography 43: 1890–1899.

Heiri, O., 2004. Within-lake variability of subfossil chironomid assemblages in shallow Norwegian lakes. Journal of Paleolimnology 30: 273–289.

Heiri, O. & A. F. Lotter, 2001. Effect of low count sums on quantitative environmental reconstructions: an example using subfossil chironomids. Journal of Paleolimnology 26: 343–350.

Heiri, O. & A. F. Lotter, 2003. 9000 years of chironomid assemblage dynamics in an Alpine lake: long-term trends, sensitivity to disturbance, and resilience of the fauna. Journal of Paleolimnology 30: 273–289.

Hofmann, W., 1978. Analysis of animal microfossils from the Großer Segeberger See. Archiv für Hydrobiologie 82: 316–346.

Hongve, D., 1975. On the ecology and distribution of *Chaoborus* (Chaoboridae, Diptera) from the upper Romerike District, south-east Norway. Norwegian Journal of Entomology 22: 49–57.

Horppila, J. & A. Liljendahl-Nurminen, 2005. Clay–turbid interactions may not cascade—a reminder for lake managers. Restoration Ecology 13: 242–246.

Hynynen, J., A. Palomäki, J. J. Meriläinen, A. Witick & K. Mäntykoski, 2004. Pollution history and recovery of a boreal lake exposed to a heavy bleached pulping effluent load. Journal of Paleolimnology 32: 351–374.

Ilyashuk, E. A., B. P. Ilyashuk, D. Hammarlund & I. Larocque, 2005. Holocene climatic and environmental changes inferred from midge records (Diptera: Chironomidae, Chaoboridae, Ceratopogonidae) at lake Berkut, southern Kola Peninsula, Russia. The Holocene 15: 897–914.

Jager, I. S. & N. Walz, 2002. *Chaoborus flavicans* (Diptera) is an oxy-regulator. Archiv für Hydrobiologie 155: 401–411.

Johnson, M. G., J. R. M. Kelso, O. C. McNeil & W. B. Morton, 1990. Fossil midge associations and the historical status of fish in acidified lakes. Journal of Paleolimnology 3: 113–127.

Kansanen, P. H., T. Jaakkola, S. Kulmala & R. Suutarinen, 1991. Sedimentation and distribution of gammaemitting radionuclides in bottom sediments of southern Lake Päijänne, Finland, after the Chernobyl accident. Hydrobiologia 222: 121–140.

Kauppila, T., T. Moisio & V. P. Salonen, 2002. A diatom-based inference model for autumn epilimnetic total phosphorus concentration and its applications to a presently eutrophic boreal lake. Journal of Paleolimnology 27: 261–273.

Korhola, A., H. Olander & T. Blom, 2000. Cladoceran and chironomid assemblages as quantitative indicators of water depth in subarctic Fennoscandian lakes. Journal of Paleolimnology 24: 43–54.

Korhola, A., K. Vasko, H. T. T. Toivonen & H. Olander, 2002. Holocene temperature changes in northern Fennoscandia from chironomids using bayesian modelling. Quaternary Science Reviews 21: 1841–1860.

Larocque, I., 2001. How many chironomid head capsules are enough? A statistical approach to determine sample size for palaeoclimatological reconstructions. Palaeogeography, Palaeoclimatology, Palaeoecology 172: 133–142.

Liljendahl-Nurminen, A., 2006. Invertebrate predation and trophic cascades in a pelagic food web—the multiple roles of *Chaoborus flavicans* (Meigen) in a clay-turbid lake. PhD thesis, University of Helsinki: 35 pp.

Liljendahl-Nurminen, A., J. Horppila, P. Eloranta, T. Malinen & L. Uusitalo, 2002. The seasonal dynamics and distribution of *Chaoborus flavicans* larvae in adjacent lake basins of different morphometry and degree of eutrohication. Freshwater Biology 47: 1283–1295.

Luoto, T. P., 2009. Subfossil Chironomidae (Insecta: Diptera) along a latitudinal gradient in Finland: development of a new temperature inference model. Journal of Quaternary Science 24: 150–158.

Luoto, T. P., L. Nevalainen & K. Sarmaja-Korjonen, 2008. Multiproxy evidence for the 'Little Ice Age' from Lake Hampträsk, southern Finland. Journal of Paleolimnology 40: 1097–1113.

Nevalainen, L., 2004. Karkkilan Arapisto-järven paleolimnologinen kehitys vesikirppujen ja piilevien kuvastamana [Paleolimnological development of Lake Arapisto in Karkkila, southern Finland, reflected by cladocerans and diatoms]. Unpublished M.Sc. thesis, Department of Geography, University of Helsinki, Finland (in Finnish).

Nevalainen, L., 2008. Parthenogenesis and gamogenesis in seasonal succession of chydorids (Crustacea, Chydoridae) in three low-productive lakes as observed with activity traps. Polish Journal of Ecology 56: 85–97.

Nevalainen, L. & K. Sarmaja-Korjonen, 2008. Intensity of autumnal gamogenesis in chydorid (Cladocera, Chydoridae) communities in southern Finland, with a focus on *Alonella nana* (Baird). Aquatic Ecology 42: 151–163.

Nevalainen, L., T. P. Luoto & K. Sarmaja-Korjonen, 2008. Late Holocene water-level changes in lake Iso Lehmälampi, southern Finland, reflected in subfossil cladocerans and chironomids. Studia Quaternaria 25: 33–42.

Ouimet, C., 2001. Implications of *Chaoborus* pupation and ecdysis in cold water. Freshwater Biology 46: 1169–1177.

Paasivirta, L., 1982a. Saarijärven Pyhä-Häkin kansallispuiston ja sen lähiympäristön metsäjärvien makroskooppinen pohjaeläimistö. Jyväskylän yliopiston biologian laitoksen tiedonantoja [Macroscopic bottom fauna in the national park of Pyhä-Häkli in Saarijärvi and in the small forest lakes in close proximity]. Bulletin of the University of Jyväskylä 29: 31–39 (in Finnish).

Paasivirta, L., 1982b. Saarijärven Pyhä-Häkin kansallispuiston ja sen lähiympäristön metsäjärvien koskikorento-, sudenkorento-, kaislakorento-, verkkosiipis- ja sääskilajisto. Jyväskylän yliopiston biologian laitoksen tiedonantoja [Plecoptera, Odonata, Megaloptera, Neuroptera and Nematocera species in the national park of Pyhä-Häkli in Saarijärvi and in the small forest lakes in close proximity]. Bulletin of the University of Jyväskylä 29: 40–49 (in Finnish).

Paasivirta, L., 1989. Pohjaeläimistötutkimuksen liittäminen järvisyvännealueiden seurantaan Vesi- ja ympäristöhallinnon monistesarja [Studies on the bottom fauna in deepwater follow-ups]. Publications of the Finland's Environmental Administration 164: 1–69 (in Finnish).

Paasivirta, L., T. Lahti & T. Perätie, 1988. Emergence phenology and ecology of aquatic and semi-terrestrial insects on a boreal raised bog in Central Finland. Holarctic Ecology 11: 96–105.

Rautio, M., 1998. Community structure of crustacean zooplankton in subarctic ponds—effects of altitude and physical heterogeneity. Ecography 21: 327–335.

Sæther, O. A., 1997. Diptera Chaoboridae, Phantom midges. In Nilsson, A. (ed.), Aquatic Insects of North Europe 2. Apollo Books, Stenstrup: 149–161.

Sarmaja-Korjonen, K., 2002. Multi-proxy data from Kaksoislammi Lake in Finland: dramatic changes in the late Holocene cladoceran assemblages. Journal of Paleolimnology 28: 287–296.

Sarmaja-Korjonen, K., M. Nyman, S. Kultti & M. Väliranta, 2006. Paleolimnological development of Lake Njargajavri, northern Finnish Lapland, in a changing Holocene climate and environment. Journal of Paleolimnology 35: 65–81.

Scholz, F. & I. Zerbst-Boroffka, 1998. Environmental hypoxia affects osmotic and ionic regulation in freshwater midgelarvae. Journal of Insect Physiology 44: 427–436.

Smol, J. P., 2002. Pollution of Lakes and Rivers: A Paleoenvironmental Perspective. Arnold Publishers, London.

Stenson, J. A. E., 1978. Differential predation by fish on two species of *Chaoborus* (Diptera, Chaoboridae). Oikos 31: 98–101.

Stenson, J. A. E., 1981. The role of predation in the evolution of morphology, behaviour and life history of two species of *Chaoborus*. Oikos 37: 323–327.

Sweetman, J. N. & J. P. Smol, 2006. Reconstructing fish populations using *Chaoborus* (Diptera: Chaoboridae) remains—a review. Quaternary Science Reviews 25: 2013–2023.

Szeroczyńska, K. & K. Sarmaja-Korjonen, 2007. Atlas of Subfossil Cladocera from Central and Northern Europe. Friends of the Lower Vistula Society, Świecie, Poland.

Thienemann, A., 1925. Die binnengewässer nittelewopas eine limnologische einfuhrung. Binnengewässer 1: 1–225.

Uutala, A. J., 1990. *Chaoborus* (Diptera: Chaoboridae) mandibles—paleolimnological indicators of the historical status of fish populations in acid-sensitive lakes. Journal of Paleolimnology 4: 139–151.

Uutala, A. J. & J. P. Smol, 1996. Paleolimnological reconstructions of long-term changes in fisheries status in Sudbury area lakes. Canadian Journal of Fisheries and Aquatic Sciences 53: 174–180.

Walker, I. R., 2001. Midges: Chironomidae and related Diptera. In Smol, J. P., H. J. B. Birks & W. M. Last (eds), Tracking Environmental Change using Lake Sediments: Zoological Indicators. Developments in Paleoenvironmental Research, Vol. 4. Kluwer Academic Publishers, Dordrecht: 43–66.

Wissel, B., W. J. Boeing & C. W. Ramcharan, 2003. Effects of water color on predation regimes and zooplankton assemblages in freshwater lakes. Limnology and Oceanography 48: 1965–1976.

Yan, N. D., R. W. Nero, W. Keller & D. C. Lasenby, 1985. Are *Chaoborus* larvae more abundant in acidified than in nonacidified lakes in Central Canada? Holarctic Ecology 8: 93–99.

Yan, N. D., W. Keller, H. J. MacIsaac & L. J. McEachern, 1991. Regulation of zooplankton community structure of an acidified lake by *Chaoborus*. Ecological Applications 1: 52–65.

Holocene climate on the Modoc Plateau, northern California, USA: the view from Medicine Lake

Scott W. Starratt

Originally published in the journal Hydrobiologia, Volume 631, No. 1, 197–211.
DOI: 10.1007/s10750-009-9811-z

Abstract Medicine Lake is a small (165 ha), relatively shallow (average 7.3 m), intermediate elevation (2,036 m) lake located within the summit caldera of Medicine Lake volcano, Siskiyou County, California, USA. Sediment cores and high-resolution bathymetric and seismic reflection data were collected from the lake during the fall of 1999 and 2000. Sediments were analyzed for diatoms, pollen, density, grain size (sand/mud ratio), total organic carbon (TOC), and microscale fabric analysis. Using both ^{14}C (AMS) dating and tephrochronology, the basal sediments were estimated to have been deposited about 11,400 cal year BP, thus yielding an estimated average sedimentation rate of about 20.66 cm/1,000 year. The lowermost part of the core (11,400–10,300 cal year BP) contains the transition from glacial to interglacial conditions. From about 11,000–5,500 cal year BP, Medicine Lake consisted of two small, steep-sided lakes or one lake with two steep-sided basins connected by a shallow shelf. During this time, both the pollen (*Abies*/*Artemisia* ratio) and the diatom (*Cyclotella*/*Navicula* ratio) evidences indicate that the effective moisture increased, leading to a deeper lake. Over the past 5,500 years, the pollen record shows that effective moisture continued to increase, and the diatom record indicates fluctuations in the lake level. The change in the lake level pattern from one of the increasing depths prior to about 6,000 cal year BP to one of the variable depths may be related to changes in the morphology of the Medicine Lake caldera associated with the movement of magma and the eruption of the Medicine Lake Glass Flow about 5,120 cal year BP. These changes in basin morphology caused Medicine Lake to flood the shallow shelf which surrounds the deeper part of the lake. During this period, the *Cyclotella*/*Navicula* ratio and the percent abundance of *Isoetes* vary, suggesting that the level of the lake fluctuated, resulting in changes in the shelf area available for colonization by benthic diatoms and *Isoetes*. These fluctuations are not typical of the small number of low-elevation Holocene lake records in the region, and probably reflect the hydrologic conditions unique to Medicine Lake.

Keywords Diatoms · Pollen · Laminated sediments · Multiproxy

Guest editors: K. Buczkó, J. Korponai, J. Padisák & S. W. Starratt
Palaeolimnological Proxies as Tools of Environmental Reconstruction in Fresh Water

S. W. Starratt (✉)
U.S. Geological Survey, 345 Middlefield Road, Menlo Park, CA 94025, USA
e-mail: sstarrat@usgs.gov

Introduction

Geologic history

Medicine Lake volcano is a large Quaternary shield volcano located about 50 km northeast of Mount

Shasta in northern California, USA. The volcano began erupting about 500,000 years ago (Donnelly-Nolan & Ramsey, 2001). At least 17 eruptions have occurred during the past 12,900 calendar years. The composition of these lavas ranges from basalt to rhyolite, with those of silicic composition dominating the last 5,000 years. Depending on their location, these eruptions affected the morphology of the Medicine Lake caldera, provided a source for some of the tephra layers found in lake sediments, and may have indirectly affected lake level through influences on groundwater flow. About 5,120 cal year BP (J. Donnelly-Nolan & M. Nathenson, personal communication), the dacite of the Medicine Lake Glass Flow erupted on the floor of the caldera. Between about 3,300 and 800 cal year BP, seven eruptions of varying compositions occurred.

Anderson (1941) was the first to study the region in some detail. In addition to discussing the eruptive history of the volcano, he briefly presents the evidence for glacial activity and the presence of lacustrine sediments that pre-date the present lake. Geomorphological evidence for the Last Glacial Maximum at Medicine Lake volcano extends down to an elevation of at least 1,890 m (Donnelly-Nolan, in press). Ice accumulated in the caldera and flowed out to the southeast through the lowest point on the caldera rim. The suggestion that there was ice as thick as 150 m on the volcano is supported by mapping (Anderson, 1941; Donnelly-Nolan & Nolan, 1986) and fluid inclusion studies (Barger, 2001).

Medicine Lake presently occupies a portion of the summit caldera (Fig. 1) of Medicine Lake volcano. Anderson (1941) noted both geomorphological and sedimentological evidences for higher lake levels during the Quaternary. He suggested that at one time the basin to the northeast had been connected to Medicine Lake, and that this connection was the result of the eruption of the Medicine Lake Glass Flow in the late Quaternary. Recent analysis indicates that the eruption of the dacite of the Medicine Lake Glass Flow occurred around 5,120 cal year BP (J. Donnelly-Nolan and M. Nathenson, personal communication). His evidence for past higher lake levels included

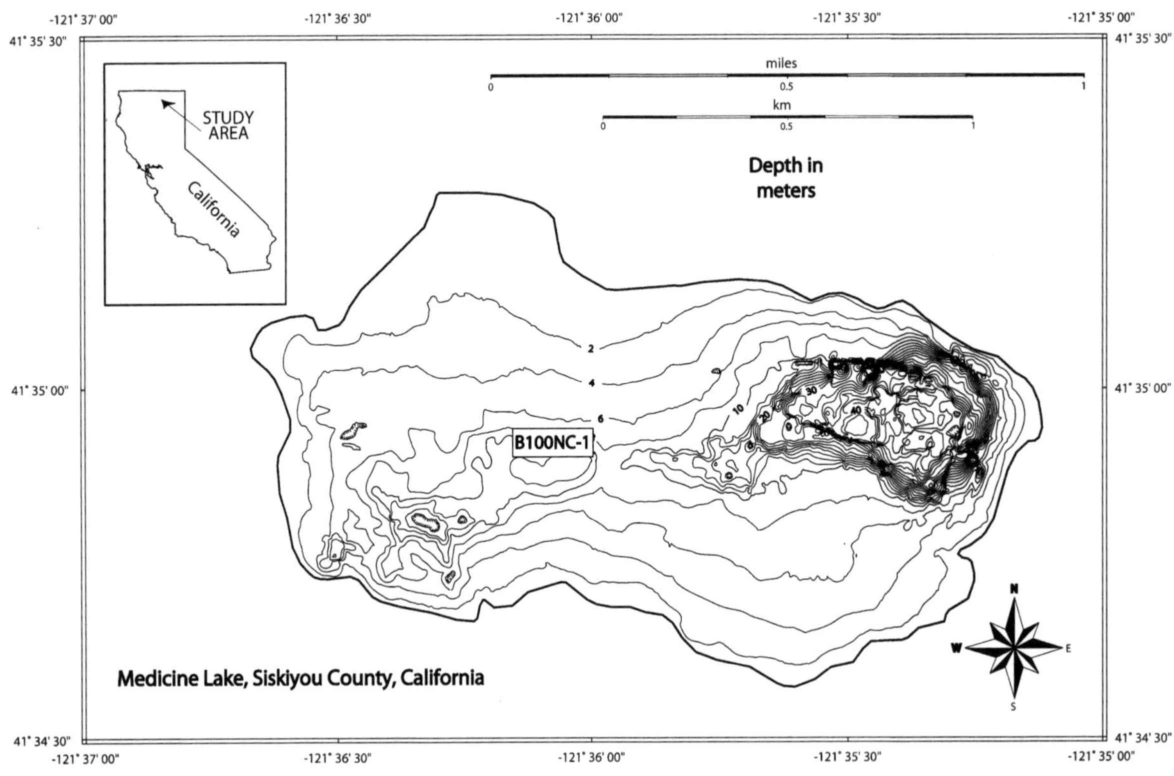

Fig. 1 Map showing the location of Medicine Lake. The lake is located in eastern Siskiyou County, California, about 50 km northeast of Mount Shasta

abandoned shorelines north of the lake and diatomaceous deposits to the south.

Hydrocast profiles

Medicine Lake is a small (165 ha), relatively shallow (average 7.3 m), medium altitude (2,036 m) lake. In September 1992, temperature, specific conductance, dissolved oxygen, and pH data were collected at Medicine Lake using a multiparameter probe (Schneider & McFarland, 1996).

The lake exhibits a strong seasonal thermocline, indicating that the present lake is either a warm monomictic or dimictic lake. The strength, duration, and timing of seasonal overturn may vary depending on the date of lake thaw, which presently ranges from late May to July. Surface temperature in September 1992 was about 15°C and approached 4°C at 25 m. The thermocline occurred between about 12 and 15 m. The dissolved oxygen profiles indicate an orthograde oligotrophic lake in which the oxygen concentrations are controlled by wind mixing with saturation at or near 100%. The highest levels of oxygen (10 mg/l) were recorded between 12 and 17 m. Specific conductance is uniform (20 μS/cm) between profiles with the exception of the profile measured in the shallowest part of the lake (90 μS/cm). The pH values were neutral to slightly acidic reflecting the increased concentration of humic acids in deeper parts of the lake.

Modern climate and vegetation

Medicine Lake lies within a semi-arid winter rainfall climate. More than 75% of the annual precipitation falls during the period from October to March. Large variations in temperature occur both diurnally and seasonally. Warm summer temperatures and low precipitation, coupled with high evapotranspiration, lead to a low effective moisture for the region. During the summer, orographic and convection processes cause sporadic late afternoon thunderstorms. Occasionally, during the late spring and summer, day-long storms descend over Medicine Lake volcano (J. Donnelly-Nolan, personal communication).

The modern forest surrounding Medicine Lake is dominated by Lodgepole pine (*Pinus contorta* Loudon). Other pine species in the area include Ponderosa (*P. ponderosa* Laws), Jeffrey (*P. jeffreyi* Grev. et Balf.), sugar (*P. lambertiana* Douglas), and western white (*P. monticola* Douglas). Red (*Abies magnifica* Andr. Murray) and white (*A. concolor* (Gordon et Glend.) Lindley) fir are found at higher elevations and on shaded north-facing slopes. Incense cedar (*Calocedrus decurrens* (Torrey) Florin) and western juniper (*Juniperus occidentalis* Hook.) are present in small numbers on dry slopes and at lower elevations. *Artemisia tridentata* Nutt. is common in open areas at lower elevations throughout the Modoc Plateau. Oaks are represented by the huckleberry oak (*Quercus vaccinifolia* Kellogg). Several species of grass and bracken fern (*Pteridium* spp.) are common in forest clearings.

Methods

In September 2000, core B100NC-1 (226 cm long; 12.6 m depth) was collected from the shallower western basin of Medicine Lake (Childs et al., 2000) using a Vibracore system (Fig. 2). Density was measured at intervals of 1 cm before the core was split. Paired samples (26) were collected for paleontological analyses. Sampling intervals were determined by variations in the stratigraphy of the core. Additional samples were taken for dating (AMS ^{14}C and tephra analyses), grain size, and total organic carbon (TOC) analyses. Additional information on the core in Starratt et al. (2003).

Three samples containing fibrous root material were dated at the CAMS facility at Lawrence Livermore National Laboratory using AMS ^{14}C. Seven tephra samples from discrete layers were analyzed using the JEOL electron microprobe at the U.S. Geological Survey (Menlo Park). Four of the samples were correlated with identified tephras.

Twenty-six samples for diatom analysis were processed using hydrogen peroxide, hydrochloric acid, and nitric acid, (Battarbee, 1986) to remove carbonate and organic matter, and sodium pyrophosphate to deflocculate the remaining clay. Approximately 50 μl of the resulting suspension was dried on a 22 × 30-mm coverslip and permanently mounted using Naphrax®. At least 300 frustules were enumerated following the method of Schrader & Gersonde (1978) using an Olympus BH-2 microscope at magnifications of 500× or 1,000×. Diatom identification was based on Lowe (1974), Gasse (1986), Krammer &

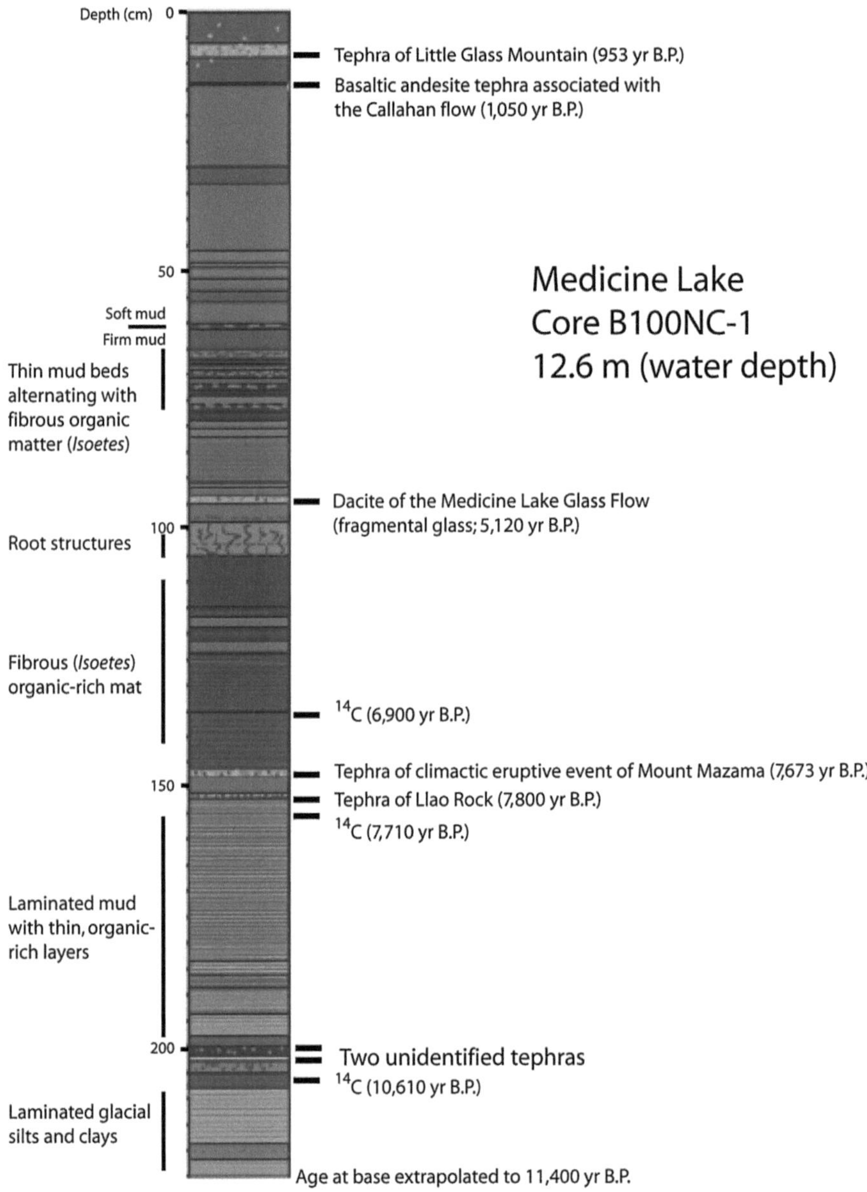

Fig. 2 Generalized stratigraphic column for Core B100NC-1. Dates from geochemical correlation with ash beds of known age and AMS ^{14}C dates. See Childs et al. (2000) for a detailed description of symbols used in stratigraphic column

Lange-Bertalot (1986, 1988, 1991a, b), Lange-Bertalot & Krammer (1987, 1989), Cumming et al. (1995), Round & Bukhtiyarova (1996), Krammer (1997a, b, 2000, 2002), and Lange-Bertalot (2001).

Twenty-six pollen samples were prepared using standard techniques (Faegri & Iversen, 1975; Wood et al., 1996), which include digestion in hydrochloric acid, potassium hydroxide, hydrofluoric acid, nitric acid, and acetolysis. Residues were stained with safranin and mounted in silicone oil, and a minimum of 300 non-aquatic pollen grains were counted at 400×. Due to the high proportion (80–90%) of *Pinus* (pine) pollen in all samples, counts were extended to include at least 100 non-*Pinus* grains.

One hundred and seventy-seven samples were analyzed for sediment grain size (sand/mud). Samples were disaggregated and sieved using a 63-μm screen. A Columetrics Incorporated coulometer was also used to analyze 66 samples for TOC content.

The interval between 184 and 168.5 cm was embedded with low-viscosity resin, polished, and analyzed using back-scattered electron imagery (BSEI; R. B. Pearce and A. E. S. Kemp, written communication).

Results

Age model

Three AMS ^{14}C dates were calibrated using CALIB 4.3 (Stuiver et al., 1998). These dates (10,610 cal year BP, 7,710 cal year BP, and 6,900 cal year BP) from depths of 206.00, 151.50, and 137.00 cm, respectively, constrain the age of the lower part of the core (Figs. 2 and 3). Four of the tephra samples can be correlated with tephras of known age, which were dated by stratigraphic relationships with units dated using AMS ^{14}C. These include two from Mount Mazama, in what is now southern Oregon (Llao Rock, 7,800 cal year BP, climactic eruptive event of Mount Mazama, 7,673 cal year BP at 151.25 and 147.50 cm, respectively (C. Bacon, personal communication)) and two from local volcanic centers (basaltic andesite tephra associated with the Callahan flow; 13.50 cm–1,050 cal year BP, tephra of Little Glass Mountain; 6.75 cm–953 cal year BP). All ages discussed in this article are in cal year BP. The three AMS ^{14}C and four tephra dates were used to construct the age model for Medicine Lake. Based on these ages, the sediment accumulation rate is approximately 20.66 cm/1,000 year (Fig. 3).

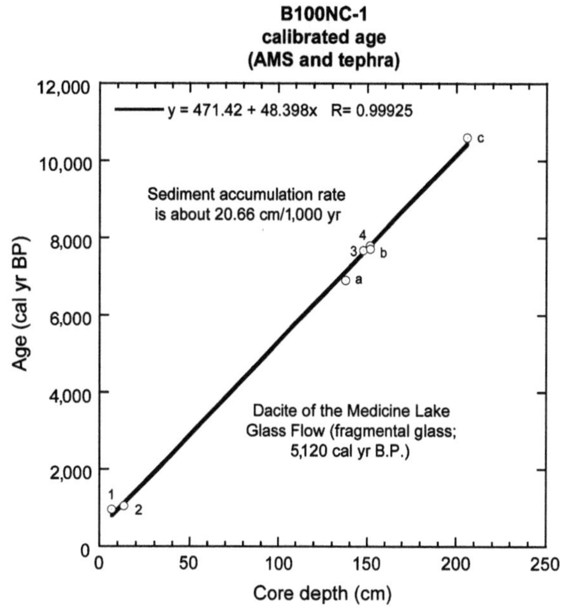

Fig. 3 Age model for Core B100NC-1 based on three AMS ^{14}C dates and correlation to four ash beds of known age. Control points: (1) tephra of Little Glass Mountain (953 year BP), (2) basaltic andesite tephra associated with the Callahan flow (1,050 year BP), (3) tephra of climactic eruptive event of Mount Mazama (7,673 year BP), (4) tephra of Llao Rock (7,800 year BP), (a) ^{14}C 6,900 year BP, (b) ^{14}C 7,710 year BP, and (c) ^{14}C 10,610 year BP

Stratigraphy

With the exception of the tephra layers, the sediments are predominantly silt and clay. From the base of the core to a depth of about 60 cm, the sediments are firm, and become less consolidated toward the top of the core. The sand fraction records the deposition of varying amounts of fine-grained mud clasts, diatoms, seeds, disseminated tephra, along with seven discrete tephra layers (Fig. 2). The organic material in some of the laminated intervals comprises the aquatic macrophyte *Isoetes* (quillwort). The tephra layers contain between 17% and 68% sand-size sediment (volcanic glass, minerals, and lithic fragments). The sediments at the bottom of the core are estimated to be about 11,400 years old. The earliest laminated interval, approximately 11,400–10,600 cal year BP, consists of silt and clay which contains diatoms, seeds, and fine-grained disseminated tephra. Two unidentified tephra layers were deposited between 10,500 and 10,250 cal year BP. The sequence from 10,250 to 8,300 cal year BP consists primarily of laminated mud with thin, organic-rich layers. Between 10,250 and 8,300 cal year BP, the sand fraction is composed mainly of rare diatoms, seeds, and abundant fine-grained mud clasts (fecal pellets). The laminated interval between 8,300 and 7,780 cal year BP is similar in composition, with abundant mud clasts and the remains of *Isoetes*. Two tephra beds at the top of the laminated sequence have been identified as the products of two closely spaced eruptive events at Mount Mazama (Llao Rock and climactic eruption). The laminated sequence deposited from 7,540 to 6,090 cal year BP consists of fibrous organic-rich mud. The sand fraction contains abundant mud clasts and diatoms with thick mats of *Isoetes* present from about 7,200 to 6,090 cal year BP. The sediments deposited between 6,090 and 4,590 cal year BP are laminated and the sand fraction contains abundant diatoms and mud clasts, as well as rare disseminated tephra. A layer of fragmental glass (tephra related to the dacite of the Medicine Lake Glass Flow) was deposited about 5,120 cal year BP. The laminated interval deposited between 4,590 and

3,380 cal year BP contains abundant *Isoetes* remains, mud clasts, and diatoms. The consistency of the sediment changes from firm to soft mud about 3,380 calendar years ago. A change to poorly defined bedding occurs gradually between 3,380 and 1,680 cal year BP with the sand fraction consisting of disseminated tephra, mud clasts, diatoms, and rare tubes (root casts). Sand-size sediments deposited over the last 1,680 calendar years are dominated by mixed black and white pumice clasts, with distinct tephra layers deposited at about 1,050 (basaltic andesite tephra associated with the Callahan flow) and 953 (tephra of Little Glass Mountain) cal year BP.

Between 184 and 168.5 cm, three of the intervals (183.3–180, 173.35–172, and 171.1–169.1 cm) evaluated using BSEI contained relatively well-preserved laminae. These laminated units are interbedded with massive or poorly laminated beds. Three types of laminae were recognized: (1) clay-diatom lamina with silt; (2) clay-diatom lamina with rare grains of silt; and (3) diatom ooze lamina. Clay-diatom lamina with silt are dominated by clay, complete diatom frustules, disarticulated diatom frustules, and diatom fragments. The diatom flora is dominated by benthic taxa, although *Cyclotella* spp. are present in small numbers. Chrysophyte stomatocysts, pollen, and spores are present. Silt (including tephra) comprises up to 10% of the lamina. Most of the boundaries of these lamina are abrupt. The second common lamina type is the clay-diatom lamina with rare grains of silt. These lamina can be further divided into two subtypes, based on the amount of clay present. In both cases, the abundance of silt is less than 10%. Lamina dominated by diatoms, composed of 50–90% diatomaceous debris, are the rarest type. Intact frustules of benthic species are the dominant component, with relatively few *Cyclotella* spp. Silt is rare (less than 10%).

Density

With the exception of two density peaks which correspond to tephra layers, the density gradually decreases from 11,400 to approximately 5,500 cal year BP. (Fig. 4). This decrease in density represents the dilution of glacially derived sediments by fine-grained organic matter. Over the last 5,500 calendar years, the density increased slightly. The lowest density intervals between 8,000 and 6,000, and

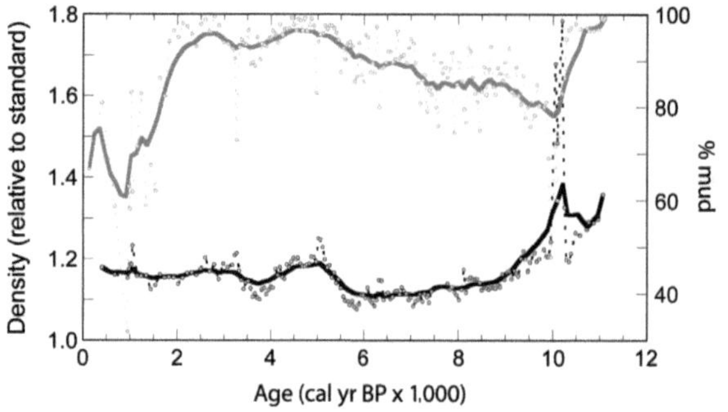

Fig. 4 Record of variation in density (relative to density calibration standard) (*black curve*) and percent of mud in sediment (*gray curve*). Curves are smoothed using the Stineman function which incorporates ±10% of the data range

4,000 and 3,500 cal year BP corresponded to intervals in which the sediment is rich in fibrous organic material (*Isoetes* root mats).

Grain size

The percentage of sediment that comprises mud (<63 μm) ranges from 60 to nearly 100% (Fig. 4). Peaks in the abundance of sand (63 μm to 2 mm) comprise airfall tephra deposits.

Between 11,400 and 10,400 cal year BP, the mud component decreased from 98% to 80%, as silt and clay derived from local glacial activity were replaced by coarser grained detrital sediment. The mud content slowly increased from 80% to 95% between 10,300 and 5,500 cal year BP, as the organic matter diluted the clastic input. The percentage of mud remained relatively stable until about 2,000 cal year BP. The decrease in mud over the last 2,000 years was due to a decrease in organic matter and the presence of two tephra layers that are dominated by sand-size clasts.

Total organic carbon (TOC)

Organic carbon represents the dominant form of carbon in Medicine Lake sediments. Peaks of seven to eight percent TOC occur between 9,000 and 5,500 and at 4,300 and 3,300 cal year BP (Fig. 5). The sediments deposited during these time periods contain a high percentage of *Isoetes* root mat material. Organic carbon is low in the lower part of the core due to the higher amount of glacial flour and tephra in the sediments. The lower amounts of organic matter over the past 2,000 cal year appear to be related to an

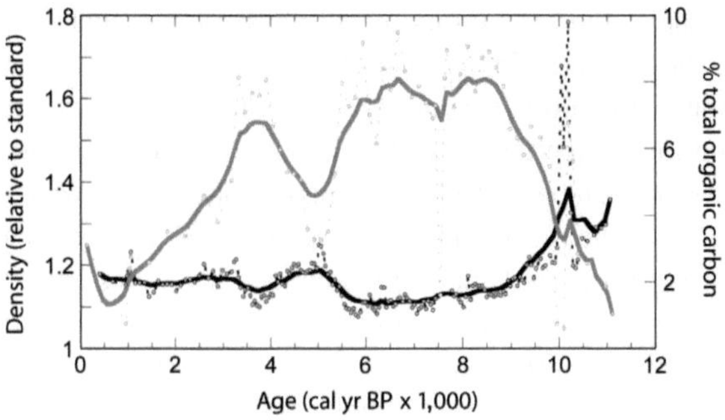

Fig. 5 Record of variation in density (relative to density calibration standard) (*black curve*) and percent total organic carbon (TOC) (*gray curve*). Curves are smoothed using the Stineman function which incorporates ±10% of the data range

increase in the abundance of sand-sized sediment, including pumice from local eruptions.

Diatoms and chrysophyte stomatocysts

Diatoms and chrysophyte stomatocysts were extracted from surface samples collected in 1999 and 2000 and from 26 samples from core B100NC-1. More than 160 species and varieties of diatoms were identified in the modern and fossil floras. Taxa accounting for at least 1% of the assemblage in any one sample are listed in Table 1. These include airborne contaminants from regional late Neogene and early Quaternary diatomite deposits. Thirty-five chrysophyte stomatocyst types were also identified. A detailed analysis (including SEM) would likely yield a significantly more diverse flora.

The benthic component of the flora is represented by more than 25 genera. These include both periphytic and mobile taxa. Several of the samples contain aerophilic taxa which suggests that there is at least some input of sediment from the surrounding watershed. *Navicula s.l.* is the dominant genus, both in terms of diversity and absolute abundance, and is used as a proxy for the relative amount of shallow shelf area in the lake. Planktonic taxa (primarily *Cyclotella pseudostelligera* and *C. stelligera*) are abundant in the sediments in the deeper part of the lake. As the environmental parameters controlling the distribution of many of the taxa found in Medicine Lake are not well understood, it is difficult to make a detailed analysis of the physical and chemical changes that have taken place over the past 11,400 calendar years.

Table 1 Diatom taxa in medicine lake sediments[a]

Achnanthes didyma Hust.
A. helvetica (Hust.) Lange-Bertalot
A. holstii Cleve
A. lanceolatum (Bréb.ex Kütz.) Grun. in Cleve and Grun.
A. levanderi Hust.
A. minutissima Kütz.
A. nitidiformis Lange-Bertalot
A. subatomoides (Hust.) Lange-Bertalot et Archibald
A. exigua Grun.
Amphora inariensis Krammer
Anomoeneis serians (Bréb.) Cleve
A. vitrea (Grun.) Ross
Aulacoseira crenulata (Ehrenb.) Thwaites
Aulacoseira spp.
Caloneis bacillum (Grun.) Cleve
C. silicula (Ehrenb.) Cleve
Cocconeiformis sp.
Cosmioneis spp.
Cyclotella ocellata Pant.
C. pseudostelligera Hust.
C. stelligera Cleve et Grun.
Diploneis finnica (Ehrenb.) Cleve
D. marginstriata Hust.
Encyonema lunata (W. Smith) D.G. Mann
E. minuta (Hilse) D.G. Mann
E. minuta var. *pseudogracilis* (Chol.) Reimer
E. silesiacum (Bleisch) D.G. Mann
Eunotia arcus Ehrenb.
E. intermedia (Krasske) Nörpel et Lange-Bertalot
E. cf. *sudetica* O. Mueller
Fragilaria brevistriata Grun. in Van Heruck
F. construens (Ehrenb.) Grun.
F. vaucheriae (Kütz.) J.B. Peterson
Frustulia rhomboides var. *capitata* (Raben.) De Toni
Gomphonema acuminatum Ehrenb.
G. parvulum (Kütz.) Kütz.
G. truncatum Ehrenb.
Hantzschia amphioxys (Ehrenb.) W. Smith
Hippodonta capitata (Ehrenb.) Lange-Bertalot, Metzeltin, et Witkowski
Luticola mutica (Kütz.) D.G. Mann
Meridion circulare Agardh
Navicula absoluta Hust.
N. amphibola Cleve
N. angusta Grun.
N. aurora Sovereign
N. bacillum Ehrenb.
N. bryophila J.B. Peterson
N. clementis Grun.
N. cocconeiformis Gregory in Greville

Table 1 continued

N. cohnii (Hilse) Lange-Bertalot
N. concentrica Carter
N. crytopcephala Kütz.
N. cryptotenella Lange-Bertalot
N. difficillima Hust.
N. digitulus Hust.
N. disjuncta Hust.
N. explanata Hust.
N. halophila (Grun.) Cleve
N. laevissima Kütz.
N. laterostrata Hust.
N. pseudoscutiformis Hust.
N. pseudoventralis Hust.
N. radiosa Kütz.
N. reinhardtii Grun.
N. rhynchocephala Kütz.
N. seminulum Grun.
Nedium iridis (Ehrenb.) Cleve
Nitzschia amphibia Grun.
N. dissipata (Kütz.) Grun.
N. fonticola Grun.
N. frustulum (Kütz.) Grun.
N. palea (Kütz) W. Smith
N. perminuta (Grun.) Peragallo
Nupela sp.
Opephora martyi Heribaud
Orthoseira roseana (Raben.) O'Meara
Pinnularia borealis Ehrenb.
P. divergens W. Smith
P. interrupta W. Smith
P. maior (Kütz.) Raben.
P. microstauron (Ehrenb.) Cleve
P. viridis (Nitzsch) Ehrenb.
Sellaphora pupula (Kütz.) Mereschkowsky (several forms)
Stauroneis anceps Ehrenb.
S. phoenicenteron (Nitzsch) Ehrenb.
Stenoprobia curvula (W. Smith) Krammer
S. delicatissima (Lewis) Van Heurck
Surirella biserata Bréb.
S. linearis W. Smith
Tabellaria fenestrata (Lyng.) Kütz.

[a] Species and varieties comprising more than one percent of the assemblage in at least one sample

The composition of the modern flora indicates a fresh water, circum-neutral to slightly alkaline, oligotrophic lake. The samples from core B100NC-1 suggest that these conditions have not changed substantially over the past 11,400 calendar years. Despite the limitations of the physical and chemical data, it is possible to construct a lake level history using the diatom flora. *Cyclotella* spp. are used as a proxy for periods when the oligotrophic lake occupied a small, steep-sided basin with little shallow shelf area. *Navicula s.l.*, the most abundant benthic genus, is used as a proxy for periods during which lake level was higher and the deep central basins were surrounded by a broad shallow shelf (less than about 5 m deep). Rühland et al. (2003) and Rühland & Smol (2005) have shown that planktonic diatoms such as *Cyclotella* spp. increase in abundance in waters deeper than 5 m. The ratio of *Cyclotella*/*Navicula* is used as a proxy for lake level (presence or absence of a shallow shelf which provided habitat for the benthic diatoms). The increase in the abundance of the obligate planktonic *Fragilaria* spp. (*Fragilaria capucina*, *F. vaucheriae*, *F. parasitica*, *Pseudostaurosira brevistriata*, and *Staurosira construens* and variants.) over the last 5,000 calendar years suggests an increase in the availability of a shallow shelf environment. The *Cyclotella*/*Navicula* ratio (Fig. 6) gradually increases from near zero at 11,400 cal year BP, to 7.5 about 6,000 cal year BP, suggesting a gradual rise in the lake level. The ratio then drops to an average of about 3.0 for the last 5,500 calendar years, with variations in the ratio that suggest fluctuations in the amount of shallow shelf available for colonization by benthic taxa.

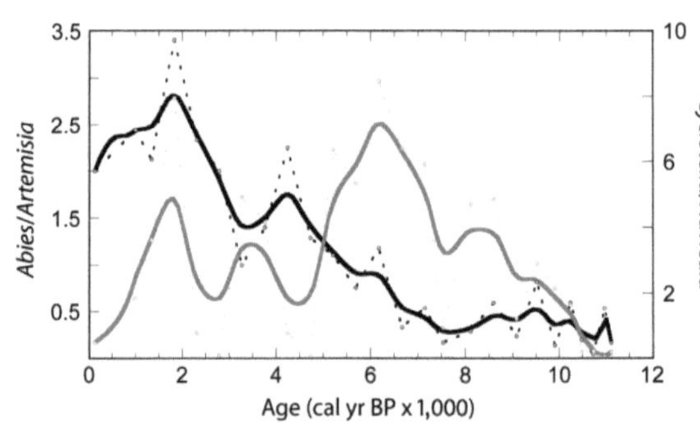

Fig. 6 Record of variations in the ratios of *Abies*/*Artemisia* (*black curve*) and *Cyclotella*/*Navicula* (*gray curve*). The *Abies*/*Artemisia* ratio is a proxy for regional changes in effective moisture. Higher values indicate greater availability of moisture. The *Cyclotella*/*Navicula* ratio is a proxy for relative lake level. Higher values indicate a deeper lake with minimal shallow shelf area. Curves are smoothed using the Stineman function which incorporates ±10% of the data range

Pollen (J. Wanket, personal communication)

Although the Medicine Lake record is dominated by *Pinus* (pine), variations in the secondary terrestrial and aquatic pollen types suggest changes in regional climate and site conditions (Starratt et al., 2003). Prior to 7,000 cal year BP, low levels of *Abies* (fir) pollen can be explained by an inconsistent snowpack, whereas relatively high levels of *Artemisia* (sagebrush) pollen suggest that sagebrush existed at higher elevations than it does today. The presence of *Pteridium* (fern) may be due to frequent disturbance by fire. The abundance of *Isoetes* (quillwort) spores may be correlated with the area available for colonization. *Isoetes bolanderi* var. *pygmaea* (Engelm.) Clute, the species presently found in Medicine Lake, is generally found in shallow lakes above 1,500 m. The spores mature in late summer.

After 7,000 cal year BP, the increasing importance of *Abies* and declining importance of *Artemisia* probably reflect a middle and late Holocene trend toward the establishment of a snowpack similar to present day levels. Shorter-term variations in the aquatic pollen record spanning two to four samples (1,000–2,000 years) may reflect variations in lake level over the past 7,000 calendar years. Periods of high *Isoetes* abundance coincide with an increased proportion of emergent aquatic pollen, which may be due to increased shallow shelf area.

The pollen data can be used to evaluate both the regional climate and variations in lake level. The *Abies/Artemesia* ratio is stable from 11,400 to about 7,000 cal year BP. The ratio increases from an average of about 0.5 to 3.0 from 7,000 to 1,500 cal year BP, and decreases to 2.0 today. The abundance of *Isoetes* follows a similar pattern with relative low numbers until about 7,000 cal year BP, and then increasing to the present. A ratio of *Abies/Artemesia* (Figs. 6, 7, 8) can be used as a proxy for effective moisture, and the abundance of *Isoetes* spores (Fig. 7) can be used as an indicator of lake level and available shallow shelf.

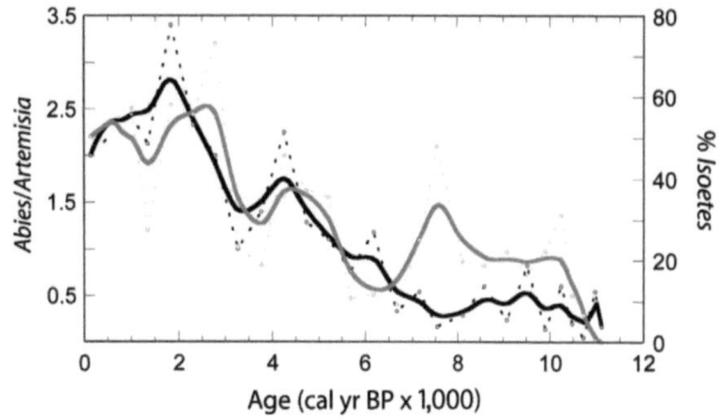

Fig. 7 Record of variations in the ratio of *Abies/Artemisia* (*black curve*) and percent abundance of *Isoetes* (*gray curve*). The *Abies/Artemisia* ratio is a proxy for regional changes in effective moisture. Higher values indicate greater availability of moisture. The *Isoetes* abundance is a proxy for relative lake level. Higher values indicate a lake with greater shallow shelf area. Curves are smoothed using the Stineman function which incorporates ±10% of the data range

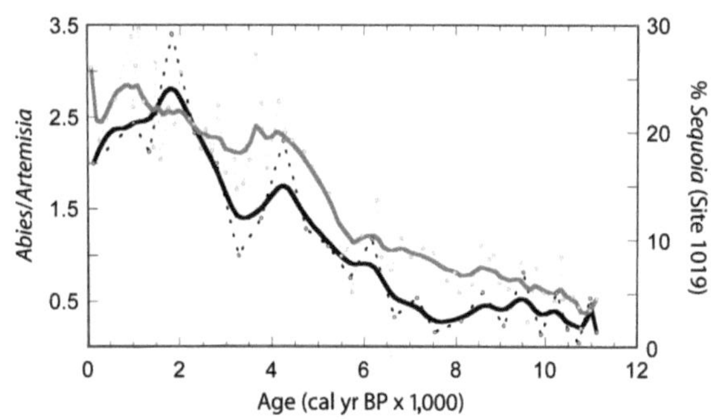

Fig. 8 Record of variations in the ratio of *Abies/Artemisia* (*black curve*) and percent abundance of *Sequoia* (*gray curve*) from Ocean Drilling Program Leg 167 Site 1019. The *Abies/Artemisia* ratio is a proxy for regional changes in effective moisture. The *Sequoia* abundance record is an indicator of the effective moisture along the northern California coast. In both cases, higher values indicate greater availability of moisture in their respective regions. Curves are smoothed using the Stineman function which incorporates ±10% of the data range

Discussion

In a recent study of ODP Site 1019, off the northern California coast, Barron et al. (2003) and Barron & Bukry (2007) used a number of paleontological and geochemical proxies to unravel the complex Holocene climate history of the region. Over the past 11,400 calendar years, the abundance of *Sequoia* pollen increased from about five to more than 25% of the palynoflora (Fig. 8). An increase in *Sequoia* pollen indicates a shift toward the climate of the present, with wet winters and cool, foggy summers (Huyer, 1983; Lynn & Simpson, 1987; Herbert et al., 2001). The *Abies/Artemesia* ratio reflects a similar trend in effective moisture. Throughout the Holocene, but particularly over the last 7,000 calendar years, the

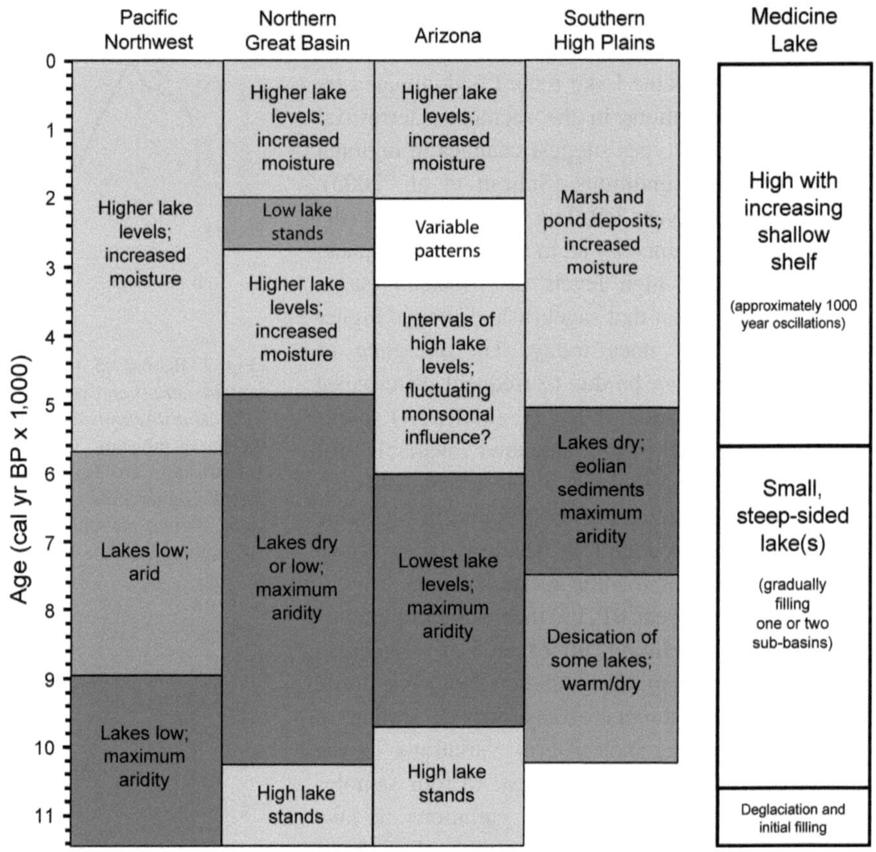

Fig. 9 Paleoclimatic interpretations of selected paleolimnological records for the United States. *Darker gray areas* indicate periods of aridity in the Pacific Northwest, northern Great Basin, Arizona, and the southern high plains. The record at Medicine Lake indicates (1) an initial period of filling following the melting of local glacial ice, (2) a period during which two small, steep-sided lakes or one lake with two steep-sided basins connected by a small shallow shelf, and (3) a period during which the lake level increased, possibility in response to a dam created by the Medicine Lake glass flow, during which the amount of shallow shelf increased, providing increased area for colonization by benthic diatoms

amount of precipitation in the form of both snow and rain has increased. This trend is similar to the trend found in other lakes in the western U.S. (Davis et al., 1976; Lindström, 1990; Rose & Lindström, 1996; Fritz et al., 2001; Benson et al., 2002) which show that the middle Holocene (to about 5,500–4,500 cal year BP) was dryer than the later part of the Holocene (Fig. 9). Changes in middle Holocene seasonality along the California coast (Bograd et al., 2002; Diffenbaugh & Sloan, 2004; Diffenbaugh et al., 2006; Diffenbaugh & Ashfaq, 2007) may have been responsible for changes in the timing and magnitude of precipitation in the basin. Potito et al. (2006) have demonstrated a causal connection between sea surface temperatures in the northeastern Pacific Ocean and temperatures at sites in the Sierra Nevada, several hundred miles south of Medicine Lake. A similar relationship between sea surface temperature and the temperature on the Modoc Plateau may have led to increased evaporation during the summer. Such changes affected lake level, sediment influx, and regional vegetation (Thompson et al., 1993; Bartlein et al., 1998; Thompson & Anderson, 2000; Harrison et al., 2003).

Using the age model discussed above, the section analyzed using BSEI was deposited between 9,400 and 8,600 cal year BP. The presence of these early Holocene laminated sediments suggests that there was less thermal mixing at that time. In their interpretation of these sediments, Pearce and Kemp (personal communication) suggested that modern dimictic lakes in northern Sweden may serve as an analog for Medicine Lake during the early Holocene (Petterson, 1996). In northern Sweden, there is a rapid change between stratified summer and winter conditions, with the thermally destratified spring and fall lasting only a few weeks. The laminae are atypical of ice-covered lakes, which are characterized by fine-grained organic-rich layers indicative of spring–summer production. The lack of these layers is due to oligotrophic conditions in Medicine Lake.

The data from Medicine Lake suggest that during the early Holocene, the basin was occupied by either two small lakes, with a surface about 10 m below the

Fig. 10 Comparison of modern lake boundary and a possible shoreline (*bold line*) for the lake at about 5,500 cal year BP

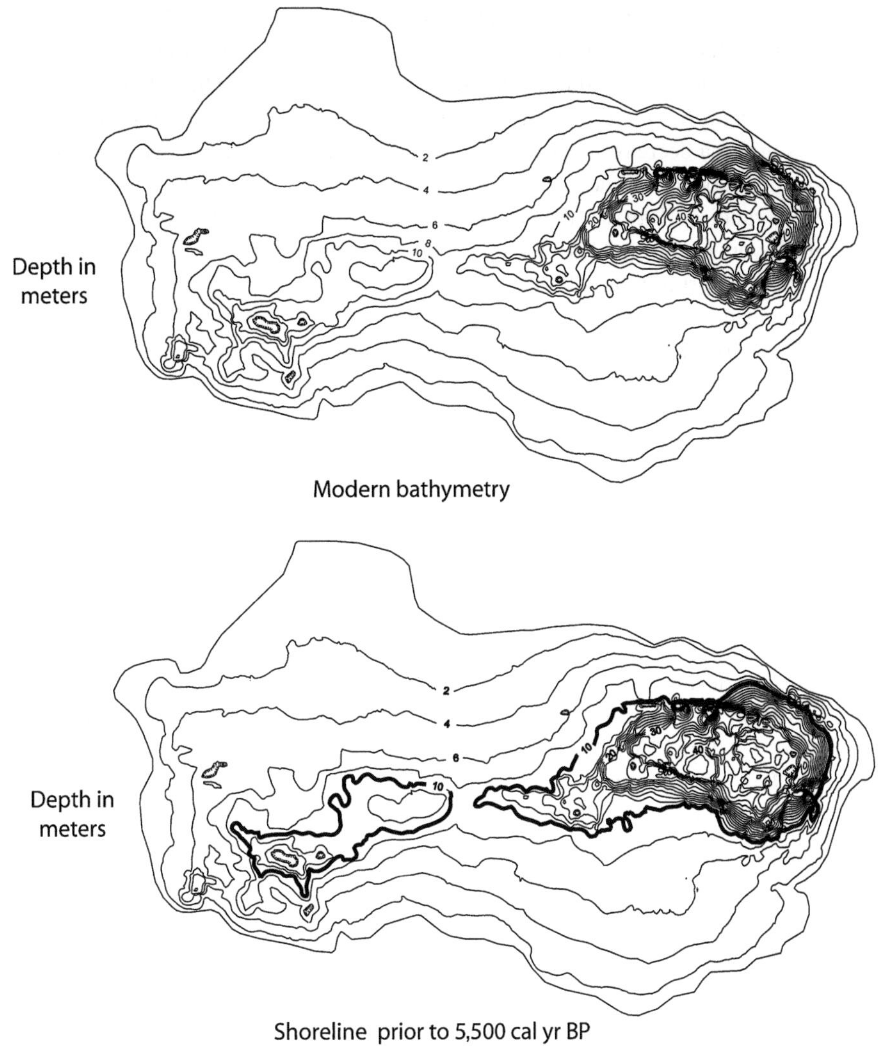

modern lake level, or that the lake was a single narrow one with two deeper basins connected by a small shallow shelf (Fig. 10). For the remainder of the Holocene, the surface area of the lake varied by several meters with climate-driven changes in effective moisture. The most recent evidence for these variations in the lake level came with a reduction in the surface area of the lake during the droughts of the twentieth century.

The ratio of *Cyclotella/Navicula*, variations in the abundance of the benthic diatom flora, and the abundance of *Isoetes* also suggest a gradual filling of one or two small, isolated basins until about 5,500 cal year BP. During this interval, the lake may behave like a small arctic lake (Smol and Cuming, 2000). In the early part of the interval, the combination of an extended season over which the surface of the lake is frozen and the steep sides of the basin results in limited production of benthic and planktonic diatoms. As the length of the ice-free season increases, the production of planktonic species increases, but the abundance of benthic taxa remains relatively low because there is little colonizable habitat. The late middle and late Holocene lake levels exhibited changes in depth of several meters, which affected the area of shallow shelf available for colonization by *Isoetes* and benthic diatoms. As the available colonizable lake bottom increases with a longer ice-free season, the *Cyclotella/Navicula* ratio decreases. *Isoetes*, which has been shown to deplete CO_2 in the water column (Sandquist & Keeley, 1990), might have had an impact on diatom abundance and diversity, particularly in shallow water. Increased evaporation in the middle Holocene may have enhanced this effect.

Between 2,000 and 1,500 cal year BP, an increase in the *Cyclotella/Navicula* ratio and the abundance of *Isoetes* suggest a decrease in available moisture. This interval corresponds to decreased fresh water flow into San Francisco Bay via the Sacramento and San Joaquin Rivers (Starratt, 2001, 2004; Malamud-Roam et al., 2006; Negrini et al., 2006), a brief decrease in the abundance of *Sequoia* at ODP Site 1019 (Barron et al., 2003), low lake levels at Mono Lake on the east side of the Sierra Nevada (Stine, 1990), and increased varve thickness in Saanich Inlet, British Columbia, which may correspond to the Roman Warm Period (Nederbragt & Thurow, 2001). A more detailed analysis of short-term regional trends in El Niño periodicity (Cole et al., 2002; Moy et al., 2002) or the Pacific Decadal Oscillation (Mantua et al., 1997) is not possible at this time due to the lack of high-resolution data.

These conclusions are supported by the results of analyses from the lakes in the Klamath basin (elevation about 1,230 m) 25 km to the north. Paleontological (pollen and diatom) and archeological data suggest that the lake levels were relatively higher in the early part of the Holocene (Bradbury et al., 2004; Hakala & Adam, 2004; Rosenbaum and Reynolds, 2004a, b). For the last 7,000 calendar years, the lake levels in the Klamath basin appear to have fluctuated in much the same way as the lake levels at Medicine Lake (Bradbury, 1991). The timing of regional (northern California and southern Oregon) low elevation lake level fluctuation is less clear, as precipitation in these basins is subject to local topographic conditions (Viau et al., 2002). The differences between the lakes on the Klamath basin and Medicine Lake may be due to the differences in moisture availability in low- and intermediate elevation lakes, changes in the morphology of the Medicine Lake volcano caldera due to magmatic movement, or some combination of these two explanations.

Conclusion

The geological and paleontological data for Medicine Lake provide evidence of both the local geological processes and the regional climatic variations that affected the Medicine Lake region since the Last Glacial Maximum. Lake levels in Medicine Lake are a function of both intrinsic and extrinsic factors.

The retreat of the ice within the Medicine Lake volcano summit caldera occurred around 11,400 years ago followed by filling of two sub-basins. Sediments deposited in Medicine Lake between 11,400 and 10,000 years ago were dominated by glacial mud. The absence of *Cyclotella* indicates that the early lake was probably less than 5 m deep. The abundance of *Artemisia* exceeds that of *Abies*, suggesting that the climate was relatively dry.

Over the next 4,000 years, the level of the lake rose, as relatively organic-rich fine-grained sediments filled the basin. The increase in the abundance of *Cyclotella* also suggests that the lake gradually deepened. The abundance of Abies in the basin also increases, indicating the presence of a deeper snowpack which existed into the late spring and summer. The increased snowpack was the primary source of the water that filled the lake during this period.

About 6,000 years ago, the lake began to flood the shallow shelf area surrounding the two sub-basins. Changes in the abundance of *Cyclotella* and benthic taxa, dominated by *Navicula* vary, indicating that the area of the flooded shelf fluctuates during this interval. The abundance of Isoetes and Abies both have a paired response to changes in the basin, both suggesting an increase in effective moisture. This increase corresponds to an increase in Sequoia pollen at ODP Site 1019, which records the establishment of modern oceanographic conditions along the northern California coast and the connection between coastal and inland climates.

The Medicine Lake record is unusual because it responds not only to local and regional climate signals, but also to changes in conditions on Medicine Lake volcano. The eruption of the dacite of the Medicine Lake Glass Flow 5,120 years ago, as well as more recent unnamed local flows may have had an impact on basin morphology and the movement of groundwater. Despite the fact that the last eruption was more than 800 years ago, recent GPS and tiltmeter data record changes in elevation, which if continued on a centennial scale could have a measurable effect on the volcano.

Acknowledgments I thank John Barron, Elmira Wan, and Dave Wahl of the U.S. Geological Survey for their detailed reviews which not only cleaned up the grammar in the text but also greatly clarified the figures. Julie Donnelly-Nolan generously provided a great deal of information on the history of Medicine Lake volcano. Age control data were provided by Charlie Bacon, Julie Donnelly-Nolan, and Manuel Nathenson, and ^{14}C AMS dates were provided by Jack McGeehin, all of the USGS. Dates on the tephra were originally determined by the USGS Tephrochronology Project. This research was supported by the Earth Surface Dynamics and Volcano Hazards Programs of the USGS.

References

Anderson, C. A., 1941. Volcanoes of the medicine lake highland, California. University of California Publications Bulletin of the Department of Geological Sciences 25: 347–422.

Barger, K. E., 2001. Fluid-inclusion studies of hydorthermal minerals from geothermal drill holes at Medicine Lake Volcano, northern California. California Geology 54: 12–21.

Barron, J. A. & D. Bukry, 2007. Development of the California Current during the past 12, 000 years based on diatoms and silicoflagellates. Palaeogeography, Palaeoclimatology, Palaeoecology 248: 333–338.

Barron, J. A., L. Heusser, T. Herbert & M. Lyle, 2003. High-resolution climatic evolution of coastal Northern California during the past 16,000 years. Paleoceanography 18. doi:10.1029/2002PA000768.

Bartlein, P. J., K. H. Anderson, P. M. Anderson, M. E. Edwards, C. J. Mock, R. S. Thompson, R. S. Webb & C. Whitlock, 1998. Paleoclimate simulations for North America over the past 21,000 years: features of the simulated climate and comparisons with paleoenvironmental data. Quaternary Science Reviews 17: 549–585.

Battarbee, R. A., 1986. Diatom analysis. In Berglund, B. E. (ed.), Handbook of Holocene Palaeoecology and Palaeohydrology. Wiley, New York: 527–570.

Benson, L., M. Kashgarian, R. Rye, S. Lund, F. Paillet, J. Smoot, C. Kester, S. Mensing, D. Meko & S. Lindström, 2002. Holocene multidecadal and multicentennial droughts affecting northern California and Nevada. Quaternary Science Reviews 21: 659–682.

Bograd, S. J., F. B. Schwing, R. Mendelssohn & P. Green-Jessen, 2002. On the changing seasonality over the North Pacific. Geophysical Research Letters 29. doi: 10.1029/2001GL013790.

Bradbury, J. P., 1991. The late Cenozoic diatom stratigraphy and paleolimnology of Tule Lake, Siskiyou Co. California. Journal of Paleolimnology 6: 205–255.

Bradbury, J. P., S. M. Coleman & R. L. Reynolds, 2004. The history of recent limnological changes and human impact on Upper Klamath Lake, Oregon. Journal of Paleolimnology 31: 151–165.

Childs, J. R., J. B. Lowenstern, R. L. Phillips, P. Hart, J. J. Rytuba, J. A. Barron, S. W. Starratt & S. Spaulding, 2000. Bathymetric, geophysical and geologic sample data from Medicine Lake, Siskiyou County, northern California. U.S. Geological Survey Open-File Report 00-043 (http://geopubs.wr.usgs.gov/open-file/of00-043/).

Cole, J. E., J. T. Overpeck & E. R. Cook, 2002. Multiyear La Niña events and persistent drought in the contiguous United States. Geophysical Research Letters 29(1647). doi: 10.1029/2001GL013561.

Cumming, B. F., S. E. Wilson, R. J. Hall & J. P. Smol, 1995. Diatoms from British Columbia (Canada) lakes and their relationship to salinity, nutrients and other limnological variables. Bibliotheca Diatomologica 31: 207 pp.

Davis, J. O., R. Elston & G. Townsend, 1976. Coastal geomorphology of the south shore of Lake Tahoe: suggestion of an altithermal lowstand. Nevada Archeological Survey Research Paper Number 6: 40–65.

Diffenbaugh, N. S. & M. Ashfaq, 2007. Response of California Current forcing to mid-Holocene insolation and sea surface temperatures. Paleoceanography 22(PA3101). doi: 10.1029/2006PA001382.

Diffenbaugh, N. S. & L. C. Sloan, 2004. Mid-Holocene orbital forcing of regional-scale climate: a case study of western North America using a high-resolution RCM. Journal of Climate 17: 2927–2937.

Diffenbaugh, N. S., M. Ashfaq, B. Shuman, J. W. Williams & P. J. Bartlein, 2006. Summer aridity in the United States: response to mid-Holocene changes in insolation and sea surface temperature. Geophysical Research Letters 33(L22712). doi: 10.1029/2006GL028012.

Donnelly-Nolan, J. M., in press. The geologic map of Medicine Lake volcano, northern California. U.S. Geological Survey Miscellaneous Investigations Map.

Donnelly-Nolan, J. M. & K. M. Nolan, 1986. Catastrophic flooding and eruption of ash-flow tuff at Medicine Lake volcano, California. Geology 14: 875–878.

Donnelly-Nolan, J. M. & D. W. Ramsey, 2001. Geologic mapping of Medicine Lake volcano, CA, USA. Eos, Transactions, American Geophysical Union. 82: F1318–F1319.

Faegri, K. & J. Iversen, 1975. Textbook of Pollen Analysis. Munksgaard, Copenhagen: 101–117.

Fritz, S. C., S. E. Metcalfe & W. Dean, 2001. Holocene climate patterns in the Americas inferred from paleolimnological records. In Markgraf, V. (ed.), Interhemispheric Climate Linkages. San Diego, Academic Press: 241–263.

Gasse, F., 1986. East African diatoms: Taxonomy, ecological distribution. Bibliotheca Diatomologica 11: 201 pp.

Hakala, K. J. & D. P. Adam, 2004. Late Pleistocene vegetation and climate in the southern Cascade Range and the Modoc Plateau region. Journal of Paleolimnology 31: 189–215.

Harrison, S. P., J. E. Kutzbach, Z. Liu, P. J. Bartlein, B. Otto-Bliesner, D. Muhs, I. C. Prentice & R. S. Thompson, 2003. Mid-Holocene climates of the Americas: a dynamical response to changes seasonality. Climate Dynamics 20: 663–688.

Herbert, T. D., J. D. Schuffert, D. Andreasen, L. Heusser, M. Lyle, A. Mix, A. C. Ravello, L. D. Stott & J. C. Herguera, 2001. Collapse of the California Current during glacial maxima linked to climate change on land. Science 293: 71–76.

Huyer, A., 1983. Coastal upwelling in the California Current system. Progress in Oceanography 12: 279–290.

Krammer, K., 1997a. Die cymbelloiden Diatomeen. Eine Monographie der weltweit bekannten Taxa. Teil 1.

Allgemeines und Encyonema part. Bibliotheca Diatomologica Band 36. J. Cramer, Stuttgart: 382 pp.
Krammer, K., 1997b. Die cymbelloiden Diatomeen. Eine Monographie der weltweit bekannten Taxa. Teil 2. Encyonema part., Encyonopsis und Cymbellopsis. Bibliotheca Diatomologica Band 37. J. Cramer, Stuttgart: 469 pp.
Krammer, K., 2000. Diatom of Europe Diatoms of the European Inland Waters and Comparable Habitats. In Lange-Bertalot, H. (ed.), The Genus Pinnularia, Vol. I. Koeltz Scientific Books, Germany: 703 pp.
Krammer, K., 2002. Diatoms of Europe. Diatoms of the European Inland Waters and Comparable Habitats, Vol. 3. Cymbella. A. R. G. Gantner Verlag K. G, Ruggell: 584 pp.
Krammer, K. & H. Lange-Bertalot, 1986. Bacillariophyceae. 1. Teil: Naviculaceae. In Ettl, H., J. Gerloff, H. Heynig & D. Mollenhauer (eds), Süsswasser flora von Mitteleuropa, Band 2/1. Gustav Fischer Verlag, Stuttgart, New York: 876 pp.
Krammer, K. & H. Lange-Bertalot, 1988. Bacillariophyceae. 2. Teil: Bacillariaceae, Epithemiaceae, Surirellaceae. In Ettl, H., J. Gerloff, H. Heynig & D. Mollenhauer (eds), Süsswasserflora von Mitteleuropa, Band 2/2. VEB Gustav Fischer Verlag, Jena: 596 pp.
Krammer, K. & H. Lange-Bertalot, 1991a. Bacillariophyceae. 3. Teil: Centrales, Fragilariaceae, Eunotiaceae. In Ettl, H., J. Gerloff, H. Heynig & D. Mollenhauer (eds), Süsswasserflora von Mitteleuropa, Band 2/3. Gustav Fischer Verlag, Stuttgart, Jena: 576 pp.
Krammer, K. & H. Lange-Bertalot, 1991b. Bacillariophyceae 4 Teil: Achnanthaceae, Kritische Ergänzungen zu Navicula (Lineolatae) und Gomphonema, Gesamtliteraturverzeichnis Teil 1–4. In Ettl, H., G. Gärtner, J. Gerloff, H. Heynig & D. Mollenhauer (eds), Süsswasserflora von Mitteleuropa, Band 2/4. Gustav Fischer Verlag, Stuttgart, Jena: 437 pp.
Lange-Bertalot, H., 2001. Diatoms of Europe. Diatoms of the European Inland Waters and Comparable Habitats. Vol. 2. Navicula sensu stricto. 10 Genera Separated from Navicula sensu lato. Frustulia. A. R. G. Gantner Verlag K.G, Ruggell: 526 pp.
Lange-Bertalot, H. & K. Krammer, 1987. Bacillariaceae, Epithemiaceae, Surirellaceae. Bibliotheca Diatomologica 15. J. Cramer, Stuttgart: 289 pp.
Lange-Bertalot, H. & K. Krammer, 1989. Achnanthes eine Monographie der Gattungen. Bibliotheca Diatomologica 18. J. Cramer, Stuttgart: 393 pp.
Lindström, S., 1990. Submerged tree stumps as indicators of middle Holocene aridity in the Lake Tahoe basin. Journal of California and Great Basin Anthropology 12: 146–157.
Lowe, R. L., 1974. Environmental requirements and pollution tolerance of freshwater diatoms. U.S. Environmental Protection Agency Report EPA-670/4-74-005, National Environmental Research Center, Cincinnati, OH: 334 pp.
Lynn, R. J. & J. J. Simpson, 1987. The California current system: the seasonal variability of its physical characteristics. Journal of Geophysical Research 92: 12947–12966.
Malamud-Roam, F. P., B. L. Ingram, M. Hughes & J. L. Florsheim, 2006. Holocene paleoclimate records from a large California estuarine system and its watershed region: linking watershed climate and bay conditions. Quaternary Science Reviews 25: 13–14, 1570–1598.
Mantua, N. J., S. R. Hare, Y. Zhang, J. M. Wallace & R. C. Francis, 1997. A Pacific interdecadal climate oscillation with impacts on salmon production. Bulletin of the American Meteorological Society 78: 1069–1079.
Moy, C. M., G. O. Seltzer, D. T. Rodbell & D. M. Anderson, 2002. Variability of El Niño/Southern Oscillation activity at millennial timescales during the Holocene Epoch. Nature 420: 162–165.
Nederbragt, A. J. & J. W. Thurow, 2001. A 6000 yr varve record of Holocene climate in Saanich Inlet, British Columbia, from digital sediment colour analysis of ODP Leg 169S cores. Marine Geology 174: 95–110.
Negrini, R. M., P. E. Wigand, S. Draucker, K. Gobalet, J. K. Gardner, M. Q. Sutton & R. M. Yohe II, 2006. The Rambla highstand shoreline and the Holocene lake-level history of Tulare Lake, California, USA. Quaternary Science Reviews 25(13–14): 1599–1618.
Petterson, G., 1996. Varved sediments in Sweden: a brief view. In Kemp, A. E. S. (ed.), Paleoclimatology and Paleoceanography from Laminated Sediments. Geological Society, London Special Publications. 116: 37–48.
Potito, A. P., D. F. Proinchu, G. M. MacDonald & K. A. Moser, 2006. A late quaternary chironomid-inferred temperature record from the Sierra Nevada, California, with connections to northeast Pacific sea surface temperatures. Quaternary Research 66: 356–363.
Rose, M. & S. Lindström, 1996. Middle Holocene decrease in the surface level of Lake Tahoe, Nevada and California. U.S. Geological Survey Circular 1119: 77–79.
Rosenbaum, J. G. & R. L. Reynolds, 2004a. Record of late Pleistocene glaciation and deglaciation in the southern Cascade Range. II. Flux of glacial flour in a sediment core from Upper Klamath Lake, Oregon. Journal of Paleolimnology 31: 235–252.
Rosenbaum, J. G. & R. L. Reynolds, 2004b. Basis for paleoenvironmental interpretation of magnetic properties of sediment from Upper Klamath Lake (Oregon): effects of weathering and mineralogical sorting. Journal of Paleolimnology 31: 253–265.
Round, F. & L. Bukhtiyarova, 1996. Four new genera based on Achnanthes (Achnanthidium), together with a redefinition of Achnanthidium. Diatom Research 11: 345–361.
Rühland, K. & J. P. Smol, 2005. Diatom shifts as evidence for recent subarctic warming in a remote tundra lake, NWT, Canada. Palaeogeography, Paleoclimatology, Palaeoecology 226: 1–16.
Rühland, K., A. Priesnitz & J. P. Smol, 2003. Paleolimnological evidence from diatoms for recent environmental changes in 50 lakes across Canadian Arctic treeline. Arctic, Antarctic, and Alpine Research 35(1): 110–123.
Sandquist, D. R. & J. E. Keeley, 1990. Carbon uptake characteristics in two high elevation populations of the aquatic CAM plant *Isoetes bolanderi* (Isoetacae). American Journal of Botany 77: 682–688.
Schneider, R. R. & W. D. McFarland, 1996. Hydrologic data and description of a hydrologic monitoring plan for Medicine Lake volcano, California. U.S. Geological Survey Open-File Report 9: 5–750.
Schrader, H. & R. Gersonde, 1978. Diatoms and silocoflagellates. Utrecht Micropaleontology Bulletin 17: 129–176.

Smol, J. P. & B. F. Cuming, 2000. Tracking long-term changes in climate using algal indicators in lake sediments. Journal of Phycology 36: 986–1011.

Starratt, S. W., 2001. Diatoms as indicators of freshwater flow variation in central California. In West, G. J., L. D. Buffaloe (eds), Proceedings of the Eighteenth Annual Pacific Climate Workshop; 18–21 March 2001; Asilomar, California. Interagency Ecological Program for the San Francisco Estuary Technical Report 69. Sacramento (CA): California Department of Water Resources: 129–144.

Starratt, S. W., 2004. Diatoms as indicators of late Holocene freshwater flow variation in the San Francisco Bay estuary, central California, U.S.A. In Poulin, M. (ed.) Proceedings of the Seventeenth International Diatom Symposium, Bristol, Biopress Limited: 371–397.

Starratt, S. W., J. A. Barron, T. Kneeshaw, R. L. Phillips, J. L. Bischoff, J. B. Lowenstern & J. A. Wanket, 2003. A Holocene record from Medicine Lake, Siskiyou County, California: preliminary diatom, pollen, geochemical, and sedimentological data. In West, G. J. & N. L. Blomquist (eds), Proceedings of the Nineteenth Annual Pacific Climate Workshop, Technical Report 71 of the Interagency Ecological Program for the San Francisco Estuary: 131–148.

Stine, S., 1990. Late Holocene fluctuations of Mono Lake, eastern California. Palaeogeography, Palaeoclimatology, Palaeoecology 78: 333–381.

Stuiver, M., P. J. Reimer, E. Bard, J. W. Beck, G. S. Burr, K. A. Hugen, B. Kromer, F. G. McCormac, J. van der Plicht & M. Spurk, 1998. INTCAL98 radiocarbon age calibration 24,000-0 cal BP. Radiocarbon 40: 1041–1083.

Thompson, R. S. & K. H. Anderson, 2000. Biomes of western North America at 18,000, 6000 and 0 C-14 yr BP reconstructed from pollen and pacrat midden data. Journal of Biogeography 27: 555–584.

Thompson, R. S., C. Whitlock, P. J. Bartlein, S. P. Harrison & W. G. Spaulding, 1993. Climatic changes in the western United States since 18,000 yr B.P. In Wright, H. E., Jr., J. E. Kutzbach, T. Webb III, W. F. Ruddiman, F. A. Street-Perrott & P. J. Bartlein (eds), Global Climate Since the Last Glacial Maximum. Minneapolis, Minnesota University Press: 468–513.

Viau, A. E., K. Gajewski, P. Fines, D.-E. Atkinson & M. C. Sawada, 2002. Widespread evidence of 1500 yr climate variability in North America during the past 14000 yr. Geology 30: 455–458.

Wood, G. D., A. M. Gabriel & J. C. Lawson 1996. Palynological techniques—processing and microscopy. In Jansonius, J. & D. C. McGregor (eds), Palynology: Principles and Applications, Vol. 1. American Association of Stratigraphic Palynologists Foundations: 29–50.

Multiproxy study of anthropogenic and climatic changes in the last two millennia from a small mire in central Poland

Mariusz Lamentowicz · Zofia Balwierz · Jacek Forysiak · Mateusz Płóciennik · Piotr Kittel · Marek Kloss · Juliusz Twardy · Sławomir Żurek · Jacek Pawlyta

Originally published in the journal Hydrobiologia, Volume 631, No. 1, 213–230.
DOI: 10.1007/s10750-009-9812-y © Springer Science+Business Media B.V. 2009

Abstract The Żabieniec kettle hole is the first peatland in central Poland analyzed quantitatively with four biotic proxies (plant macrofossils, pollen, testate amoebae and chironomids) to reconstruct the past environmental change. Palaeoecological data were supported by historical and archaeological records. We focused on autogenic vegetation change and human impact in relation to climatic effects. The aims of our study were (a) to describe the development history of the mire during the last 2,000 years, (b) to date and reconstruct the anthropogenic land-use changes and (c) to discuss a possible climatic signal in the peat archive. The combination of proxies revealed dramatic shifts that took place in the peatland since the Roman Period. Żabieniec was a very wet telmatic habitat until ca. AD 600. Then, the water table declined, and the site transformed into a *Sphagnum*-dominated mire. This dry shift took place mainly during the Early Medieval Period. Human impact was gradually increasing, and it was particularly emphasized by deforestation since AD 1250 (beginning of the Late Medieval Period). Consequently, surface run-off and aeolian transport

Guest editors: K. Buczkó, J. Korponai, J. Padisák & S. W. Starratt
Palaeolimnological Proxies as Tools of Environmental Reconstruction in Fresh Water

M. Lamentowicz (✉)
Department of Biogeography and Palaeoecology, Faculty of Geosciences, Adam Mickiewicz University, Dzięgielowa 27, 61-680 Poznan, Poland
e-mail: mariuszl@amu.edu.pl

M. Lamentowicz
Ecosystem Boundaries Research Unit, Wetlands Research Group Station 2, Swiss Federal Institute for Forest Snow and Landscape Research (WSL), 1015 Lausanne, Switzerland

M. Lamentowicz
Laboratoire des Systèmes Ecologiques Station 2, Ecole Polytechnique Fédérale de Lausanne (EPFL), 1015 Lausanne, Switzerland

Z. Balwierz · J. Forysiak · P. Kittel · J. Twardy
University of Łódź, Faculty of Geographical Sciences, Narutowicza 88, 90-139 Lodz, Poland

M. Płóciennik
University of Łódź, Faculty of Biology and Environment Protection, Banacha 12/16, 90-237 Lodz, Poland

M. Kloss
Centre for Ecological Research, Polish Academy of Sciences, Dziekanów Leśny, 05-092 Łomianki, Poland

S. Żurek
Department of Quaternary Studies and Nature Conservation, University of Humanistic and Nature Science of Kielce, Institute of Geography, Świętokrzyska 15, 25-406 Kielce, Poland

J. Pawlyta
Department of Radioisotopes, Silesian University of Technology, Krzywoustego 2, 44-100 Gliwice, Poland

from the exposed soils caused the eutrophication of the mire. Furthermore, chironomids and testate amoebae reveal the beginning of a wet shift ca. AD 1350. Openness considerably increased in the Late Medieval and the Modern Periods. The highest water table during the last 1,000 years was recorded between AD 1500 and 1800. This wet event is connected with deforestation, but it could be also associated with the Little Ice Age. Our study shows plant succession in the Żabieniec peatland, which can be explained with the recent landscape transformation. However, such changes are also possibly linked with the major climatic episodes during the last two millennia, such as the Medieval Warm Period and the Little Ice Age.

Keywords Central Poland · Kettle hole · Testate amoebae · Pollen · Macrofossils · Human impact · Climate · Environmental archaeology

Introduction

Peatlands are very sensitive to climate change and human impact (Blackford, 1993; Charman, 2002; Rydin & Jeglum, 2006). They record the hydrological and trophic variations within the peatland ecosystem as well as changes in the catchment area (Bragg, 2002). Many records from peat archives have proven to be useful for reconstructing the climatic changes (Schoning et al., 2005; Booth et al., 2006; Oksanen & Valiranta, 2006; Chambers et al., 2007; Charman & Blundell, 2007). However, other studies have concentrated mainly on the signal of human impact (Warner et al., 1989; Chambers & Charman, 2004; Lamentowicz et al., 2007; Amesbury et al., 2008).

Wetland ecosystems function under various kinds of anthropogenic stress. Their present state is a result of a long history of changes. Disturbance was gradually increasing along with economic development in central Poland during the last 2,000 years. This period was also very important because of the phases of considerable climatic changes such as the Medieval Warm Period and Little Ice Age. Recovery of a disturbed peatland ecosystem is often not possible after it has passed a critical threshold. Therefore, it is important to assess the present condition and naturalness of peatlands. Moreover, the importance of human influence and climate should be estimated to understand the state of a mire ecosystem and the methods available for active nature conservation (Willis et al., 2007).

One of the major aims of late Quaternary palaeoecology is to answer the question: what is the proportion of the climatic to anthropogenic signal revealed in proxies? Late Holocene deposits are particularly challenging for palaeoclimatic inferences (Yeloff et al., 2007; van der Linden et al., 2008). Human activity not only transformed the forest vegetation in many parts of the world but also triggered indirect physical and chemical changes in peatlands. Since the Early Medieval period, the vegetation of European forests was considerably transformed (Ralska-Jasiewiczowa & Latałowa, 1996; Williams, 2000; Ralska-Jasiewiczowa et al., 2004). Deforestations that progressed since the Medieval Age caused increased run-off and abrupt hydrological fluctuations. Furthermore, most of the European peatlands were drained (Joosten & Clarke, 2002), which disturbed or eliminated the climatic signal from the peat archive. Consequently, multidisciplinary studies made at many sites are necessary to reconstruct spatially past environmental changes.

Recently, most palaeoenvironmental peatland studies have focused on ombrotrophic mires (Yeloff & Mauquoy, 2006; Hughes et al., 2007; Sillasoo et al., 2007; Blundell et al., 2008), and a minority has been carried out in fens and kettle-hole mires (Lagerback & Robertsson, 1988; Warner, 1993; Yu, 2006; Lamentowicz et al., 2008b), which can also provide interesting palaeoenvironmental data.

A typical feature of a kettle-hole peatland is the presence of a telmatic fen–bog transition in the stratigraphic sequence. The autogenic terrestrialization makes the reconstruction of a palaeoclimatic signal very difficult, although it is not impossible to detect it in subsequent stages of growth. Żabieniec in Central Poland is an example of a peatland where *Sphagnum* appeared and dominated in the following successive hydroseral stages. This may result from a natural trajectory to acidification (Hughes & Barber, 2004) or even land-use change in the surroundings (Lamentowicz et al., 2007). Multiproxy palaeoenvironmental studies from peatlands with relation to historical and archaeological sources are not common in this part of Europe. They are particularly rare in the central part of our country, located in periglacial conditions during the last Ice Age. This is because most wetlands are concentrated within the former range of the last Vistulian (=Weichselian) Glaciation in the northern half of Poland, where a majority of studies were carried out.

The Żabieniec kettle hole is the first peatland in Poland and this part of Europe analyzed with four biotic proxies (plant macrofossils, pollen, chironomids and testate amoebae), which we compare with archaeological records in this study. Results of Cladocera and analyses of diatoms are currently being prepared for future publication. We focused mainly on the period of the last 2,000 years to track the increasing human disturbance that should have been recorded in the peatland. It was assumed that the most important hydrological and trophic change in the studied mire was initiated along with the beginning of the deforestation in the mire catchment in the Medieval Period. For this reason, the aims of our study were (a) to describe the development history of the mire during the last 2,000 years, (b) to date and reconstruct the anthropogenic land-use change and (c) to discuss a possible climatic signal in the analyzed peat archive.

Study site

The Żabieniec mire is situated in Central Poland, 25 km east of Łódź, in an area of morainic upland (Fig. 1). The mire is located in a closed oval basin, which is 1.5-km long, in a watershed between the rivers Mrożyca (to the west) and Mroga (to the east). This basin was formed during the ice melting of the Warta Glacier as a melt-out depression with a highly varied bottom configuration. Żabieniec is situated in the central part of this glacial depression. The mire area is ca. 2.5 ha. Peat and lake sediments were deposited within this depression probably from the end of the Warta (=Saalian) Glaciation until the upper Plenivistulian (=late Weichselian). Biogenic deposits of the Late Vistulian and Holocene were formed only in the deepest part of the depression, i.e. in the Żabieniec mire basin.

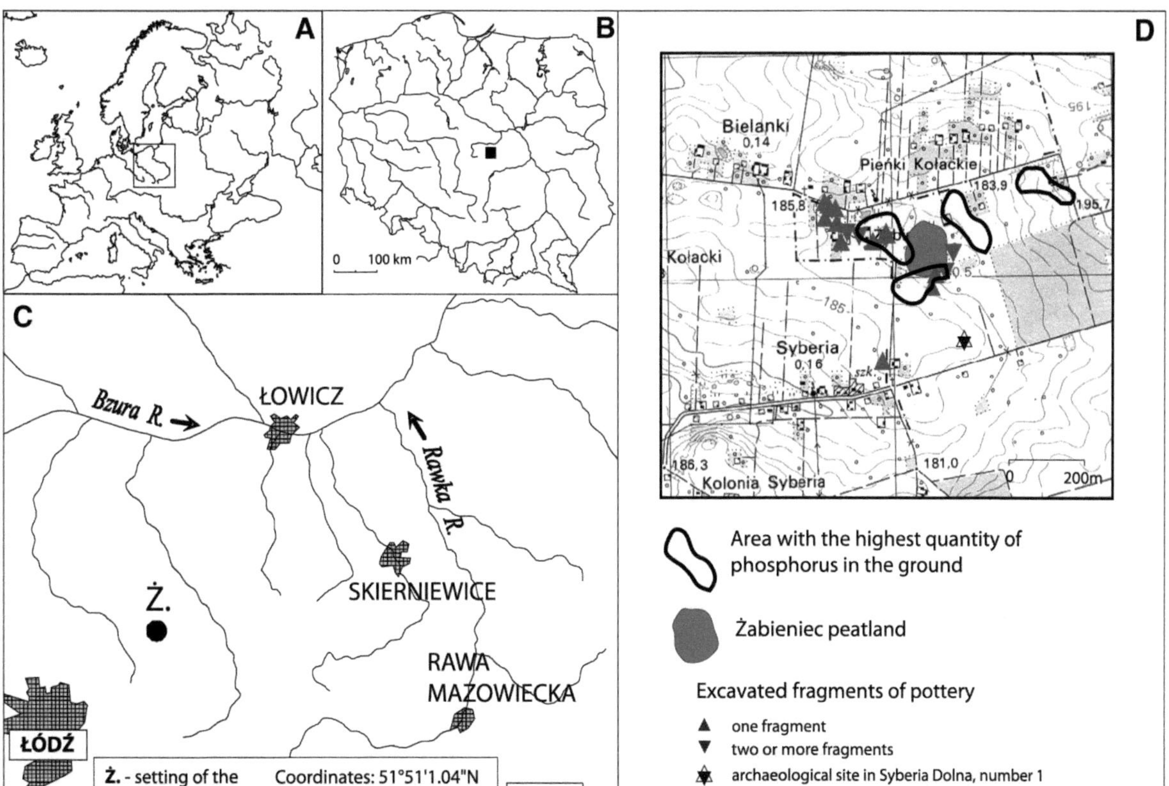

Fig. 1 Setting of the study site at the background of **A** Europe, **B** Poland, **C** central Poland and **D** map showing archaeological sites around the peatland

Climate in this region of Poland is rather continental and drier than in the northern part of the country, which is more influenced by the wet air masses from the Baltic and northern Atlantic. According to the climatic typology of Poland, this area has a transitional climate (Woś, 1999).

The average annual temperature in 1931–1989 was 7.7°C, while the mean of the warmest month (July) was 18°C and that of the coldest month (January) was −3.3°C. However, these annual and monthly temperatures are highly variable, which is probably caused by the influence of oceanic and continental climate. Average annual precipitation in 1951–1989 was 590 mm, but in this case also, a high variability (ca. 30%) was observed (Kłysik, 2001).

Methods

Fieldwork

One core (Z-2) was sampled from the central part of the mire (Fig. 2). In addition, a 60-cm peat monolith was extracted for the study of the last millennium. The main core was taken with a piston corer. Furthermore, a geological survey of the mire was done with a Russian sampler (diameter—50 mm). Samples for phosphorus analyses were collected by geological drilling at depths of ca. 90–100 cm. Distance between drillings was 50–100 m. One hundred samples were collected for the analysis. Phosphorus analysis was realized according to the methodology described by Eidt (1973) and Arrhennius (1950).

Laboratory

An age-depth model for the peat profile was constructed by radiocarbon dating of the organic material (Table 1). Samples prefixed with 'Poz' in the laboratory code were dated with accelerator mass spectrometry (AMS), whereas all the other samples were dated with radiometric methods. All the AMS dates were obtained on the basis of several *Sphagnum* stems carefully selected for radiocarbon dating, avoiding any rootlets and other contamination (Table 1). All the other dates were obtained on the basis of bulk

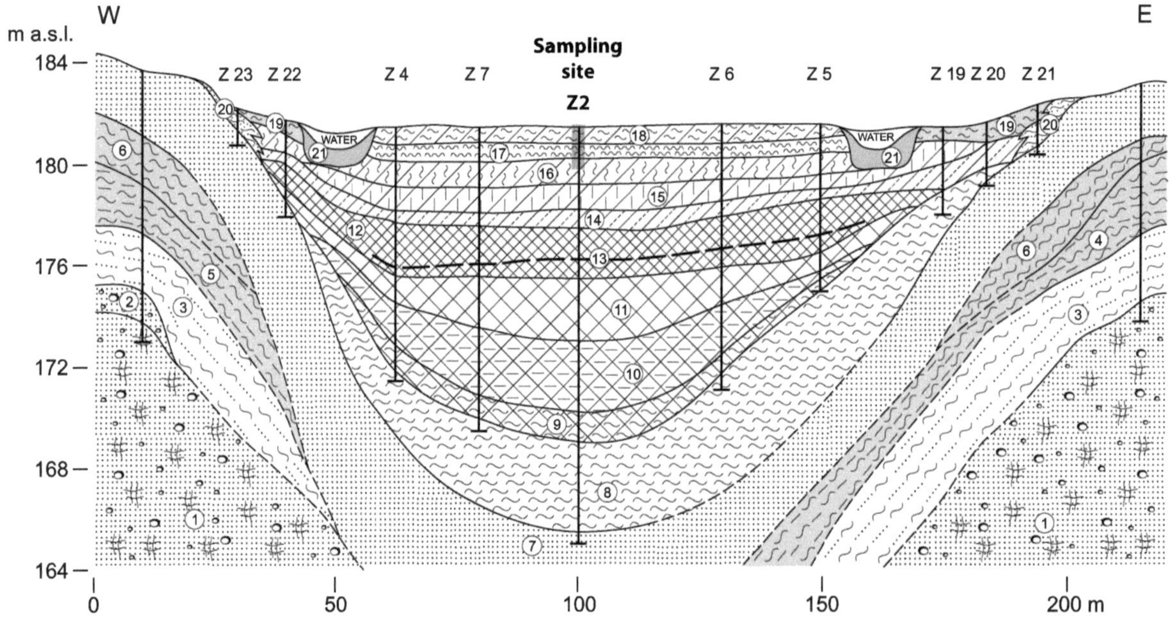

Fig. 2 Geological profile (WE) of the Żabieniec peatland. Warta Glaciation (=Saalian): 1—glacial till, 2—fluvioglacial sand and gravel, and 3—mineral silt; Eem Interglacial: 4—organic silt; Vistulian (=Weichselian): 5—mineral-organic silt with fine sand, 6—organic-mineral silt, 7—sand, 8—mineral and organic silt, 9—gyttja with silt, 10—detritus-clay gyttja; Holocene: 11—fine detritus gyttja, 12—coarse detritus gyttja, 13—brown-moss peat, 14—sedge-moss peat, 15—*Sphagnum* peat, 16—sedge-moss peat, 17—*Sphagnum cuspidatum* peat, 18—sedge-moss peat and brown moss peat, 19—organic-mineral silt with detritus, 20—silt, 21—deposits disturbed by exploitation

Table 1 Results of radiocarbon dating

Sample name	Depth (cm)	^{14}C age (BP)	Calibrated age (95.40% probability)
Poz-19019	45	645 ± 50	1276 AD (95.4%) 1405 AD
Poz-17117	60	1085 ± 30	894 AD (95.4%) 1016 AD
Poz-17116	120	1475 ± 35	535 AD (95.4%) 652 AD
Lod-1391	305	3010 ± 50	1410 BC (95.4%) 1114 BC
Lod-1390	375	3690 ± 50	2271 BC (0.8%) 2259 BC
			2206 BC (94.6%) 1937 BC
Poz-23658	436	4580 ± 35	3499 BC (18.2%) 3436 BC
			3379 BC (42.8%) 3318 BC
			3292 BC (0.2%) 3290 BC
			3274 BC (0.5%) 3268 BC
			3236 BC (33.8%) 3108 BC
Poz-23638	531	6270 ± 40	5325 BC (86.4%) 5204 BC
			5168 BC (9.0%) 5076 BC
Poz-23639	826	9130 ± 50	8530 BC (1.2%) 8519 BC
			8474 BC (94.2%) 8255 BC
Poz-23640	1,036	11,860 ± 60	11921 BC (95.4%) 11609 BC
Poz-23659	1,290	14,120 ± 70	15286 BC (95.4%) 14450 BC

organic matter of peat. Both AMS and radiometric methods used the standard alkali–acid–alkali pre-treatment procedure. Sample pre-treatment, graphite target preparation and ^{14}C concentration measurements for 'Poz' samples were done in the AMS facility of the Poznań Radiocarbon Laboratory. Radiometric dating for 'Lod' prefixed samples were done in the Łódź Radiocarbon Laboratory.

For the analysis of testate amoebae, 4-cm^3 samples were taken in 2.5-cm intervals for the upper 60 cm and in 5-cm intervals for the section of 60–125 cm. Samples were prepared according to the sieving and back-sieving procedure described by Hendon and Charman (1997). Testate amoebae were identified and counted to a total of 150 individuals per sample, at a magnification of 200–400×. Their identification was based on the available literature (Grospietsch, 1958; Ogden & Hedley, 1980; Charman et al., 2000; Clarke, 2003; Mitchell, 2003).

For pollen analysis, 2-cm^3 samples were taken in 5-cm intervals and treated with 10% KOH and then acetolyzed for 3 min (Berglund & Ralska-Jasiewiczowa, 1986; Faegri & Iversen, 1989). Minerogenic matter was removed with hydrofluoric acid. In general, a minimum of 500 pollen grains of trees and shrubs (arboreal pollen = AP) was counted in each sample, but in samples where their frequency was very low, all the pollen grains from the two slides 22 × 22 mm^2 were identified. The sum of arboreal and non-arboreal pollen was the basis of percentage calculation (AP + NAP = 100%).

The material for plant macrofossil analysis was composed of samples taken in 10-cm intervals, and the sample volume was 20 cm^3. The material was rinsed with water onto sieves with mesh sizes of 0.25 and 0.5 mm. The residue was identified under a stereoscopic microscope at a magnification of 10–100×. One slide from each sample was examined under a microscope at 200–400× to determine the peat composition. Macrofossils were identified with the use of available guides (e.g. Katz et al., 1965, 1977; Berggren, 1968; Tobolski, 2000; Mauquoy & van Geel, 2007).

Chironomid fossil analysis was performed according to the methods described by Brooks et al. (2007) using a 63-μm mesh size sieve. In samples where a small number of head capsules were present, kerosene floatation on a bigger volume of sediment (up to 80 cm^3) was used for the extraction of chironomid subfossils (Rolland & Larocque, 2007). Head capsules were identified by using mainly the taxonomic keys by Wiederholm (1983), Klink & Moller Pillot (2003) and Brooks et al. (2007). Water quality classification based on the proportions of tolerant and intolerant taxa follows Wilson and Ruse (2005).

Numerical analyses and graphical presentation

Conventional radiocarbon dates were calibrated by using OxCal 4.05 (Bronk Ramsey, 2001) and the IntCal04 calibration curve (Reimer et al., 2004). Five-year smoothing was chosen for the calibration. Probability distributions for 10 calibrated dates were used for the estimation of an age-depth model. In addition, the calendar year 2006 was attributed to the top of the core. The age-depth model was built by using the P_Sequence deposition model procedure of OxCal 4.05 (Bronk Ramsey, 2008). Parameter *k* of the P_Sequence function was set to 0.1, and the interpolation parameter was set to 2. The resulting expected values modelled for each dated depth were then interpolated by the Akima spline algorithm (Akima, 1970) to give the date for each centimetre of depth (Fig. 3). In addition, normalized probability distributions for each calibrated radiocarbon date used for age-depth modelling are also presented in Fig. 3. The grey shaded area in Fig. 3 represents the 68% probability range of the model.

For the summary diagram, general zones were delimited mainly on the basis of archaeological

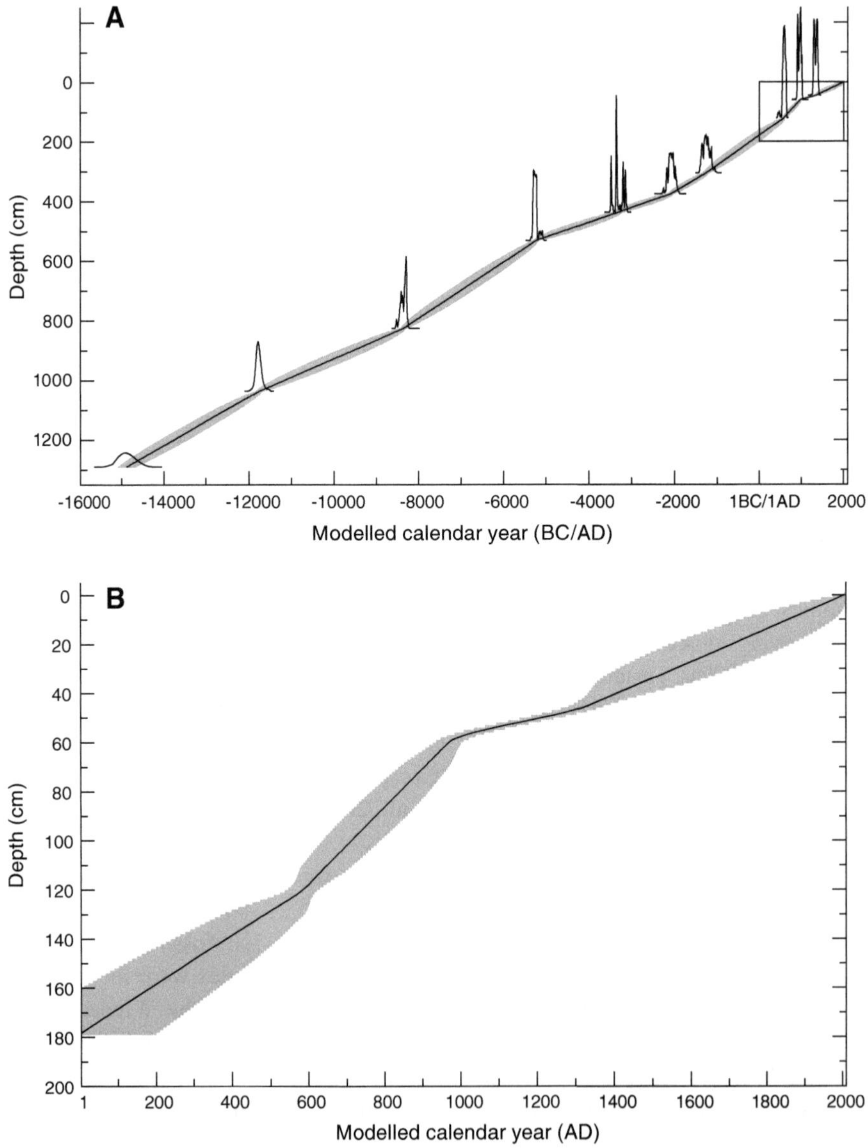

Fig. 3 Age-depth model, (**A**) entire profile, (**B**) last 2,000 years

periods (Kaczanowski & Kozłowski, 1998), but an additional zone Z6 was created because of a significant, abrupt change in all proxies.

In order to reconstruct quantitatively the water table and pH in the mire, we applied a training set consisting of 123 surface samples taken from natural *Sphagnum* mires in Pomerania (Lamentowicz & Mitchell, 2005; Lamentowicz et al., 2008b), western Poland (Lamentowicz et al., 2008a) as well as in Żabieniec in 2005. In the publications mentioned above, the performance of four models was tested: partial least squares (PLS), weighted averaging (WA), tolerance down-weighted weighted averaging (WAtol) and weighted averaging partial least squares (WA-PLS), by using the program C2 (Juggins, 2003). The root mean square error of prediction (RMSEP) was calculated by using the jackknifing cross-validation procedure (Crowley, 1992). For depth to the water table (DWT) (range of modern samples min = −3 cm; max = 70 cm, SD = 14 cm), the model with the lowest RMSEP was WAtol, with a maximum prediction bias of 8.9 cm and a jacknived root mean square error of prediction (RMSEP(jack)) of 4.3 cm. For pH, the model with the lowest RMSEP was WA-PLS (range of modern samples min = 3.14; max = 7.27, SD = 1.11), the maximum bias of the prediction was 0.5 pH units and the RMSEP(jack) was 0.41 pH units (Lamentowicz et al., 2008b; Lamentowicz & Mitchell, 2005). Reconstruction of the general trophic status of the peatland was based on the quantitative pH reconstruction as well as qualitative attributes of testate amoebae and plant macrofossils data.

Results

Chronology and peat accumulation rates

The age-depth model presented in Fig. 3 shows that accumulation rate throughout the history of the peat bog was approximately constant. The average accumulation rate calculated by linear approximation of the age-depth model curve is about 0.74 mm/year. However, some small variations of accumulation rate can be observed. Small decreases in the accumulation rate during 12000–7000 BC and 5000–2000 BC as well as a small increase during 8000–5000 BC are apparent.

Biological records

Proxies analyzed in this study allowed reconstructing the history of the peatland and its surroundings, spanning the last two millennia. Initially, we developed a zonation for each proxy. However, because zones for the particular proxies did not remain in agreement, we presented the development history in the context of human impact, which is a useful criterion from the point of view of the aim of this study. Low agreement between zonations might be caused by different resolution of sampling, especially in the case of pollen analysis.

Detailed diagrams of each proxy are presented in separate figures for plant macrofossils (Fig. 4), pollen (Fig. 5), chironomids (Fig. 6) and testate amoebae (Fig. 7).

A summary diagram with a short interpretation is presented in Fig. 8. The development history of the Żabieniec mire was divided into six stages (Z1–Z6) according to the archaeological periods.

History of the Żabieniec peatland

Roman Period, ca. 0 BC–AD 375 (Z1) (Figs. 4, 5, 6, 7)

During this period, the Żabieniec peatland was a mesotrophic–telmatic habitat with shallow water, as indicated by brown mosses, *Scheuchzeria palustris*, *Carex* spp., and remains of *Pediastrum* and *Botryococcus* (Fig. 4–5). Between AD 100 and AD 200, *Sphagnum* spp. of the section *Sphagnum* became more abundant, whereas *Scheuchzeria palustris* and Cyperaceae declined, which suggests a temporal acidification (van Breemen, 1995; Tallis, 1983). At about AD 400, the peatland became mesotrophic again, which was probably connected with a rise in the water table. All proxies show relatively stable hydrological conditions. Testate amoebae were absent. Human impact was insignificant, but infrequent pollen of *Rumex acetosa*, *Plantago lanceolata* and cereals (including *Secale*) shows human presence (Behre, 1981) in more distant areas.

Chironomid analysis indicates that Żabieniec in the Roman Period was a mire with a shallow water body overgrown with macrophytes. The dominant species was *Paratendipes nudisquama*. *Paratendipes* larvae often occur in small standing water bodies like

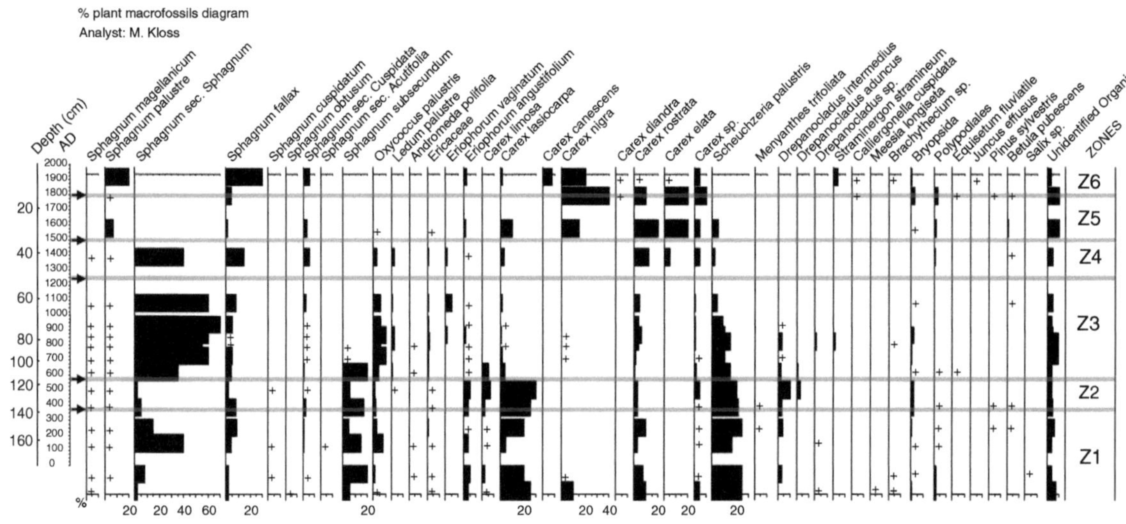

Fig. 4 Percentage plant macrofossil diagram. Zonation based on archaeological periods

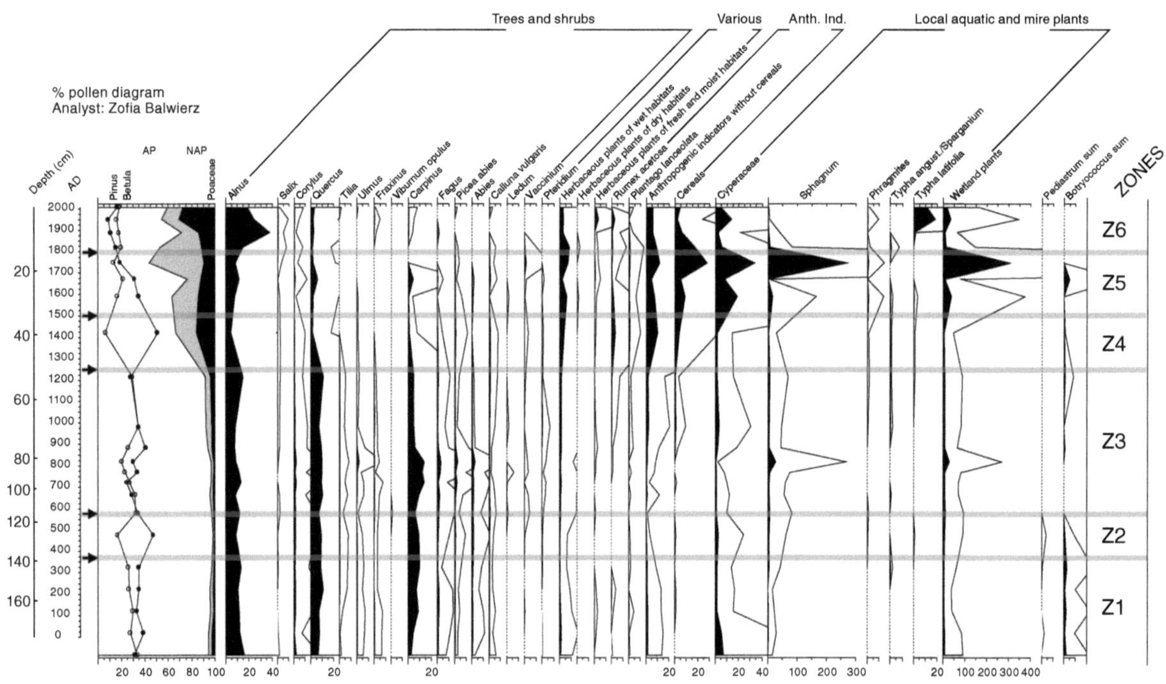

Fig. 5 Percentage pollen diagram. Zonation based on archaeological periods

ponds and bogs (Wiederholm, 1983). Chironomid communities were relatively rich in taxa, with high dominance of macrophyte-related taxa characteristic of small acid water bodies with dense vegetation such as *Monopelopia tenuicalcar* and *Lauterborniella agrayloides* (ecological preferences after Klink & Moller Pillot (2003) and Vallenduuk & Moller Pillot (2007)).

Migration period, ca. AD 375–568 (Z2)
(Figs. 4, 5, 6, 7)

This stage was characterized by indicators of wet and mesotrophic habitats. *Scheuchzeria palustris* and *Carex* spp. dominated among vascular plants. *Pediastrum* and *Botryococcus* increased, representing open water surface. There was no decline in AP. During this

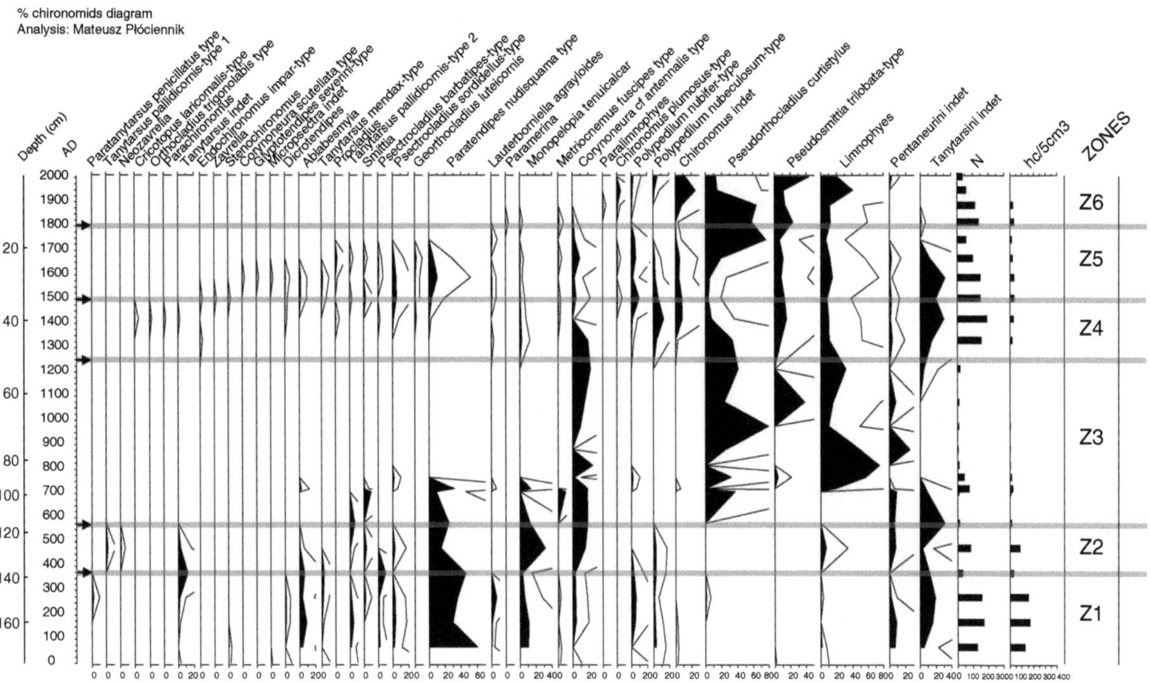

Fig. 6 Percentage chironomid diagram. Zonation based on archaeological periods

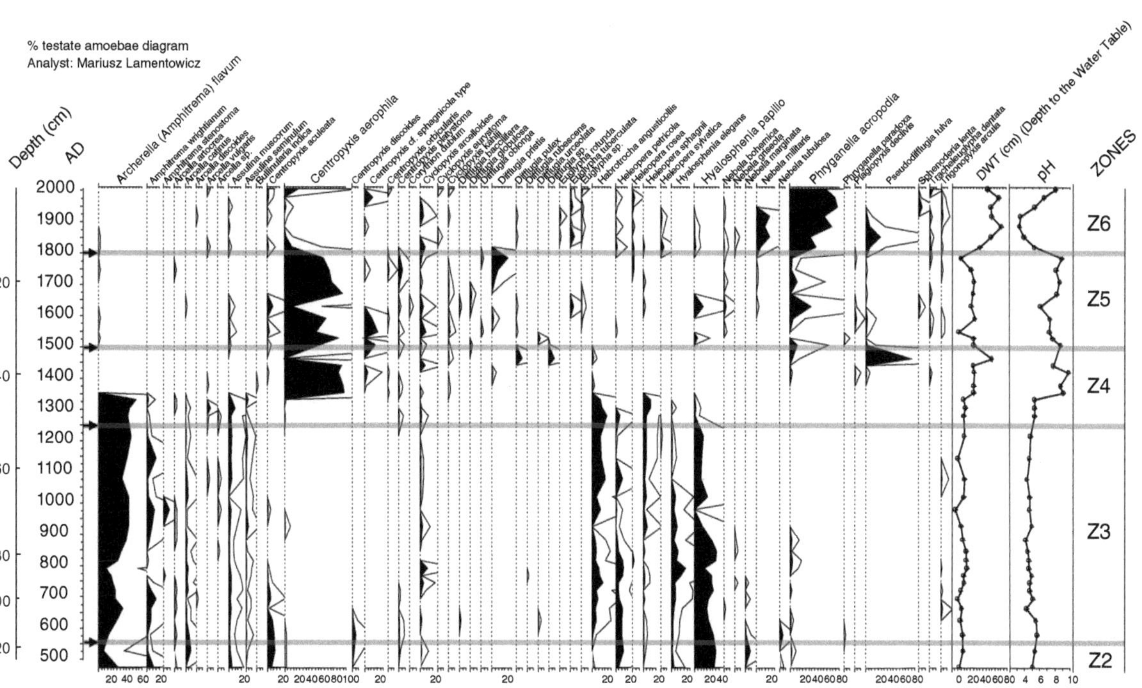

Fig. 7 Percentage testate amoebae diagram. Zonation based on archaeological periods

stage (AD 500), the first testate amoebae appeared, mainly oligotrophic taxa (Lamentowicz & Mitchell, 2005; Lamentowicz et al., 2008b) (*Amphitrema wrightianum*, *Archerella* (*Amphitrema*) *flavum*) and taxa typical for mesotrophic–telmatic conditions (Lamentowicz & Mitchell, 2005; Lamentowicz et al.,

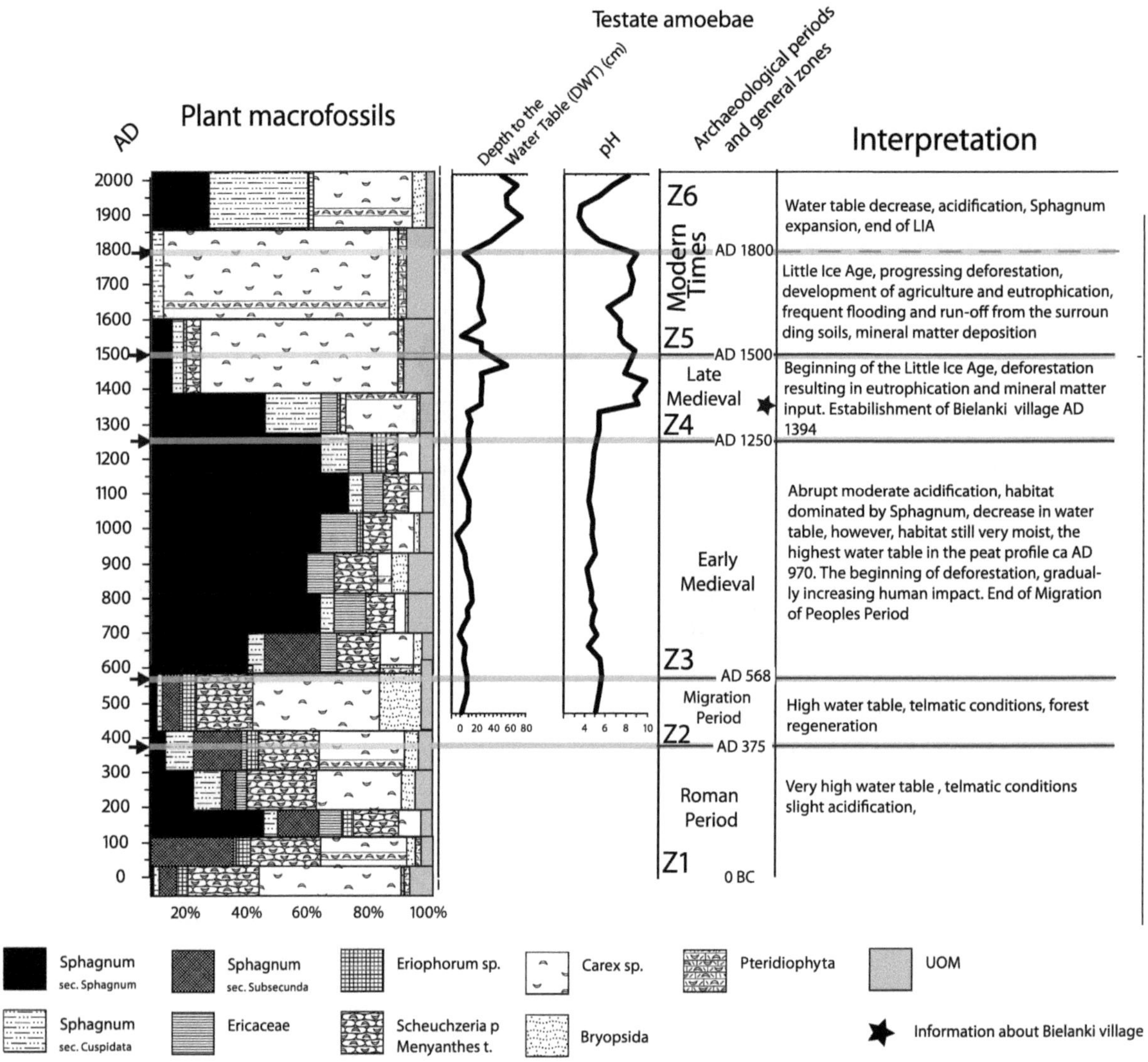

Fig. 8 Comparison of plant macrofossils data with quantitative reconstructions based on testate amoebae. Macrofossils show local habitat changes. Environmental variables from testate amoebae are based on regional transfer function. Other proxies are summarized in the Results and Discussion sections. First information about location of Bielanki village is also mentioned and the zonation is based on Polish archaeological periods

2008b) (such as *Centropyxis aculeata* and *C. discoides*). Testate amoebae appeared abruptly, which suggests their rapid establishment. Their presence was connected with an increasing abundance of *Sphagnum*. The resolution of plant macrofossil analysis does not make it possible to determine the exact date of *Sphagnum* expansion, but remains of testate amoebae reveal that it was not a gradual process.

The composition of chironomid assemblages changed and diversity decreased, which indicates a shift in the ecosystem. The terrestrial *Smittia* appeared (Brooks et al., 2007), the domination of *Paratendipes nudisquama* decreased, while shares of some acidophilic taxa—*Monopelopia tenuicalcar*, *Limnophyes*—briefly increased. Furthermore, *Corynoneura* cf. *antennalis* increased; this species is usually associated with streams but frequent also in acidic waters (Klink & Moller Pillot, 2003). Larvae in this morphotype in Żabieniec seem to be associated with acidic conditions.

Early Medieval Period, ca. AD 568–1250 (Z3) (Figs. 4, 5, 6, 7)

From ca. AD 600, oligotrophic *Sphagnum* peatland developed. This is roughly concurrent with the end of the Migration Period. However, during the entire Early Medieval Period, anthropogenic pollen indicators were low, but gradually increasing. There were no direct changes in the peatland surroundings, but distant transport of some pollen taxa (*Rumex acetosa*, *Plantago lanceolata* and cereals) represents the Early Medieval economy. Between AD 700 and AD 850, *Carpinus* forest regenerated. *Carpinus* shows a constant decline since AD 850, along with an increase in pollen indicators of human activity. Furthermore, the proportion of NAP gradually increased, representing landscape openness. The testate amoebae assemblage was composed mainly of species characteristic for the moist *Sphagnum* lawn (Lamentowicz & Mitchell, 2005; Lamentowicz et al., 2008b), e.g. *Amphitrema flavum* and *Hyalosphenia papilio*; however, an important moisture indicator *Amphitrema wrightianum* (Charman et al., 2000; Lamentowicz & Mitchell, 2005; Lamentowicz et al., 2008b) shows wetter episodes during ca. AD 690, ca. AD 1000 and ca. AD 1150. Reconstructed values of depth to the water table fluctuate between 0 and 20 cm. Average pH was around 4.5. These stable hydrological conditions in the peatland were maintained until ca. AD 1350.

From ca. AD 750 to ca. AD 1250, only four chironomid taxa were present. The abundance of larve was low from ca. AD 750 to ca. AD 1200, which may indicate water table decrease and *Sphagnum* expansion. A rapid decline and disappearance of nearly all shallow-water taxa (including *Paratendipes nudisquama*) up to AD 750 and their replacement by semiterrestrial ones tolerant to low pH such as *Pseudorthocladius curtistylus* and *Limnophyes* (Klink & Moller Pillot, 2003; Brooks et al., 2007), may indicate semiterrestrial conditions and coincide with those having a pH below 5.0 as inferred by testate amoebae (Fig. 8). Only the percentage of *Corynoneura* cf. *antennalis* increased. This species is not a terrestrial one (Klink & Moller Pillot, 2003) which suggests the presence of at least temporally open water table on the mire surface.

Late Medieval Period, AD 1250–1500 (Z4) (Figs. 4, 5, 6, 7)

Historically, year ca. AD 1250 marks the end of the Early Medieval Period, but the most important transition in the peatland record is dated to ca. AD 1350. Up to this date, conditions were relatively stable, with rapid peat accumulation by *Sphagnum*. The beginning of the Late Medieval Period is marked by ca. AD 1250, and the probable deforestation in the immediate vicinity of the Żabieniec mire is marked by ca. AD 1350, which might have triggered a change in the Żabieniec peatland vegetation and peat composition. The closeness of the deforestation can be estimated indirectly by testate amoebae and chironomids, as the pollen record possesses a lower sampling resolution. Peat-forming vegetation became dominated by Cyperaceae, indicating habitat eutrophication (Fig. 8). The curve of depth to the water table shows a decrease in the water table, with the lowest value ca. AD 1450, but pH increased to unusual value (>9). This very high pH value is hard to explain; it may be caused by the co-occurrence of minerotrophic indicators in fossil testate amoebae data. However, it reveals the instability of the peatland ecosystem, caused by change in the surroundings. Pollen data show an increase in human impact and NAP percentage, and a decline in pollen of all trees. Furthermore, an increase of anthropogenic indicators is apparent, e.g. of cereals, *Rumex acetosa* and *Plantago lanceolata*. Remains of *Botryococcus* represent increased wetness. However, this peak is visible in only one sample. This wet shift is also reflected in the pH curve, representing eutrophication. Testate amoebae show a dramatic change in the community composition. *Centropyxis aerophila* reached abundances over 90% from ca. AD 1350 when the peat consisted of mineral matter, whereas before ca. AD 1350, nearly no mineral particles were recorded in the peat. Most testate amoebae occurring during this phase represent the genera *Centropyxis*, *Cyclopyxis* and *Difflugia*. At about AD 1400, the highest value of pH was recorded. The presence of *Phryganella acropodia* indicates short dry shifts (Mitchell et al. 2001).

Non-biting midges also revealed a sudden shift in the mire in the Late Medieval Period, associated with human influence on the bog. Human impact increased the abundance and diversity of chironomid communities, with the maximum ca. AD 1510. Most of the 31 taxa from the Late Medieval Period were represented also in the Roman Period and during the Migration Period. The factors that seemed to have determined the composition of assemblages in this period were pH and moderate eutrophication. Also, the water level was higher than that in the Early Medieval Period, which was crucial for the existence of rich assemblages of littoral species and decreased the dominance of terrestrial ones. While semiterrestrial taxa typical for acidified habitats (e.g. *Pseudorthocladius curtistylus*) and *Corynoneura* cf. *antennalis* were less abundant, others considered as eurytopic, occurring in neutral waters (Klink & Moller Pillot, 2003; Brooks et al., 2007) (*Psectrocladius sordidellus*-type, *Dicrotendipes*, *Polypedilum nubeculosum*-type) were more abundant. Lower percentage of intolerant taxa (following the classification in Wilson & Ruse, 2005) than that in the Roman Period indicates higher trophy coinciding with higher pH inferred from testate amoebae.

Modern period, ca. AD 1500–1800 (Z5) (Figs. 4, 5, 6, 7)

During this phase, the decline in NAP and the increase in anthropogenic indicators continued. Local vegetation remained dominated by Cyperaceae, with an even lower percentage of *Sphagnum*. Cereals reached over 30% of the total pollen during ca. AD 1750, and such high percentages should be interpreted as indicating the cultivation in the direct surroundings of the peatland. The landscape became more and more open, as all tree taxa declined in abundance through the interval. Macrofossils and pollen of wetland plants, including *Botryococcus*, indicate shallow water conditions. Also, testate amoebae reveal a high water table and high pH. *Centropyxis aerophila* was accompanied by other species present at lower abundance, e.g. *Centropyxis* cf. *sphagnicola* type, *C. platystoma*, *Cyclopyxis arcelloides* and *Phryganella acropodia*.

Abundant chironomid assemblages existed to AD 1670, whereas later only 10 mainly acidophilic and terrestrial ones remained. The percentage of *Pseudorthocladius curtistylus* sharply increased during ca. AD 1700, indicating the return of the mire habitat to conditions preceding human influence on the mire in the Late Medieval Period.

Modern period, ca. AD 1800–2006 (Z6) (Figs. 4, 5, 6, 7)

The last phase of Żabieniec mire development is characterized by the reappearance of *Sphagnum*, afforestation and an abrupt increase in *Alnus*. Human indicators and cereals slightly decreased. From ca. AD 1800, a second complete transformation of the habitat took place. Among testate amoebae, *Phryganella acropodia*, which prefers dry conditions, reached a very high percentage (>70%). This reflects a decrease in the water table. Also, *Nebela militaris*, a good dry indicator according to the existing transfer function of Lamentowicz et al. (2008c), confirmed this dry shift. *Typha latifolia* appeared during ca. AD 1900 in the peatland margins and has persisted there up to the present days.

Chironomids reveal a similar successional shift from the aquatic to terrestrial environment. Disappearance of nearly all littoral taxa and the dominance of *Pseudorthocladius curtistylus*, later also the terrestrial taxa *Pseudosmittia trilobata*-type and *Limnophyes* (Klink & Moller Pillot, 2003), may indicate a low water level and terrestrial character of the bog. *Polypedilum* and *Chironomus*, which occur in samples from ca. AD 1890–2006, may indicate at least temporary water pools on the mire in the last century of the Żabieniec history, as these taxa are eurytopic and not terrestrial (Brooks et al., 2007). The number of head capsules was at first higher than those in the previous sections, but later it declined. Diversity of the assemblages decreased to ca. AD 1890 as well, and then slightly increased.

Archaeological data

In the immediate vicinity of the Żabieniec mire, only one archaeological site was discovered during an archaeological surface survey (Fig. 1D). It is the site Syberia Dolna no. 1, located ca. 350 m north of the mire, on the surface of a western slope of a dry valley. Results of an archaeological study in 2007 did not confirm the occurrence of any relics at this site. In the area of 150 km^2, in the surroundings of the mire, 84 archaeological sites (with 119 archaeological relics) have been registered. Half of them date to the Late Medieval Period and Modern Period, and only five to the Early Medieval Period (not earlier than the eleventh century). In the group of 13 prehistoric sites, five date to the Mesolithic, two to the Mesolithic or the Neolithic, 13 to the Bronze Age, seven to the Late Bronze Age/Early Iron Age, one to the Early Iron Age (Hallstatt Period), one to the Hallstatt Period/La Tene Period and 10 to the Roman period. Most of these were documented by few archaeological relics. The sites are located mainly in or near valleys of the rivers Mroga and Mrożyca. Most of the sites are up to 5 km away from the mire (AZP, unpublished data).

In the area of the Żabieniec village and of the nearby Bielanki village, we uncovered only few fragments of pottery during an intensive surface survey, dated to the Modern Period and probably to the Late Medieval. In the same area, we recorded the highest quantity of phosphorus in the ground, detected by a field method in the surface layer (90–100-cm thick). Simultaneously, in preliminary archaeological excavations, no evidence of human activity was found. The oldest historical sources were recorded in the close vicinity of the Żabieniec mire two villages, Kołacinek and Bielanki. Both were noted in the Early Middle Ages—Bielanki in 1394 (in *Księgi Łęczyckie*), and Kołacinek in 1257 and 1334 as the Kuyavian Duke's possessions. In the sixteenth century, the former village existed within the borders of the Brzeziny parish as Bylanowo and in 1576 as Bilianovo (noblemen's possessions). In records from the nineteenth century, we can find the name Bielanki. In the sixteenth century, Kołacinek was found among noblemen's possessions as well (Zajączkowski & Zajączkowski, 1966).

The nineteenth century cartographic sources (i.e. Gilly's Map 1803 and the so-called *Topographic Map of the Polish Kingdom* 1839) show continuous woodlands in the surroundings of Żabieniec. In the late nineteenth century (according to Gilly's Map), we can find open areas only in proximity to the rivers Mroga and Mrożyca, near the village Wola Cyrusowa (north of Żabieniec), and probably in the vicinity of Bielanki. The Żabieniec village was present in AD 1825. The *Topographic Map of the Polish Kingdom* presents the increasing deforestation in the first half of the nineteenth century, in the surroundings of Bielanki as well. The Żabieniec mire was still forested at that time.

In the nineteenth century, in the area of the Brzeziny district, most arable land was moderately fertile, suitable for growing rye and potatoes. In this region, the extensive woodlands were administered or possessed by the state. In 1820–1853, the forest area declined in this region by ca. 30%. In 1820, woodlands occupied 40% of the district area, and in 1853, only 28% (Ohryzko-Włodarska, 1972).

Based on the results of former (AZP, unpublished data) and our archaeological field research, we can conclude that the human impact near the Żabieniec mire was insignificant almost until just prior to the Modern Period. Only a few relics, which are uncovered by archaeological research and dated to the Late Medieval Period, have been discovered. Older settlements occupied the areas close to the Mroga and Mrożyca river valleys.

Discussion

The multiproxy approach to the study of the Żabieniec peat archive allowed us to look at many aspects of the past change of the peatland and the surrounding landscape. This site may be regarded as an important reference point for higher resolution studies. All proxies give a very clear and sharp signal of abrupt changes in the peatland ecosystem and its surroundings. Two main questions arose during the investigation: (1) How and when did the anthropogenic land-use change affect the autogenic processes, and (2) How did the climate modify human activities and the natural signal provided by proxies?

Land-use change and autogenic processes

Our study shows that the direct human impact appeared in the Late Medieval Period, although settlements existed in the Bronze Age and the Iron Age in the nearby river valleys of Mroga and Mrożyca. Only one archaeological site was discovered in the immediate neighbourhood of the mire. Despite the late human influences, the peatland ecosystem completely changed since ca. AD 1350 together with the transformation of the landscape.

The Żabieniec mire is a classical example of plant succession in a former lake, which has progressed since the Late Glacial. In our study, we concentrated on the last stages of the terrestrialization process. During most of its history, this site was not directly disturbed by human activity. Human impact began to affect the peatland ecosystem quite recently. Deforestation and development of agriculture may lead to various trophic states and various types of vegetation.

One very important study was realized on floating bogs of southern Ontario by Warner et al. (1989). These authors showed an influence of deforestations on peatland ecosystems (water table fluctuations and vegetation change). Magyari et al. (2001) also interpreted the transition to higher mire water table as at least partly induced by gradually intensifying human activity in northeastern Hungary. The authors state that the periodic supply of nutrients together with human-induced water table increases may have delayed the autogenic succession. Other example of human impact to peatlands (vegetation change) was provided by Rybníček and Rybníčková (1974).

What is more important is that we can incorrectly believe that present peatland systems are on their natural path of development (Warner, 1996). The Żabieniec peatland (and possibly most other peatlands in central Poland) represents altered ecosystems, disturbed in the past by, e.g. draining, agriculture and exploitation. However, past deforestation has been an underestimated factor, because no precise palaeoenvironmental data of the recent peat deposits are available for most of the sites.

The habitat in Żabieniec was very wet and telmatic until ca. AD 600. Then, the water table decreased and the site transformed into a *Sphagnum*-dominated mire. This drying took place during the Early Medieval Period and might be interpreted as a decrease in the water table leading to oligotrophication.

The strongest evidence for the gradual increase in Human impact on the region was post-AD 1350 deforestation (beginning of the Late Medieval Period). Consequently, run-off and aeolian transport from exposed soils caused eutrophication which can be tracked through changes in pH. Geochemical results (increased values of magnesium, iron, potassium, zinc, as well as decrease of organic matter) obtained from the same core (Borówka et al., unpublished data) confirm our assumptions of soil erosion. Furthermore, chironomids and testate amoebae also clearly responded to the change in AD 1350. *Centropyxis aerophila* domination indicates minerotrophic conditions. The shell morphology of this taxon allows living in mineral soil (Foissner, 1987, 2000); therefore, its dominance can be an indicator of mineral deposition on a peatland surface. The date of ca. AD 1350 may be connected with the first mention about the Bielanki village in AD 1394 when forest exploitation became more intensive in the direct vicinity of the mire. Openness increased considerably through the Late Medieval and Modern Periods. During these periods, intensive development of agriculture was observed in central Poland, mainly in uplands (Twardy, 2008).

Other Polish studies of the recent peat cover (past 1–3 millennia) show how multidirectional peatland development can result in the development of different types of peatlands. One study from the Tuchola Forest (Lamentowicz et al., 2007) revealed an opposite (to the present study) response of peatland ecosystem to deforestations. In that case, forest cutting resulted in acidification and *Sphagnum* expansion during the last 200 years. This kettle-hole peatland is located in a sandy outwash plain covered by pine forest, where run-off from the acid soils led to pH decrease, which promoted *Sphagnum* establishment. Human impact appeared much later in this peatland than in the Żabieniec mire. Unfortunately, there is a shortage of high-resolution studies from this part of Europe to compare with the results from Żabieniec.

Climate—human or autogenic change?

Despite very pronounced human impact, it is probable that the Żabieniec peatland has also responded to climatic change. Until AD 600, the peatland was very wet, and then *Sphagnum* expanded (testate amoebae also increase rapidly in abundance at this time). This could have been a result of autogenic tendencies of the peatland to oligotrophication (Zobel, 1988) or of a decrease in the water table, caused by climatic change (Hughes & Barber, 2003, 2004). The location of this peatland in an area of continental climate influences suggests that temperature might be the most important parameter governing the peatland hydrology (Schoning et al., 2005; Charman, 2007), and the increase in temperature during the Medieval Warm Period may have influenced the water table. At present, the surface of the Żabieniec mire is flooded in wet years. Our proxies show that such flooding occurred also in the past. A good example for comparison is the previously mentioned kettle-hole peatland in Tuchola (Lamentowicz et al., 2008b), where such flooding took place in the past and is still observed today. However, the pattern of changes in Tuchola is different, as this mire acidified much earlier, ca. 5000 BP. In the case of kettle holes such as the ones in Żabieniec and Tuchola, acidification is not a synonym of ombrotrophication and it may have depended on soil leaching during the Holocene.

We suggest that during most of the history of Żabieniec, climate was a very important factor, but the signal was more recognizable as terrestrialization progressed. The dry shift during ca. AD 600 and relatively stable hydrological conditions were probably related to climate. However, this was too early for the Medieval Warm Period (which is usually dated to AD 800–1300), although during this time also, many other areas of the world experienced drought episodes (Bradley et al., 2003). Nevertheless, it is possible that in this part of Europe, the situation was different, and the Medieval Warm Period started earlier. Having no data from this part of Poland, we can compare our results with those obtained from northern Poland, from a Baltic bog in Stążki (Lamentowicz et al., 2009). At this site, the water table remained high until AD 1000–1100, and later, it decreased and became very unstable. The last part of the Early Medieval Age and the Late Medieval Age were very dry at Stążki, and this multiproxy study shows that at the beginning of the Little Ice Age, the record of climate change may have been modified human impact.

In the case of Żabieniec, intensified human impact was synchronous with that of the Little Ice Age (LIA). This causes difficulties in identifying the climate signal. Pronounced human impact occurs during the LIA, which is dated differently in various parts of the world. The LIA was recorded in many environmental archives in Europe (Mauquoy et al., 2002; Matthews, 2005; Weckström et al., 2006; Blass et al., 2007; van der Linden et al., 2008), and it is commonly dated to AD 1550–1850 (Bradley & Jones, 1992). In Poland, this important event is not well documented, but records of it may be more common than we suppose. A comparison of our wet shift dated to ca. 1350 AD can also be made with other European data such as those of Magny (2004) for central Europe, who dates a final phase of lake level increase to AD 1394.

The major hydrological shift at Żabieniec at AD 1350 corresponds to the Wolf minimum, suggesting that the shift was in response to reduced solar activity. The impact of the Maunder minimum (Shindell et al., 2001) is well documented because all proxies show that a very wet period occurred between AD 1500 and AD 1800. It is possible that climate was the decisive factor for human settlement in the vicinity of Żabieniec that is located on the morainic plateau. Formerly, settlements were only recorded in the river valleys. Due to the increase in wetland areas, people may have been forced to search for more suitable places for colonization.

The openness of the vegetation significantly increased in the period AD 1800–2006, which is well documented in the historical sources, but the water table decreased in the peatland. This intriguing dry shift may be interpreted as the end of the Little Ice Age. Until the early twentieth century, no peatland exploitation took place which indicates that any changes in the peatland were due to climate variability. This also confirms that not only did deforestation influence the water table in the peatland, but also that climate played a crucial role in the past. We excluded autogenic change as the reason of water table decrease because there are no *Sphagnum* hummocks in the mire surface. At present, *Sphagnum fallax* dominates in the moss layer. This species tolerates a very wide range of trophic conditions, and even a high input of phosphorus may not be disturbing (Limpens et al., 2003).

Acknowledgements The study was supported by a grant from the Polish Ministry of Science and Higher Education, No. 2P04E02228, 'Changes in the Environment of the Łódź Hills (Wzniesienia Łódzkie) during the Vistulian (Weichselian) and Holocene in the light of interdisciplinary palaeoecological research of the Żabieniec mire' (Principal Investigator: Jacek Forysiak). Mariusz Lamentowicz's activity was funded by the above-mentioned grant, as well as, a second grant from the Polish Ministry of Science and Higher Education, No. 2PO4G03228, 'Climatic changes in Pomerania (northwestern Poland) in the last millennium, based on the multiproxy high-resolution studies'. We thank Milena Obremska for her help in pollen diagram preparation and Sylwia Ufnalska for improving the English language. We also thank the three anonymous referees for their valuable comments.

References

Akima, H., 1970. A new method of interpolation and smooth curve fitting based on local procedures. Journal of the Association for Computing Machinery 17: 589–602.

Amesbury, M. J., D. J. Charman, R. M. Fyfe, P. G. Langdon & S. West, 2008. Bronze Age upland settlement decline in southwest England: testing the climate change hypothesis. Journal of Archaeological Science 35: 87–98.

Arrhennius, O., 1950. Főrhistorisk bebyggeise antydd genom kemisk analys. Fornvännen 45: 59–62.

Behre, K. E., 1981. The interpretation of anthopogenic indicators in pollen diagrams. Pollen and Spores 23: 225–245.

Berggren, G., 1968. Atlas of Seeds and Fruits of North-European Plant Species (Sweden, Norway, Denmark, East Fennoscandia and Iceland) with Morphological

Descriptions. Part 2, *Cyperaceae*. Swedish Natural Science Research Council, Stockholm.
Berglund, B. E. & M. Ralska-Jasiewiczowa, 1986. Pollen analysis and pollen diagrams. In Berglund, B. E. (ed.), Handbook of Holocene Paleoecology and Paleohydrology. Wiley, Chichester-Toronto: 455–484.
Blackford, J., 1993. Peat bogs as sources of proxy climatic data: past approaches and future research. In Chambers, F. M. (ed.), Climate Change and Human Impact on the Landscape. Chapman and Hall, New York: 47–56.
Blass, A., M. Grosjean, A. Troxler & M. Sturm, 2007. How stable are twentieth-century calibration models? A high-resolution summer temperature reconstruction for the eastern Swiss Alps back to AD 1580 derived from proglacial varved sediments. The Holocene 17: 51–63.
Blundell, A., D. J. Charman & K. Barber, 2008. Multiproxy late Holocene peat records from Ireland: Towards a regional palaeoclimate curve. Journal of Quaternary Science 23: 59–71.
Booth, R. K., M. Notaro, S. T. Jackson & J. E. Kutzbach, 2006. Widespread drought episodes in the western Great Lakes region during the past 2000 years: Geographic extent and potential mechanisms. Earth and Planetary Science Letters 242: 415–427.
Bradley, R. S. & P. D. Jones, 1992. When was the "Little Ice Age"? In Mikami, T. (ed.), International Symposium on the Little Ice Age Climate. Tokyo Metropolitan University, Tokyo, Japan: 1–4.
Bradley, R. S., K. R. Briffa, J. E. Cole, M. K. Hughes & T. J. Osborn, 2003. The Climate of the Last Millennium. In Alverson, K., R. Bradley & T. Pedersen (eds.), Paleoclimate, Global Change, and the Future. Springer-Verlag, Berlin, Heidelberg, New York: 105–142.
Bragg, O. M., 2002. Hydrology and peat-forming wetlands in Scotland. The Science of the Total Environment 294: 111–129.
Bronk Ramsey, C., 2001. Development of the Radiocarbon Program OxCal. Radiocarbon 43: 355–363.
Bronk Ramsey, C., 2008. Deposition models for chronological records. Quaternary Science Reviews 27(1–2): 42–60. doi:10.1016/j.quascirev.2007.01.019.
Brooks, S. J., P. G. Langdon & O. Heiri, 2007. The identification and use of Palaearctic chironomidae larvae in palaeoecology. QRA Technical Guide No. 10. Quaternary Research Association, London.
Chambers, F. M. & D. J. Charman, 2004. Holocene environmental change: contributions from the peatland archive. The Holocene 14: 1–6.
Chambers, F. M., D. Mauquoy, S. A. Brain, M. Blaauw & J. R. G. Daniell, 2007. Globally synchronous climate change 2800 years ago: proxy data from peat in South America. Earth and Planetary Science Letters 253: 439–444.
Charman, D. J., 2002. Peatlands and Environmental Change. Wiley, Chichester: 301 pp.
Charman, D. J., 2007. Summer water deficit variability controls on peatland water-table changes: implications for Holocene palaeoclimate reconstructions. The Holocene 17: 217–227.
Charman, D. & A. Blundell, 2007. A new European testate amoebae transfer function for palaeohydrological reconstruction on ombrotrophic peatlands. Journal of Quaternary Science 22: 209–221.
Charman, D. J., D. Hendon & W. A. Woodland, 2000. The Identification of Testate Amoebae (Protozoa: Rhizopoda) in Peats. Technical Guide No. 9. Quaternary Research Association, London: 147 pp.
Clarke, K. J., 2003. Guide to Identification of Soil Protozoa—Testate Amoebae. Freshwater Biological Association, Ambleside: 40 pp.
Crowley, P. H., 1992. Resampling methods for data analysis in computation-intensive ecology and evolution. Annual Review of Ecology and Systematics 23: 405–447.
Eidt, R., 1973. A rapid chemical field test for archaeological site surveying. American Antiquity 38: 207–209.
Faegri, K. & J. Iversen, 1989. Textbook of Pollen Analysis. Wiley, Chichester-Toronto.
Foissner, W., 1987. Soil protozoa: fundamental problems. Ecological significance, adaptations in ciliates and testaceans, bioindicators, and guide to literature. Progress in Protistology 2: 69–212.
Foissner, W., 2000. The *Centropyxis aerophila* complex (Protozoa: Testacea). Acta Protozoologica 39: 257–273.
Grospietsch, T., 1958. Wechseltierchen (Rhizopoden). Kosmos, Stuttgart: 1–86.
Hendon, D. & D. J. Charman, 1997. The preparation of testate amoebae (Protozoa: Rhizopoda) samples from peat. The Holocene 7: 199–205.
Hughes, P. D. M. & K. E. Barber, 2003. Mire development across the fen-bog transition on the Teifi floodplain at Tregaron Bog, Ceredigion, Wales, and a comparison with 13 other raised bogs. Journal of Ecology 91:253–264.
Hughes, P. D. M. & K. E. Barber, 2004. Contrasting pathways to ombrotrophy in three raised bogs from Ireland and Cumbria, England. The Holocene 14: 65–77.
Hughes, P. D. M., S. H. Lomas-Clarke, J. Schulz & P. Jones, 2007. The declining quality of late-Holocene ombrotrophic communities and the loss of Sphagnum austinii (Sull. ex Aust.) on raised bogs in Wales. The Holocene 17: 613–625.
Joosten, H. & D. Clarke, 2002. Wise Use of Mires and Peatlands—Background and Principles Including a Framework for Decision-making. International Mire Conservation Group and International Peat Society, Saarijärvi.
Juggins, S., 2003. C2 User guide. Software for Ecological and Palaeoecological Data Analysis and Visualisation. University of Newcastle, Newcastle upon Tyne, UK: 69 pp.
Kaczanowski, P. & J. K. Kozłowski, 1998. Great History of Poland, Vol. 1. The Oldest History of Poland Area (to VII Century). Fogra, Kraków: 382 pp.
Katz, N. J., S. V. Katz & M. G. Kipiani, 1965. Atlas and Keys of Fruits and Seeds Occurring in the Quaternary Deposits of the USSR. Nauka, Moskva: 367 pp.
Katz, N. J., S. V. Katz & E. I. Skobiejeva, 1977. Atlas of Macrofossils from Peats. Nedra, Moskva: 370 pp.
Klink, A. G. & H. K. M. Moller Pillot, 2003. Chironomidae Larvae. Key to the Higher Taxa and Species of the Lowlands of Northwestern Europe 1.0 CD-ROM. ETI.
Kłysik, K., 2001. Climatic conditions. In Liszewski, S. (ed.), Monograph of Łódź Voivodeship. Łódzkie Towarzystwo Naukowe, Łódź. (in Polish).
Lagerback, R. & A. N. N. M. Robertsson, 1988. Kettle holes: stratigrafical archives for Weichselian geology and

palaeoenvironment in nothernmost Sweden. Boreas 17: 439–468.

Lamentowicz, M. & E. A. D. Mitchell, 2005. The ecology of testate amoebae (Protists) in Sphagnum in north-western Poland in relation to peatland ecology. Microbial Ecology 50: 48–63.

Lamentowicz, M., K. Tobolski & E. A. D. Mitchell, 2007. Palaeoecological evidence for anthropogenic acidification of a kettle-hole peatland in northern Poland. The Holocene 17: 1185–1196.

Lamentowicz, Ł., M. Lamentowicz & M. Gąbka, 2008a. Testate amoebae ecology and a local transfer function from a peatland in western Poland. Wetlands 28: 164–175.

Lamentowicz, M., M. Obremska & E. A. D. Mitchell, 2008b. Autogenic succession, land-use change, and climatic influences on the Holocene development of a kettle hole mire in Northern Poland (Northern Poland). Review of Palaeobotany & Palynology 151: 21–40.

Lamentowicz, M., A. Cedro, G. Miotk-Szpiganowicz, E. A. D. Mitchell, J. Pawlyta & T. Goslar, 2008c. Last millennium palaeoenvironmental changes from a Baltic bog (Poland) inferred from stable isotopes, pollen, plant macrofossils and testate amoebae Palaeogeography, Palaeoclimatology. Palaeoecology 265: 93–106.

Lamentowicz, M., K. Milecka, M. Gałka, A. Cedro, J. Pawlyta, N. Piotrowska, Ł. Lamentowicz & W. O. van der Knaap, 2009. Climate- and human-induced hydrological change since AD 800 in an ombrotrophic mire in Pomerania (N Poland) tracked by testate amoebae, macro-fossils, pollen, and tree-rings of pine. Boreas 38: 214–229.

Limpens, J., B. M. Hilde & F. Berendse, 2003. Expansion of *Sphagnum fallax* in bogs: striking the balance between N and P availability. Journal of Bryology 25: 83–90.

Magny, M., 2004. Holocene climate variability as reflected by mid-European lake-level fluctuations and its probable impact on prehistoric human settlements. Quaternary International 113: 65–79.

Magyari, E., P. Sumegi, M. Braun, G. Jakab & M. Molnar, 2001. Retarded wetland succession: anthropogenic and climatic signals in a Holocene peat bog profile from north-east Hungary. Journal of Ecology 89: 1019–1032.

Matthews, J. A., 2005. 'Little Ice Age' glacier variations in Jotunheimen, southern Norway: a study in regionally controlled lichenometric dating of recessional moraines with implications for climate and lichen growth rates. The Holocene 15: 1–19.

Mauquoy, D. & B. van Geel, 2007. Mire and Peat Macros. In A., E. S. (ed.), Vol. 3. p. 2315–2336, Elsevier.

Mauquoy, D., B. van Geel, M. Blaauw & J. van der Plicht, 2002. Evidence from northwest European bogs shows 'Little Ice Age' climatic changes driven by variations in solar activity. The Holocene 12: 1–6.

Mitchell, E. A. D., 2003. Identification keys for testate amoebae. http://wslar.epfl.ch/mitchell/edward/Identification_keys/Keys.htm.

Mitchell, E. A. D., W. O. van der Knaap, J. F. N. Leeuwen, A. Buttler, B. G. Warner & J. M. Gobat, 2001. The palaeoecological history of the Praz-Rodet bog (Swiss Jura) based on pollen, plant macrofossils and testate amoebae (Protozoa). The Holocene 11: 65–80.

Ogden, C. G. & R. H. Hedley, 1980. An Atlas of Freshwater Testate Amoebae. British Museum (Natural History) and Oxford University Press, (London and Oxford), London: 222 pp.

Ohryzko-Włodarska, C., 1972. Transformation of the Village of the Polish Kingdom in Years 1846–1870. An Example of Brzeziny District. Ludowa Spółdzielnia Wydawnicza, Warszawa: 364 pp. (in Polish).

Oksanen, P. O. & M. Valiranta, 2006. Palsa mires in a changing climate. Suoseura 57: 33–43.

Ralska-Jasiewiczowa, M. & M. Latałowa, 1996. Poland. In Berglund, B. E., H. J. B. Birks, M. Ralska-Jasiewiczowa & H. E. Wright Jr (eds.), Palaeoecological Events During the Last 15000 years: Regional Syntheses of Palaeoecological Studies of Lakes and Mires in Europe. Wiley, Chichester: 403–472.

Ralska-Jasiewiczowa, M., M. Latałowa, K. Wasylikowa, K. Tobolski, T. Madeyska, H. E. Wright & C. Turner Jr, 2004. Late Glacial and Holocene History of Vegetation in Poland Based on Isopollen Maps. W. Szafer Institute of Botany, Polish Academy of Sciences, Kraków.

Reimer, P. J., M. G. L. Baillie, E. Bard, A. Bayliss, J. W. Beck, C. J. H. Bertrand, P. G. Blackwell, C. E. Buck, G. S. Burr, K. B. Cutler, P. E. Damon, R. L. Edwards, R. G. Fairbanks, M. Friedrich, T. P. Guilderson, A. G. Hogg, K. A. Hughen, B. Kromer, G. McCormac, S. Manning, C. B. Ramsey, R. W. Reimer, S. Remmele, J. R. Southon, M. Stuiver, S. Talamo, F. Taylor, J. van der Plicht & C. Weyhenmeyer, 2004. IntCal04 terrestrial radiocarbon age calibration, 0–26 cal kyr BP. Radiocarbon 46: 1029–1058.

Rolland, N. & I. Larocque, 2007. The efficiency of kerosene flotation for extraction of chironomid head capsules from lake sediments samples. Journal of Paleolimnology 37: 565–572.

Rybníček, K. & E. Rybníčková, 1974. The origin and development of waterlogged meadows in the central part of the Sumava Foothills. Folia Geobotanica & Phytotaxonomica. 9: 45–70.

Rydin, H. & J. Jeglum, 2006. The Biology of Peatlands. Oxford University Press, Oxford: 343 pp.

Schoning, K., D. J. Charman & S. Wastegård, 2005. Reconstructed water tables from two ombrotrophic mires in eastern central Sweden compared with instrumental meteorological data. The Holocene 15: 111–118.

Shindell, D. T., G. A. Schmidt, M. E. Mann, D. Rind & A. Waple, 2001. Solar forcing of regional climate change during the Maunder Minimum. Science 294: 2149–2152.

Sillasoo, Ü., D. Mauquoy, A. Blundell, D. Charman, M. Blaauw, J. R. G. Daniell, P. Toms, J. Newberry, F. M. Chambers & E. Karofeld, 2007. Peat multi-proxy data from Männikjärve bog as indicators of late Holocene climate changes in Estonia. Boreas 36: 20–37.

Tallis, J. H., 1983. Changes in wetland communities. In Gore, A. J. P. (ed.), Mires: Swamp, Bog, Fen and Moor. Elsevier, Amsterdam: 311–347.

Tobolski, K. 2000. Przewodnik do oznaczania torfów i osadów jeziornych (A guide for identification of peat and lake sediments). Wydawnictwo Naukowe PWN, Warszawa, 508 pp.

Twardy, J., 2008. Transformacja rzeźby centralnej części Polski Środkowej w warunkach antropopresji. Transformation of the Relief of the Central Poland in Conditions of the Human Impact. Wydawnictwo Uniwersytetu Łódzkiego, Łódź. (in Polish).

Vallenduuk, H. J. & H. K. M. Moller Pillot, 2007. Chironomidae Larvae of the Netherlands and Adjacent Lowlands. General Ecology and Tanypodinae. KNNV Publishing, Zeist, The Netherlands: 144 pp.

van Breemen, N., 1995. How *Sphagnum* bogs down other plants. Trends in Ecology and Evolution 10: 270–275.

van der Linden, M., J. Barke, E. Vickery, D. J. Charman & B. van Geel, 2008. Late Holocene human impact and climate change recorded in a North Swedish peat deposit. Palaeogeography, Palaeoclimatology, Palaeoecology 258: 1–27.

Warner, B. G., 1993. Palaeoecology of floating bogs and landscape change in the Great Lakes drainage basin of North America. In Chambers, F. M. (ed.), Climate Change and Human Impact on the Landscape. Chapman and Hall, New York: 237–245.

Warner, B. G., 1996. Vertical gradients in peatlands. In Mulamoottil, G., B. G. Warner & A. E. McBean (eds.), Wetlands. Environmental Gradients, Boundaries, and Buffers. Lewis Publisher, Waterloo: 45–66.

Warner, B. G., H. J. Kubiw & K. I. Hanf, 1989. An anthropogenic cause for quaking mire formation in southwestern Ontario. Nature 340: 380–384.

Weckström, J., A. Korhola, P. Erästö & L. Holmström, 2006. Temperature patterns over the past eight centuries in Northern Fennoscandia inferred from sedimentary diatoms. Quaternary Research 66: 78–86.

Wiederholm, T., 1983. Chironomidae of the Holarctic Region. Keys and Diagnoses. Larvae, Vol. 1. Entomological Society of Lund, Sweden: 457 pp.

Williams, M., 2000. Dark ages and dark areas: global deforestation in the deep past. Journal of Historical Geography 26: 28–46.

Willis, K. J., M. B. Araujo, K. D. Bennett, B. Figueroa-Rangel, C. A. Froyd & N. Myers, 2007. How can a knowledge of the past help to conserve the future? Biodiversity conservation and the relevance of long-term ecological studies. Phil Trans R Soc 362: 175–186.

Wilson, R. S. & L. P. Ruse, 2005. A guide to the identification of genera of chironomid pupal exuviae occurring in Britain and Ireland (including common genera from Northern Europe) and their use in monitoring lotic and lentic fresh waters. Freshwater Biological Association Special Publications 13, 176 pp.

Woś, A., 1999. Klimat Polski. Climate of Poland. Wydawnictwo Naukowe PWN, Warszawa: 302 pp. (in Polish).

Yeloff, D. & D. Mauquoy, 2006. The influence of vegetation composition on peat humification: implications for palaeoclimatic studies. Boreas 35: 662–673.

Yeloff, D., P. Broekens, J. Innes & B. v. Geel, 2007. Late Holocene vegetation and land-use history in Denmark: a multi-decadally resolved record from Lille Vildmose, northeast Jutland. Review of Palaeobotany and Palynology 146: 182–192.

Yu, Z., 2006. Holocene carbon accumulation of fen peatlands in boreal western Canada: a complex ecosystem response to climate variation and disturbance. Ecosystems 9: 1278–1288.

Zajączkowski, S. & S. M. Zajączkowski, 1966. Materiały do słownika geograficzno-historycznego dawnych ziem łęczyckiej i sieradzkiej do 1400 roku, Część I. (Materials for geographical–historical dictionary of old Łęczyce and Sieradz lands until AD 1400. Part I (Abramowice – Mzurki)) Łódź (in Polish).

Zobel, M., 1988. Autogenic succession in boreal mires–a review. Folia Geobotanica & Phytotaxonomica 23: 417–445.

Sedimentary multiproxy response to hydroclimatic variability in Lagunillo del Tejo (Spain)

Lidia Romero-Viana · M. Rosa Miracle · Charo López-Blanco · Estela Cuna · Gloria Vilaclara · Jordi Garcia-Orellana · Brendan J. Keely · Antonio Camacho · Eduardo Vicente

Originally published in the journal Hydrobiologia, Volume 631, No. 1, 231–245.
DOI: 10.1007/s10750-009-9813-x

Abstract Lagunillo del Tejo is a small groundwater-fed sinkhole lake in the karst region of the Iberian Range (central-eastern Spain), which undergoes significant lake level fluctuation in response to rainfall variability. The aim of this study is to understand the record of water level fluctuations in Lagunillo del Tejo over the last two-and-a-half centuries. This information could be used in future studies to interpret longer sedimentary sequences. We analysed photosynthetic pigments, diatoms and cladoceran remains in sediment sequences recovered from the deepest part of the lake. The paleoecological proxies traced two different communities which have switched their prevalence during the past: (1) a planktonic community of algae, including diatoms, chlorophytes, cryptophytes and cyanobacteria, and phototrophic bacteria associated with higher lake level and water column seasonal stratification; (2) a littoral community with the higher levels of macrophyte pigments and associated epiphytic diatoms and chydorids, all of which indicate lower lake level. The levels of coherence between different proxies, each having an independent mechanistic link to lake-level variability, enhance the reliability of palaeolimnological inferences. The high-resolution stratigraphical data from the upper part of the core was compared with lake-level inferences from instrumental rainfall series (1859–2005) to establish the correspondence between Lagunillo del Tejo sediment sequences and climate record.

Guest editors: K. Buczkó, J. Korponai, J. Padisák & S. W. Starratt
Palaeolimnological Proxies as Tools of Environmental Reconstruction in Fresh Water

L. Romero-Viana (✉) · M. R. Miracle · C. López-Blanco · A. Camacho · E. Vicente
Department of Microbiology and Ecology, University of Valencia, 46100 Burjassot, Valencia, Spain
e-mail: lidia.romero@uv.es

E. Cuna · G. Vilaclara
Facultad de Estudios Superiores Iztacala, Universidad Nacional Autónoma de México, CP-54090 Mexico, Mexico

J. Garcia-Orellana
Departament de Física – Institut de Ciència i Tecnologia Ambientals, Universitat Autònoma de Barcelona, 08193 Bellaterra, Spain

J. Garcia-Orellana
School of Marine and Atmospheric Sciences, State University of New York, Stony Brook, NY 11794-5000, USA

B. J. Keely
Department of Chemistry, University of York, Heslington, York Y010 5DD, UK

Keywords Iberian Peninsula · Lake-level fluctuation · Cladoceran remains · Photosynthetic pigments · Diatoms · Lake sediments

Introduction

The past few decades have witnessed the development of a large variety of paleolimnological approaches to track past water-depth changes, such as geomorphological evidence (Digerfeldt, 1986; Benson et al., 1991), changes in sediment lithology (Almquist-Jacobson, 1995; Rodó et al., 2002; Magny et al., 2007) and a wide spectrum of paleoecological techniques, including diatoms (i.e. Fritz, 1990; Battarbee et al., 2001). In spite of inherent difficulties, most of them related to the chronology of sedimentary sequences, such as changes in sediment accumulation rates and hiatuses due to sediment erosion during desiccation periods (Verschuren, 2003), numerous studies have confirmed the suitability of these methodologies.

In semi-arid and semi-humid regions, lake-level changes may be closely linked to climatic variability, particularly in closed basins where the links between climate and lake level are controlled by the balance between precipitation and evapotranspiration, but may also include groundwater regimes (Dearing, 1997). In this study, we analysed three short sediment records from Lagunillo del Tejo (central-eastern Spain) recovered between 2003 and 2008. Lagunillo del Tejo is a small karstic closed-basin lake sensitive to Mediterranean hydrological regimes and is fed mainly by groundwater. Because of its morphological features, the planktonic community is segregated from a plant-associated community inhabiting a highly developed littoral ring of macrophytes. During the last three decades, maximum depth has varied from 11 to 3.5 m with a clear trend towards lower lake levels. On the other hand, limnological surveys during this period (Vicente & Miracle, 1984 and unpublished authors' observations) have indicated that it lodged a changing algal and phototrophic bacterial community that we suspect as being directly related to lake-level fluctuation (Romero-Viana et al., 2009). Moreover, the recent lowering has resulted in a reduction of the macrophyte ring.

The aim of this study is to understand how Lagunillo del Tejo has recorded water-level fluctuations over the last two-and-a-half centuries, knowledge that could be used to interpret longer sedimentary sequences. Lake-level fluctuations may have an overriding effect on lacustrine ecology, mainly due to their structural role in determining the spatial and temporal extension and functioning of the planktonic–littoral and aquatic–terrestrial transition zones (Coops et al., 2003). Therefore, we have analysed photosynthetic pigments, diatoms and cladoceran remains as potential tracers of changes in the relative significance of planktonic-to-littoral communities. The high-resolution stratigraphical data obtained in this study and the comparison of qualitative lake-level inferences with instrumental rainfall series (1859–2005) provided strong evidence that Lagunillo del Tejo preserves a reliable climate record.

Site description

Lagunillo del Tejo (Fig. 1) is one of the seven doline lakes formed by dissolution of Cenomanian and Turonian dolostones that sub-horizontally overlie impermeable Cenomanian marls in Cañada del Hoyo (Cuenca, Spain).The lake, fed mainly by groundwater, is subject to marked water-level fluctuations. At the time of the corings during May 2003, May 2005 and March 2008, Lagunillo del Tejo had a maximum depth of 8, 6 and 4 m, respectively. In May 2003 its diameter was 72 m. The lake is monomictic, which is thermally stratified from May to November, and an anoxic zone develops over the years with high water level (Vicente & Miracle, 1984 and unpublished data). The waters are bicarbonate rich, with a pH around 9 in the epilimnetic waters, and a conductivity around 600 μS cm^{-1}.The order of major ion concentration is $HCO_3^- \gg SO_4^{2-} > Cl^-$ and $Mg^{++} \gg Ca^{++} > Na^+$. However, during thermal stratification, pH decreases to 7.5–7, and conductivity may reach 900–1000 μS cm^{-1} in the anoxic hypolimnion.

Phototrophic bacterial populations were reported to occur in the anoxic layer (Vicente & Miracle, 1984). The phototrophic algal biomass included diatoms and chlorophytes that grew in the epilimnion and metalimnion, and dense cyanobacterial and cryptophyta populations that developed at the oxic–anoxic interface. Littoral macrophytic vegetation included *Potamogeton pectinatus*, as main species in the inner ring and *Myriophyllum spicatum, Polygonum amphibium*, and *Chara* spp. (Cirujano, 1995),

Fig. 1 Geographical location of Lagunillo del Tejo (*black arrow*), bathymetry of Lagunillo del Tejo (May 2003) and location of the coring site. The three pictures show the progressive decrease of lake level from May 2003 to May 2006

structuring a more diverse community in the outer ring near the shoreline. In spring time, macrophytes may be also covered by filamentous metaphyton (*Spirogyra* and other green alga). During the driest periods when the lake-level drops to 4-m of maximum depth, the littoral zone become a single macrophyte ring composed mainly of *P. pectinatus* while the outer ring completely dry up.

The study area is characterised by a Mediterranean climate with a typical seasonal pattern of very dry, hot summers and cooler, rainier winters. Total annual rainfall is 525 ± 123 mm (mean of instrumental data series 1950–2003 from the nearby town of Cuenca). Regional winter precipitation, contributing at 50% of the total amount, is highly correlated with the phase of the North Atlantic Oscillation (NAO) (Romero-Viana et al., 2008). The annual mean evapotranspiration in the lake area is approximately 130 mm with monthly maxima over 200 mm in summer and monthly minima below 50 mm in winter. Mean monthly temperature ranges from 5.6°C in the coldest month (January) to 25°C in the warmest month (July). Monthly temperatures variations can be quite extreme and differences between day and night are also very important, especially in summer, indicating the continental character of its climate.

Materials and methods

Sediment coring and sampling

Sediment cores were recovered from a securely moored raft (fixed with cables to shore elements) in the central and the deepest point of Lagunillo del Tejo on three different occasions, namely, May 2003, May 2005, and March 2008. The first two were recovered using a Phleger gravity corer (Kahl Scientific Instruments) of 3.5-cm diameter and 80-cm length. The cores, extracted in a methacrylate cylinder, were immediately protected from light by wrapping in foil and stored in a cold chamber. The 2003 core CN-1 was sliced into 2–3-mm sections sealed in sterile "Whirlpack" bags and conserved at −20°C in darkness until pigment and diatom analyses. The 2005 core CN-2 was sliced into 0.5-cm sections in the uppermost 5 cm, and into 1-cm sections from 5 cm to the core bottom. The samples placed in plastic bags were stored until radionuclide analyses. Finally, 2008 core CN-3 was recovered by means of a chamber corer (Eijkelkamp) of 4-cm diameter and 50-cm length in two steps. The lithology of the CN-3 core was described in the field. The sediment core was deposited in a plastic half-cylinder, wrapped tightly with film and stored in a refrigerator until sliced into 1-cm sections for cladoceran analysis. Unlike the gravity corer, the chamber corer type allowed the retrieval of uncompressed sediment.

Analytical methods

Water content and density were measured in the three cores. Water content was determined by oven-drying aliquots of wet sediment for 2 h at 105°C. Density was calculated as wet sediment weight normalised by the known volume of wet sediment aliquots. The organic matter content was determined in CN-1 and CN-3 from dried samples by loss-on-ignition for 6 h at 460°C (APHA, 1992) and expressed as percentage of dry matter. Sediment dating involved ^{210}Pb, ^{137}Cs and ^{226}Ra measurements carried out on core CN-2 at Universitat Autònoma de Barcelona (UAB). Determination of ^{210}Pb activities was accomplished through the measurement of its daughter nuclide, ^{210}Po, following the methodology described by Sánchez-Cabeza et al. (1998). In brief, after addition of a given amount of ^{209}Po as internal tracer, sediment aliquots of 200–300 mg of each sample were totally dissolved in acid medium by using an analytical microwave oven. Polonium isotopes were plated onto pure silver discs and counted with PIPS α-spectrometers (CANBERRA, Model PD-450.18 AM). Determinations of ^{226}Ra (via ^{214}Pb through its 351 keV emission line) and ^{137}Cs were done by γ-spectrometry, using a high-purity well-type Ge detector (CANBERRA).

Samples for photosynthetic pigment analysis of CN-1 sediment core were extracted with acetone. Acetone extracts were treated with diazomethane to methylate free acid groups (Airs et al., 2001), dried under a stream of N_2 and stored at 4°C until liquid chromatography (HPLC) analysis was performed as described by Airs et al. (2001; method A). The mobile phase gradient used a mixture of four solvents: NH_4Ac (0.01 M), MeOH, MeCN and EtAc. A detailed description of these analyses as well as the LC-MS methods used to confirm pigment identification are given in Romero-Viana et al. (2009). Pigment concentrations were expressed in micrograms per gram of organic matter ($\mu g\ g_{om}^{-1}$).

Diatom analyses were carried out on high resolution (2–3 mm) sediment samples of CN-1 core. The same samples used for acetone pigment extraction were dried before weighing 0.5 g, which were digested in a hot (<100°C) oxidant/acid mixture (HCl, H_2O_2 and HNO_3) to eliminate organic matter and carbonates (Battarbee, 1986). All the samples were repeatedly settled, poured off and rinsed. A known fraction of the resultant slurries was dried onto coverslips, and mounted with Naphrax®. At least 400 valves were counted in samples with adequate diatom abundance and preservation, with a phase contrast Zeiss Microscope, at 1000× magnification. In samples relatively devoid of diatoms, counting was limited to fewer valves, which revealed the dominant taxa, despite information loss. Diatoms were identified to the lowest possible taxonomic level using a scanning electron microscopy (JEOL JSM-6380LB). Previous centrifugation for pigment extraction resulted in a high proportion of diatom breakage, which put an extra difficulty in identification; nevertheless, it was yet possible to identify them, because many individuals remained unaltered, and valve counting on broken valves was performed on central pieces that allowed recognition of taxa. Taxonomic and autoecological information for diatom taxa were obtained from several sources, including Hustedt (1930, 1959–1966), Cholnoky (1968), Lowe (1974), Gasse (1986), Krammer & Lange-Bertalot (1986, 1988, 1991a, b), Krammer (1997a, b, 2000, 2002), Lange-Bertalot & Krammer (1987, 1989), Lange-Bertalot (2001), Round et al. (1991) and Round & Bukhtiyarova (1996).

For cladoceran analysis, CN-3 sediment samples of 1 cm^3 were heated in 10% KOH solution (<100°C) using a thermostatic heating plate on an orbital shaking for 30 min. After the first 5 min of treatment, ultrasonic waves were applied for 30 s to enhance cleaning. The samples were then sieved though a 40-μm mesh and the residue was transferred from the sieve back into the beaker with a stream of water from a wash bottle. Some drops of glycerol–safranin were added. The samples were counted in a Petri dish by means of an inverted microscope Olympus U-PMTVC. A minimum of 200 remains of the most abundant species were counted. The total number of individuals was estimated as the maximum count of head shields, post abdomens or caparaces. Identifications and ecological characteristics of the species found were obtained from Frey (1959, 1962), Margaritora (1985), Alonso (1996) and Szeroczynska & Sarmaja-Korjonen (2007). Major stratigraphic diatoms and subfossil cladocera zones were identified using a constrained cluster analysis CONISS performed with the computer program TILIA v. 2.02 (Grimm, 1987).

Results

Lithology and sediment core physical characteristics

The cores showed dark mud-reduced layers, R zones, alternating with brown oxidised ones, O zones (Fig. 2). Some layers of more coarse material, included in O zones, were observed at some particular depths (O3 and O4) along the core. Distinct colour changes and also water and organic content were used as criteria for core correlation. The cores, CN-1 and CN-2, which were recovered with a gravity corer, showed similar depth profiles of water content, density and LOI, and therefore, CN-2 data have not been plotted in Fig. 2. The described zones in cores CN-1 and CN-2 shared similar thickness, but CN-3 core, which was retrieved without compression, showed greater thickness. The O zones were characterised by higher density and lower water and organic content. Water content values in R zones ranged from 40% to 60% and values were above 60% at some depths (Fig. 2).

Chronological model

The unsupported ^{210}Pb (Fig. 3A) data showed the exponential curve expected from radioactive decay that is interrupted around the 4 g cm^{-2} (4.75 cm) because of the presence of a sediment interval (O5 zone) of detritic material from 5 to 12 m of depth, where ^{210}Pb activity remains roughly constant and slightly over supported ^{226}Ra. This anomalous 210Pb profile is detailed in Romero-Viana et al. (2009, Fig. 4). However, total ^{210}Pb activity slightly increases after 13-cm depth before reaching a secular equilibrium with the supported ^{226}Ra that could be set

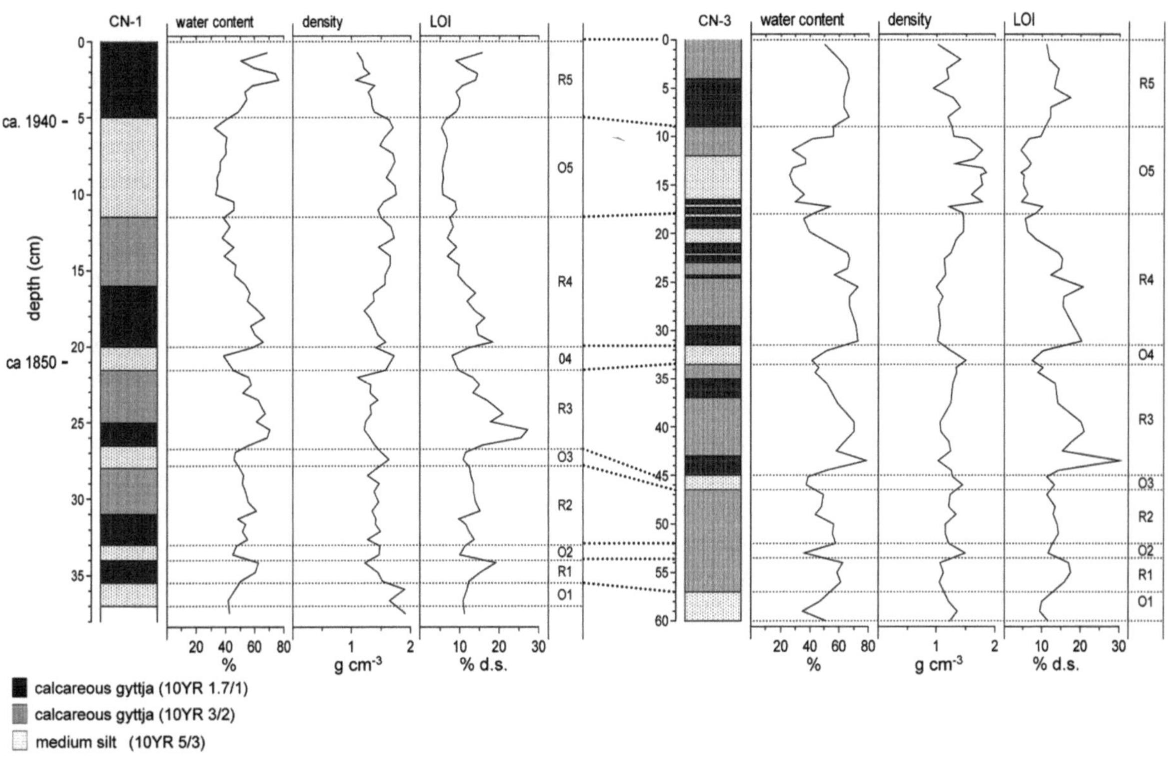

Fig. 2 Physical characteristics of Lagunillo del Tejo sediment cores, CN-1 and CN-3; lithology, water content (%), density (g cm^{-3}) and LOI (% of dry sediment). *Dashed lines* represent core correlations of the main sedimentary units. Some age indications are shown at the *left*

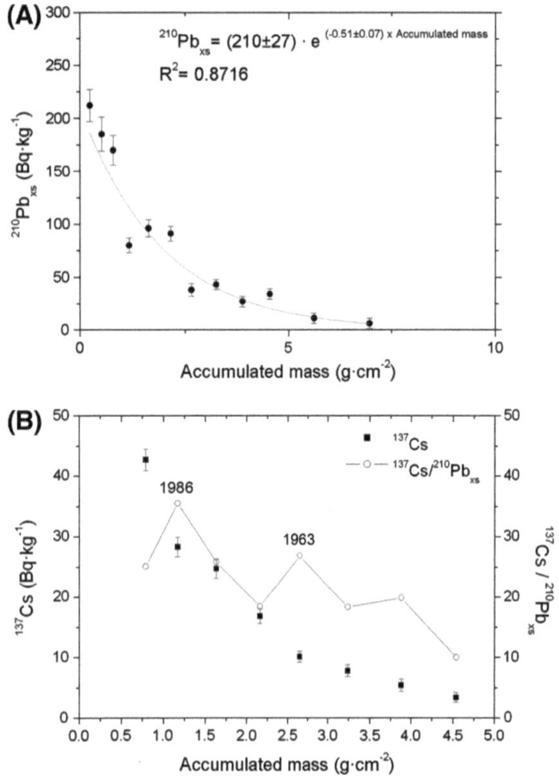

Fig. 3 Radionuclide activities (Bq kg^{-1}) from core CN-2. **A** Regression line of unsupported ^{210}Pb versus accumulated mass (from sediment surface to 10 g cm^{-2}). **B** ^{137}Cs (*black squares*), and ^{137}Cs to ^{210}Pb excess ratio in the uppermost 5 g cm^{-2}

at around 20 cm of depth. Chronology for each section was determined by using CRS model (Appleby & Oldfield, 1978; Appleby, 2001) without considering the allochthonous material from 5 to 12 depth interval. Mean accumulation rate in the non-perturbed areas was 64 mg cm^{-2} yr^{-1} (0.8 mm yr^{-1}).

The maximum ^{137}Cs concentration was observed at the top of the core, decreasing to 5-cm depth, and did not show any subsurface maximum. ^{137}Cs activity versus depth (Fig. 3B) did not present a well-resolved peak because of the 1963 fallout maximum from the atmospheric testing of nuclear weapons (1963), or because of the Chernobyl accident (1986). Moreover, dilution events could distort the ^{137}Cs profile and under these circumstances, the ^{137}Cs/^{210}Pb ratio is used as a better tool to validate the ^{210}Pb dating. Figure 3B presents the ^{137}Cs/^{210}Pb ratio against the ^{210}Pb-derived chronology, showing two well-defined maxima at 1.75 and 3.25 cm, which may correspond to 1986 and 1963, respectively. These ^{137}Cs-derived ages agree with the ^{210}Pb chronology, discarding a possible mobility of ^{137}Cs with depth.

Photosynthetic pigments profiles, diatom and cladoceran assemblages

Up to 60 different compounds were identified in Lagunillo del Tejo sediment core (see Romero-Viana et al., 2009, for a detailed description). Among them, specific carotenoids were used as tracers of algal and bacterial populations (Rowan, 1989; Jeffrey & Vesk, 1997). The depth profiles of specific carotenoids (Fig. 4) derived from planktonic algal populations such as cryptophyceae, chlorophyceae, and also cyanobacteria, followed similar patterns, but cyanobacterial signal was remarkably higher in the lowermost part of zone R4 and upper part of R3 (23–18-cm depth). On the other hand, specific carotenoid for macrophytes showed opposite depth profiles to those of planktonic algae, reaching maxima at the bottom of the R3 zone and in the middle of R4, where the planktonic signal was less apparent. Among non-specific carotenoids which probably have more than one principal source (diatoms, Chrysophyceae, dinoflagellates), diadinoxanthin had a significant positive correlation ($r = 0.56$, $n = 34$, $P < 0.01$) with diatom valve concentrations, which confirms diatoms as being the main source. Purple and green sulphur bacteria derivatives were also present throughout the sedimentary sequence. Pigment derivatives from purple sulphur bacteria were coincident with increases in the abundances of chlorins and planktonic algal and cyanobacterial carotenoids, except in R3 zone, where the signal from chromatiaceae and zeaxanthin were not coincident. The analytical techniques employed confirm the presence of a cluster of compounds identified as steryl chlorin esters (SCEs). Steryl chlorin esters are biotransformation products formed by zooplankton herbivory (Harris et al., 1995; Harradine et al., 1996). Biomarkers of grazing activity were present throughout the core, pointing to some relationship with macrophyte carotenoids (e.g. maximum SCEs concentration at the bottom of the R3 zone coincides with the macrophyte carotenoid maximum).

In the high-resolution (2–3 mm) samples from CN-1 sediment core, 34 diatom taxa (26 genera) were identified (Fig. 5). Absolute diatom concentrations were low, ranging from negligible at 31-cm depth, to 19.72 × 10^{6} valves g^{-1} dry sediment (d.s.) at 2-cm

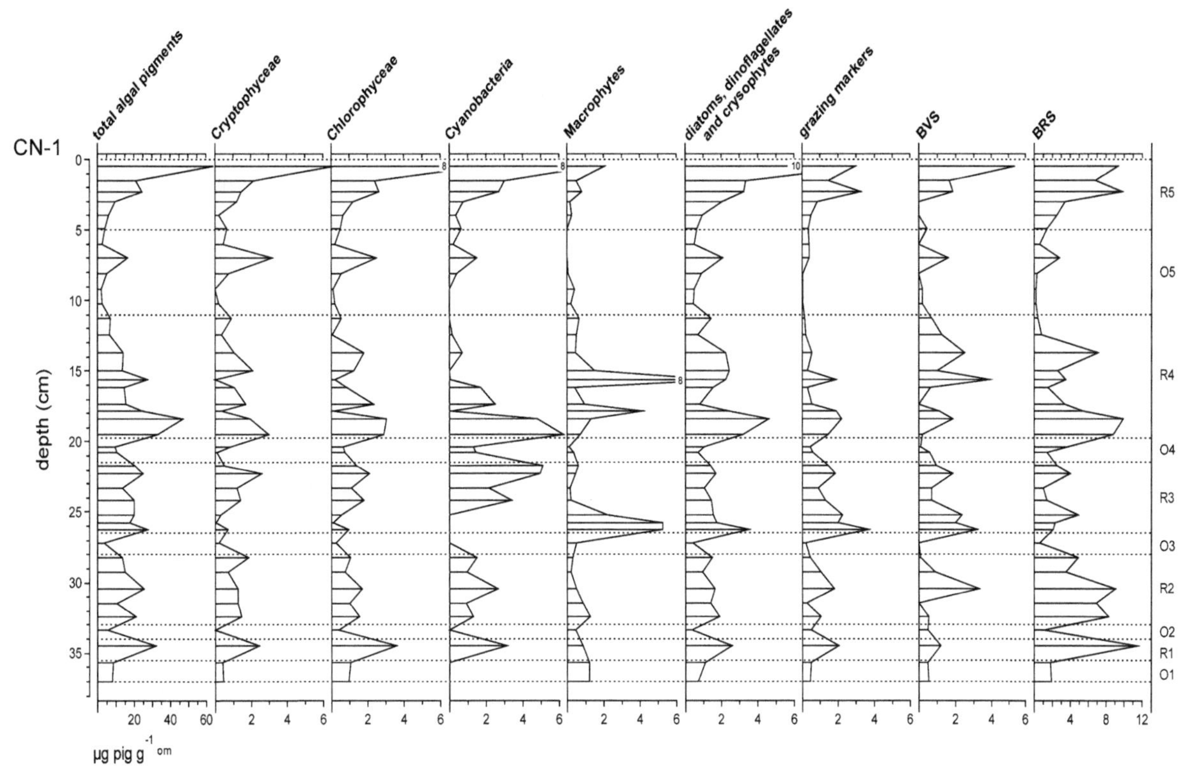

Fig. 4 Stratigraphic profiles of photosynthetic pigments ($\mu g\ g_{om}^{-1}$) from core CN-1: Total algal pigments including chlorin derivatives and carotenoids, Cryptophyceae markers: alloxanthin + crocoxanthin, Chlorophyceae markers: lutein + crytoxanthin, Cyanophyceae markers: zeaxanthin + echinenone, Macrophyte markers: lutein-5,6-epoxy + lactucaxanthin, Diatoms, dinoflagellates and crysophytes markers: diadinoxanthin + fucoxanthin + diatoxanthin, Steryl chlorin esters (SCEs) as grazing markers, green sulfur bacteria (GSB) markers: bchl *d* homologues + chlorobactene and purple sulfur bacteria (PSB) markers: bchl *a* derivatives + okenone

depth. Diatom assemblages were mainly dominated by peri-epiphytic genera, and characterised by mesotrophic and alkaliphylic taxa. Based on CONISS analysis, the Lagunillo del Tejo sequence can be divided into five main diatom zones (DZ). DZ-1 and DZ-2 (37–19 cm) had a relatively low number of valves ($0.04 \times 10^6 - 0.9 \times 10^6$ valves g^{-1}d.s.), and were highly heterogeneous. The first zone DZ-1 (37–31 cm) showed an alternation of planktonic (*Aulacoseira* spp.) and peri-epiphytic taxa (*Sellaphora pupula, Nitzschia frustulum* var. *perminuta* and *Navicula* aff. *radiosa*). The next zone, DZ-2 (31–19 cm), suggests important changes in environmental conditions, showing fluctuations from epiphytic to planktonic diatoms, such as *Cyclotella*, which became abundant in specific periods, whereas *Aulacoseira granulata* disappeared. In addition, the aerophilous species *Luticola mutica* and *Hantzschia amphioxys* became relatively important. The upper three zones, DZ-3, DZ-4 and DZ-5 (from 19 to 0 cm) were dominated mostly by peri-epiphytic species (*Amphora lybica, Anomoeoneis sphaerophora, Navicula* aff. *radiosa, Sellaphora pupula* and *Achnanthidium minutissimum*). Zone DZ-3 (19–8.5 cm) displays higher absolute valve numbers than before ($1.4 \times 10^6 - 16.7 \times 10^6$ valves g^{-1}d.s.) having prominent proportions of *Navicula* aff. *radiosa* and *Achnanthidium minutissimum*; on the other hand, DZ-4 (8.5–4.5 cm) showed again low valve numbers ($0.2 \times 10^6 - 1.8 \times 10^6$ valves g^{-1} d.s.), which was characterised by aerophilous species and a peak in *A. spahaerophora*. The highest absolute valve numbers ($13.6 \times 10^6 - 19.7 \times 10^6$ valves g^{-1}d.s.) are found in zone DZ-5 (4–0 cm), having *Achnanthidium minutissimum*, *Epithemia adnata* and *Cocconeis placentula* as the main diatom taxa.

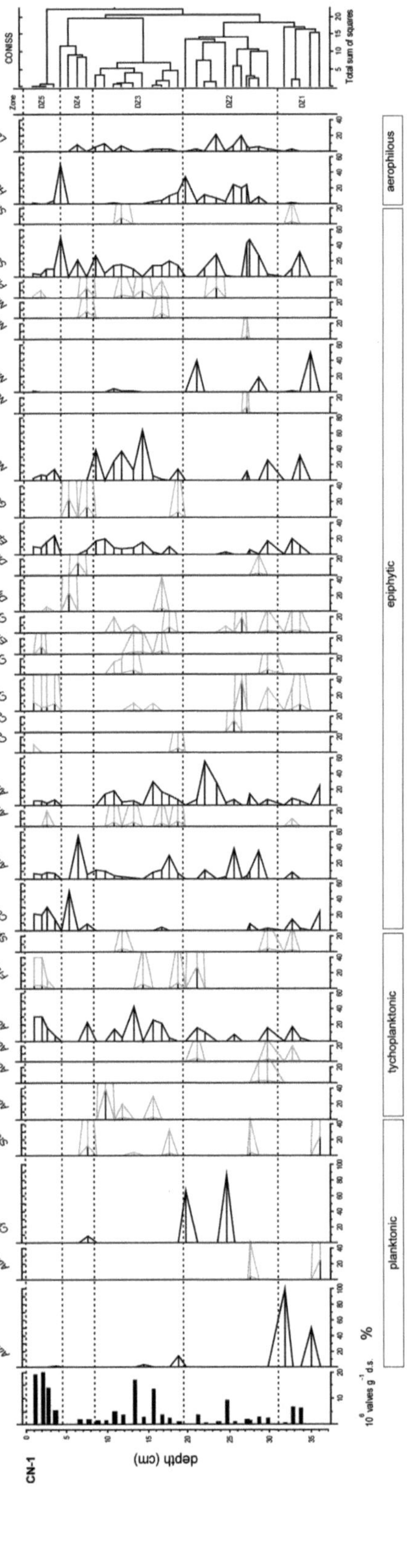

Fig. 5 Total number of valves of diatoms in core CN-1 and relative abundances of each species of planktonic, tychoplanktonic, epiphytic and aerophilous diatoms. *Grey lines* represent a 10× exaggeration scale. Diatom zones (DZ) are based on the indicated constrained cluster analysis

In core CN-3, cladoceran rests of 16 littoral–benthonic and one planktonic species (*Daphnia*) were identified. There is an alternation of ubiquitous and tolerant species *Chydorus sphaericus*, *Alona rectangula* and *Alona guttata*, and more specialised plant-associated species *Graptoleberis testudinaria* and *Acroperus angustatus*. Figure 6 shows the relative abundance of the more abundant taxa and the four main assemblage zones (CZ). In this case, profiles summarise the abundance in 1-cm sections. Zone CZ-1 (60–44 cm) was characterised by the dominance of *C. sphaericus* and *A. rectangula*, with a lower presence of *G. testudinaria* and *A. angustatus*. It is the zone which has lower *Daphnia* remains. In its upper part, a subzone can be distinguished where *G. testudinaria* was not found. The next zone, CZ2 (44–21 cm) was marked by an increase of the relative abundance of *G. testudinaria*, which together with *C. sphaericus* were the predominant species. The presence of *Alonella excisa* was limited to a restricted section at the bottom this zone. Zone CZ-3 (21–11 cm) was distinguished by a drastic decrease of total number of remains and the increase of the relative abundance of the planktonic species, *D. Longispina*, with the disappearance of species associated to plants such as *G. testudinaria, A. angustatus* and *P. aduncus*. The top zone, CZ-4 (11–0 cm) was characterised by a decrease towards the top of *D. longispina*, the dominance of *Chydorus sphaericus* and reappearance of plant-associated species, with the recovery of the high relative abundances of *G. testudinaria* and *A. angustatus*. Figure 7 shows the summarised coherent stratigraphic results obtained by the different paleoecological proxies studied and plotted in an estimated time scale according to the above described chronological model and ^{210}Pb data. In O1 zone (lower part of diatom zone DZ-1), both planktonic and epiphytic diatom taxa and pigment markers for submerged vegetation and phototrophic organisms, such as cryptophytes and cyanobacteria, were present in similar proportions. Afterwards, the percentage of

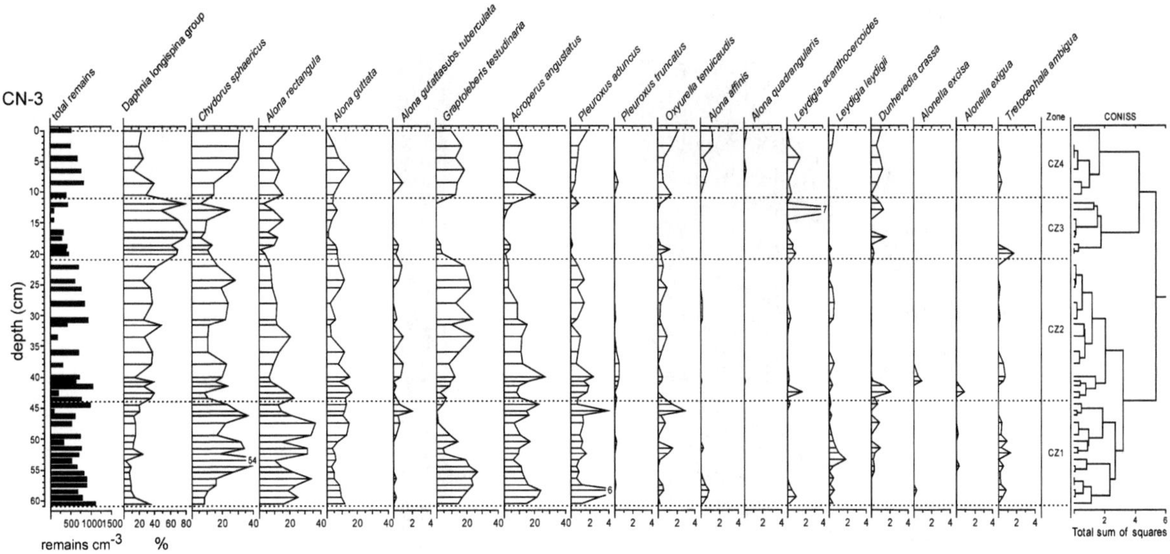

Fig. 6 Total numbers of cladorean subfossils and relative abundance of each cladocera species in CN-3 core. Zones (CZ) are based on the indicated constrained cluster analysis

epiphytic diatom taxa increased notably, whereas planktonic algae and cyanobacteria contribution almost disappeared in O2 zone (middle part of DZ-1), which is suggestive of a progressive decrease in lake level. The percentage of planktonic diatom taxa reached the highest level at the bottom half of R2 zone (upper part of DZ-1) and the percentage of cyanobacteria, chlorophyta, and cryptophyta carotenoids increased significantly with respect to other specific carotenoids through this zone. The oxidised O3 zone marks an abrupt change separating more or less the two-first diatom zones and two-first cladoceran zones. Two periods could be clearly differentiated in R3 zone (upper part of DZ-2 and lower part of CZ-2); the dominance of pigment markers of macrophytes and aerophilous and epiphytic diatom taxa, as well as tolerant benthic cladoceran taxa defined the bottom of R3. By contrast, the proportions of planktonic and tychoplanktonic diatom taxa and specific pigments of cyanobacteria and cryptophyceae populations, as well as planktonic cladocera, increased at the top of R3 zone. Notably, aerophilous diatom taxa reappeared at O4 zone with high planktonic diatom and cladoceran relative abundances, whereas epiphytic diatom taxa were almost absent. Zone O4 also separates the diatom zones DZ-2 and DZ-3 and the cladoceran subzones of CZ-2. The R4 zone (DZ-3 and upper subzone of CZ-2) was mainly composed by peri-epiphytic diatom taxa that can become temporarily tychoplanktonic, and displays a proportional mixture of littoral and planktonic pigment markers, as well as important proportions of specialised phytophilous cladoceran (*G. testudinaria*); the whole zone indicated a relatively stable shallow lake with good macrophyte development. Zone O5 (DZ-4 and CZ-3) is characterised by low number of rests; in this zone, *D. longispina* showed a notable increase of its percentage with respect to total cladocera, but with a decrease in total number of its remains. At R5, the total number of rests and pigment signal from planktonic algae increased and all seemed to indicate a rise of water level. The community of phytophilous cladocerans recovered, but *Daphnia* remained low, probably due to fish introduction.

Discussion

Comprehensive paleoecological analysis of sedimentary sequences based on the limnological information, acquired over the past few decades in Lagunillo del Tejo, suggest past changes in water depth can be traced by comparing planktonic *versus* littoral biological proxies, as well as by changes in sediment characteristics. The coherence between different proxies, each

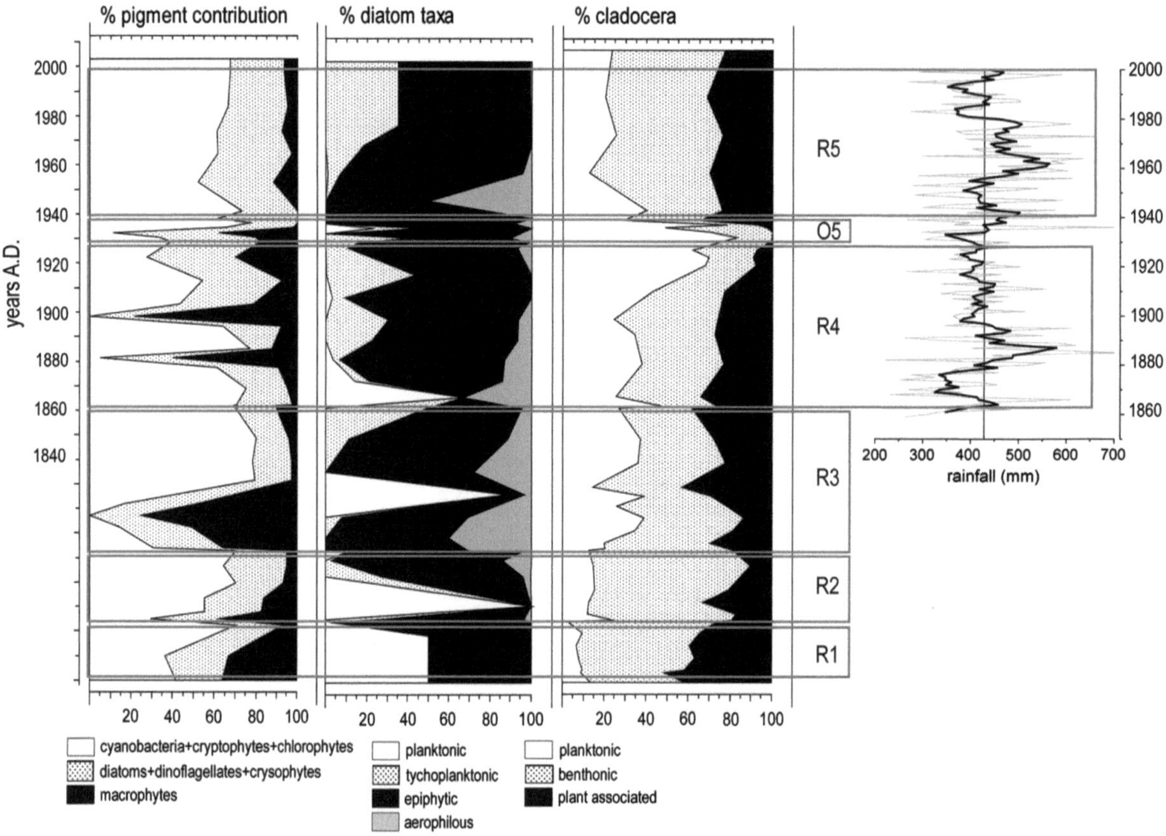

Fig. 7 Comparison between biological proxies temporal scaled and rainfall variability. From *right to left*, relative abundances of main specific carotenoids of different algal populations of core CN-1, relative abundances of diatom and cladocera groups of core CN-1 and CN-3, respectively, and annual rainfall values (*grey line*) with a centred 5-year moving average (*black line*) from Madrid (Spain) during the period 1859–2003. The *grey lines* and zone names are included to facilitate the comparison between climatic and sedimentary signals

having an independent mechanistic link to lake-level variability, enhances the reliability of paleolimnological inferences.

Lagunillo del Tejo is very sensitive to hydrological regime because of its basin topography, morphological features and karstic origin, according to the optima criteria proposed by Digerfeldt (1986). Due to the particular morphology of the sink hole, small fluctuations of a few meters in maximum depth cause drastic changes in total water volume and relative depth which affect directly the biotic community and the relationship between planktonic and littoral assemblages, changing the ecological pond structure. For example, the 4-m level decrease observed between 2003 and 2008 generated a drastic reduction to 25% in the total lake volume, as well as a change in mean depth from 3.7 to 2.1 m and in relative depth from 11.5 to 7.5 m. On the other hand, the bacterial pigment derivatives throughout the sedimentary column suggest the presence of seasonal thermal stratification and associated bottom anoxic layer which contributes to the preservation of the sedimentary signal. Coring in the central part of the lake ensures sediment accumulation from the water column and also transported sediment from shallower depths due to convection currents. (Håkanson & Jansson, 1983; Hilton et al., 1986). Therefore, the paleoecological record is able to provide strong lines of evidence for past lake-level fluctuations.

Paleoecological proxies recorded in the sediments trace two different ecological communities which have switched in relative importance during the past centuries in response to lake-level variability; (1) a planktonic group of algal populations comprising chlorophytes, cryptophytes, cyanobacteria, and phototrophic bacteria populations associated to higher lake level and water column temporal stratification; and (2) a littoral community with benthic and epiphytic alga and macrophytes, indicators of lower level. Furthermore, cladoceran assemblages provide additional information about the communities cited above. On one hand, phytophilous chydorids (*Graptoleberis testudinaria, Acroperus angustatus* and *Pleuroxus aduncus*) increase their abundances in association with macrophyte expansion under stable conditions. By contrast, higher water levels allow the development of planktonic conditions, indicated by higher abundances of *Daphnia longispina* remains in the sediment. The estimation of water-level changes using the planktonic/benthic ratio in cladoceran assemblages has been very useful in many studies (Hyvärinen & Alhonen, 1994; Sarmaja-Korjonen & Alhonen, 1999; Amsinck et al., 2003; Hoffmann, 2003; Khorola et al., 2005). However, as noted by some authors (Frey, 1976, 1986; Verschuren et al., 2000; Jeppesen et al., 2001; Khorola et al., 2005), changes in the planktonic/littoral ratio of Cladocera should be interpreted with caution since other processes could also cause fluctuations in this proportion. In the uppermost part of the sediment (over the last 50 years), the decrease of *Daphnia* may be due to recent fish introduction. We must also consider that in Lagunillo del Tejo, periods of rapid drought affecting peripheral zones, after lake level drops, provoked the reduction of the macrophyte cover by loss of its outer zone. The sediment shows this event as a rise in the relative abundance of *Daphnia longispina* as a consequence of the almost complete disappearance of the specialised phytophilous species.

The high percentages of specific pigments of macrophyte vegetation, as well as epiphytic diatoms and phytophilous cladocera, suggest the presence of an important littoral community throughout the sediment sequence. During periods of low water level in Lagunillo del Tejo (e.g. zone R4 and the bottom of zone R3), primary producers such as macrophytes and associated phototrophic populations, mostly epiphytic diatoms, had an important role, whereas in these macrophyte-dominated periods, phytoplankton played a minor role in the biotic movement of carbon (Padisák & Reynolds, 2003). The low density and high organic matter content in these zones indicate the autogenic character of the sediment with insignificant proportions of transported materials from surrounding shore lands.

In contrast, increases in pelagic phototrophic community pigments may be interpreted as rises of lake level (e.g. the top of zone R3 and the bottom of zone R2). During the periods of higher water level, the lake became easily stratified during most of the year leading to the establishment of anaerobic deep water layers. Then just above the oxic–anoxic interface, dense plates of filamentous cyanobacteria and/or cryptomonads could develop, overlying photosynthetic bacteria that colonised the anoxic zone. We observed dense growths of the mentioned organisms in the oxic–anoxic interface of this lake since 1980, the year of the first studies in it. We observed, as well, that there were not such growths when the maximum depth of this lake was less than 5–6 m, and so it was not stratified. In mountain lakes, diadinoxanthin-to-alloxanthin pigment ratio, tracers of diatoms and cryptophyta populations, respectively, have been used as indication of the relative importance of benthic versus planktonic primary production, which could, in turn, be related to water column transparency and/or to lake-level fluctuations (Buchaca & Catalán, 2007).

The sedimentary signal recorded in Lagunillo del Tejo has shown two different temporal patterns of water-level fluctuation. Paleoecological proxies suggest past periods of either low or high lake-level and lithological characteristics, and physical properties in the sediments suggest rapid events of lake-level fluctuation. These events were recorded in the oxidised "O zones" which suggest rapid events of lake lowering. Named "O zones" (O1–O5) were characterised by high density and lower water content, lower LOI and the reduction of the absolute abundances of diatom frustules and cladoceran subfossiles. Rapid lake-level decline causes increased sedimentation rates of coarse mineral materials from surrounding lakeside, which dilute the organic matter contents, the pigments signal and remains of organisms. Moreover, the organic content also decreases when water level is low due to ready exposure to oxidation (Rowan et al., 1992). Furthermore, the increase in the relative abundance of aerophilous diatoms, in some occasions together with planktonic diatom taxa, at these depths suggest occasional transport from subaerial, yet humid shoreline materials to the central area, linked to rapid lowering of waters (e.g. just after zones O4 and O5 coinciding with reductions of rainfall values around 1860 and 1930, respectively, Fig. 7). On the other hand, phytophilous cladocerans, such as the specialised *G. testudinaria*, are reduced when rapid changes in water level deteriorate the outer ring of macrophytes. As mentioned earlier, when this occured, the sedimentary sequence showed a decrease of all the species, but the relative abundance of the planktonic species (*Daphnia longispina*) increased relative to phytophilous species. Another indication of these events was the relative increase in "O zones" of *Leydigia acanthocercoides* and *Dunhevedia crassa*, which are well adapted to temporary habitats and may be indicators of fluctuations in the shore line. The least-affected species during these periods are *Chydorus sphaericus* and *Alona rectangula* which have a great capacity of adaptation (Alonso, 1996).

The observed plateau in ^{210}Pb activity (perturbations of exponential decay activity) indicates changes in the sediment accumulation (Garcia-Orellana et al., 2006; Romero-Viana et al., 2009). This irregularity of the ^{210}Pb activity profile was observed in the zone O5, characterised by higher density and lower organic matter content; therefore, it may be attributed to the dilution effect related to an increase in detrital material. Even though variations in the ^{226}Ra concentration could be used as indicator of changes in sediment composition due to the presence of rapid sedimentation (turbidites or floods), the fact that ^{226}Ra concentration is roughly constant (Romero-Viana et al., 2009) along the profile shows that the allochthonous material come from the surroundings of the lake. The above-average thickness of zone O5 may be due to the construction of agricultural terraces in the steep walls of Lagunillo del Tejo doline in the second quarter of the twentieth century, at the end of a persistent dry period (Fig. 7). The terraces, now abandoned, are still present. The chronological model indicates that the allochthonous material was introduced into the Lagunillo lake between 1926 and 1940. Taken together, the anthropogenic activity and drought could have had a synergistic effect, increasing sedimentation rates in this zone O5.

The lake-level variability inferred over the last two centuries in Lagunillo del Tejo could be related to instrumental data (1860–2003) from Madrid (200 km) from the study site, but highly correlated with the available instrumental series (1950–2003) from Cuenca. Although the Iberian Peninsula is characterised by a high spatial climatic heterogeneity, comparative analysis is possible since the central region of Spain is above the influence of similar seasonal rainfall patterns (Rodríguez-Puebla et al., 1998). The rainfall series (Fig. 7) shows two significant dry periods: an extreme drought (1870) after a wet decade around 1850 documented in many localities in Spain (Font-Tullot, 1988; Sousa and Garcia-Murillo, 2003; Rodrigo & Barriendos, 2008) and a lengthy dry period during the first three decades of twentieth century. We speculate that Lagunillo del Tejo reached a high level at the top of zone R3 corresponding to the documented positive anomaly around 1850. However, aerophilous diatoms at the bottom of zone R4 suggest rapid lake-level drop around the driest years of 1870. The littoral community sedimentary signal suggests a progressive lake-level decrease reaching probably the lowest level around 1880. The level was low, but the lake was not dry since evidences of non-integrity or discontinuity of the sediment record were not observed during this period, which was the most negative anomaly phase of the last 140 years. Conversely, the positive anomalies registered around 1890 and the planktonic signal suggest that the pond returned to a higher level. It is therefore likely that the maximum depth between 1880 and 1890 was even higher than 11 m of maximum depth as reported by Vicente & Miracle (1984) during 1980–1981. This high level phase at the end of the nineteenth century has been also reported in lake Gallocanta (NE Spain) (Rodó et al., 2002). Despite notable differences between Gallocanta and Lagunillo del Tejo (e.g. watershed area, chemical composition of water and morphological features), both record the wettest period over the last 140 years. The prevalence of the littoral community compared to the planktonic proxy signal suggests a low lake level corresponding to the dry period observed over the first three decades of twentieth century.

Recently, the number of studied Iberian lacustrine sedimentary sequences has increased notably (e.g. Pérez-Obiol & Julià, 1994; Giralt et al., 1999; Reed et al., 2001; Luque & Julià, 2002; Rodó et al., 2002; Riera et al., 2004; Valero-Garcés et al., 2004; González-Sampériz et al., 2006; Morellón et al., 2007; Valero-Garcés et al., 2008; Martín-Puertas et al., 2008). These new paleodata have contributed to the paleoclimatic characterisation of the western area of the Mediterranean basin during the Holocene. The

Lagunillo del Tejo record has been shown to provide a suitable natural archive, which records both high and low-frequency climate variability. A high-resolution climate proxy record cannot be taken uncritically to represent a high resolution record of past climate variability (Verschuren, 2003). However, with this study focused on the relationships between paleoecological proxies and lake hydrology, we have been able to show that Lagunillo del Tejo has preserved a continuous and high-resolution climate record. We attempted this study with the idea that by comparing sedimentary proxies with recorded meteorological conditions over the last two centuries, we could increase the temporal resolution of paleoclimatic data. This could be further used as a starting point to understand longer sedimentary sequences to infer regional hydro-climatic variability and thus contribute to the regional and global network of accurate climate reconstructions.

Acknowledgements We are very grateful to the INM (Spanish Institute of Meteorology) which kindly provided the meteorological data. Lidia Romero held a scholarship from Generalitat Valenciana, and Charo Lopez is granted by the Spanish Ministry of Education and Science. This study was financed by CICyT grant REN2002-03272 CLI and CGL2005-04040 from the Spanish Ministry of Education and Science to M. Rosa Miracle and Eduardo Vicente, respectively.

References

Airs, R. L., J. E. Atkinson & B. J. Keely, 2001. Development and application of a high resolution chromatographic method for the analysis of complex pigment distributions. Journal of Chromatography 917: 167–177.

Almquist-Jacobson, H., 1995. Lake-level fluctuations at Ljustjärnen, central Sweden and their implications for the Holocene climate of Scandinavia. Palaeogeography Palaeoclimatology Palaeoecology 118: 269–290.

Alonso, M., 1996. Fauna Ibérica: Branchiopoda. Museo de Ciencias Naturales. Consejo Superior de Investigaciones Científicas (CSIC), Madrid: 486 pp.

Amsinck, S. L., E. Jeppesen & D. Ryves, 2003. Cladoceran stratigraphy in two shallow brackish lakes with special reference to changes in salinity, macrophyte abundance and fish predation. Journal of Paleolimnology 29: 495–507.

APHA-AWWA-WEF, 1992. Standard Methods for the Examination of Water and Wastewater. American Public Health Association, Washington, DC: 1268 pp.

Appleby, P. G., 2001. Chronostratigraphic techniques in recent sediments. In Last, W. M. & J. P. Smol (eds), Tracking Environmental Change Using Lake Sediments: Basin Analysis, Coring and Chronological Techniques, Vol. 1. Kluwer Academic Publishers, Dordrecht: 171–203.

Appleby, P. G. & F. Oldfield, 1978. The calculation of ^{210}Pb dates assuming a constant rate of supply of unsupported ^{210}Pb to the sediment. Catena 5: 1–8.

Battarbee, R. W., 1986. Diatom analysis. In Berglund, B. E. (ed.), Handbook of Holocene Palaeoecology and Palaeohydrology. Wiley, Chichester: 527–570.

Battarbee, R. W., V. J. Jones, R. J. Flower, N. G. Cameron, H. Bennion, L. Carvalho & S. Juggins, 2001. Diatoms. In Smol, J. P., H. J. B. Birks & W. M. Last (eds), Tracking Environmental Change Using Lake Sediments: Terrestrial, Algal, and Siliceous Indicators, Vol. 1. Kluwer Academic Publishers, Dordrecht: 155–202.

Benson, L. V., P. A. Meyers & R. J. Spencer, 1991. Change in size of Walker Lake during the past 5000 years. Palaeogeography Palaeoclimatology Palaeoecology 81: 189–214.

Buchaca, T. & J. Catalán, 2007. Factors influencing the variability of pigments in the surface sediments of mountain lakes. Freshwater Biology 52: 1365–1379.

Cholnoky, B. J., 1968. Die Okologie der Diatomeen in Binnengewassern. Cramer, Lehre: 699 pp.

Cirujano, S., 1995. Flora y vegetacion de las lagunas y humedales de la provincial de Cuenca. Consejo Superior de Investigaciones Científicas, Madrid.

Coops, H., M. Beklioglu & T. L. Crisman, 2003. The role of water-level fluctuations in shallow ecosystems—workshop conclusions. Hydrobiologia 506–509: 23–27.

Dearing, J. A., 1997. Sedimentary indicators of lake-level changes in the humid temperate zone: a critical review. Journal of Paleolimnology 18: 1–14.

Digerfeldt, G., 1986. Studies on past lake-level fluctuation. In Berglund, B. E. (ed.), Handbook of Holocene Palaeoecology and Palaeohydrology. Wiley, Chichester: 127–143.

Font-Tullot, I., 1988. Historia del clima de España. Cambios climáticos y sus causas. Instituto Nacional de Meteología, Madrid: 297 pp.

Frey, D. G., 1959. The taxonomic and phylogenetic significance of the head pores of Chydoridae (Cladocera). International Revue der Gesamen Hydrobiologie 44: 27–50.

Frey, D. G., 1962. Supplement to: the taxonomic and phylogenetic significance of the head pores of Chydoridae (Cladocera). International Revue der Gesamen Hydrobiologie 47: 603–609.

Frey, D. G., 1976. Interpretation of Quaternary paleoecology from Cladocera and midges, and prognosis regarding usability of other organisms. Canadian Journal of Zoology 54: 2208–2226.

Frey, D. G., 1986. Cladocera analysis. In Berglund, B. E. (ed.), Handbook of Holocene Palaeoecology and Palaeohydrology. Wiley, Chichester: 667–692.

Fritz, S. C., 1990. Twentieth-century salinity and water-level fluctuations in Devils Lake, North Dakota: test of a diatom-based transfer function. Limnology and Oceanography 41: 882–889.

Garcia-Orellana, J., E. Gracía, A. Vizcaino, P. Masque, C. Olid, F. Martínez-Ruiz, E. Pinero, J. A. Sánchez-Cabeza & J. Danobeitia, 2006. Identifying instrumental and historical earthquake records in the SW Iberian margin using Pb-210 turbidite chronology. Geophysical Research Letters 33(24): L24601.

Gasse, F., 1986. East African diatoms: Taxonomy, Ecological Distribution. Bibliotheca Diatomologica. J. Cramer, Berlin, Stuttgart: 201 pp.

Giralt, S., F. Burjachs, J. R. Roca & R. Julià, 1999. Late Glacial to early Holocene environmental adjustment in the Mediterranean semi-arid zone of the Salines playa-lake (Alicante, Spain). Journal of Paleolimnology 21: 449–460.

González-Sampériz, P., B. L. Valero-Garcés, A. Moreno, G. Jalut, J. M. García-Ruiz, C. Martí-Bono, A. Delgado-Huertas, A. Navas, T. Otto & J. J. Dedoubat, 2006. Climate variability in the Spanish Pyrenees during the last 30,000 yr revealed by the El Portalet sequence. Quaternary Research 66: 38–52.

Grimm, E. C., 1987. CONISS: a FORTRAN 77 program for stratigraphically constrained cluster analysis by the method of incremental sum of squares. Computers and Geosciences 13: 13–35.

Håkanson, L. & M. Jansson, 1983. Principles of Lake Sedimentology. Springer, Berlin: 177–212.

Harradine, P. J., P. G. Harris, R. N. Head, R. P. Harris & J. R. Maxwell, 1996. Steryl chlorin esters are formed by zooplankton herbivory. Geochimica et Cosmochimica Acta 60: 2265–2270.

Harris, P. G., J. F. Carter, R. N. Head, R. P. Harris, G. Eglington & J. R. Maxwell, 1995. Identification of chlorophyll transformation products in zooplankton faecal pellets and marine sediment extracts by liquid chromatography/mass spectrometry atmospheric pressure chemical ionization. Rapid Communications in Mass Spectrometry 9: 177–183.

Hilton, J., J. P. Lishman & P. V. Allen, 1986. The dominant processes of sediment distribution and focusing in a small, eutrophic, monomictic lake. Limnology and Oceanography 31: 125–133.

Hoffmann, W., 2003. Cladocerans and chironomids as indicators of lake level changes in north temperate lakes. Journal of Paleolimnology 19: 55–62.

Hustedt, F., 1930. Die Süsswasserflora Mitteleuropas. In Pascher, A. (ed.), Bacillariophyta (Diatomeae), Heft 10, 2nd edn. Verlag von Gustav Fischer, Germany: 466 pp.

Hustedt, F., 1959–1966. Die Kieselalgen. In Rabenhorst, L. (ed.), Kryptogamen-flora von Deutschland, Osterreich und der Schweiz. Akademische Verlagsgesselsschaft, Greest & Porting K. G., Leipzig: 816 pp.

Hyvärinen, H. & P. Alhonen, 1994. Holocene lake level changes in the Fennoscandian tree-line region, western Finnish Lapland: diatom and cladoceran evidence. The Holocene 4: 251–258.

Jeffrey, S. W. & M. Vesk, 1997. Introduction to marine phytoplankton and their pigments signature. In Jeffrey, S. W., R. F. C. Mantoura & S. W. Wright (eds), Phytoplankton Pigments in Oceanography. UNESCO Publishing, Paris: 37–84.

Jeppesen, E., P. Leavitt, L. De Meester & J. P. Jensen, 2001. Functional ecology and palaeolimnology: using cladoceran remains to reconstruct anthropogenic impact. Trends in Ecology and Evolution 16: 191–198.

Khorola, A., M. Tikkanen & J. Weckström, 2005. Quantification of the Holocene lake-level changes in Finnish Lapland by means of a cladocera-lake depth transfer model. Journal of Paleolimnology 34: 175–190.

Krammer, K., 1997a. Die cymbelloiden Diatomeen. Eine Monographie der weltweit bekannten Taxa. Teil 1. Allgemeines und Encyonema part. Bibliotheca Diatomologica, Band 36. J. Cramer, Stuttgart: 382 pp.

Krammer, K., 1997b. Die cymbelloiden Diatomeen. Eine Monographie der weltweit bekannten Taxa. Teil 2. Encyonema part. Encyonopsis und Cymbellopsis. Bibliotheca Diatomologica, Band 37. J. Cramer, Stuttgart: 469 pp.

Krammer, K., 2000. Diatom of Europe. Diatoms of the European inland waters and comparable habitats. In Lange-Bertalot, H. (ed.), The Genus Pinnularia, Vol. 1. Koeltz Scientific Books, Germany: 703 pp.

Krammer, K., 2002. Diatoms of Europe. Diatoms of the European inland waters and comparable habitats, Vol. 3. Cymbella. A.R.G. Gantner Verlag K.G., Ruggell: 584 pp.

Krammer, K. & H. Lange-Bertalot, 1986. Bacillariophyceae. 1. Teil: Naviculaceae. In Ettl, H., J. Gerloff, H. Heynig & D. Mollenhauer (eds), Süsswasser flora von Mitteleuropa, Band 2/1. Gustav Fischer Verlag, Stuttgart, New York: 876 pp.

Krammer, K. & H. Lange-Bertalot, 1988. Bacillariophyceae. 2. Teil: Bacillariaceae, Epithemiaceae, Surirellaceae. In Ettl, H., J. Gerloff, H. Heynig & D. Mollenhauer (eds), Süsswasserflora von Mitteleuropa, Band 2/2. VEB Gustav Fischer Verlag, Jena: 596 pp.

Krammer, K. & H. Lange-Bertalot, 1991a. Bacillariophyceae. 3. Teil: Centrales, Fragilariaceae, Eunotiaceae. In Ettl, H., J. Gerloff, H. Heynig & D. Mollenhauer (eds), Süsswasserflora von Mitteleuropa, Band 2/3. Gustav Fischer Verlag, Stuttgart, Jena: 576 pp.

Krammer, K. & H. Lange-Bertalot, 1991b. Bacillariophyceae. 4. Teil: Achnanthaceae, Kritische Ergänzungen zu Navicula (Lineolatae) und Gomphonema, Gesamtliteraturverzeichnis Teil 1–4. In Ettl, H., G. Gärtner, J. Gerloff, H. Heynig & D. Mollenhauer (eds), Süsswasserflora von Mitteleuropa, Band 2/4. Gustav Fischer Verlag, Stuttgart, Jena: 437 pp.

Lange-Bertalot, H., 2001. Diatoms of Europe. Diatoms of the European Inland Waters and Comparable Habitats. Vol. 2. Navicula sensu stricto. 10 Genera separated from Navicula sensu lato. Frustulia. A.R.G. Gantner Verlag K.G., Ruggell: 526 pp.

Lange-Bertalot, H. & K. Krammer, 1987. Bacillariaceae, Epithemiaceae, Surirellaceae. Bibliotheca Diatomologica, Vol. 15. J. Cramer, Stuttgart: 289 pp.

Lange-Bertalot, H. & K. Krammer, 1989. Achnanthes eine Monographie der Gattungen. Bibliotheca Diatomologica, Vol. 18. J. Cramer, Stuttgart: 393 pp.

Lowe, R. L., 1974. Environmental requirements and pollution tolerance of freshwater diatoms. U.S. Environmental Protection Agency Report EPA-670/4-74-005, National Environmental Research Center, Cincinnati, OH: 334 pp.

Luque, J. A. & R. Julià, 2002. Lake sediment response to land-use and climate change during the last 1000 years in the oligotrophic Lake Sanabria (Northwest of Iberian Peninsula). Sedimentary Geology 148: 343–355.

Magny, M., J. L. Beaulieu, R. Drescher-Schneider, B. Vannière, A. V. Walter-Simonnet, Y. Miras, L. Millet, G. Bossuet, O. Peyron, E. Brugiapaglia & A. Leroux, 2007. Holocene climate changes in the central Mediterranean as recorded by lake-level fluctuations in Lake Accesa (Tuscany, Italy). Quaternary Science Reviews 26: 1736–1758.

Margaritora, F. G., 1985. Fauna d'Italia v. XXIII: Cladocera. Calderini (ed.), Bologna, Italy: 399 pp.

Martín-Puertas, C., B. L. Valero-Garcés, M. P. Mata, P. González-Sampériz, R. Bao, A. Moreno & V. Stefanova, 2008. Arid and humid phases in southern Spain during the last 4000 years: the Zoñar lake record, Córdoba. Holocene 18: 1–15.

Morellón, M., B. L. Valero-Garcés, A. Moreno, P. González-Sampériz, P. Mata, O. Romero, M. Maestro & A. Navas, 2007. Holocene palaeohydrology and climate variability in Northeastern Spain: the sedimentary record of Lake Estanya (Pre-Pyrenean range). Quaternary International 181: 15–31.

Padisák, J. & C. S. Reynolds, 2003. Shallow lakes: the absolute, the relative, functional and pragmatic. Hydrobiologia 506–509: 1–11.

Pérez-Obiol, R. & R. Julià, 1994. Climatic change on the Iberian Peninsula recorded in a 30000-yr pollen record from Lake Banyoles. Quaternary Research 41: 91–98.

Reed, J. M., A. C. Stevenson & S. Juggins, 2001. A multi-proxy record of Holocene climatic change in southwestern Spain: the Laguna de Medina, Cádiz. Holocene 11: 707–719.

Riera, S., G. Wansard & R. Julià, 2004. 2000-year environmental history of a karstic lake in the Mediterranean Pre-Pyrenees: the Estanya lakes (Spain). Catena 55: 293–324.

Rodó, X., S. Giralt, F. Burjachs, F. A. Comín, R. G. Tenorio & R. Julià, 2002. High-resolution saline lake sediments as enhanced tools for relating proxy paleolake records to recent climatic data series. Sedimentary Geology 148: 203–220.

Rodrigo, F. S. & M. Barriendos, 2008. Reconstruction of seasonal and annual rainfall variability in the Iberian peninsula (16th–20th centuries) from documentary data. Global and Planetary Change 63: 243–257.

Rodríguez-Puebla, C., A. H. Encinas, S. Nieto & J. Garmendia, 1998. Spatial and temporal patterns of annual precipitation variability over the Iberian Peninsula. International Journal of Climatology 18: 299–316.

Romero-Viana, L., R. Julià, A. Camacho, E. Vicente & M. R. Miracle, 2008. Climate signal in varved sediment: La Cruz (Spain) a case study. Journal of Paleolimnology 40: 703–714.

Romero-Viana, L, B. J. Keely, A. Camacho, E. Vicente & M. R. Miracle, 2009. Photoautotrophic community changes in Lagunillo del Tejo (Spain) in response to lake-level fluctuation: two centuries of sedimentary pigment records. Organic Geochemistry 40: 376–386.

Round, F. & L. Bukhtiyarova, 1996. Four new genera based on Achnanthes (Achnanthidium), together with a redefinition of Achnanthidium. Diatom Research 11(2): 345–361.

Round, F., R. M. Craword & D. G. Mann, 1991. The Diatoms Biology and Morphology of the Genera. Cambridge University Press, Cambridge: 747 pp.

Rowan, K. S., 1989. Photosynthetic Pigments of Algae. Cambridge University Press, Cambridge: 334 pp.

Rowan, D. J., J. Kalff & J. B. Rasmussen, 1992. Profundal sediment organic content and physical character do not reflect lake trophic status, but rather reflect inorganic sedimentation and exposure. Canadian Journal of Fisheries and Aquatic Sciences 49: 1431–1438.

Sánchez-Cabeza, J. A., P. Masqué & I. Ani-Ragolta, 1998. ^{210}Pb and ^{210}Po analysis in sediments and soils by microwave acid digestion. Journal of Radioanalytical and Nuclear Chemistry 227: 1–2.

Sarmaja-Korjonen, K. & P. Alhonen, 1999. Cladocera and diatom evidence of lake-level fluctuations from a Finnish lake and the effect of aquatic-moss layers on microfossil assemblages. Journal of Paleolimnology 22: 277–290.

Sousa, A. & P. Garcia-Murillo, 2003. Changes in the wetlands of Andalusia (Doñana Natural Park, SW Spain) at the end of the Little Ice Age. Climatic Change 58: 193–217.

Szeroczynska, K. & K. Sarmaja-Korjonen, 2007. Atlas of Subfossil Cladocera from Central and Northern Europe. Friends of the lower Vistula Society, Swiecie (Poland): 84 pp.

Valero-Garcés, B. L., P. González-Sampériz, A. Navas, J. Machín, A. Delgado-Huertas, J. L. Peña-Monné, C. Sancho-Marcén, T. Stevenson & B. Davis, 2004. Paleohydrological fluctuations and steppe vegetation during the last glacial maximum in the central Ebro valley (NE Spain). Quaternary International 122: 43–55.

Valero-Garcés, B. L., A. Moreno, A. Navas, P. Mata, J. Machín, A. Delgado-Huertas, P. González-Sampériz, A. Schwalb, M. Morellón, H. Cheng & R. L. Edwards, 2008. The Taravilla lake and tufa deposits (Central Iberian Reange, Spain) as paleohydrological and paleoclimatic indicators. Palaeogeography, Palaeoclimatology, Palaeoecology 259: 136–156.

Verschuren, D., 2003. Lake-based climate reconstruction in Africa: progress and challenges. Hydrobiologia 500: 315–330.

Verschuren, D., J. Tibby, K. Sabbe & N. Roberts, 2000. Effects of depth, salinity, and substrate on the invertebrate community of a fluctuating tropical lake. Ecology 81: 164–182.

Vicente, V. & M. R. Miracle, 1984. Distribution of photosynthetic organisms in a temporal stratified karstic pond near Cuenca, Spain. Verhandlungen International Verein Limnologie 22: 1704–1710.

Basin elevation and salinity changes: late Holocene development of two freshwater lakes at the Karelian White Sea coast, northwest Russia as reflected in their sediments

Mirko Dreßler · Manuela Schult · Michael Schubert · Jessica Buck

Originally published in the journal Hydrobiologia, Volume 631, No. 1, 247–266.
DOI: 10.1007/s10750-009-9814-9

Abstract In this paleolimnological multi-proxy study, anthropogenic influences and salinity changes of the northwest Russian lakes Erchovskye Ozero East (ESE) and Erchovskye Ozero West (ESW) were assessed using diatoms, pollen and other microfossils from ^{210}Pb-dated sediment cores. Both lakes are situated on the Fennoscandian Shield in direct vicinity to the White Sea coast, a region that is still subject to isostatic land uplift processes today. The analyses showed that both lakes evolved from typical saltwater-dominated coastal waters to freshwater lakes within the last two centuries. Salinity was found to be the determining environmental factor for the aquatic biota whereas no significant anthropogenic influences on the lakes could be detected. The decreasing salinity in both lakes was caused by basin elevation and isolation due to isostatic rebound of the Fennoscandian Shield. A hypothetical land uplift rate of 5.7 mm yr^{-1} was calculated based on data from the sediment cores analysed.

Guest editors: K. Buczkó, J. Korponai, J. Padisák & S. W. Starratt
Palaeolimnological Proxies as Tools of Environmental Reconstruction in Fresh Water

M. Dreßler (✉) · J. Buck
Department of Bio-Science, Institute for Botany, University of Rostock, Wismarsche Str. 8, 18051 Rostock, Germany
e-mail: mirko.dressler@gmx.de

M. Schult
Institut für Botanik und Landschaftsökologie, AG Moor- und Paläoökologie, EMAU Greifswald, Universität Greifswald, 17487 Greifswald, Germany

M. Schubert
Department Analytik, Helmholtz-Zentrum für Umweltforschung – UFZ, Permoserstr. 15, 04318 Leipzig, Germany

Keywords Diatoms · Pollen · Salinity changes · Land uplift · Basin elevation · Northeastern Europe · Multi-proxy study · Paleolimnology

Introduction

The postglacial land uplift of Fennoscandia has been intensely studied (Pan & Sjöberg, 1999), extensively describing shoreline displacements over the last 10,000 years caused by isostatic land uplift (Påsse & Andersson, 2005). Thus, a good understanding of coastline changes due to land uplift processes has been obtained for the Baltic region (Berglund et al., 2005). However, little data exists for the eastern part of the Fennoscandian Shield such as the Karelian White Sea coast.

Postglacial uplift not only changes coastlines but also commonly leads to basin elevation and isolation, changing marine bays into freshwater lakes (Weller & Derksen, 1979). These saltwater–freshwater transitions are accompanied by drastic changes in various environmental parameters, such as pH and salinity, which affect habitat conditions for aquatic organisms (Miettinen et al., 2007).

Diatoms as organisms inhabiting either freshwater or marine habitats are very sensitive to change in environmental variables. The major factors affecting the distribution of diatoms are salinity (Ryves et al., 2004), trophic conditions (Dreßler et al., 2007) and pH (Cameron et al., 1999). Thus, diatoms are frequently used in palaeolimnology for tracking changes in salinity and pH, and therefore, for defining shoreline displacement and sea level fluctuations in coastal areas (Kjemperud, 1980). Diatoms from sediments in transition zones between marine and lacustrine ecosystems are suited for these analyses (Hedenström & Risberg, 1999).

Aquatic and shoreline vegetation is also sensitive to changes in salinity (Garde et al., 2004). However, human activity can influence the vegetation of a lake and its catchment (Dreßler et al., 2006). Thus, in contrast to analyses of diatoms, which are mainly indicators of water conditions (Miettinen et al., 2005), analyses of pollen flora from sediments mainly give information about natural and anthropogenic processes in the catchment area of a given lake (Moore et al., 1991).

In the present palaeolimnological study, we analysed diatoms, pollen and other microfossils from ^{210}Pb- and ^{137}Cs-dated sediment cores of two recent freshwater lakes situated in the eastern part of the Fennoscandian Shield in the Karelian region of the White Sea coast.

Both of the study lakes are situated in close vicinity of the present shoreline and are only slightly above the present level of the White Sea. Taking into account the proximity to the marine coast and the uplift rate of the Fennoscandian Shield, that for today's Karelia is approximately 4 mm yr^{-1} (Pan & Sjöberg, 1999), it can be assumed that the basins of both lakes were part of the White Sea several hundred of years ago. This assumption is supported by geological analyses that found ca. 500-year-old clam layers of marine origin below the lacustrine sediments in both basins (Natalia Tschewtschenko, Pers. comm.). However, the specific development history of the lakes over the past 500 years and the timing of basin isolation from the sea are unknown.

Detailed knowledge of shoreline displacement, primary succession and anthropogenic impact on the White Sea coast does not exist for the last centuries. Therefore, we aimed to answer the following questions: (1) how did the temporal development of the lakes proceed with regard to salinity and anthropogenic influence? and (2) do both of the selected lakes represent a potentially sensitive system for analyses of land uplift processes in the coastal region of the White Sea?

Study site

Both of the analysed lakes are called Erchovskye Ozera in Russian, hereafter referred to as Erchovskye Ozero East (ESE) and Erchovskye Ozero West (ESW). Both study sites are oligotrophic shallow freshwater lakes with maximum depths of 2 m and a relatively small common catchment area of approximately 12 km^2. They are situated in the northwestern part of European Russia in the region of northern Karelia (66°32′N, 33°04′E; Fig. 1). Southeast of the Kindo peninsula, the lakes form a western extension of the marine bay Kislaya Guba (Fig. 1). The lakes are connected with the sea via a small stream. The lakes are situated 1.6 m (ESE) and 2.3 m (ESW) above the sea level (Fig. 1). With the tidal range at the Kindo peninsula of 1.3–1.8 m (average 1.6 m), at high tide the ESE is still subject to effects of seawater incursions.

The climate in the study area is continental with marine influence (Filatov et al., 2005) and the yearly precipitation averages about 700 mm. The coldest month is January with mean temperatures from −9°C to −12°C, the hottest month is July, with mean temperatures of 14°C to 17°C. The lakes are covered with ice for 5–6 months each winter, usually from November until May (Filatov et al., 2005).

The lakes are situated in a depression formed by tectonic processes during the early Holocene. Geological analyses revealed 500-year-old marine shell beds below lacustrine sediments of both lakes (Natalia Tschewtschenko, Pers. Comm.), indicated that the lake basins most probably still were part of the White Sea 500 years ago.

The bedrock of the catchment area is very variable and mainly consists of 2.0–1.8 billion-year-old metamorphic rock (gneiss), with quartz, potash feldspar, biotite, garnet and amphiboles.

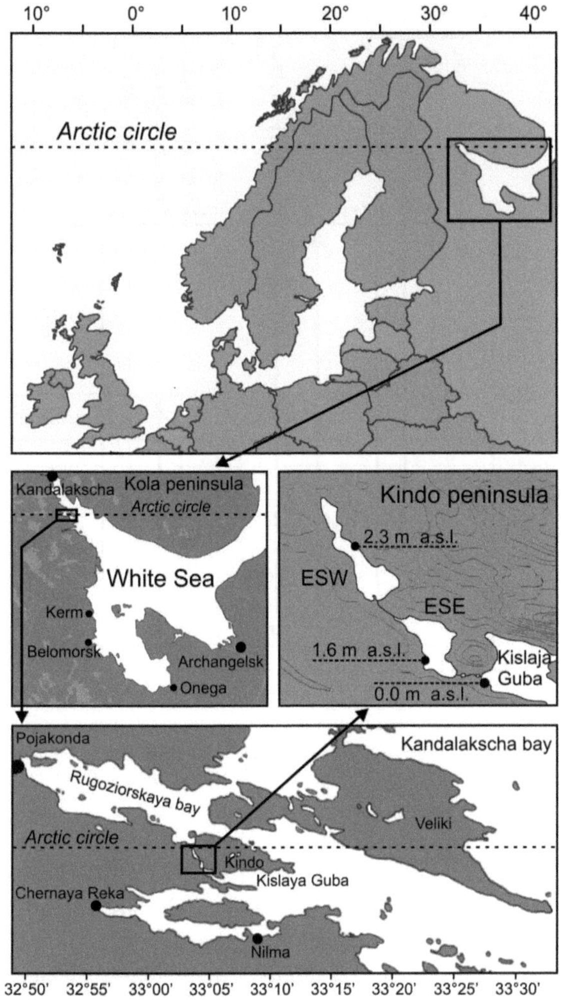

Fig. 1 Geographical position of the study lakes ESE and ESW at the White Sea coast near the Kindo peninsula in the north-western part of European Russia

The study area is situated in the northern taiga, i.e. the boreal coniferous forest zone. Almost the entire catchment is covered by woodland with raised bogs and forest on waterlogged soil covering 20 and 30% of the area, respectively. Conifers, mainly *Pinus sylvestris* L. and *Picea abies* Karsten, constitute 90% of the forests (Hyvärinen 1975) and 90% of the deciduous trees consist of *Betula*, *Populus* and *Alnus*. Shorelines of both lakes are rocky with very scarce macrophyte cover. *Carex*, *Nuphar lutea* (L.) Smith, *Nymphaea alba* L., *Lysimachia thyrsiflora* L., *Menyanthes trifoliata* L., *Hippuris vulgaris* L., *Utricularia*, *Phragmites australis* (Cavanilles) Trinius and *Myriophyllum* are dominant.

Methods

Sediment coring and sampling

The sediment studied originates from two cores (ESE: 52 cm; ESW: 48 cm) taken in August 2004 from the deepest part of each lake (2 m water depth) using a Mondsee corer (UWITEC, Austria). For sub-sampling, the sediment was divided into 1 cm sections in the field and placed in 50 ml plastic bottles. For diatom analyses, measurements of ^{210}Pb and ^{137}Cs activities and the determination of loss on ignition, samples were taken at 2 cm intervals. Samples for pollen analyses were taken at 4 cm intervals.

Chronology

Sediment dating was performed at the Centre for Environmental Research (UFZ) Leipzig-Halle, Germany. Sample dry weights ranged from 0.4 to 2.0 g. Some samples (ESW 48, 46 cm; ESE 52 cm) were heavier.

The activity concentrations of $^{210}Pb_{ex}$ and ^{137}Cs were measured using two ORTEC-Gamma-X HPGe Coaxial low-energy n-type detectors based on their distinct gamma emission energies of 46.5 and 661 keV, respectively (Appleby et al., 1992). Spectral analysis was performed with the software GAMMAW®. In the case of $^{210}Pb_{ex}$ sediment chronology was calculated applying two different models leading to different results. The Constant Initial Concentration (CIS) model assumes a constant initial excess ^{210}Pb concentration for the sediment deposited over time. The Constant Rate of Supply model (CRS) assumes a constant ^{210}Pb input at the sediment surface but permits the sediment supply to vary (Appleby & Oldfield, 1983). Validation of the applicability of the particular models for dating of the cores taken was done by comparing the respective results to the data achieved by means of the ^{137}Cs method.

Diatom analysis

Sediment for diatom analyses (2 cm^3) was treated with HCl, H_2O_2, H_2SO_4 and $KMnO_4$ following Kalbe & Werner (1974) and mounted on permanent slides with Naphrax®. Light microscopic analysis was accomplished by means of a Zeiss Axioplan, equipped with an oil immersion Plan-Apochromat objective (magnification 1,000×, numerical aperture 1.4). In each sample, 400–600 diatom valves were counted. Diatom species identification was according to Krammer & Lange-Bertalot (1986–1991, 2004), Lange-Bertalot & Moser (1994), Lange-Bertalot et al. (1996), Lange-Bertalot & Metzeltin (1996), Lange-Bertalot & Genkal (1998), Krammer (1997a, b, 2000, 2002, 2003), Witkowski et al. (2000) and Lange-Bertalot (2001).

The salinity history of the lakes was assessed using the salinity optima of diatom species based on Hustedt's halobous system (Hustedt, 1953, 1957) including modifications of Simonsen (1962). Hustedt divided diatom species into four salinity groups: polyhalobous (salinity >30‰), mesohalobous (salinity 0.2–30‰), oligohalobous (taxa having their optimum in brackish water and freshwater, but tolerating brackish water) and halophobous taxa (exclusively freshwater taxa).

Simonsen (1962) also divided marine and brackish water species into three upper-levels, and then subdivided oligohalobous and polyhalobous taxa further depending on their salinity tolerance into the subgroups oligo-, meio-, meso-, pleio- and holoeuryhalobic, with oligohalobic indicating a small and holoeuryhalobic a large range of tolerance (Fig. 2).

In addition, to show biological effects of salinity changes, the halobios index according to Ziemann (1971) was calculated. This index is based on diatoms occurring in inland saline waters (Ziemann, 1971, 1982). Thus, several typical polyhalobous (oceanic) species, which may be present at the White Sea coast, are not included. However, because many diatoms included in Ziemann's index also occur in saline habitats of brackish and marine coasts (Pankow et al., 1990; Witkowski et al., 2000), this index is suitable for showing general trends in salinity change in coastal areas.

Ziemann's index not only considers assemblage composition but also estimates abundances of different species, using abundance classes from 1 (very rare) to 6 (abundant). Thus, the relative frequency of occurrence of a particular diatom taxon has to be converted into one of Ziemann's (1982) classes of relative abundance (Hofmann, 1997). The halobios index (Ziemann, 1971, 1982) ranges from −100 (extremely low salt content, generally acidic freshwater) to +100 (typical saltwater).

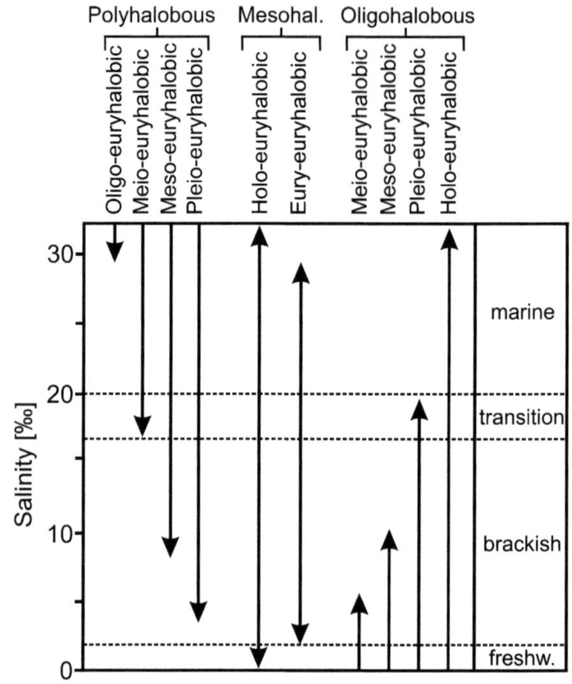

Fig. 2 Salinity tolerance of marine and brackish water species based on Simonsen (1962), modified according to Pankow et al. (1990). Mesohal. = mesohalobous, freshw. = freshwater

Sediment zonation was calculated with a stratigraphically constrained incremental sum of squares cluster analysis (CONISS; Grimm, 1987) implemented in the computer program Tilia 2 (Grimm, 1990) and based on diatom assemblage changes (square-root transformed relative abundance of diatoms). The statistical significance of diatom assemblages zones (DAZ) was assessed by a broken-stick model (Bennett, 1996) implemented in the computer program PSIMPOLL 4.25 (Bennett, 2005).

Diatom data (relative abundance) were analysed numerically as response or dependent variables with multivariate statistical analyses using the computer program CANOCO 4.5 (Ter Braak & Śmilauer, 2002). Detrended correspondence analyses (DCA) (Ter Braak, 1987) were used to choose between linear response models such as principal components analysis (PCA) or unimodal models, such as correspondence analysis (CA) (Ter Braak & Prentice, 1988; Lepš & Šmilauer, 2003). Detrending was done by segments, and non-linear rescaling resulted in standard deviation (SD) units for the first axis (Hill & Gauch, 1980).

Pollen and other microfossil analysis

Pollen samples (0.5 cm^3) were treated using standard methods (Faegri & Iversen, 1975). At least 1,000 pollen grains per sample were counted and identified using a ZEISS Axiolab light microscope (magnification 500×). Pollen and other microfossil data were expressed as percentages of the total sum of terrestrial plant pollen. Pollen species identification was according to Punt (1976), Punt & Clarke (1980, 1981), Punt et al. (1988, 1996), Moore et al. (1991), Punt & Blackmore (1991) and Beug (2004). Sediment zonation and the significant number of pollen assemblage zones (PAZ) were calculated in the same way as for the diatoms.

In addition to pollen and diatoms, also chrysophyte cysts, marine Dinophyceae, foraminifera, *Cymatiosphaera*, *Botryococcus*, *Pediastrum* and *Pediastrum boryanum* (Turpin) Meneghini were counted. Numbers of chrysophyte cysts are expressed as percentages of the total sum of diatoms. The identification of these microfossils was according to Bakker & van Smeerdijk (1982), van Geel (2001), Jankovská & Komárek (2000) and Komárek & Jankovská (2001).

Results

Chronology

Because of increasing uncertainties of the measured activity concentrations with increasing sediment depth, the $^{210}Pb_{ex}$-method could only be applied to depths of <26 cm (ESW) and <16 cm (ESE). Since a reliable interpretation of the detected $^{210}Pb_{ex}$ activities, represented by the differences between ^{210}Pb and ^{226}Ra activity concentrations, is only possible if relative errors are below 100%, no results for deeper sediment layers can be given.

The vertical distribution of ^{137}Cs differs significantly between the two cores and only in the sediment of ESW significant maxima are identifiable (Fig. 3). The most recent ^{137}Cs maximum in the core of ESW is at 3–4 cm depth (Fig. 3). A second, less distinct ^{137}Cs maximum occurs at 11–12 cm depth. The evaluation of the ^{137}Cs data indicates that the CRS-model is the best suited method for ^{210}Pb-excess dating of the ESW sediment core. According to the CRS-model, the oldest datable sample of the ESW sediment core (26 cm) represents an age of 136.7 ± 54.7 years (Fig. 3).

In the core of ESE, relatively constant ^{137}Cs activities of approximately 150 Bq kg^{-1} were detected down to a depth of 12 cm. Below depth range 12–17 cm, this activity values dropped to 20 Bq kg^{-1} (Fig. 3). As no distinctive ^{137}Cs maxima that could be attributed to historic fallout events were found, a reliable dating of the core was not possible using ^{137}Cs, thus excluding validation of the $^{210}Pb_{ex}$-dating results on basis of ^{137}Cs data. Nevertheless, samples down to a depth of 16 cm were dated according to $^{210}Pb_{ex}$ data to obtain a rough approximation of the time scale. According to the CRS-model the oldest $^{210}Pb_{ex}$-datable sample of the ESE sediment core (16 cm) revealed an age of 149.0 ± 59.6 years (Fig. 3).

Diatoms of lake ESW

In the 24 samples of the ESW core, 157 diatom taxa were identified, 12 of them reached relative abundances above 5% in at least one sample (Fig. 4). The number of species per sample ranges from 8 to 37. Out of a total of 157 species, 122 were used for the calculation of the Ziemann halobios index. Three different local diatom assemblage zones (DAZ) were identified based on variations in salinity: DAZ-ESW-1 (47–35 cm), DAZ-ESW-2 (33–17 cm) and DAZ-ESW-3 (15–1 cm) (Fig. 4).

In the oldest sediments of DAZ-ESW-1 (47 and 45 cm), meso-euryhalobic polyhalobous diatoms (Fig. 5) as *Paralia sulcata* (Ehrenberg) Cleve constitute the dominating group with 38% relative abundance (Figs. 4 and 5). Pleio-euryhalobic polyhalobous diatoms (Fig. 5) as *Cocconeis scutellum* Ehrenberg constituted 14% (Fig. 4). Meio-euryhalobic polyhalobous diatoms such as *Rhabdonema minutum* Kützing amounted to 8.8% (Figs. 4 and 5) whereas *Diploneis smithii* (Brebisson) Cleve (pleio-euryhalobic polyhalobous) (not shown on the diagram) only to 2.2%.

Above 45 cm, meso-euryhalobic oligohalobous taxa, mainly *Staurosira pinnata* Schmidt, Kamenik, Lange-Bertalot & Klee (Figs. 4 and 5) and *Staurosira brevistriata* (Grunow) Grunow, are the dominating group with 50–69% of relative abundance (Figs. 4 and 5). Meio-meso-euryhalobic oligohalobous diatoms, only represented by *Staurosira* aff. *venter* (Ehrenberg) Cleve & Möller (Fig. 4), dominate this section with 20–45% (Fig. 5). Freshwater (halophobous)

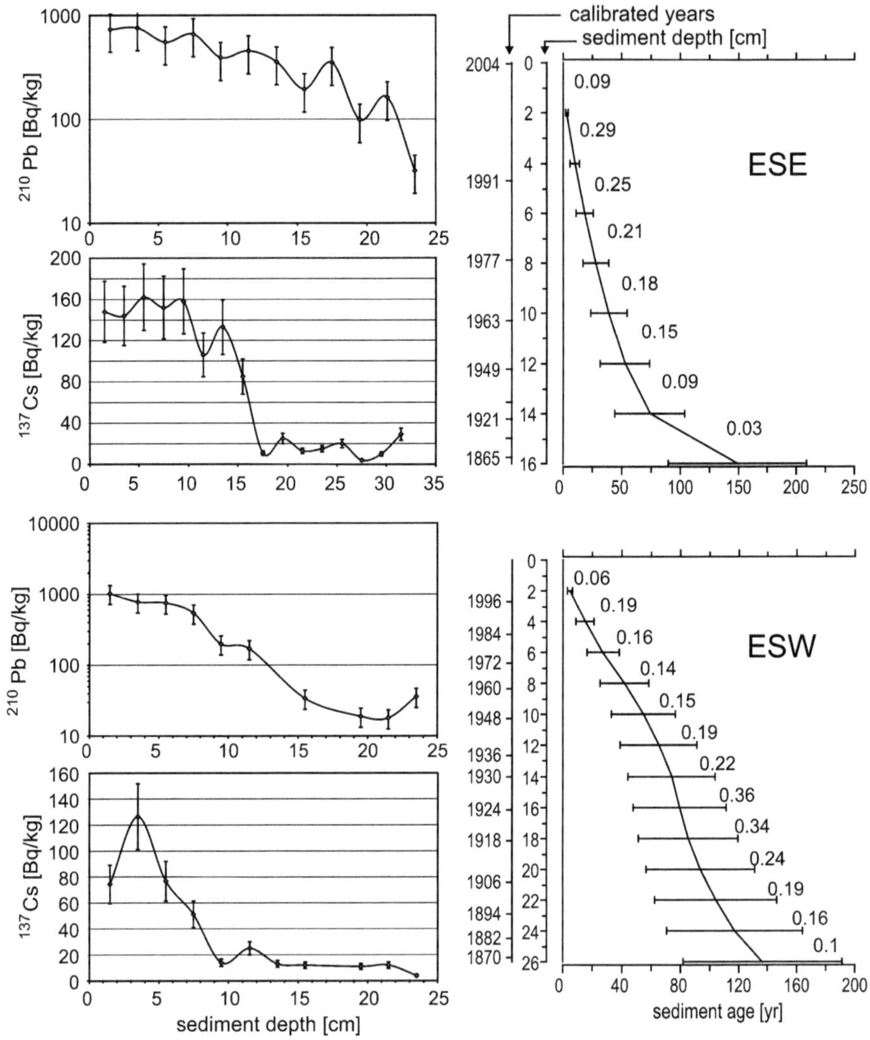

Fig. 3 ^{210}Pb and ^{137}Cs profiles of the lakes ESE (*above*) and ESW (*below*) are shown *on the left side*. The ^{210}Pb chronology (*on the right side*) was calculated using the CRS-model. Error bars represent 95% confidence intervals. Sedimentation rates (*on the right side*) are given as mm yr^{-1}

diatoms account for 0–3%. Mesohalobous taxa constitute <2% in DAZ-ESW-1. The halobios index fluctuates around +45 throughout DAZ-ESW-1 except for the oldest sample (47 cm) where a value of +55 was calculated (Fig. 5).

In DAZ-ESW-2, abundance of polyhalobous diatoms (e.g. *P. sulcata*, *C. scutellum* and *R. minutum*) further decreases (Fig. 5). Instead, DAZ-ESW-2 is dominated by *S.* aff. *venter*, a meio-meso-euryhalobic oligohalobous taxon accounting for 50–86% (Fig. 4). The content of meso-euryhalobic oligohalobous taxa including *S. pinnata* and *S. brevistriata* fluctuates around 10%, decreasing to 6% at the transition to DAZ-ESW-3 (Figs. 4 and 5). Holo- and pleio-euryhalobic oligohalobous diatoms occur for the first time at a depth of 27 cm with contents of 2.7 and 13%, respectively (Fig. 5). In the same samples, mesohalobous diatoms represented by, e.g. *Fragilaria fasciculata* (Agardh) Lange-Bertalot and *Navicula gregaria* Donkin appear, but constitute less than 3% of the assemblage (not shown).

The transition to DAZ-ESW-3 is marked by increasing importance of meio-euryhalobic oligohalobous taxa. The abundance of halophobous taxa, e.g. *Fragilaria exigua* Grunow, *Tabellaria flocculosa* Kützing and *Achnanthes subatomoides* (Hustedt) Lange-Bertalot & Archibald, increased continuously reaching 15.5% at 17 cm (Figs. 4 and 5). The halobios index oscillate around +45 between 35 and 23 cm. Between 21 and 19 cm, the halobios index further declines and reaches values representing well-buffered freshwater. At 17 cm, a value of

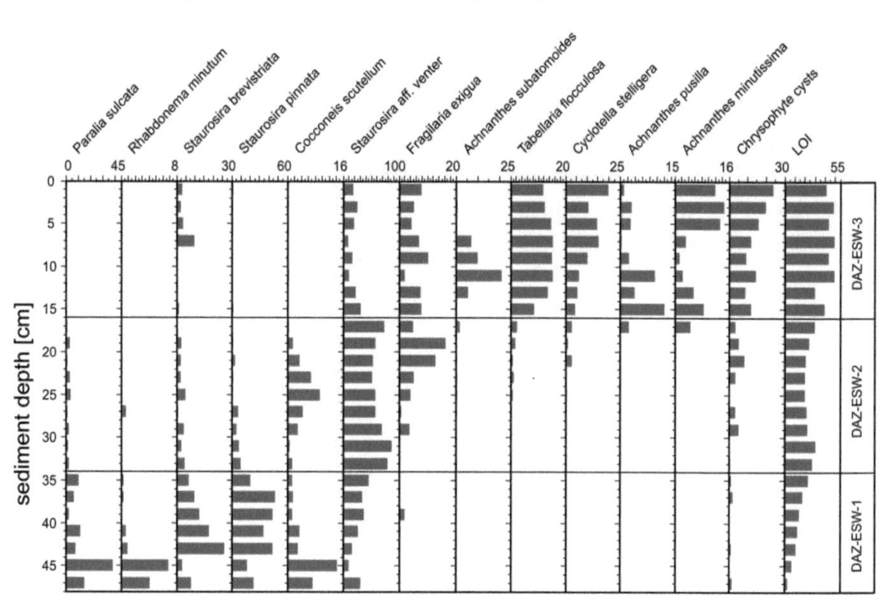

Fig. 4 Main diatom taxa (%), chrysophyte cysts (% of all diatoms), LOI (%) and diatom assemblage zones (DAZ) of lake ESW. Zonation was defined by CONISS

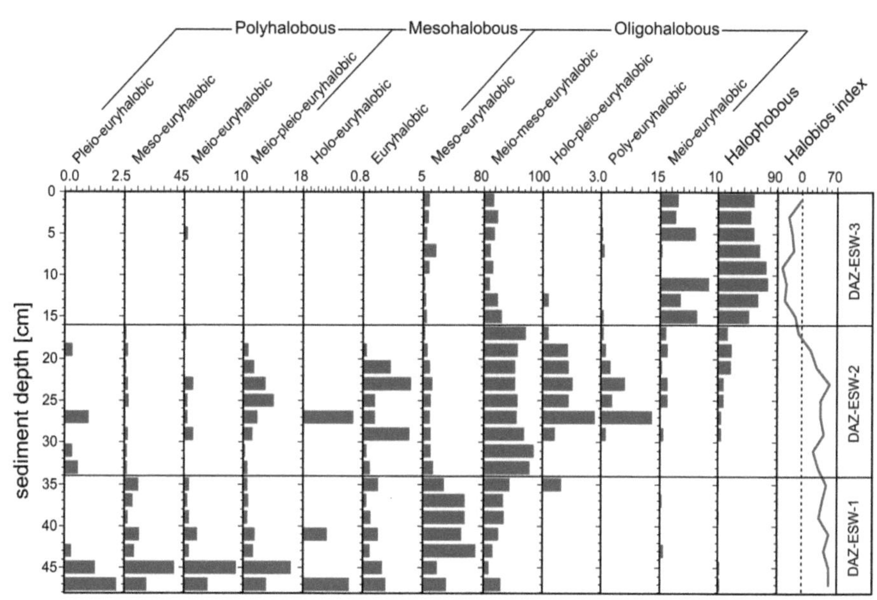

Fig. 5 Relative abundance of diatom groups (polyhalobous, mesohalobous, oligohalobous and halophobous) of lake ESW based on salinity optima according to Hustedt (1953, 1957). Subdivision of the diatom groups (oligo-, meio-, meso-, pleio- and holoeuryhalobic) according to Simonsen (1962). The halobios index (*on the right side*) was calculated according to Ziemann (1971, 1982)

−6 (Fig. 5) indicates weakly buffered freshwater conditions.

In DAZ-ESW-3, the previously common polyhalobous diatoms can be detected only sporadically (<1%, Fig. 5), while mesohalobous taxa disappear from the core. Instead, this zone is dominated by halophobous species (e.g. *T. flocculosa*, *Cyclotella stelligera* Cleve & Grunow, *A. subatomoides* and *F. exigua*) (Figs. 4 and 5). These taxa reach their maximum (75%) at 11 cm. In DAZ-ESW-3, *Achnanthes minutissima* Kützing, an oligohalobous species with a maximum abundance in freshwater, but tolerating brackish water, and the halophobous *Achnanthes pusilla* (Grunow) De Toni (Fig. 4), *Achnanthes suchlandtii* Hustedt, and *Eunotia sudetica* O. Müller (not shown) are recorded for the first time in the sediment of ESW and reach abundances exceeding 5%. Of the oligohalobous taxa, meso-euryhalobic (e.g. *S. pinnata* and *S. brevistriata*) reach 5–17% while meio-meso-euryhalobic diatoms (e.g. *S.* aff. *venter*) range between 10 and 30% (Figs. 4 and 5). The content of meio-euryhalobic oligohalobous diatoms vary between 6 and 8% in the older parts of DAZ-ESW-3, then decrease to <1% between 9 and 7 cm before they rise again in the youngest sediments to 2–6% (Fig. 5). The halobios index gives negative values of −12 to −39 throughout DAZ-ESW-3.

The length of the first gradient of the detrended correspondence analysis (DCA) performed with diatom samples of lake ESW is 3.7 standard deviation (SD) units. This gradient, if greater than 4 SD units, reflects the unimodal distribution of species in the downcore dataset, thereby warranting the use of unimodal methods in further analyses (Ter Braak & Šmilauer, 2002). Due to the relatively short gradient, principal components analysis (PCA) (linear method) has been used for lake ESW (Fig. 6). The gradient length of the second axis is 1.6 SD. Eigenvalues of the first axis are 0.559, i.e. this axis explains 56% of the variation in the diatom flora. The second axis explains 30% of the variation. The PCA groups the ESW-samples into three zones (Fig. 6), which correspond to the diatom assemblage zones calculated by the cluster analysis (Fig. 4).

Along the first PCA-axis, the opportunistic taxa *S.* aff. *venter* can be found (Fig. 6). Along the second axis halophobous taxa are arranged towards samples of DAZ-ESW-3 (Fig. 6). Polyhalobous diatoms are also scattered along the second axis, but in the direction towards samples of DAZ-ESW-1. Thus, the second axis presumably corresponds to salinity changes, whereas the first axis is consistent with the so-called disturbance gradient.

Diatoms of lake ESE

In the 26 samples from the ESE core, 132 diatom taxa were identified. Twelve of them reach relative abundances exceeding 3% in at least one sample (Fig. 7). The number of species per sample ranges from 20 to 41 and the diatom-halobios index was calculated using 116 out of the 132 species identified. Changes in the diatom distribution allowed us to distinguish three local diatom assemblage zones (DAZ): DAZ-ESE-1 (51–37 cm), DAZ-ESE-2 (35–17 cm) and DAZ-ESE-3 (15–1 cm) (Fig. 7).

Meso-euryhalobic polyhalobous diatoms, mainly *Paralia sulcata*, constitute 20% in DAZ-ESE-1 at 51 cm (Figs. 7 and 8). Other polyhalobous species (e.g. *Cocconeis scutellum*) do not exceed 3% in this sample (Fig. 8). In the following sediment samples,

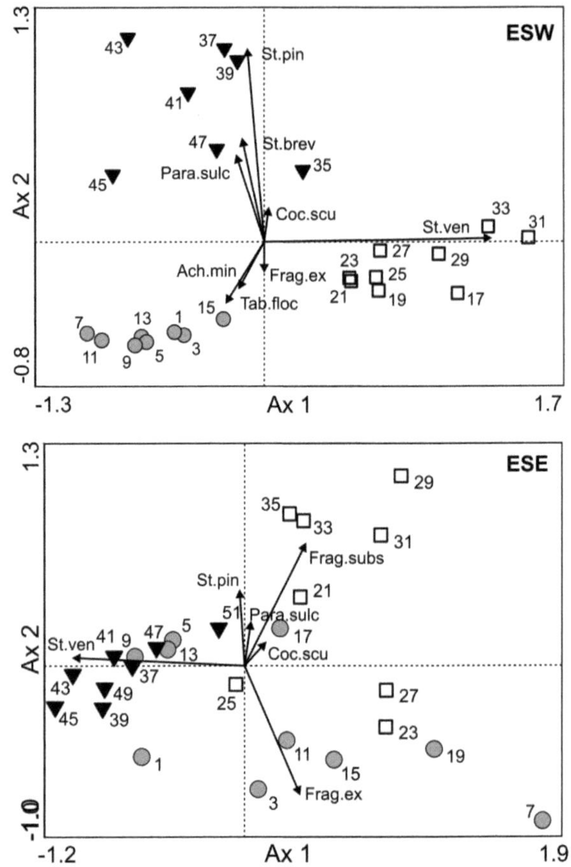

Fig. 6 PCA plots of the lake ESW and ESE diatom samples. *Arrows* indicate important diatom taxa. Diatom assemblage zones (DAZ) defined using CONISS. *Triangles*: samples belonging to DAZ-1, *Quadrates*: samples belonging to DAZ-2, *Circles*: samples belonging to DAZ-3. Ach.min = *Achnanthes minutissima*; Coc.scu = *Cocconeis scutellum*; Frag.ex = *Fragilaria exigua*; Frag.subs = *Fragilaria subsalina*; Para.sulc = *Paralia sulcata*; St.brev = *Staurosira brevistriata*; St.pin = *Staurosira pinnata*; St.ven = *Staurosira* aff. *venter*; Tab. floc = *Tabellaria flocculosa*

percentages of all polyhalobous taxa decrease, reaching values of only 1–2% at the transition to DAZ-ESE-2 (Fig. 8). The dominant taxon throughout DAZ-ESE-1 is *Staurosira* aff. *venter* (56–75%; Fig. 7). Mesohalobous diatoms (e.g. *Fallacia cryptolyra* (Brockmann) Stickle & Mann) and meso-euryhalobic oligohalobous taxa contribute to 1–4% and 5–15%, respectively. Neither freshwater species nor meio-euryhalobic oligohalobous diatoms reach contents exceeding 1%. The halobios index indicates saltwater with moderate to high salinities throughout DAZ-ESE-1.

Fig. 7 Main diatom taxa (%), chrysophyte cysts (% of all diatoms), LOI (%) and diatom assemblage zones (DAZ) of lake ESE. Diatom assemblage zones (DAZ) defined using CONISS

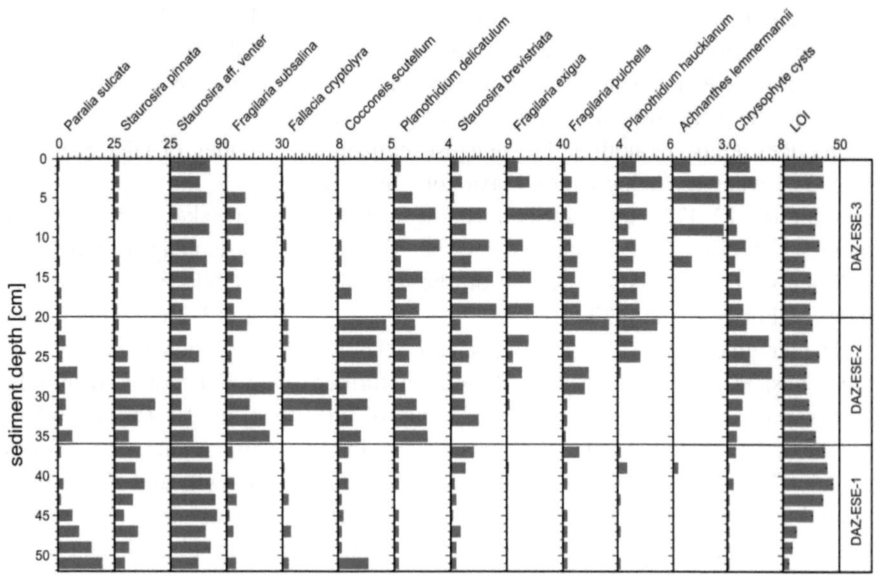

Fig. 8 Relative abundance of diatom groups (polyhalobous, mesohalobous, oligohalobous and halophobous) of lake ESE based on salinity optima according Hustedt (1953, 1957). Subdivision of the diatom groups (oligo-, meio-, meso-, pleio- and holoeuryhalobic) according to Simonsen (1962). The halobios index (*on the right side*) was calculated according to Ziemann (1971, 1982)

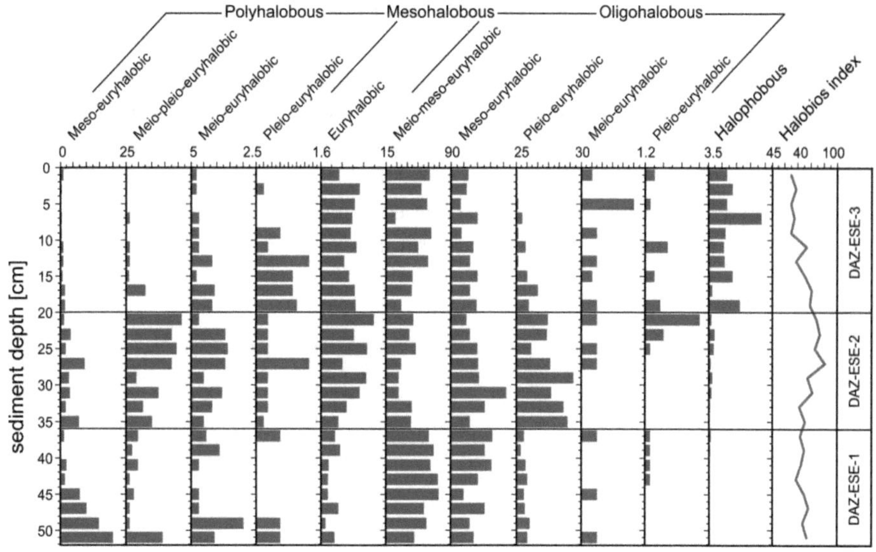

In contrast to the oldest sediments, meso-euryhalobic polyhalobous taxa contribute <10% in DAZ-ESE-2 (Fig. 8). The abundances of other polyhalobous taxa on the other hand increase slightly in this zone, together accounting for 3–6% (Fig. 8). Euryhalobic mesohalobous taxa (e.g. *Planothidium hauckianum* (Grunow) Round & Bukhtiyarova and *F. cryptolyra*) increase in abundance and fluctuate around 10% (Figs. 7 and 8). *Staurosira* aff. *venter* is still the dominant taxon, however, at 40–50%, its abundance is lower than in DAZ-ESE-1. Meso- and pleio-euryhalobic oligohalobous taxa (e.g. *Staurosira pinnata*, *S. brevistriata* and *Fragilaria subsalina* (Grunow) Lange-Bertalot) both reach relatively high percentages, with 20% each (Fig. 8). Also freshwater species (*Fragilaria exigua*, *Navicula rhynchotella* Lange-Bertalot and *Tabellaria flocculosa* (not shown)) are more abundant in DAZ-ESE-2 than in the previous zone, but never exceed 10% (Fig. 8). The halobios index fluctuates considerably, varying between +50 and +80 (saltwater conditions with high salinities).

In DAZ-ESE-3, the content of all polyhalobous taxa decreases to <2% (Fig. 8). Euryhalobic mesohalobous diatoms (e.g. *P. hauckianum*) usually exceed 10% (Fig. 8). *Staurosira* aff. *venter* dominates this zone as well, again reaching values of approximately 70% (Fig. 7). The pleio-euryhalobic oligohalobous *F. subsalina* does not occur at all in the youngest sediments (Fig. 7), whereas halophobous taxa on average account for 12–17%, temporarily reaching a maximum of 37.5% (Figs. 7 and 8). In the younger sediments of DAZ-ESE-3, values of the halobios index stabilise around +30.

The length of the first DCA axis of lake ESE is 1.5 SD. Thus, PCA was the method chosen for further multivariate analyses. The gradient length of the second axis is 0.9 SD. The first axis of the PCA explains 51% of the variation in the diatom flora, the second 31%. The PCA groups, the ESE-samples into three zones (Fig. 6), which correspond to the diatom assemblage zones calculated earlier by the cluster analysis (Fig. 7). However, transitions between the zones are not as distinct as in lake ESW, but rather blurred.

Staurosira aff. *venter* can be found along the first axes of the PCA (Fig. 6). Along the second axes close to the centres of the coordinate systems, halophobous taxa are arranged towards samples of DAZ-ESE-3 (Fig. 6). Polyhalobous diatoms are also scattered along the second axes, but in the direction towards samples of DAZ-ESE-1. Thus, the second axes presumably reflect salinity changes, whereas the first axes are consistent with the so-called disturbance gradient.

Pollen of lake ESW

In lake ESW, two significantly different pollen assemblages zones (PAZ) can be identified by CONISS: PAZ-ESW-1 (45–21 cm) and PAZ-ESW-2 (17–1 cm) (Fig. 9).

Arboreal pollen (AP) dominates throughout both pollen zones with relative abundances of 96–99% (Fig. 9). *Pinus* (~50%, not shown), is the most important genus, followed by *Betula* (~30%, not shown) and *Picea* (6–12%, not shown). *Ulmus* and *Alnus* also occur regularly, but never reach contents exceeding 1 and >2%, respectively (not shown). *Juniperus communis* L. occurs with relative abundances of 1–3.3% throughout PAZ-ESW-1 (Fig. 9), while pollen of *Salix*, *Quercus* and *Corylus* can be found only sporadically and always contribute <1% each (not shown).

In PAZ-ESW-1, the relative abundance of non-arboreal upland pollen (NAP-upland; pollen of terrestrial plants, which do not grow in water or wetlands) does not exceed 2% (Fig. 9). Cyperaceae occur regularly but relative abundances are <3%

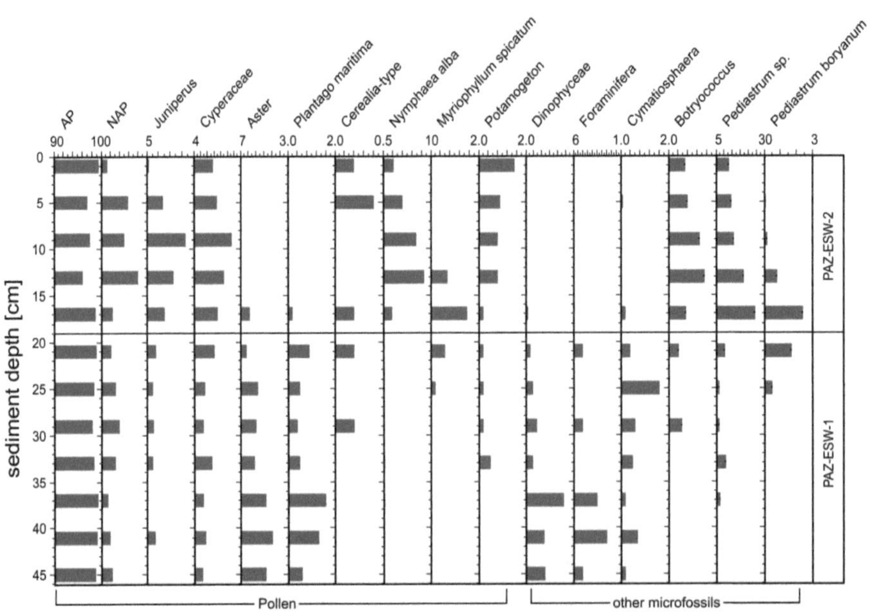

Fig. 9 Main pollen and other microfossils for lake ESW expressed as percentages of the total sum of terrestrial plants. Pollen assemblage zones (PAZ) (*on the right*) defined using CONISS. AP = arboreal pollen; NAP = non-arboreal pollen (upland)

throughout PAZ-ESW-1. Pollen of *Potamogeton* (Fig. 9) can be found for the first time in a depth of 33 cm and the share of all aquatic plants (referring to the sum of all terrestrial plants; *Nymphaea alba*, *Myriophyllum spicatum* L., *Potamogeton*, *Nuphar* and *Sparganium*) does not exceed 1% in PAZ-ESW-1. Cerealia-type pollen reached relative abundances of 0.2% in depths of 29 and 21 cm (Fig. 9). Furthermore, pollen of *Plantago maritima* L. (0.4–2%), *Aster* (0.3–2%; Fig. 9) and *Rumex* (<0.5%, not shown) can be found regularly in PAZ-ESW-1.

In PAZ-ESW-2, abundances of AP decrease slightly from 97% to 93% between 13 and 5 cm (Fig. 9). Simultaneously, the content of NAP-upland increases from 1% to 4% and thus, reaches its maximum (Fig. 9). At 9 cm, Cyperaceae (Fig. 9) reaches its maximum abundance of 5.5%, before these values drop again to 2.7% at 1 cm. Cerealia-type pollen fluctuate throughout PAZ-ESW-2 and reach their highest contents in depths of 13 cm (0.9%) and 5 cm (0.5%; Fig. 9). Relative abundances of aquatic plants increase up to a depth of 13 cm (10%) before they decline again to 5% in the younger sediments. The abundance of halotolerant plants as *P. maritima* decrease significantly in PAZ-ESW-2 (Fig. 9), while pollen of *N. alba* reach abundances between 2 and 8.5%. *Rumex* (not shown) and *Potamogeton* (Fig. 9) occur regularly, also in the youngest sediments, but never exceed 2%. Between 29 and 9 cm, pollen of *Myriophyllum spicatum* can be found, reaching a maximum (1.5%) at the transition between PAZ-ESW-1 and 2 (Fig. 9).

Pollen of lake ESE

In lake ESE, CONISS also identified two significantly different pollen assemblage zones: PAZ-ESE-1 (52–22 cm) and PAZ-ESE-2 (21–1 cm; Fig. 10). As in lake ESW, AP are the dominant group throughout the ESE sediment core (97.7–99.6%, Fig. 10), with *Pinus* (not shown) being the most abundant species (consistently >55%, not shown), followed by *Betula* and *Picea* (<20 and <10%, respectively, not shown). Pollen of *Alnus* (1–3.5%) and *Corylus* (<1%) (not shown) also occur regularly in PAZ-ESE-1, while those of *Salix*, *Quercus*, *Ulmus* (not shown) and *Juniperus* (Fig. 10) can be found only sporadically and with low abundances. NAP-upland are represented with ~1% in the beginning of PAZ-ESE-1 (Fig. 10). Between 31 and 22 cm their abundance increases to >2%, leading to a decline of AP to <98%.

Cyperaceae occur regularly with a maximum abundance of <3% in PAZ-ESE-1 (Fig. 10). *Potamogeton* on the other hand can only be detected once

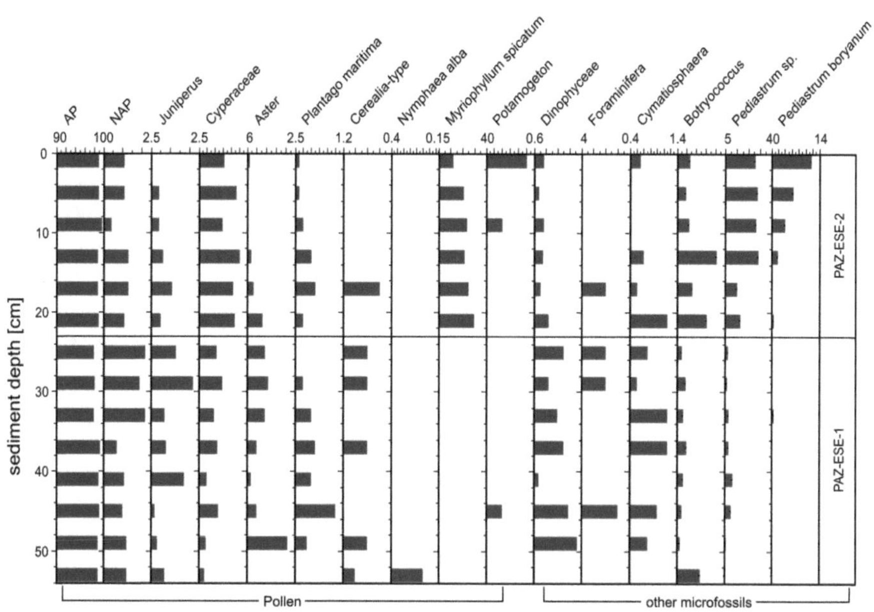

Fig. 10 Main pollen and other microfossils for lake ESE expressed as percentages of the total sum of terrestrial plants. Pollen assemblage zones (PAZ) (*on the right*) were set according to CONISS. AP = arboreal pollen; NAP = non-arboreal pollen (upland)

at 52 cm (Fig. 10). The content of all aquatic plants (*Nymphaea alba*, *Myriophyllum spicatum*, *Potamogeton*, *Nuphar* and *Sparganium*) consistently is <1% throughout this zone. Also Cerealia-type pollen occurs only erratically with abundances of $\leq$0.2% (Fig. 10). Relative abundances of *Plantago maritima* and *Aster* fluctuate throughout PAZ-ESE-1, reaching values of $\leq$1 and $\leq$2.4%, respectively (Fig. 10).

In PAZ-ESE-2, Cyperaceae reach slightly higher abundances (>3%) than in the previous zone (Fig. 10). Cerealia-type pollen can only be found in one sample of PAZ-ESE-2 (14 cm), where they contribute 0.3% (Fig. 10). The share of all aquatic plants, on the other hand, increases to >20% in PAZ-ESE-2, with *N. alba* alone reaching abundances of up to 8% (Fig. 10).

Other microfossils of lake ESW

With relative abundances of 0.0–1.6%, the Chrysophyte cysts are negligible in DAZ-ESW-1. In DAZ-ESW-2, their content increases to 5–8% (Fig. 4). This increase of Chrysophyte cysts continues in DAZ-ESW-3, resulting in relative abundances of >20% in the youngest sediments (Fig. 4).

Of the microfossils counted in pollen samples, indicators of saline conditions as marine Dinophyceae, foraminifera and *Cymatiosphaera* can be found almost exclusively in PAZ-ESW-1, with their contribution exceeding 5, 1 and 2%, respectively (Fig. 9). At the transition to PAZ-ESW-2, *Pediastrum boryanum* reaches abundances of >2% before it disappear again (Fig. 9). Within PAZ-ESW-2, the freshwater taxa *Botryococcus* (2–4%) and *Pediastrum* (8–25%) dominate.

Other microfossils of lake ESE

Chrysophyte cysts (<1.3%) are negligible in DAZ-ESE-1. Again, their contents increase in DAZ-ESE-2 and reach values of 6.4 and 5.9% in depths of 27 and 23 cm, respectively (Fig. 10). In contrast to lake ESW, however, chrysophyte cysts decrease again in DAZ-ESE-3 to values of 1–4% (Fig. 10).

Sea water indicators including marine Dinophyceae, foraminifera and *Cymatiosphaera* also dominant in PAZ-ESE-1 (<4, <0.4 and <1.3%, respectively, Fig. 10) but occur sporadically with low abundances in PAZ-ESE-2, too. Similar to lake ESW, the freshwater taxa *Botryococcus* (<5%), *Pediastrum* (10–40%) and *Pediastrum boryanum* (0–11%) can be found regularly in PAZ-ESE-2.

Loss on ignition

In lake ESW, the organic matter content of the sediment, expressed as loss on ignition (LOI), increases in the course of DAZ-ESW-1 from 7% to 23% (Fig. 4). In DAZ-ESW-2, it averages ca. 25% before it further increases at the transition to DAZ-ESW-3 to 40% and then remains at this level throughout DAZ-ESW-3 (Fig. 4).

In lake ESE, the LOI is 12% at the beginning of DAZ-ESE-1 but than increases rapidly to its maximum of 44% (Fig. 7). In DAZ-ESE-2, LOI amounts to 20–30%. In DAZ-ESE-3, it increases slightly and fluctuates between 20 and 35% (Fig. 7).

Discussion

Chronology

Increasing measurement uncertainties with increasing sediment depth do not allow reliable ^{210}Pb-dating of the sediment cores below 26 cm (approx. AD 1870) in lake ESW and 16 cm (ca. AD 1865) in lake ESE. Partly, these measurement uncertainties may be due to low counting rates based on low sample weight (Appleby, 2001). Furthermore, the large tidal range, which occurs in the study area, may have contributed to the errors in dating of the older sediments: tides can cause substantial disturbances in the sedimentation process due to resuspension, redeposition and dissolution especially of coastal and estuarine sediments (Appleby & Oldfield, 1983). This would explain why the older sediments, which where subject to saltwater inundation due to tidal impacts, were undatable, whereas in the younger, freshwater-dominated sediments, where sedimentation processes were less disturbed, had relatively small ^{210}Pb-dating errors.

Furthermore, using different models as CRS and CIC, reliable dating based on the half-life of ^{210}Pb (22.3 yr) is only possible for the last ~150 years (Appleby, 2001). As the deepest datable sediments of this study are ~130 years (ESW) and ~135 (ESE) years old, it can be assumed that dating of deeper sediments would not have been possible also due to this methodological limit.

Moreover, it has to be considered that both ESW and ESE are shallow lakes. The sediment of shallow lakes regularly is subject to disturbance of the sedimentation process due to bioturbation, and reworking by winds and currents (Conley, 2004). This could have added to the measurement uncertainties.

Results of ^{210}Pb-dating based on different models (CRS, CIC) commonly are evaluated using ^{137}Cs (Andersen et al., 2000), a radioactive isotope, which exclusively originates from anthropogenic sources. This method measures ^{137}Cs accumulations in sediments from known historical emission events (Blais et al., 1995). In this study, only the vertical ^{137}Cs distribution of lake ESW gave useable results. Here, the most recent ^{137}Cs maximum (in a depth of 3–4 cm) presumably is attributable to the Chernobyl disaster of 1986 (Andersson & Roed, 1994). A second, less distinctive ^{137}Cs maximum in a depth of 11–12 cm possibly marks the first fallout-maximum from 1959, which resulted from above-ground nuclear weapons tests (Appleby, 2001). The stratigraphic position of these two events indicates that the CRS-model of ^{210}Pb-excess dating is the most suitable method to date the sediment core of ESW.

In the sediment of lake ESE, no distinct ^{137}Cs-maxima occur. Partly, this may be due to sediment disturbances. Additionally, caesium can migrate within the sediment (Olsson, 1986). And furthermore, it has to be considered that samples were analysed in 2 cm intervals, thus sharp ^{137}Cs-maxima have been overlooked.

Salinity changes of lake ESW

In the sediments of lake ESW, cluster analysis and PCA identify three diatom assemblage zones. The oldest sediments (DAZ-ESW-1) were deposited in saltwater, the most recent ones (DAZ-ESW-3) in freshwater and the intermediate DAZ-ESW-2 represents the transition between saltwater and freshwater conditions.

Diatoms that have their salinity optimum at 17–30‰ and, thus, are characteristic of saltwater and brackish water (e.g. *Rhabdonema minutum* and *Cocconeis scutellum*) can be found in DAZ-ESW-1. The dominant species in this zone is *Paralia sulcata*, which is typical for marine, sandy, coastal tidal flats. With a salinity optimum of 8–30‰ it is perfectly adapted to tidal dynamics (McQuiod & Nordberg, 2003). The halobios index (+40 to +60) indicate moderate to high salt concentrations (Ziemann, 1971, 1982).

At the top of DAZ-ESW-1, the salinity starts decreasing as indicated by the increasing importance of oligohalobous diatoms and the simultaneously decreasing abundances of polyhalobous taxa (Ziemann, 1971). However, despite the consistently decreasing salinity, saltwater conditions still dominate throughout most of DAZ-ESW-2. Only at the end of this zone mesohalobous (*Diatoma tenuis*) and polyhalobous diatoms (e.g. *C. scutellum*) finally disappear. Accordingly, the halobios index corresponds to weakly buffered freshwater at the end of DAZ-ESW-2.

The striking feature of DAZ-ESW-2 is the dominance (50–85%) of *Staurosira* aff. *venter*, a benthic and tychoplanktonic diatom, which is known to reach high abundances in salinities up to 10‰ (Pankow et al., 1990). This small species shows high reproduction rates and a wide ecological amplitude, which makes it a very opportunistic species that outcompetes other diatoms under changing environmental conditions, especially in climates with short vegetation periods as found in arctic or alpine regions (Lotter et al., 1999). This competitiveness of *S.* aff. *venter* has been documented for changes in ice-covering (Lotter & Bigler, 2000) and trophic level (Schmidt et al., 2004) as well as for changing salinity (Denys, 1990) as found in DAZ-ESW-2.

In DAZ-ESW-3, change towards freshwater reached its final stage as poly- and mesohalobous taxa disappear while halophobous species increase. From this point, freshwater species including *Tabellaria flocculosa* and *Achnanthes subatomoides* occur. These are characteristic for acidic to slightly alkaline zones in oligothrophic, electrolyte-poor waters (Hofmann, 1994). The halobios index corresponds to weakly buffered, acidic freshwater conditions (Ziemann, 1997). This is consistent with recent environmental settings.

The strong correlation ($r^2 = 0.85$) between the changing halobios index, which reflects the biological effects of salinity changes, and the sample scores of the first DCA axis indicates that the changing salinity was the environmental variable responsible for the development of the diatom flora in lake ESW.

The cluster analysis performed with the pollen data distinguishes two zones in the sediment of lake ESW. The border between these pollen zones is situated in a depth between 20 and 16 cm. This corresponds to the transition from DAZ-ESW-2 to DAZ-ESW-3, which lies between 17 and 15 cm. Thus, PAZ-ESW-1 largely correlates to the salt- and brackish water-dominated DAZ-ESW-1 and DAZ-ESW-2, whereas PAZ-ESW-2 corresponds to the freshwater-dominated DAZ-ESW-3. Hence, it can be concluded that the pollen data also clearly reflect changes in the salinity.

Accordingly, the occurrence of pollen of halotolerant species, such as *Aster tripolium* and *Plantago maritima* in PAZ-ESW-1, indicates typical marine shoreline vegetation as characteristic also for recent drift lines in the adjacent White Sea bays (Filatov et al., 2005).

At the transition to PAZ-ESW-2, pollen abundances of these halotolerant species decline while those of freshwater species, such as Cyperaceae and *Potamogeton*, increase. Furthermore, in the transition zone between PAZ-ESW-1 and PAZ-ESW-2, also pollen of *Myriophyllum spicatum* becomes abundant. The occurrence of this species, which prefers carbonate-rich waters, suggests electrolyte-rich freshwater (Casper & Krausch, 1981). Thus, despite already prevailing freshwater conditions, occasional saltwater influences presumably persisted well into PAZ-ESW-2 leading to an accumulation of electrolytes and thereby creating a suitable habitat for *M. spicatum*.

During the course of PAZ-ESW-2, pollen abundances of characteristic freshwater species such as Cyperaceae and *Potamogeton* increases. Furthermore, pollen of *Nymphaea alba*, a species that does not tolerate saltwater (Ellenberg, 1992), appear. Thus, the pollen analysis results also suggest that freshwater conditions prevail in PAZ-ESW-2.

Similar to diatom and pollen data, also the other microfossils indicate salinity changes in the sediment of lake ESW: In DAZ-ESW-1 and DAZ-ESW-2 (i.e. in PAZ-ESW-1) mainly marine organisms such as Dinophyceae, foraminifera, and *Cymatiosphaera* (Bakker & van Smeerdijk, 1982) occur. At the transition to PAZ-ESW-2, their abundances decrease (in concordance with the beginning freshwater conditions). And also the appearance of chrysophyte cysts, *Pediastrum* and *Botryococcus* in PAZ-ESW-2 indicates decreasing salinity as *Pediastrum* and *Botryococcus* are found mainly in waters of moderate salinity with neutral or slightly alkaline pH or in freshwater (Florin, 1977; van Geel, 2001).

Salinity changes of lake ESE

The cluster analysis performed with the diatom data of lake ESE shows similar results to that of lake ESW, dividing the sediment into three distinctive zones. Here, too, a tendency of decreasing salinity is indicated. However, salinity changes are not as pronounced as in lake ESW. Thus, although results of the PCA support the tripartition of the sediment, it is not as distinct as in lake ESW and transitions between the zones are rather blurred.

The beginning of DAZ-ESE-1 is dominated by the polyhalobous *P. sulcata*, which indicates brackish or marine conditions. However, this species declines in abundance through DAZ-ESE-1 while other poly-, meso- and, mainly, oligohalobous taxa occur. Accordingly, the halobios index corresponds to saltwater conditions with high to moderate salt concentrations (Ziemann, 1971, 1982).

In the following DAZ-ESE-2, polyhalobous taxa, such as *Cocconeis scutellum* and *Fragilaria subsalina*, increase. Furthermore, mesohalobous species regularly reach higher abundances than in DAZ-ESE-1, resulting in a higher halobios index.

In DAZ-ESE-3, the decreasing importance of mesohalobous taxa and the simultaneous increase of halophobous diatoms, which for the first time reach more than 37%, indicate stronger freshwater influences (Ziemann, 1971, 1982). Also, the halobios index points out to a development towards well-buffered freshwater conditions. However, in contrast to lake ESW, the halobios index does not reach negative values. Thus, also the youngest sediments of lake ESE presumably were subject to marine influences, as affirmed by the persisting abundances of oligo-, meso- and (partly) poly-halobous species (Ziemann, 1997).

Therefore, in contrast to lake ESW, the establishment of freshwater conditions is not completed yet in lake ESE. No stable conditions have been reached. Instead, the diatom flora and the halobios index indicate that constant dynamic changes occurred. This is supported by the dominance of the opportunistic *Staurosira* aff. *venter*, which reaches abundances of 10–80% throughout the sediment core of lake ESE. The occurrence of other small *Fragilaria* and *Staurosira*-species (*S. pinnata*, S. *brevistriata*, *F. subsalina* and *F. pulchella*) in lake ESE, too, can be interpreted as indicator of dynamic processes as these species previously were found to occur in areas, where isostatic land uplift caused the transformation of marine bays into freshwater lakes (Stabell, 1985).

As in lake ESW, the cluster analysis of pollen from lake ESE indicates the division of the sediment into two zones. Again, PAZ-ESE-1 largely corresponds to the salt- and brackish water-dominated DAZ-ESE-1 and DAZ-ESE-2 whereas PAZ-ESE-2 corresponds to DAZ-ESE-3, which is characterised by increasing freshwater influence (transition between PAZ-ESE-1 and PAZ-ESE-2: between 24 and 20 cm; transition between DAZ-ESE-2 and DAZ-ESE-3: between 21 and 19 cm).

The pollen spectrum for lake ESE equals that of lake ESW, showing the tendency to shift from a typical marine shoreline vegetation with halotolerant plants, such as *Aster tripolium* and *Plantago maritima*, towards freshwater conditions. However, like the diatoms, the pollen also indicates that the final stage of weakly buffered, acidic freshwater is not yet reached in lake ESE: *Myriophyllum spicatum* (characteristic of carbonate-rich waters) prevails well into the youngest sediments whereas the halophobous species *Nymphaea alba* did not become established. Thus, the pollen data reflect electrolyte-rich freshwater with marine influences in the youngest sediments of lake ESE.

The analysis of other microfossils supports this incomplete change from salt- to freshwater environment: In PAZ-ESE-1 (i.e. DAZ-ESE-1 and DAZ-ESE-2), marine organisms, such as Dinophyceae, foraminifera and *Cymatiosphaera* (Bakker & van Smeerdijk, 1982), prevail. However, some freshwater taxa (e.g. chrysophyte cysts, *Botryococcus*, and *Pediastrum*) already occur, indicating slight freshwater influences. In PAZ-ESE-2, the abundances of freshwater species (*Botryococcus*, *Pediastrum* and *P. boryanum*) increase. Simultaneously, marine taxa decline but do not vanish completely. Similar to results of the pollen and diatom analyses, this indicates a tendency towards freshwater conditions, which are not yet fully established.

Anthropogenic impact in the catchment area

According to the pollen analyses, the forest in the catchment area of both lakes is dominated by *Pinus*, *Picea* and *Betula*. Pollen data thus reflect typical natural conditions of the boreal zone (Hyvärinen, 1975) and correspond to pollen accumulations of the last few centuries similar to lakes of this region which have not been subject to anthropogenic influences (Kremenetski et al., 1999; Solovieva et al., 2005).

In both lakes studied, AP are extremely abundant throughout the sediment core and only once (ESW: 13–5 cm, ESE: 32–24 cm) decline slightly while NAP-upland pollen increase. These changes in the pollen composition could be due to forest clearance in the vicinity of the lakes. Nevertheless, AP abundances still exceed 96% throughout both sediment cores, suggesting a more or less closed forest cover with no clearly detectable anthropogenic influence during the period investigated (Aaby, 1994).

Furthermore, forest clearance (decreasing arboreal pollen) and associated erosion processes in the vicinity of lakes normally cause increased input of mineral matter into the lakes (Pederson et al., 2005). However, the loss on ignition (LOI), an indicator of high mineral inputs and erosion processes (Dean, 1974; Heiri et al., 2001), does not correlate to the (anyway little) changes in AP abundances in the sediment of the study lakes. Instead, in the sediment of the study lakes the LOI seems to be more dependent on salinity changes as decreasing tidal influence (and decreasing salinity) presumably led to lower resuspension and redeposition of organic matter. Therefore, especially in lake ESW, the LOI increases with decreasing marine influences (decreasing tidal influence and salinity) due to undisturbed organic sedimentation, but does not indicate any human activity (forest clearance and erosion processes) in the area surrounding our study lakes.

Furthermore, results of the pollen analysis suggest non-existent or at most marginal anthropogenic influences in the catchment area as Cerealia-type pollen occur only sporadically and in low abundance. Their appearance can either be explained by pollen drift from distant regions (Prentice, 1985) or by the occurrence of wild grass pollen, which are difficult to distinguish from pollen of cultivated grasses (Behre, 2007). The identification of pollen from wild and cultivated grasses is especially difficult when only few pollen grains are present (Tweddle et al., 2005) as is the case in the sediments of both lakes. Furthermore, identification of grass pollen is particularly problematic in coastal areas as wild grasses here produce very large pollen grains, which cannot be distinguished from those of cultivated grasses (Joly et al., 2007). Hence, the few large Cerealia-type pollen grains detected in the sediments most probably are not only from cultivated species but also from species growing in shoreline habitats.

It was not possible to assess the impact of settlements in the vicinity of the lakes. Instead, all changes detected in the pollen composition can be explained by natural processes, i.e. mainly changes in salinity.

Land uplift and the development of both lakes

The analyses of diatoms, pollen and other microfossils show that both lakes underwent a similar history over the past few centuries, with salinity the controlling environmental factor for the aquatic biota. There was no evidence of significant anthropogenic impacts.

Lake ESW evolved from saltwater to dystrophic freshwater conditions whereas in lake ESE, which is located closer to the sea, the development from saltwater to freshwater conditions has only reached the stage of well-buffered freshwater. Here, isolation of the lake from the sea is not yet complete, and the lake presumably is still subject to marine influences, especially during high tides or storm surges.

Results of the paleolimnological analyses of the youngest sediments (ESW: dystrophic freshwater; ESE: well-buffered freshwater with saltwater influences) agree with recent conditions in the lakes: In the years 2000 and 2004, only freshwater diatoms were found in lake ESW while freshwater and saltwater diatoms occurred in lake ESE (Dreßler, 2004). Analyses of the years 2000, 2004 and 2006 (Dreßler, unpublished) determined pH-values of <6.5 and a salinity of 0‰ in lake ESW and pH-values of 7.0–9.2 and a salinity of 0–10‰ in lake ESE.

Both lakes are located on the Fennoscandian Shield, an area of isostatic land uplift with recent uplift rates of 4 mm yr^{-1} at the Karelian White Sea coast (Filatov et al., 2005). The lakes are situated in direct vicinity to the sea; ESW in an altitude of 2.3 m, ESE in an altitude of 1.6 m above mean tide level. As tidal range at the coast next to the studied lakes varies between 1.3 and 1.8 m (average: 1.6 m), the flood exceeds mean tide level on average by 0.8 m. Thus, saltwater impacts should occur only occasionally in lake basins that are situated higher than 0.8 m above mean tide level. During the progression of land uplift processes, marine influences should decrease further and the character of these lakes should shift continuously from salt- to freshwater.

However, high tides and storm surges can exceed 0.8 m above mean tide, which results in occasional saltwater influx into lake ESE. In lake ESW, which is situated in higher elevation and further away from the coast, the last saltwater influence was identified at a sediment depth of approximately 21 cm, a zone dated at 104.4 ± 41.8 yr by ^{210}Pb-dating. Conditions in this zone of lake ESW (decreasing salinity and the establishment of freshwater conditions) equal recent conditions in lake ESE. Thus, it can be assumed that in the past, lake ESW was located as high as or slightly higher above sea level than ESE is today (i.e. 1.6–1.8 m). Hence, the basin of lake ESW presumably rose 0.5–0.7 m within the last 104.4 ± 41.8 yr which corresponds to an uplift rate of approximately 5.7 mm yr^{-1}.

However, this calculated rate of 5.7 mm yr^{-1} can only be considered a rough approximation of the true land uplift rate as several sources of error possibly limit its accuracy. For instance results of ^{210}Pb-dating have large error margins and it is hard to assess which minimum elevation above sea level effectively prevents any saltwater influx also during high tides and surges. Furthermore, the analogy between conditions in a sediment depth of 21 cm in lake ESW and recent conditions in lake ESE is an estimation based solely on similarities in diatom, pollen and other microfossil data.

Another problem with assessing basin isolation through diatom assemblage changes is the occurrence of fresh surface water in near shore areas (Granskog et al., 2005a), especially in sheltered bays and at freshwater inflows (Granskog et al., 2005b). Such pronounced vertical salinity gradients easily develop in calm weather, probably preferentially during spring snowmelt. It may be only a very thin surface freshwater stratum, but this can be a significant phenomenon for diatoms and other algae (Jiang et al., 1998; Ulanova & Snoeijs, 2006; Kaartokallio et al., 2007). Due to this temporary vertical salinity gradient, the first clear indicators of freshwater influence (halophobous and wide-range salinity-tolerant diatoms) may appear before basin isolation has started. Thus, the calculated dates of the isolation event may be a little too old, leading to a slightly overestimated land uplift rate.

Nevertheless, the estimated uplift rate is supported by values calculated by previous studies: Filatov et al. (2005) give a recent uplift rate of 4 mm yr^{-1} for the Karelian coast. Pan & Sjöberg (1999) determined recent uplift rates of the Fennoscandian Shield to range from 1 to 11 mm yr^{-1} (standard errors of 1–3 mm yr^{-1}) using GPS-data from Baltic Sea level project GPS campaigns in 1993 and 1997. Påsse & Andersson (2005) calculated uplift rates of 2–3 mm yr^{-1} based on empirical data. Thus, despite relatively large sources of error the land uplift rate calculated for lakes ESW is in good agreement with results of previous studies and can be used as approximation for further analysis.

Basin elevation and associated transition from marine to freshwater environments caused by land uplift is not unique to our study lakes. Many investigations of basin elevation covering much longer time spans were carried out in Sweden (e.g. Risberg et al., 2005), Norway (e.g. Kjemperud, 1980) and Finland (e.g. Miettinen, 2004). These studies show that processes similar to those occurring in our study lakes during the last two centuries, took place much earlier during the Holocene in other regions of the Fennoscandian Shield. Based on diatom, pollen and spore data, for example, it was calculated that the Getsjö archipelago (south central Sweden) of the Yoldia Sea passed the transition stage to freshwater around 10,000 years BP (Florin, 1977).

Conclusions

The results of diatoms, pollen and other microfossils found in ^{210}Pb-dated sediment cores do not indicate any anthropogenic impact on the two northwest Russian lakes ESE and ESW, situated at the Karelian White Sea coast. Instead, salinity was found to be the determining environmental factor for the aquatic biota as both lakes evolved from typical saltwater-dominated coastal waters to freshwater lakes within the two last centuries.

In lake ESW, this led to the establishment of electrolyte-poor, dystrophic freshwater conditions, whereas in lake ESE, which is situated lower above sea level and closer to the coast, the process is not yet complete. Here, well-buffered freshwater conditions with significant saltwater impacts still prevail today.

The decreasing salinity in both lakes presumably was caused by basin elevation and isolation due to isostatic land uplift of the Fennoscandian Shield. A hypothetical land uplift rate of approximately 5.7 mm yr^{-1} was calculated based on data from the sediment cores analysed.

Acknowledgements We gratefully acknowledge Helmut Winkler, Robert Dietrich and Unn Klare (University of Rostock) for logistic help and sediment coring assistance. We also gratefully acknowledge Hanns-Christian Treutler (Environmental Research Centre Leipzig-Halle) for lead and caesium measurements and Natalia Tschewtschenko (Moscow State University) for helpful comments and geological information about the Kindo peninsula. We also like to thank Unn Klare for translating the manuscript, Regine Paschen (University of Rostock) for assistance in the laboratory, Georgij Gennadievich Norvikov (White Sea Biological Station of the Moscow State University) for providing the research opportunities at the Kindo peninsula and two anonymous reviewers for very helpful comments on an earlier draft of this article.

References

Aaby, B., 1994. NAP-percentages as an expression of cleared areas. In Frenzel, B. (ed.), Evaluation and land surfaces cleared from forests in Roman Iron Age and the time of migrating German tribes based on regional pollen diagrams. Paläoklimaforsch 12:13–27.

Andersson, K. G. & J. Roed, 1994. The behaviour of Chernobyl Cs-137, Cs-134 and Ru-106 in undisturbed soil—implications for external radiation. Journal of Environmental Radioactivity 22: 183–196.

Andersen, T. J., O. A. Mikkelsen, A. L. Moller & M. Pejrup, 2000. Deposition and mixing depths on some European intertidal mudflats based on Pb-210 and Cs-137 activities. Continental Shelf Research 20: 569–1591.

Appleby, P. G., 2001. Chronostratigraphic techniques in recent sediments. In Last, W. M. & J. P. Smol (eds), Developments in paleoenvironmental research, Vol. 1: Tracking environmental changes in lake sediments: physical and chemical techniques. Kluwer Academic Publishers, Dordrecht: 171–203.

Appleby, P. G. & F. Oldfield, 1983. The assessment of ^{210}Pb dates from sites with varying sediment accumulation rates. Hydrobiologia 103: 29–35.

Appleby, P. G., N. Richardson & P. J. Nolan, 1992. Self-absorption corrections for well-type germanium detectors. Nuclear Instruments and Methods in Physics Research Section B 71: 228–233.

Bakker, M. & D. G. Van Smeerdijk, 1982. A palaeoecological study of a late Holocene section from "Het Ilperveld", western Netherlands. Review of palaeobotany and palynology 36: 95–163.

Behre, K.-E., 2007. Evidence for Mesolithic agriculture in and around central Europe? Vegetation History and Archaeobotany 16: 203–219.

Bennett, K. D., 1996. Determination of the number of zones in a biostratigraphical sequence. New Phytologist 132: 155–170.

Bennett, K. D., 2005. Documentation for psimpoll 4.25 and pscomb 1.03. C programs for plotting pollen diagrams and analysing pollen data. Quaternary Geology, Department of Earth Sciences, Uppsala University, Uppsala, Sweden.

Berglund, E. B., P. Sandgren, L. Barnekow, G. Hannon, H. Jiang, G. Skog & S. Y. Yu, 2005. Early Holocene history of the Baltic Sea, as reflected in costal sediments in Blekinge, southeastern Sweden. Quaternary International 130: 111–139.

Beug, H.-J., 2004. Leitfaden der Pollenbestimmung für Mitteleuropa und angrenzende Gebiete. Verlag Dr. Friedrich Pfeil, Munchen.

Blais, J. M., J. Kalff, R. J. Cornett & R. D. Evans, 1995. Evaluation of ^{210}Pb dating in lake sediments using stable Pb, Ambrosia pollen, and ^{137}Cs. Journal of Paleolimnology 13: 169–178.

Cameron, N. G., H. J. B. Birks, V. J. Jones, F. Berge, J. Catalan, R. J. Flower, J. Garcia, B. Kawecka, K. A. Koinig, A. Marchetto, P. Sánchez-Castillo, R. Schmidt, M. Šiško, N. Solovieva, E. Štefková & M. Toro, 1999. Surface-sediment and epilithic diatom pH calibration sets for remote European mountain lakes (AL:PE Project) and their comparison with the Surface Waters Acidification Programme (SWAP) calibration set. Journal of Paleolimnology 22: 291–317.

Casper, S. J. & D. H. Krausch, 1981. Pteridophyta und Anthophyta, 2. Teil. In Ettl, H., J. Gerloff & H. Heynig (eds), Süßwasserflora von Mitteleuropa 24. VEB Gustav Fischer, Jena.

Conley, D. J., 2004. Molten: monitoring long-term trends in eutrophication and nutrients in the costal zone: Creation of guidelines for the evaluation of background conditions, anthropogenic influence and recovery. Deliverable 12 Management Tool: 2–14.

Dean, W. E., 1974. Determination of carbonate and organic matter in calcareous sediments and sedimentary rocks by loss on ignition: comparison with other methods. Journal of Sedimentary Petrology 44: 242–248.

Denys, L., 1990. *Fragilaria* blooms in the Holocene of the western coastal plain of Belgia. In Simola, H. (ed.), Proceedings of the tenth international diatom symposium. Koeltz, Koenigstein: 397–406.

Dreßler, M. 2004. Limnic periphytic Diatoms (Bacillariophyceae) on various substrates in different freshwater systems in the north west of Russia: a short report. Proceedings of the Pertsov White Sea Biological Station. 7th International Conference 9: 56–65.

Dreßler, M., U. Selig, W. Dörfler, S. Adler, H. Schubert & T. Hübener, 2006. Environmental changes and the migration period in northern Germany as reflected in the sediments of Lake Dudinghausen. Quaternary Research 66: 25–37.

Dreßler, M., T. Hübener, S. Görs, P. Werner & U. Selig, 2007. Multi-proxy reconstruction of trophic state, hypolimnetic anoxia and phototrophic sulphur bacteria abundance in a dimictic lake in Northern Germany over the past 80 years. Journal of Paleolimnology 37: 205–219.

Ellenberg, H., 1992. Zeigerwerte von Pflanzen in Mitteleuropa. Goltze, Göttingen.

Faegri, K. & J. Iversen, 1975. Textbook of pollen analysis. Hafner Press, New York.

Filatov, N., D. Pozdnyakov, O. M. Johannessen, L. H. Pettersson & L. Bobylev, 2005. White Sea: Its marine environment and ecosystem dynamics influenced by global change. Springer, Chichester.

Florin, M. B., 1977. Late glacial and pre-boreal vegetation in southern central Sweden. II. Pollen spore and diatom analyses. Striae 5: 1–60.

Garde, L. M., J. M. Nicol & J. G. Conran, 2004. Changes in vegetation patterns on the margins of a constructed wetland after 10 years. Ecological Management & Restoration 5: 111–117.

Granskog, M. A., H. Kaartokallio, D. N. Thomas & H. Kuosa, 2005a. Influence of freshwater inflow on the inorganic nutrient and dissolved organic matter within coastal sea ice and underlying waters in the Gulf of Finland (Baltic Sea). Estuarine, Coastal and Shelf Science 65: 109–122.

Granskog, M. A., J. Ehn & M. Niemelä, 2005b. Characteristics and potential impacts of under-ice river plumes in the seasonally ice-covered Bothnian Bay (Baltic Sea). Journal of Marine Systems 53: 187–196.

Grimm, E. C., 1987. CONISS: A Fortran 77 Program for stratigraphically constrained cluster analysis by the method of incremental sum of squares. Computer & Geosciences 13: 13–35.

Grimm, E. C., 1990. TILIA and TILIA • GRAPH: PC spreadsheet and graphics software for pollen data. INQUA - Commission for the Study of the Holocene, Working Group on Data-Handling Methods Newsletter 4: 5–7.

Hedenström, A. & J. Risberg, 1999. Early Holocene shore-displacement in southern central Sweden as recorded in elevated basins. Boreas 28: 490–504.

Heiri, O., A. F. Lotter & G. Lemcke, 2001. Loss on ignition as a method for estimating organic and carbonate content in sediments: reproducibility and comparability of results. Journal of Paleolimnology 25: 101–110.

Hill, M. O. & H. G. Gauch, 1980. Detrended correspondence analysis, an improved ordination technique. Vegetatio 42: 47–58.

Hofmann, G., 1994. Aufwuchs- Diatomeen und ihre Eignung als Indikatoren der Trophie. Bibliotheca Diatomologica 30. Cramer in der Gebrüder Borntraeger Verlagsbuchhandlung, Berlin, Stuttgart.

Hofmann, G., 1997. Diatom communities in the rivers Werra and Ulster (Germany) and their response to reduced salinity. Limnologica 27: 77–84.

Hustedt, F., 1953. Die Systematik der Diatomeen in ihren Beziehungen zur Geologie und Ökologie nebst einer Revision des Halobiensystems. Svensk Botanisk Tidskrift 47: 509–519.

Hustedt, F., 1957. Die Diatomeenflora des Fluss- Systems der Weser im Gebiet der Hansestadt Bremen. Abhandlungen des Naturwissenschaftlichen Verein, Bremen 34: 181–440.

Hyvärinen, H., 1975. Absolute and relative pollen diagrams from northernmost Fennoscandia. Fennia 142: 5–23.

Jankovská, V. & J. Komárek, 2000. Indicative value of *Pediastrum* and other coccal green algae in palaeoecology. Folia Geobotanica 35: 59–82.

Jiang, H., N.-O. Svensson & S. Björck, 1998. Meltwater Discharge to the Skagerrak–Kattegat from the Baltic Ice Lake during the Younger Dryas Interval. Quaternary Research 49: 264–270.

Joly, C., L. Barillé, M. Barreau, A. Mancheron & L. Visset, 2007. Grain and annulus diameter as criteria for distinguishing pollen grains of cereals from wild grasses. Review of Palaeobotany and Palynology 146: 221–233.

Kaartokallio, H., H. Kuosa, D. N. Thomas, M. A. Granskog & K. Kivi, 2007. Biomass, composition and activity of organism assemblages along a salinity gradient in sea ice subjected to river discharge in the Baltic Sea. Polar Biology 30: 183–197.

Kalbe, L. & H. Werner, 1974. Das Sediment des Kummerower Sees. Untersuchungen des Chemismus und der Diatomeenflora. Internationale Revue der gesamten Hydrobiologie und Hydrographie 59: 755–782.

Kjemperud, A., 1980. Diatom changes in sediments of basins possessing marine/lacustrine transitions in Frosta, Nord-Trøndelag, Norway. Boreas 10: 27–38.

Komárek, J. & V. Jankovská, 2001. Review of the green algal genus *Pediastrum*: implication for pollen—analytical research. Bibliotheca Phycologica 108: 1–27. Stuttgart.

Krammer, K., 1997a. Die cymbelloiden Diatomeen. Eine Monographie der weltweit bekannten Taxa, Teil 1 Allgemeines und *Encyonema* part. Bibliotheca Diatomologica 36: 1–382. Cramer, Berlin Stuttgart.

Krammer, K., 1997b. Die cymbelloiden Diatomeen. Eine Monographie der weltweit bekannten Taxa, Teil 2 *Encyonema* part. *Encyonopsis* and *Cymbellopsis*. Bibliotheca Diatomologica 37: 1–469. Cramer, Berlin Stuttgart.

Krammer, K., 2000. The Genus *Pinnularia*. In Lange-Bertalot, H. (ed.), Diatoms of Europe 1. A.R.G. Gantner, Ruggell.

Krammer, K., 2002. Cymbella. In Lange-Bertalot, H. (ed.), Diatoms of Europe 3. A.R.G. Gantner, Rugell.

Krammer, K., 2003. Cymbopleura, Delicata, Navicymbula, Gomphocymbellopsis, Afrocymbella. In Lange-Bertalot, H. (ed.), Diatoms of Europe 4. A.R.G. Gantner, Ruggell.

Krammer, K., & H. Lange-Bertalot H., 1986–1991, 2004. Süßwasserflora von Mitteleuropa, Bacillariophyceae, Band 2.1., 2.2., 2.3. and 2.4., Fischer Verlag, Stuttgart.

Kremenetski, C. V., T. Bottger, F. W. Junge & A. G. Tarasov, 1999. Late- and postglacial environment of the Buzuluk area, middle Volga region, Russia. Quaternary Science Reviews 18: 1185–1203.

Lange-Bertalot, H., 2001. *Navicula* sensu stricto, 10 Genera separated from *Navicula* sensu lato, *Frustulia*. In Lange-Bertalot, H. (ed.), Diatoms of Europe 2. A.R.G. Gantner, Ruggell.

Lange-Bertalot, H. & S. I. Genkal, 1998. Diatoms from Siberia I, Islands in the Arctic Ocean (Ygorsky-Shar Strait) second corrected printing. In Lange-Bertalot, H. (ed.), Iconographia diatomologica 6. A.R.G. Gantner, Ruggell.

Lange-Bertalot, H. & D. Metzeltin, 1996. Annotated diatom micrographs. In Lange-Bertalot, H. (ed.), Iconographia diatomologica 2. Koeltz Scientific Books, Königstein.

Lange-Bertalot, H. & G. Moser, 1994. Brachysira, Monographie der Gattung. Bibliotheca Diatomologica 29: 1–212. Cramer, Berlin Stuttgart.

Lange-Bertalot, H., K. Külbs, T. Lauser, M. Nörpel-Schempp & M. Willmann, 1996. Annotated diatom micrographs. In Lange-Bertalot, H. (ed.), Iconographia diatomologica 3. Koeltz Scientific Books, Königstein.

Lepš, J. & P. Šmilauer, 2003. Multivariate analysis of ecological data using CANOCO. Cambridge University Press, Cambridge.

Lotter, A. F. & C. Bigler, 2000. Do diatoms in the Swiss Alps reflect the length of ice-cover? Aquatic Science 62: 125–141.

Lotter, A. F., R. Pienitz & R. Schmidt, 1999. Diatoms as indicators of environmental change near arctic and alpine treeline. In Stoermer, E. & J. P. Smol (eds), The diatoms: applications for the environmental and earth science. Cambridge University Press, Cambridge.

McQuiod, M. R. & K. Nordberg, 2003. The diatom *Paralia sulcata* as an environmental indicator species in costal sediments. Estuarine, Costal and Shelf Science 56: 339–354.

Miettinen, A., 2004. Holocene sea-level changes and glacio-isostasy in the Gulf of Finland, Baltic Sea. Quaternary International 120: 91–104.

Miettinen, J. O., M. Kukkonen & H. Simola, 2005. Hindcasting baseline values for water colour and total phosphorus concentration in lakes using sedimantary diatoms—implications for lake typology in Finland. Boreal Environment Research 10: 31–43.

Miettinen, A., H. Jansson, T. Alenius & G. Haggrén, 2007. Late Holocene sea-level changes along the southern coast of Finland, Baltic Sea. Marine Geology 242: 27–38.

Moore, P. D., J. A. Webb & M. E. Collinson, 1991. Pollen analysis. Blackwell Scientific Publications, Oxford.

Olsson, I. U., 1986. Radiometric dating. In Berglund, B. E. (ed.), Handbook of Holocene palaeoecology and palaeohydrology. Wiley, Chichester.

Pan, M. & L. E. Sjöberg, 1999. Estimating present-day post-glacial rebound and horizontal movements in Fennoscandia by represented GPS campaigns in 1993 and 1997. Geophysical Research Letters 26: 771–774.

Pankow, H., V. Kell, N. Wasmund & B. Zander, 1990. Ostsee-Algenflora. Gustav Fischer, Jena.

Påsse, T. & L. Andersson, 2005. Shore-level displacement in Fennoscandia calculated from empirical data. Geological Society of Sweden 127: 253–268.

Pederson, D. C., D. M. Peteet, D. Kurdyla & T. Guilderson, 2005. Medieval warming, little ice age, and European impact on the environment during the last millennium in the lower Hudson Valley, New York, USA. Quaternary Research 63: 238–249.

Prentice, I. C., 1985. Pollen representation, source area, and basin size: toward a unified theory of pollen analysis. Quaternary Research 23: 76–86.

Punt, W., 1976. The northwest European pollen flora, I. Elsevier, Amsterdam.

Punt, W. & S. Blackmore, 1991. The northwest European pollen flora, VI. Elsevier, Amsterdam.

Punt, W. & G. C. S. Clarke, 1980. The northwest European pollen flora, II. Elsevier, Amsterdam.

Punt, W. & G. C. S. Clarke, 1981. The northwest European pollen flora, III. Elsevier, Amsterdam.

Punt, W., S. Blackmore & G. C. S. Clarke, 1988. The northwest European pollen flora, V. Elsevier, Amsterdam.

Punt, W., S. Blackmore & P. P. Hoen, 1996. The northwest European pollen flora, VII. Elsevier, Amsterdam.

Risberg, J., G. Alm & T. Goslar, 2005. Variable isostatic uplift patterns during the Holocene in southeast Sweden, based on high-resolution AMS radiocarbon datings of lake isolations. The Holocene 15: 847–857.

Ryves, D. B., A. L. Clarke, P. G. Appleby, S. L. Amsinck, E. Jeppersen, F. Landkildehus & N. J. Anderson, 2004. Reconstructing the salinity and environment of the Limfjord and Vejlerne Nature Reserve, Denmark, using a diatom model for brackish lakes and fjords. Canadian Journal of Fisheries and Aquatic Sciences 61: 1988–2006.

Schmidt, R., C. Kamenik, H. Lange-Bertalot & R. Klee, 2004. *Fragilaria* and *Staurosira* (Bacillariophyceae) from sediment surfaces of 40 lakes in the Austrian Alps in relation to environmental variables, and their potential for palaeoclimatology. Journal of Limnology 63: 171–189.

Simonsen, R., 1962. Untersuchungen zur Systematik und Ökologie der Bodendiatomeen der westlichen Ostsee. Internationale Revue der gesamten Hydrobiologie, Systematische Beihefte 1: 1–144.

Solovieva, N., P. Tarasov & G. MacDonald, 2005. Quantitative reconstruction of Holocene climate from the Chuna Lake pollen record, Kola Peninsula, north-west Russia. The Holocene 15: 141–148.

Stabell, B., 1985. The development and succession of taxa within the diatom genus *Fragilaria* Lyngbye as a response to basin isolation from the sea. Boreas 14: 273–286.

Ter Braak, C. J. F., 1987. CANOCO—a FORTRAN program for canonical community ordination by [partial] [detrended] [canonical] correspondence analysis, principal components analysis and redundancy analysis. TNO Institute of Applied Computer Science, Statistics Dept, Wageningen. Report 871TI A 1 1.

Ter Braak, C. J. F. & T. C. Prentice, 1988. A theory of gradient analysis. Advances in Ecological Research 18: 271–317.

Ter Braak, C. J. F. & P. Śmilauer, 2002. CANOCO reference manual and CanoDraw for windows user's guide. Software for canonical community ordination (version 4.5). Microcomputer Power, Ithaca.

Tweddle, J. C., K. J. Edwards & N. R. J. Fieller, 2005. Multivariate statistical and other approaches for the separation of cereal from wild Poaceae pollen using a large Holocene dataset. Vegetation History and Archaeobotany 14: 15–30.

Ulanova, A. & P. Snoeijs, 2006. Gradient responses of epilithic diatom communities in the Baltic Sea proper. Estuarine, Coastal and Shelf Science 68: 661–674.

Van Geel, B., 2001. Non-pollen palynomorphs. In Smol, J. P., H. J. B. Birks & W. M. Last (eds), Tracking environmental change using lake sediments. Terrestrial, algae, and siliceous indicators. Kluwer Academic Publishers, Dordrecht: 99–119.

Weller, M. W. & D. V. Derksen, 1979. The geomorphology of Teshekpuk Lake in relation to coastline configuration of Alaska's Coastal Plain. Arctic 32: 152–160.

Witkowski, A., H. Lange-Bertalot & D. Metzeltin, 2000. Annotated diatom micrographs, diatom flora of marine coasts 1. In Lange-Bertalot, H. (ed.), Iconographia diatomologica 7. A.R.G. Gantner Verlag, Ruggell.

Ziemann, H., 1971. Die Wirkung des Salzgehaltes auf die Diatomeenflora als Grundlage für eine biologische Analyse und Klassifikation der Binnengewässer. Limnologica 8: 505–525.

Ziemann, H., 1982. Indikatoren für den Salzgehalt der Binnengewässer-Halobiensystem. Biologische Wirkung des Salzgehaltes. In Ausgewählte Methoden der Wasseruntersuchung Bd. 2. Biologische, mikrobiologische und toxikologische Methoden. 2. Aufl.: 89–95 und 108–114, Jena.

Ziemann, H., 1997. The influence of different ion ratios on the biological effect of salinity in running waters of Thuringia (Germany). Limnologica 27: 19–28.

An approach to the recent environmental history of Pilica Piaski spring (southern Poland) using diatoms

Agata Z. Wojtal · Andrzej Witkowski · Burkhard Scharf

Originally published in the journal Hydrobiologia, Volume 631, No. 1, 267–277.
DOI: 10.1007/s10750-009-9815-8 © Springer Science+Business Media B.V. 2009

Abstract Diatom floras from springs have received little attention until recently, despite the fact that springs provide specific conditions that cannot be found in any other aquatic system, and remain of great importance in terms of general environmental changes. Due to pollution of the Kraków-Częstochowska upland aquifers, the chemical composition of spring waters in the area is anthropogenically modified, and far from the natural state. In order to generate a baseline diatom flora for springs, a paleolimnological study was applied as one of the best recognized and applied methods used to track long-term environmental changes worldwide. Little is known, however, about the direction and nature of such changes in aquatic environments fed by springs. The present article focuses on shifts in diatom assemblages preserved in sediments collected from a small pond, situated close to several spring outlets. They were interpreted as a record of environmental changes that had taken place during the last century. For most of the history of the pond—as recorded in this 84 cm long core—the diatom assemblage was dominated by small *Fragilaria* spp. The major shifts in species composition began in a core depth of 65 cm with a decline in *Fragilaria construens* var. *venter* (Ehrenberg) Grunow and *Fragilaria pinnata* Ehrenberg complexes, and a concurrent increase in *Achnanthidium minutissimum* (Kützing) Czarnecki. The second change was recorded at the depth of 45 cm with a sharp decline of *A. minutissimum*, which again was replaced by the small *Fragilaria* spp. In recent years, further changes in the diatom assemblage occurred, with a notable increase in *Cyclotella delicatula* Hustedt. The assemblage shifts recorded at this site appear to be consistent with environmental changes triggered by land use (e.g. agriculture intensity) and/or possible changes in spring water discharge.

Keywords Springs · Bacillariophyta · Palaeoenvironmental research · Bioindication · Eutrophication

Guest editors: K. Buczkó, J. Korponai, J. Padisák & S. W. Starratt
Palaeolimnological Proxies as Tools of Environmental Reconstruction in Fresh Water

A. Z. Wojtal (✉)
W. Szafer Institute of Botany, Polish Academy of Sciences, Lubicz 46, 31-512 Krakow, Poland
e-mail: a.wojtal@botany.pl

A. Witkowski
Department of Paleoceanology, University of Szczecin, Mickiewicza 18, 70-383 Szczecin, Poland

B. Scharf
Ellhornstr. 21, 28195 Bremen, Germany

Introduction

Diatoms are known to be excellent environmental indicators as they are common in aquatic environments, respond rapidly to changing conditions, and different species often have distinct environmental optima (Stoermer & Smol, 1999; Keatley et al., 2006). The composition of siliceous frustules in sediments can directly reflect the floristic composition in the past, and also indirectly reflects water characteristics such as pH, nutrient status, salinity (Battarbee, 1986), and flow velocity. Unlike many other freshwater (fossil or sub-fossil) records from lakes, mires, and ponds, data from spring sediments are very scarce (e.g. Williams & Williams, 1997; Filippi et al., 2008). Moreover, while numerous studies of freshwater environments have focused on the use of diatoms for assessing water quality parameters in streams, rivers, and lakes, these kinds of data are relatively sparse from springs (Cantonati, 2001; Battegazzore et al., 2004; Wojtal, 2006). However, interest in European diatom floras of springs has increased considerably during the last decades (e.g., Reichardt, 1981, 1994; Kadłubowska, 1985; Sabater & Roca, 1990, 1992; Rakowska, 1996; Cantonati, 1998; Werum, 2001a; Werum & Lange-Bertalot, 2004; Poulíčková et al., 2005; Cantonati et al., 2006; Żelazna-Wieczorek & Mamińska, 2006) for several reasons. Thermal stability and continuous water flow in springs are one of the most essential characteristics. Fluctuations of chemical and physical conditions in springs are minimized, thus allowing the study of interactions between limited numbers of species and environmental factors (Round, 1981; Filippi, 2007). Additionally, they remain as biodiversity refuges in anthropogenically influenced areas (Cantonati, 2001; Werum, 2001b). Their biodiversity patterns, therefore, can be applied to understand diatoms distribution, in response to geology (e.g. Cantonati, 1998; Werum, 2001a; Werum & Lange-Bertalot, 2004), water chemistry, temperature, altitude (e.g., Sabater & Roca, 1990; Cantonati et al., 2006), flow velocity (e.g., Sabater & Roca, 1990), and human impact (Wojtal, 2006). The sedimentary records from spring-fed areas are also expected to be highly sensitive indicators of environmental changes (Smol, 2002).

The springs of the Krakowsko-Częstochowska upland are threatened habitats because of significant human pressures such as agriculture and urbanization. They are very vulnerable to pollution because of the karst-fissured geological character of the area, which allows rapid flow of surface water into groundwater systems through natural conduits. Moreover, once an aquifer is contaminated, it may remain polluted for decades (Smol, 2002). One of the most severe environmental problems in the area is eutrophication of these systems caused by high levels of nitrates derived from agricultural run-off and wastes. A steady rise in nitrate levels was observed in most springs in the whole area over the period of 1974–1999 (Baścik et al., 2001). This trend has continued into the twenty-first century (Wojtal, 2006). The total water mineralization (amount of total dissolved solids) in the Pilica Piaski spring increased over 40% within last 25 years (Baścik et al., 2001).

The evaluation of anthropogenic impacts on these springs and developing an effective conservation management strategy require calibration studies and knowledge of the natural state of these freshwater ecosystems. Samples of the bottom sediments from a spring-fed pond in the Pilica Piaski were retrieved, in order to evaluate the diatom assemblages which may represent baseline conditions for springs of the Krakowsko-Częstochowska upland. This article describes changes in the diatom assemblages of the Pilica Piaski spring-fed pond, and attempts to reconstruct changes in the spring water characteristics over a period of at least one century using the bioindication potential of diatoms.

Materials and methods

The Pilica Piaski spring is situated in the Krakowsko-Częstochowska upland, in southern Poland (50°28′N and 19°41′E), at an elevation of about 264 m a.s.l. The water-bearing layer consists of Jurassic (Malm) marl and platy limestone which yields over 40 l per second (Baścik et al., 2001). Most of the Krakowsko-Częstochowska upland springs belong to the HCO_3 -Ca category. Water temperature, conductivity, and pH were measured using an ELMETRON CC-101 and CP-103 conductivity and pH meters, respectively. Other parameters were determined using ion chromatography (Dionex 100), atomic absorption spectrophotometry, and flame AAS techniques. The physical and chemical characteristics of the water in the Pilica Piaski spring-fed pond is given in the analysis of July 2005, as (outflow): temperature—9.5°C, conductivity—372 $\mu S\ cm^{-1}$; pH—7.2; oxygen—9.2 $mg\ l^{-1}$, Cl^-—13.1 $mg\ l^{-1}$; NO_3^-—28.4 $mg\ l^{-1}$, PO_4^{3-}—<0.0,000 $mg\ l^{-1}$; SO_4^{2-}—21.4 $mg\ l^{-1}$; alkalinity—3.2 mval l^{-1}; Ca^{2+}—61.3 $mg\ l^{-1}$; Na^+—5.1 $mg\ l^{-1}$;

K^+—2.4 mg l^{-1}, whereas in the spring-fed pond temperature was c.a. 20°C, and conductivity amounted to 540 μS cm^{-1}.

The material was retrieved from a shallow pond (40–80 cm in depth) which is fed by several, small outlets, situated at the foot of a steep 6–8 m high scarp, and a few meters from the sampling site. The bottom of several small outlets was imbricated with limestone gravel, while the pond bottom was covered by thick layer of mud, due to the low-flow velocity. The sampling site is isolated from the surrounding area by the scarp and swampy zone, of a few meters in width.

In July 2005, during a very hot summer period, when daytime air temperature exceeded 30°C, an 84 cm sediment core was taken at depth ca. 0.8 m, a few meters from the outflows of the spring using a Mondsee-gravity corer with a diameter of 60 mm. The core was divided lengthwise into two halves, one of which was used for diatom analysis, while the other was stored. The 1 cm^3 samples for diatom analysis were taken at 1-cm intervals immediately after retrieval. The core was predominantly homogenous brown-grey clayey mud, turning dark brown in the uppermost part due to a large amount of organic matter. The remains of terrestrial vascular plants (leaves and stem fragments) observed from the depth of 56 cm can originate from vegetation surrounding the pond. A small addition of fine grained sand was observed at 64–62 cm and 45–44 cm. Coarse sand occurred at 58–46 cm.

Diatom samples were analyzed every 5 cm. In order to reconstruct the environmental conditions during the deposition of the sediments studied, diatoms were grouped according to their environmental requirements. The diatom assemblage zones (DAZ) were identified using TILIA (software written by E.C. Grimm, Illinois State Museum, 1995) dendrograms which utilized the sum of least squares after being transformed with square root. Shannon-Weaver diversity was calculated as follows:

$$H' = -\sum_{i=1}^{s} \frac{n_i}{n} \log_2 \frac{n_i}{n},$$

where n_i = number of individual diatoms of the species i, n = total number of individual counts, $\frac{n_i}{n}$ = relative abundance of the species i.

The ^{14}C dating was performed at the Poznań Radiocarbon Laboratory. The samples for diatom analysis were treated with 10% HCl, washed several times with distilled water, and thereafter boiled in concentrated acids (H_2SO_4, HNO_3), in order to remove organic matter. After washing several times with distilled water, the cleaned diatom material was air-dried on coverslips and mounted in Naphrax®. Observations of the diatoms were performed with a Nikon Optiphot microscope equipped with a Plan ×100 oil-immersion lens and differential interference contrast (DIC) optics. The identification of diatoms was based mainly on the following literature: Krammer & Lange-Bertalot, 1986, 1988, 1991a, 1991b; Lange-Bertalot & Metzeltin, 1996; Krammer, 1997a, 1997b, 2000, 2002, 2003; Lange-Bertalot, 1993, 2001; Lange-Bertalot et al., 2003, and numerous other more specific taxonomic publications. The relative abundance of small *Fragilaria* spp., mainly *Fragilaria construens* var. *venter* and *Fragilaria pinnata* was aggregated in diatom analysis, to avoid possible identification errors in LM, particularly during counting, when frustules in girdle view and of very small valve size made precise identification very difficult if not impossible.

Results

The accumulation of sediments, based on ^{14}C dating of the remains of terrestrial vascular plants, from a depth of 38 and 10 cm, was estimated as covering approximately 50 years (107.85 ± 0.35 and 106.58 ± 0.34 pMC, respectively). The lower part of the core consisted of more compact and pale sediments, whereas more recent sediments were darker, with the inclusion of remains of vascular plants. No laminations were observed.

The diatom analysis of the PIL1 core was based on 18 samples. A total of 136 diatom taxa were identified (11 Centrales and 125 Pennales), but only seven were considered common (i.e., present with a relative abundance exceeding 10% in at least one core depth interval). The following taxa were included in this group: *Fragilaria construens* var. *venter*, *F. pinnata*, *F. capucina* var. *capucina* Desmazières, *F. gracilis* Østrup, *F. capucina* var. *rumpens* (Kützing) Lange-Bertalot, *F. capucina* var. *vaucheriae* (Kützing) Lange-Bertalot, and *Achnanthidium minutissimum*. Other taxa with an abundance of ca. 5% were *Planothidium lanceolatum* (Brébisson) Round and

Bukhtiyarova, *Planothidium frequentissimum* (Lange-Bertalot) Round and Bukhtiyarova, *Cyclotella delicatula* Hustedt, *Gomphonema pumilum* (Grunow) Reichardt and Lange-Bertalot (sensu lato), and *Fragilaria ulna* var. *acus* (Kützing) Lange-Bertalot.

The dominant taxa were considered as tychoplanktonic of periphytic origin with respect to habitat, including e.g., *Fragilaria capucina* var. *capucina*. However, at the depth interval of 84–50 cm, the dominant taxa were periphytic forms including *A. minutissimum*. The flora was dominated by alkaliphilous and circumneutral diatoms. The analysis of salinity tolerances showed a strong dominance of fresh–brackishwater taxa including *F. construens* var. *venter* and *F. pinnata* complexes (Figs. 1, 2). It is notable that there are very few freshwater forms, and that those that are present prefer high levels of dissolved minerals in the water. The analysis of trophic tolerances showed a dominance by meso-eutraphentic taxa in the core depth of 85–65 cm. Their content decreases in successive interval as they are replaced by oligo- to eutraphentic (eurytraphentic) taxa. Finally, their relative abundance increases again, up to 97.4% from the core depth of 45 cm.

Changes in diatom species composition, and hence in proportions between particular diatom ecological groups indicate the presence of two local diatom assemblage zones (DAZ I–DAZ II) in the core. Additional division of DAZ I into two sub-zones, namely DAZ I A–B and DAZ II into three subzones, namely DAZ II A–C is possible. Particular zones and sub-zones were distinguished based on the following environmental variables: changes in habitat categories, water mineralization (salinity), pH, and trophic level. The dominants in the core included *F. capucina*, *A. minutissimum*, *P. frequentissimum*, *P. lanceolatum*, and unresolved small *Fragilaria* spp., mainly *F. construens* var. *venter* and *F. pinnata* complexes.

DAZ I A

Two subzones were distinguished within DAZ I. Sub zone DAZ I-A includes the depth interval of 85–65 cm. Tychoplanktonic forms which are taxa of periphytic origin, with respect to habitat were dominant in this subzone. Their abundance decreases from 76.6% at 84 cm to 16.4% at 65 cm. The only exception in this case is sample at a depth of 75 cm, in which they reach 72.9%. On the contrary, the abundance of periphytic taxa is increasing in this interval, however, at depth of 75 cm, they decrease to 17%. Alkaliphilous forms are dominant. However, their relative abundance decreases upward in the core as they are replaced by circumneutral forms. Analysis of trophic preferences showed an increasing abundance of taxa with a very broad tolerance spectrum covering oligo- to eutrophic waters. Their content increases from 25% to 67.2% upward the core. A strong decrease is observed in meso-eutraphentic taxa at a depth of 65 cm (from 71.5% at 84 cm to 9.5% at 65 cm).

DAZ I B

Subzone DAZ I B spans a depth range of 65 to 45 cm. The habitat category is dominated by periphytic taxa. Their content (ca. 50%) does not show any distinct changes in the whole sub-zone. However, an increase is observed in tychoplankton of periphytic origin. These reach 16.4% at the lower limit of DAZ I B and increase to 95.7% in the upper limit of the sub-zone. In terms of pH characteristics, the circumneutral taxa are dominant, and their content ranges between 50% and 72%.

A strong decrease in the abundance of meso-eutraphentic taxa was observed compared with the preceding subzone. These reach a maximum relative abundance of 23.8% at a core depth of 50 cm. Taxa with very broad tolerance spectra, inhabiting oligo- up to eutrophic waters are dominant. However, their content decreases towards the sub-zone upper limit. Mesotraphentic and eutraphentic taxa reach higher abundances in this sub-zone than in the preceding one.

DAZ II

In DAZ II, three sub-zones were distinguished.

DAZ II A

Sub-zone DAZ II A includes the sediment interval from 45 to 25 cm. There is a strong increase in tychoplaktonic taxa of periphytic origin compared to DAZ I. Their content decreases, however, towards the sub-zone upper limit from 97.7% at 45 cm to

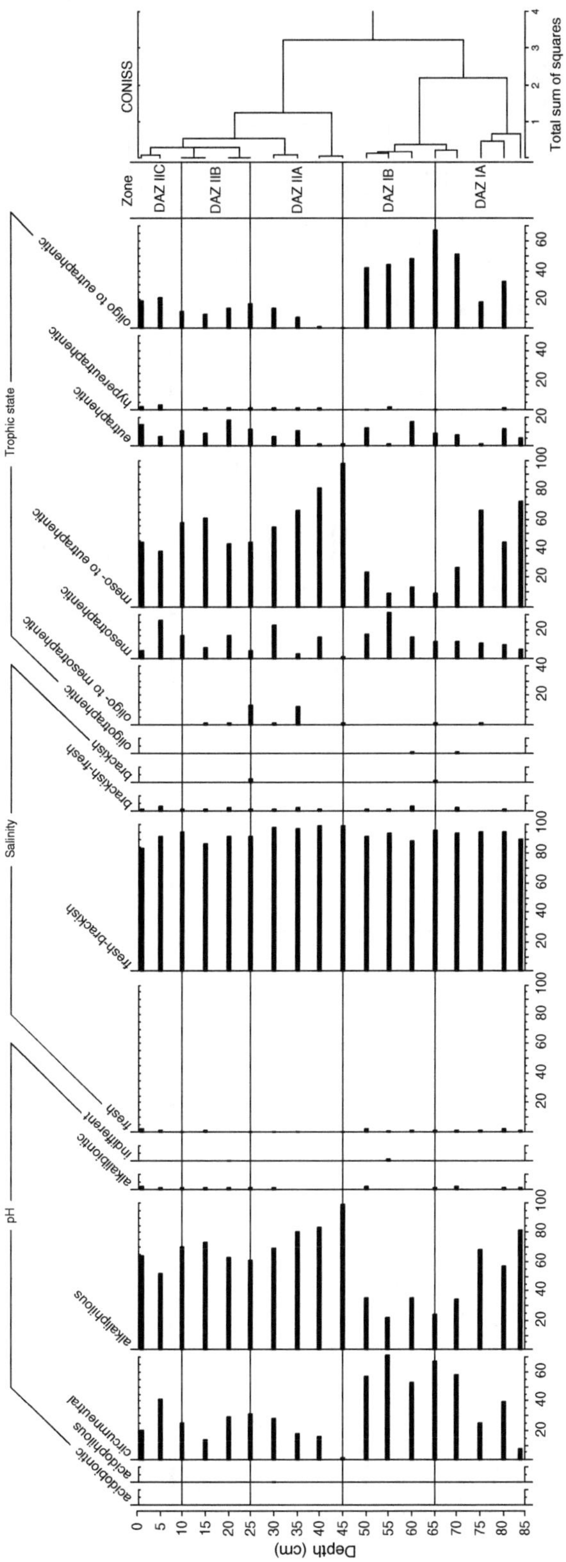

Fig. 1 Stratigraphic distribution of diatom ecological groups in core PIL-1

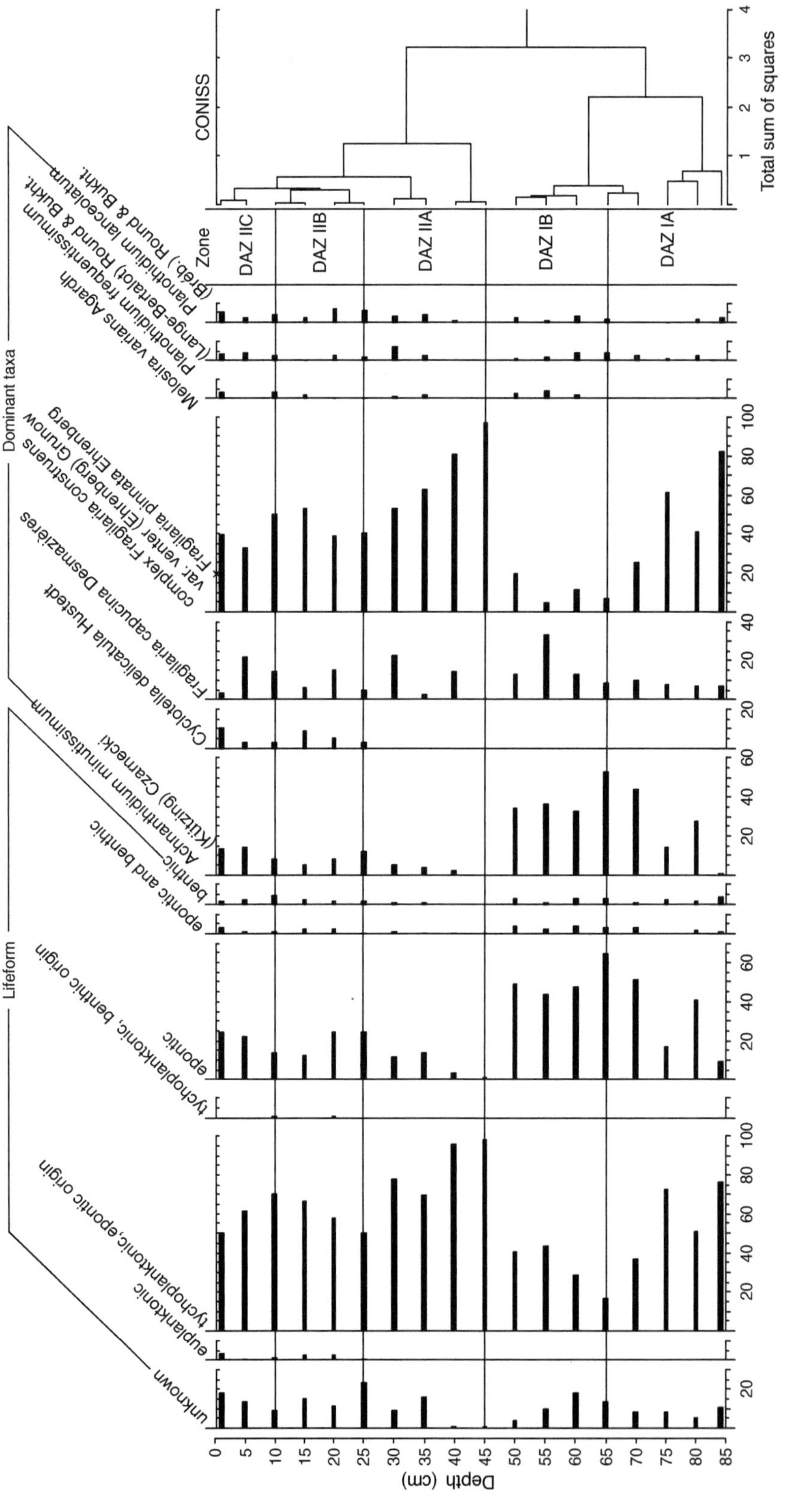

Fig. 2 Stratigraphic distribution of the most abundant taxa and of some ecological groups in core PIL-1

50.3% at 25 cm. The analysis of pH preferences showed strong differences in abundance of particular groups in comparison to DAZ I. Alkaliphilous taxa are dominant. These reach relative abundance of 98.8% at 45 cm and show gradual decrease to 60.9% at 25 cm. The minimum (0.9%) abundance of circumneutral forms occurs in this zone, however, their content increases to 31.6% at 25 cm. Despite this increase, their content is much lower than in DAZ I A and DAZ I B. Meso-eutraphentic forms are dominant among trophic groups. They reach the highest relative abundance for the whole core in this sub-zone. Their abundance decreases, however, gradually from 97.4% at 45 cm to 44.2% at 25 cm. Meso- and eutraphentic taxa remain relatively stable between 10% and 20% in this interval. The relative abundance of taxa with a very broad range of trophic tolerance (oligo- to eutraphentic) increases from 1.7% at 45 cm to 16.7% at 25 cm.

DAZ II B

Sub-zone DAZ II B encompasses a core depth range of 25–10 cm. The habitat category is dominated by the increasing abundance of tychoplanktonic taxa of periphytic origin, whereas the abundance of periphytic (epontic) taxa decreases. Truly planktonic diatoms (e.g., *Asterionella formosa* Hassall and *Aulacoseira ambigua* (Grunow) Simonsen) were represented by single valves).

The relative abundance of alkaliphilous diatoms was stable and reached ca. 70%, when compared with circumneutral forms which decrease. Diatom assemblages are dominated by freshwater–brackish forms, while solitary valves of brackish water species appear at a core depth of 25 cm.

Meso-eutraphentic forms increase in abundance from 44.2% at 25 cm to 57.1% at 15 cm. In addition, mesotraphentic, eutraphentic, and oligo-eutraphentic forms appear in low numbers (maximum up to 16%).

DAZ II C

Sub-zone DAZ II C includes the uppermost part (10–1 cm) of the core. Tychoplanktonic forms of periphytic origin are present in smaller numbers. They decrease from 70.6% at 10 cm to 50% at 1 cm. Periphytic forms increase, however, towards their maximum abundance in the core. True planktonic taxa appear in this sub-zone, and these are primarily represented by *C. delicatula* Hustedt.

Alkaliphilous forms increase in abundance compared to DAZ II B. The number of *A. minutissimum* increases. Fresh-brackish water forms are dominant amongst halobous groups, while brackish–freshwater forms (ca. 3%) appear at a core depth of 5 cm. Meso-eutraphentic species are the dominant trophic form. Their content decreases from 57.1% in the core lower limit to 43.9% in the core uppermost sample. In addition, mesotraphentic, eutraphentic, and oligo- to eutraphentic forms also occur and all reach rather similar abundance.

Discussion

The Krakowsko-Częstochowska upland springs are fed by the largest aquifers in Poland. Extensive human modification (e.g., pollution, increasing water consumption, coal mining, melioration works, and agricultural practices) have diminished the water resources of this system and indirectly damaged and reduced the number of springs supplied by them. The major threat to the springs of the Krakowsko-Częstochowska upland is eutrophication originating from unsustainable agricultural practices and improper wastewater management. The most threatened springs are situated in the area covered by sands, where the pollutants can easily percolate into the groundwater. The most important threat to water quality in the area are nitrates, as their concentration exceeds maximal allowable limits for drinking waters in Poland (Dojlido, 1987).

The absence of orthophosphate ions at detectable concentrations, by means of applied methods, can be related to biological consumption and long-term burial of phosphorus (Dodds, 2003), as precipitation of insoluble Ca-phosphates (Ca–P) is the dominant phosphorous transformation at pH greater than 7.0 (Vymazal, 1995). Fluctuations of the groundwater table influence the total ion content, including that of calcium, due to the retention time in conduits connecting the aquifer and its springs. Additional anthropogenic impact (e.g., agricultural modification, pollution) can enhance the rate of the changes in hydrochemical background.

Despite the fact that springs are the places where groundwater outflows, rarely, as in the case of Pilica

Piaski, there are spring-fed ponds which allow for sediments accumulation. The results of ^{14}C dating of terrestrial vascular plants from the depth of 38 and 10 cm did not provide precise chronologies for that part of the core, whose age could be estimated for approximately 50 years. Supposing that even high sedimentation rates the age of "bottom", samples could represent at least 100-year old diatom assemblage remains. The application of more precise dating methods might furnish with better data.

Sediment samples represent an integrated sample from all habitats within the pond and outlets supplying it, as well as an integration of several seasons of sediment accumulation. The lack of lamination within the core is presumably related to very few seasonal changes in springs due to their thermal stability. The absence of laminae may also be due to bioturbation or sediment resuspension by the wind.

The diatom flora from the Pilica Piaski presents an assemblage with relatively high diversity (H' up to 4.06), which is dominated by circumneutral and/or alkaliphilous, fresh–brackishwater, meso-eutraphentic, and oligo- to eutraphentic taxa that are common constituents of the area limnocrens (Rakowska, 1996; Wojtal, 2006). Calculated H' values were generally similar to the reported from springs of the southern, calcareous part of Pyrenees, with slow-moving to sometimes fast waters at the outflows, and high ionic strength (Sabater & Roca, 1990), but higher than in epilithic diatom assemblages in springs of the Krakowsko-Częstochowska upland (Wojtal, 2006). The occurrence of several diatom species clearly corresponds to diatom assemblages in springs of carbonate-rich water (e.g., Sabater & Roca, 1992; Cantonati, 1998; Werum, 2001a; Werum & Lange-Bertalot, 2004; Cantonati et al., 2006), on the other hand, the lack of such common diatoms as *Diatoma mesodon* (Ehrenberg) Kützing, *Karayevia clevei* (Grunow) Round and Bukhtiyarova, or very scarce species such as *Achnanthidium pyrenaicum* Hustedt is probably related to their preferences of fast flowing water. The composition of diatom assemblages from the Pilica Piaski spring-fed pond is the most similar to the one reported from Madonina Val Lomasona in Italy (Filippi et al., 2008), probably due to most analogous environmental conditions.

The composition of diatom assemblages changed dramatically at depths of 65 and 45 cm (Figs. 1, 2). A marked shift in species composition was recorded at this time, but it is likely that the diatom assemblage began to change even earlier (70 cm and 50 cm). Potential causes contributing to the recent shift in diatom assemblages include changes in water discharge, anthropogenic eutrophication, artificial nitrogen deposition, and/or changes of alkalinity. The relatively stable abundance of fresh–brackishwater diatoms for most of the spring-fed pond's history could be related to the natural ion status of the spring. Changes in diatom tolerances from circumneutral to alkaliphilous may suggest increased natural calcium leaching from the bedrock, but is more likely related to or accompanied by changes in the general nutrient status. The most distinctive shift in diatom composition, from diatoms tolerating broad range of trophic levels (DAZ I) towards meso-eutraphentic (DAZ II) suggests an increase of nutrient concentration. The changes in the composition of the diatom assemblages between DAZ IA and DAZ IB can be related to an increase in discharge as an addition of sand was observed in samples at 64–46 cm core depth, and was accompanied by a shift of alkaliphilous, benthic diatoms, e.g., *F. construens* var. *venter* and *F. pinnata* into circumneutral tychoplanktonic taxa of periphytic (epontic) origin, e.g., *A. minutissimum*.

Anthropogenic nutrient deposition (Baścik et al., 2001) may be a factor driving diatom assemblage change in the most recent Pilica Piaski sediments (beginning from the depth of 25 cm). The increase of centric diatoms [*C. delicatula* Hustedt, *Puncticulata radiosa* (Lemmermann) Håkansson, *Discostella pseudostelligera* (Hustedt) Houk & Klee, *Skeletonema* sp., *Stephanodiscus hantzschii* Grunow, and *S. neoastraea* Håkansson and Hickel] in the topmost part of the core could reflect increasing eutrophication.

Unfortunately, the most abundant diatoms, *F. construens* var. *venter* and *F. pinnata* complexes are also an important component of diatom assemblages found in a wide range of aquatic environments including freshwater (e.g., Bennion et al., 2001; Rühland et al., 2003; Schmidt et al., 2004, brackish–fresh and brackish waters (e.g., Witkowski, 1994; Witkowski et al., 2000); remain understudied. According to Lowe (1996), *F. construens* var. *venter* is one of the most important diatoms in waters of low light, low turbulence, and narrow temperature range, conditions that correspond well with those in Pilica Piaski locality. According to Tuchman (1996),

F. construens possess the ability to resume photosynthetic activity upon illumination, after being buried below the photic zone in shallow marine sediments for an extended period of time. Small *Fragilaria* spp. were also regarded as poor indicators for the trophic status as they are distributed along the whole length of TP gradient in the northwest European training set (Bennion et al., 2001). Their distribution in a wide range of aquatic environments can be, however, caused by difficulties in precise identification due to the need to resolve their diagnostic characters with a scanning electron microscope (Morales, 2002), and consequently the autecology of this group is only superficially understood. Another important constituent, *A. minutissimum*, is one of the most important species in the calcareous part of Pyrenees where it is found in slow moving to sometimes fast water at spring sources (Sabater & Roca, 1990). It is identified as a resource specialist, that is, a species capable of securing resources that are present at concentrations limited to sympatric species within lentic periphyton communities, and as a phosphorus specialist (Lowe, 1996). Also autecology of other important species, such as *C. delicatula* is very poorly known (Wojtal & Kwandrans, 2006).

Conclusions

Diatom assemblages have changed markedly within twentieth century in the Pilica Piaski spring-fed pond, indicating possible changes in pH, conductivity, and nutrient levels. Changes in current velocity could be regarded as one of the most important factors driving shifts in the diatom assemblage structure. The direction of environmental change would be confirmed by more detailed diatom analysis, especially of small *Fragilaria* representatives, ^{210}Pb dating, and chemical sediments analyses. Additional paleolimnological studies in this region will help to refine and improve future interpretations.

Acknowledgments The authors acknowledge the valuable suggestions of anonymous reviewers. We are grateful to Professor Wojciech Chełmicki and MSc. Maria Baścik (Dept. of Hydrology, Jagiellonian University) for discussions and suggestions concerning springs in the Krakowsko-Częstochowska upland, Dr. Elżbieta Wilk-Woźniak for logistic support during field works, and MSc. Agnieszka Kierzek Szczecin University for preparation of figures. The authors highly appreciate the linguistic correction of the manuscript by Dr. Collin Archibald. The senior author (A. Z. Wojtal) expresses special gratitude to her husband Piotr for his continuous help and encouragement during the research and preparation of this work. The work was partly supported by the Polish Ministry of Science and Higher Education for 2008–2011 (grant N304 092834), and 2004–2007 (grant P04G 09826).

References

Baścik, M., W. Chełmicki, A. Korska, J. Pociask-Karteczka & J. Siwek, 2001. Źródła Wyżyny Krakowsko-Wieluńskiej i Miechowskiej. Zmiany w latach 1973–2000. Springs of the Kraków-Wieluń and Miechów Uplands. Changes between 1973 and 2000 (in Polish). Uniwersytet Jagielloński, Instytut Geografii i Gospodarki przestrzennej, Zakład Hydrologii, Kraków: 1–127.

Battarbee, R. W., 1986. Diatom analysis. In Berglund, B. E. (ed.), Handbook of Holocene Palaeoecology and Palaeohydrology. John Wiley & Sons, Chichester, New York, Brisbane, Toronto, Singapore: 527–570.

Battegazzore, M., A. Morisi, B. Gallino & S. Fenoglio, 2004. Environmental quality evaluation of alpine springs in NW Italy using benthic diatoms. Diatom Research 19: 149–165.

Bennion, H., P. Appleby & G. L. Phillips, 2001. Reconstructing nutrient histories in the Norfolk Broads, UK: implications for the role of diatom-total phosphorus transfer functions in shallow lake management. Journal of Paleolimnology 26: 181–204.

Cantonati, M., 1998. Diatom communities of springs in the Southern Alps. Diatom Research 13: 201–220.

Cantonati, M., 2001. The diatom communities of the liverwort *Chiloscyphus polyanthus* var. *rivularis* in a mountain spring-fed stream in the Adamello-Brenta Regional Park, Northern Italy. In Jahn, R., J. P., Kociolek, A. Witkowski & P. Compère (eds), Lange-Bertalot-Festschrift. Studies on Diatoms. A.R.G. Gantner Verlag K.G., Ruggell: 353–368.

Cantonati, M., R. Gerecke & E. Bertuzzi, 2006. Springs of the Alps—sensitive ecosystems to environmental change: from biodiversity assessments to long-term studies. Hydrobiologia 562: 59–96.

Dodds, W. K., 2003. The role of periphyton in phosphorus retention in shallow freshwater aquatic systems. Journal of Phycology 39: 840–849.

Dojlido, J. R., 1987. Chemia wody. Water chemistry (in Polish). Arkady, Warszawa: 136–145.

Filippi, M. L., 2007. Paleo-ecological studies on springs: strategic approach and methods. In Cantonati, M., E. Bertuzzi & D. Spitale (eds), The Spring Habitat: Biota and Sampling Methods. Museo Tridentino di Science Naturali, Trento: 299–306.

Filippi, M. L., N. Angeli, L. Marziali, A. H. I. Matouk, E. Bertuzzi, V. Lencioni & M. Cantonati, 2008. Paleolimnological techniques applied to springs habitats: challenges and limits as highlighted by the CRENODAT project (southeastern Alps, Italy). Verhandlungen der Internationalen Vereinigung für Limnology 30: 329–332.

Kadłubowska, J. Z., 1985. Untersuchungen der Stetigkeit der Diatomeengesellschaften aus der Saltzquelle, den Limnokrenen und dem Moortümpel. Verhandlungen der Internationalen Vereinigung für Limnology 22: 2834–3837.

Keatley, B. E., M. S. V. Douglas & J. P. Smol, 2006. Early-20th century environmental changes inferred using subfossil diatoms from a small pond on Melville Island, N.W.T., Canadian high Arctic. Hydrobiologia 553: 15–26.

Krammer, K., 1997a. Die cymbelloiden Diatomeen. Eine Monographie der weltweit bekannten Taxa. Teil 1. Allgemeines und *Encyonema* Part. Bibliotheca Diatomologica 36. J. Cramer, Berlin, Stuttgart.

Krammer, K., 1997b. Die cymbelloiden Diatomeen. Eine Monographie der weltweit bekannten Taxa. Teil 2. *Encyonema* part., *Encyonopsis* and *Cymbellopsis*. Bibliotheca Diatomologica 37. J. Cramer, Berlin, Stuttgart.

Krammer, K., 2000. The genus *Pinnularia*. In Lange-Bertalot, H. (ed.), Diatoms of Europe. 1. A.R.G. Gantner Verlag K.G., Vaduz.

Krammer, K., 2002. *Cymbella*. In Lange-Bertalot, H. (ed.), Diatoms of Europe. 3. A.R.G. Gantner Verlag K.G., Ruggell.

Krammer, K., 2003. *Cymbopleura, Delicata, Navicymbula, Gomphocymbellopsis, Afrocymbella*. In Lange-Bertalot, H. (ed.), Diatoms of Europe. 4. A.R.G. Gantner Verlag K.G., Ruggell.

Krammer, K. & H. Lange-Bertalot, 1986. Bacillariophyceae. 1. Naviculaceae. In Ettl, H., J. Gerloff, H. Heyning & D. Mollenhauer (eds), Süβwasserflora von Mitteleuropa 2. G. Fischer Verlag, Stuttgart, New York.

Krammer, K. & H. Lange-Bertalot, 1988. Bacillariophyceae. 2. Bacillariaceae, Epithemiaceae, Surirellaceae. In Ettl, H., J. Gerloff, H. Heyning & D. Mollenhauer (eds), Süβwasserflora von Mitteleuropa 2. G. Fischer Verlag, Stuttgart, NewYork.

Krammer, K. & H. Lange-Bertalot, 1991a. Bacillariophyceae 3. Centrales, Fragilariaceae, Eunotiacae. In Ettl, H., J. Gerloff, H. Heyning & D. Mollenhauer (eds), Süβwasserflora von Mitteleuropa. 2. Gustaw Fischer Verlag, Stuttgart, Jena.

Krammer, K. & H. Lange-Bertalot, 1991b. Bacillariophyceae. 4. Achnanthaceae, Kritische Ergänzungen zu *Navicula* (Lineolatae) und *Gomphonema*, Gesamtliteraturverzeichnis. In Ettl, H., G. Gärtner, J. Gerloff, H. Heyning & D. Mollenhauer (eds), Süβwasserflora von Mitteleuropa. 2. G. Fischer Verlag, Stuttgart, Jena.

Lange-Bertalot, H., 1993. 85 new taxa and much more than 100 taxonomic clarifications supplementary to Süβwasserflora von Mitteleuropa 2(1–4). Bibliotheca Diatomologica 27. J. Cramer, Berlin, Stuttgart.

Lange-Bertalot, H., 2001. Diatoms of the European inland waters and comparable habitats. *Navicula* sensu stricto, 10 genera separated from *Navicula* sensu lato, *Frustulia*. In Lange-Bertalot, H. (ed.), Diatoms of Europe. 2. A.R.G. Gantner Verlag K.G., Vaduz.

Lange-Bertalot, H. & D. Metzeltin, 1996. Oligotrophie-Indikatoren. 800 Taxa in drei ökologisch diversen Seen-Typen. In Lange-Bertalot, H. (ed.), Iconographia Diatomologica. Annotated Diatom Micrographs 2, Ecology-Diversity-Taxonomy. Koeltz Scientific Books, Koenigstein.

Lange-Bertalot, H., P. Cavacini, N. Tagliaventi & S. Alfinito, 2003. Diatoms of Sardinia. Rare and 76 new species in rock pools and other ephemeral waters. In Lange-Bertalot, H. (ed.), Iconographia Diatomologica. Annotated Diatom Micrographs 12, Biogeography-Ecology-Taxonomy. A.R.G. Gantner Verlag K.G., Ruggell.

Lowe, R. L., 1996. Periphyton patterns in lakes. In Stevenson, R. J., M. L. Bothwell & R. L. Lowe (eds), Algal Ecology—Freshwater Benthic Ecosystems. Academic Press, San Diego: 57–76.

Morales, E. A., 2002. Studies in selected fragilarioid diatoms of potential indicator value from Florida (USA) with notes on the genus *Opephora* Petit (Bacillariophyceae). Limnologica 32: 102–113.

Poulíčková, A., P. Hašler & M. Kitner, 2005. Cyanobacteria and algae. In Paulíčková, A., M. Hájek & K. Rybníček (eds), Ecology and Paleoecology of Spring Fens of the West Carpathians. Palacký University Olomouc, Academy of Sciences of the Czech Republic, Masaryk University Brno, Olomouc.

Rakowska, B., 1996. Diatom communities occurring in Niebieskie Źródła near Tomaszów Mazowiecki, Central Poland (1963–1990). Fragmenta Floristica et Geobotanica 41: 639–655.

Reichardt, E., 1981. Die Diatomeenflora quelliger und sumpfiger Standorte in der Umgebung von Treulichtlingen. Nova Hedwigia 34: 487–519.

Reichardt, E., 1994. Zur Diatomeenflora (Bacillariophyceae) tuffabscheidender Quellen und Bäche im Südlichen Frankenjura. Berichte der Bayerischen Botanischen Gesellschaft 64: 119–133.

Round, F. E., 1981. The Ecology of Algae. Cambridge University Press, Cambridge.

Rühland, K., A. Pienitz & J. P. Smol, 2003. Paleolimnological evidence from diatoms for recent environmental changes in 50 lakes across Canadian Arctic Treeline. Arctic, Antarctic, and Alpine Research 35: 110–123.

Sabater, S. & J. R. Roca, 1990. Some factors affecting distribution of diatom assemblages in Pyrenean springs. Freshwater Biology 24: 493–507.

Sabater, S. & J. R. Roca, 1992. Ecological and biogeographical aspects of diatom distribution in Pyrenean springs. British Phycological Journal 27: 203–213.

Schmidt, R., C. Kamenik, H. Lange-Bertalot & R. Klee, 2004. *Fragilaria* and *Staurosira* (Bacillariophyceae) from sediment surfaces of 40 lakes in the Austrian Alps in relation to environmental variables, and their potential for palaeoclimatology. Journal of Limnology 63: 171–189.

Smol, J. P., 2002. Pollution of Lakes and Rivers: A paleoenvironmental perspective. Hodder Publishers, New York; Oxford University Press, Arnold, London.

Stoermer, E. F. & J. P. Smol (eds), 1999. The Diatoms: Applications for the Environmental and Earth Sciences. Cambridge University Press, Cambridge.

Tuchman, N. C., 1996. The role of heterotrophy in Algae. In Stevenson, R. J., M. L. Bothwell & R. L. Lowe (eds), Algal Ecology—Freshwater Benthic Ecosystems. Academic Press, San Diego: 299–319.

Vymazal, J., 1995. Algae and Element Cycling in Wetlands. Lewis Publishers, Boca Raton.

Werum, M., 2001a. Die Kieselalgengesellschaften in Quellen: Abhängigkeit von Geologie und anthropogener Beeinflussung in Hessen. Bundesrepublik Deutschland, Wiesbaden.

Werum, M., 2001b. Diatomeen in Quellen hessischer Mittelgebirge: Gefahrdung nach Roter Liste in Korrelation zu anthropogenen Eingriffen und Geologie. In Jahn, R., J. P. Kociolek, A. Witkowski & P. Compère (eds), Lange-Bertalot-Festschrift. Studies on Diatoms. A.R.G. Gantner Verlag K.G., Ruggell: 369–381.

Werum, M. & H. Lange-Bertalot, 2004. Diatoms in springs from Central Europe and elsewhere under the influence of hydrogeology and anthropogenic impacts. In Lange-Bertalot, H. (ed.), Iconographia Diatomologica, Annotated Diatom Monographs 13, Ecology-Hydrology-Taxonomy. A.R.G. Gantner Verlag K.G., Ruggell.

Williams, N. E. & D. D. Williams, 1997. Paleoecological reconstruction of natural and human influences on groundwater outflows. In Boon, P. J. & D. L. Howell, (eds), Freshwater Quality: Defining The Indefinable? Scottish Natural Heritage, Edinburgh: 172–180.

Witkowski, A., 1994. Recent and fossil diatom flora of the Gulf of Gdańsk, Southern Baltic Sea. Bibliotheca Diatomologica 28. J. Cramer, Berlin, Stuttgart.

Witkowski, A., H. Lange-Bertalot & D. Metzeltin, 2000. Diatom flora of marine coasts I. In Lange-Bertalot, H. (ed.), Iconographia Diatomologica 7, Annotated Diatom Monographs, Diversity-Taxonomy-Identification. A.R.G. Gantner Verlag K.G., Ruggell.

Wojtal, A. Z., 2006. Use of diatoms for monitoring springs (Southern Poland). In Ács, É., K. T. Kiss, J. Padisák & K. E. Szabó (eds), Use of Algae for Monitoring Rivers. Hungarian Algological Society, Balatonfüred: 179–183.

Wojtal, A. Z. & J. Kwandrans, 2006. Diatoms of the Wyżyna Krakowsko-Częstochowska upland (S Poland)–Coscinodiscophyceae (Thalassiosirophycidae). Polish Botanical Journal 51: 177–207.

Żelazna-Wieczorek, J. & M. Mamińska, 2006. Algoflora and vascular flora of a limestone spring in the Warta river valley. Acta Societatis Botanicorum Poloniae 75: 131–143.

Diatom-inferred trophic history of IJsselmeer (The Netherlands)

Holger Cremer · Frans P. M. Bunnik · Emiliya P. Kirilova · Eddy H. R. R. Lammens · André F. Lotter

Originally published in the journal Hydrobiologia, Volume 631, No. 1, 279–287.
DOI: 10.1007/s10750-009-9816-7

Abstract IJsselmeer was formed in 1932 through the closure of the Afsluitdijk that separated the artificial lake from the former Zuiderzee estuary. The palaeoecology of IJsselmeer was studied on a 63-cm-long sediment core. Lithology and microfossil data, particularly the diatom flora, clearly show the transition from the marine Zuiderzee into the freshwater IJsselmeer. Trophic conditions in IJsselmeer since 1932 have been inferred by qualitative and quantitative diatom-based approaches: by plotting the distribution of trophic categories based on published trophic indicator values, by a canonical correspondence analysis (CCA) yielding relative total phosphorus (TP) changes and by applying a transfer function in order to calculate TP concentrations. All three approaches indicate that IJsselmeer initially was meso- to eutrophic. A first hypertrophic period is indicated for the mid-1940s, likely due to internal loading. After 1960, the phosphorus load steadily increased and TP in IJsselmeer reached highest concentrations (ca. 150 μg l^{-1}) in the 1980s as confirmed by monitoring data since 1975. The monitored data show that the TP concentration continuously decreased after 1985 due to successful environmental protection measures. This trend is not (or not yet) evidenced by the diatom data and thus, the diatom-inferred TP concentration.

Keywords Palaeolimnology · Lake sediments · Diatoms · Phosphorus · Eutrophication

Guest editors: K. Buczkó, J. Korponai, J. Padisák & S. W. Starratt
Palaeolimnological Proxies as Tools of Environmental Reconstruction in Fresh Water

H. Cremer (✉) · F. P. M. Bunnik
TNO Built Environment and Geosciences, Geological Survey of the Netherlands, Princetonlaan 6, 3584 CB Utrecht, The Netherlands
e-mail: holger.cremer@tno.nl

E. P. Kirilova · A. F. Lotter
Institute of Environmental Biology, Laboratory of Palaeobotany and Palynology, Utrecht University, Budapestlaan 4, 3584 Utrecht, The Netherlands

E. H. R. R. Lammens
Rijkswaterstaat Waterdienst, P. O. Box 17, 8200 Lelystad, The Netherlands

Introduction

Lake IJsselmeer is a main target area for the implementation of the European Union's (EU's) Water Framework Directive (WFD) in the Netherlands. As an artificial, so-called heavily modified water body (HMWB), the lake is exposed to various man-made pressures including shoreline regulation, high nutrient loads and overexploitation of fish (Lammens et al., 2008). The WFD requires that all European Union member states determine reference conditions for their aquatic ecosystems to provide a baseline against which the effects of past and present activities should be measured. EU member states are obliged to undertake measures to achieve the formulated goals before 2015 (European Union, 2000). Reference conditions are subsequently used to classify the ecological status of an aquatic ecosystem which is defined by the (widely accepted) deviation from the reference conditions and ranges from 'good' through 'moderate' and 'poor' to 'bad'. Palaeolimnology is able to assists the implementation of the WFD through the determination of reference conditions for aquatic ecosystems (Bennion & Battarbee, 2007). Through deciphering the natural histories of ecosystems, reference conditions are indicated by the apparent 'natural state' of a water body, which prevailed in periods when human pressures were less drastic than today. Several studies (i.e. Taylor et al., 2006 and references therein) have demonstrated the value of palaeolimnology for WFD related issues.

Here, we describe the environmental history of the past 75 years of IJsselmeer based on a multiproxy palaeolimnological study with emphasis on the development of the trophic state. Various microfossils were initially used to document the transition from the former marine Zuiderzee towards the fresh IJsselmeer in 1932. Changes in trophic state of the past 75 years are reconstructed using diatoms and three different approaches. These analyses should reveal distinct changes of the total phosphorus concentration in IJsselmeer during the relatively short time span of its existence. The article also attempts to clarify the possibilities and limitations of using algae, particularly diatoms, for paleoenvironmental reconstructions in young and artificial water bodies that, in addition, are strongly affected by human activities.

Study area

IJsselmeer was created in 1932 by the separation of the former Zuiderzee estuary from the North Sea through the construction of a dam, the Afsluitdijk (Fig. 1). After its formation, IJsselmeer changed within a few years into a freshwater lake. In the years subsequent to completion, the size of the lake was reduced by land reclamation projects through which a number of border lakes were created. The construction of another dam, the Houtribdijk, in 1975 divided IJsselmeer into two lakes: IJsselmeer (ca. 1125 km^2) to the north and Markermeer (ca. 650 km^2) to the south of the newly formed dam (Fig. 1). The IJssel River, a tributary of the Rhine River, is the main freshwater source of IJsselmeer. Today, the IJsselmeer region is a strongly regulated, multifunctional area with various recreational, industrial, natural, nautical and agricultural interests.

IJsselmeer is a large, buffered and still eutrophic to hypertrophic lake with moderate chlorinity. The average and maximum water depths of IJsselmeer are 4.5 and 7 m, respectively. Regular measurements of the phosphorus concentration started in the mid-1970s when the lake was already hypertrophic (total phosphorus (TP) concentration of 200–300 $\mu g\ l^{-1}$; Lammens et al., 2008). Since 1985, the phosphorus load constantly decreased to present-day concentration of ca. 100 $\mu g\ l^{-1}$ (De Leeuw et al., 2006). Improvements over the past two decades are also indicated by changes in total nitrogen and chlorophyll *a* (De Leeuw et al., 2006).

Materials and methods

Sediment core IJ6 (Length: 63 cm) was recovered with a UWITEC gravity corer in spring 2006 in the western part of IJsselmeer (Fig. 1) at 420 cm water depth. A total of 45 samples from this core were analyzed for diatoms and palynological indicators (dinoflagellates, foraminifera, green algae and pollen).

Sediment samples for diatom analysis were freeze-dried and 0.5–2 g of this material was treated consecutively with HCl (30%) and H_2O_2 (35%) in order to remove the carbonates and organic matter. Sedimentation trays (Battarbee, 1973) and the highly refractive mountant Naphrax ($RI = 1.74$) were used to produce permanent slides. Diatom slides were analyzed at 945× magnification with a Leica DM2500 microscope equipped with differential interference contrast.

Palynological slides were prepared by treating the samples consecutively with sodium pyrophosphate (15 g/l), HCl (30%) and acetolysis in order to disperse the samples and to dissolve carbonates and organic matter. In a final step, the samples were centrifuged with heavy liquid to eliminate sand and silt particles and to concentrate pollen and other microfossils. Palynological slides were mounted using glycerin.

Diatom-based reconstructions of the trophic history of IJsselmeer were carried out with a qualitative,

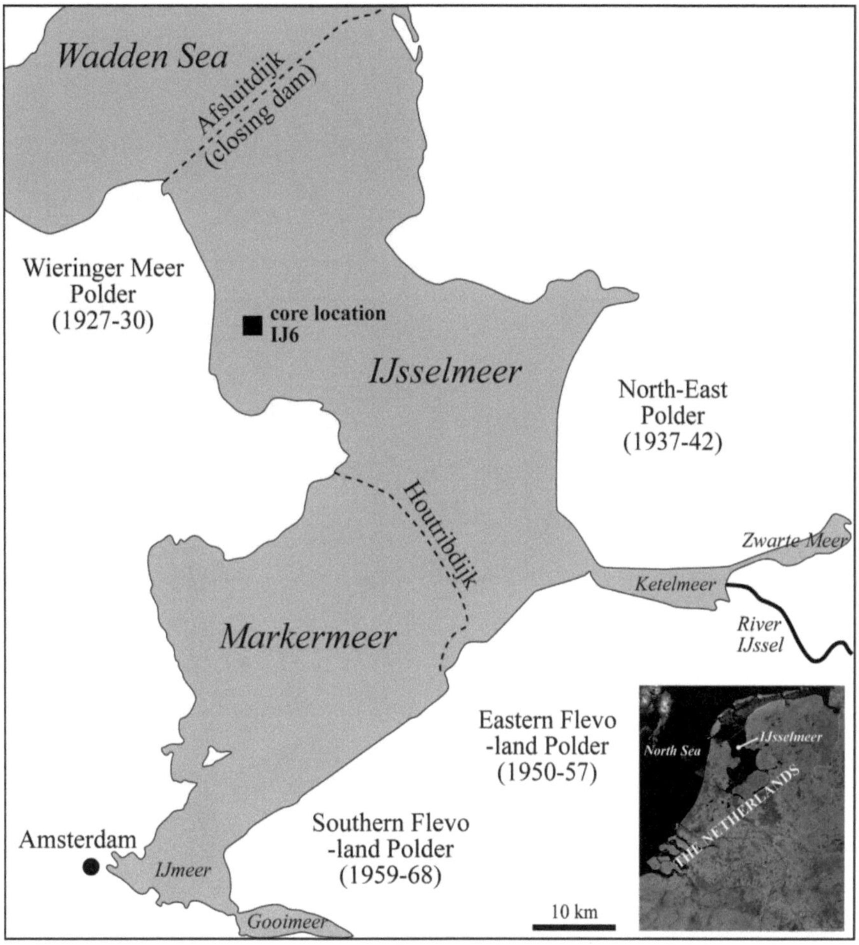

Fig. 1 Lakes and dams in the IJsselmeer area. The location of core IJ6 is indicated by a *black square*

semiquantitative and quantitative method. First, an estimation of the trophic state was performed by calculating the frequency of diatom-based trophic state categories. For this approach, trophic indicator values for diatom species according to the classification of Van Dam et al. (1994) were used. Second, a canonical correspondence analysis (CCA; Ter Braak, 1986) was performed with a modern diatom training set and TP as the only constraining environmental variable. The 28 fossil diatom assemblages of core IJ6 were used as passive samples in the CCA. The available European diatom TP datasets (the Combined TP dataset was used) of European Diatom Database Initiative (EDDI; http://craticula.ncl.ac.uk/EDDI/jsp) were used as modern training set (345 lakes; TP range 2–1,189 $\mu g\ l^{-1}$, mean 95 $\mu g\ l^{-1}$). The CCA, performed with the program CANOCO 4.52 (Ter Braak & Šmilauer, 1998), resulted in first CCA axis scores of the passive fossil samples that represent relative total phosphorus (TP) concentrations. Thirdly, a diatom-based quantitative TP reconstruction was performed using the available European diatom TP training sets of EDDI and a Modern Analogue Technique (MAT) transfer function as implemented in the software package C2 Version 1.5 (Juggins, 2007) to calculate absolute TP concentrations in IJsselmeer for the past 70 years. We used chi-squared distance as a dissimilarity coefficient and the weighted average of the 10 closest analogues.

Results

Zuiderzee to IJsselmeer transition

The transition from marine conditions of the Zuiderzee to the freshwater lake IJsselmeer is indicated by both the lithology and micropaleontology in core IJ6 (Fig. 2). The transition is indicated by a change from

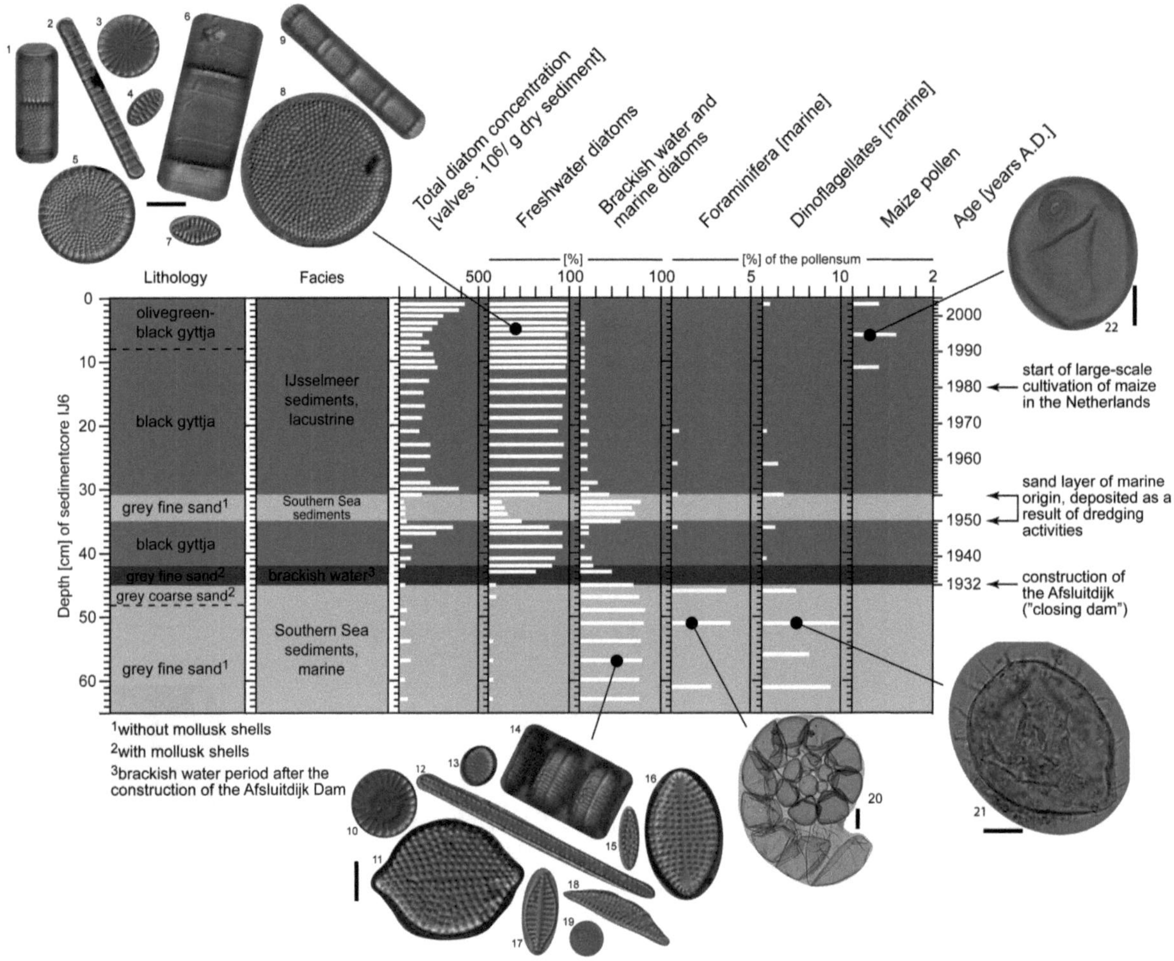

Fig. 2 Lithology, facies and abundance of diatoms, foraminifera, dinoflagellates and maize pollen in sediment core IJ6 from IJsselmeer. Micrographs. *Freshwater diatoms*: 1. *Aulacoseira ambigua* (Grunow) Simonsen. 2. *Diatoma tenuis* Agardh. 3. *Stephanodiscus hantzschii* Grunow. 4. *Staurosirella pinnata* (Ehrenberg) Williams et Round. 5. *Stephanodiscus medius* Håkansson 6. *Stephanodiscus binderanus* (Kützing) Krieger. 7. *Staurosira venter* (Ehrenberg) Cleve et Möller. 8. *Actinocyclus normanii* (Gregory) Hustedt. 9. *Aulacoseira subarctica* (O. Müller) Simonsen. *Marine and brackish water diatoms*: 10. *Cyclotella meneghiniana* Kützing. 11. *Raphoneis amphiceros* (Ehrenberg) Ehrenberg. 12. *Thalassionema nitzschioides* (Grunow) Mereschkowski. 13. *Delphineis minutissima* (Hustedt) Simonsen. 14. *Paralia sulcata* (Ehrenberg) Cleve. 15. *Cymatosira belgica* Grunow. 16. *Delphineis surirella* (Ehrenberg) Andrews. 17. *Planothidium delicatulum* (Kützing) Round et Bukhtiyarova. 18. *Campylosira cymbelliformis* (A. Schmidt) Grunow. 19. *Thalassiosira proshkinae* Makarova. *Foraminifera*: 20. Chitinous inner linings. *Marine dinoflagellates*: 21. *Operculodinium centrocarpum* (Deflandre et Cookson) Wall. *Pollen*: 22. *Zea mays* (maize). Scale bars: 10 μm

coarse sand with mollusk shells through fine sand without mollusk shells to a black gyttja. The grey fine sand layer at 45–42 cm core depth represents the brackish water period following the closure of the Afsluitdijk. The transition is also documented by the occurrence of marine diatoms, foraminifera and dinoflagellates. Shells of these protists are common below 45 cm core depth and disappear in the gyttja where freshwater diatoms constitute the main microfossil group (Fig. 2). Marine diatoms are also predominant in a sand layer at 35–31 cm core depth suggesting that this layer is of marine origin. The sand was assumingly deposited during dredging activities, which are common in IJsselmeer.

Chronology

The chronology of core IJ6 is based on the marine–freshwater transition in 1932 following the construction of the Afsluitdijk. The post-1932 period is represented in this core by the upper 45 cm of sediment, which indicates a mean sedimentation rate of 5.6 mm/year. This estimate is supported by the occurrence of *Zea mais* (maize) pollen in the sediment. In the Netherlands, large scale maize cultivation started in the 1980s. The first maize pollen in core IJ6 was found in 11 cm core depth. Applying a mean sedimentation rate of 5.6 mm/year this would correspond to the year 1986. Accordingly, the marine sand layer at 35–31 cm core depth was deposited in 1950.

Zuiderzee and IJsselmeer diatom flora

Figure 3 shows the most frequent marine and freshwater diatoms documenting the Zuiderzee and IJsselmeer diatom floras, respectively (Cremer & Bunnik, 2006). The Zuiderzee sediments were mainly characterized by *Cymatosira belgica* Grunow and *Delphineis minutissima* (Hustedt) Simonsen. *Planothidium delicatulum* (Kützing) Round et Bukhtiyarova, *Thalassiosira levanderi* Van Goor, *T. proschkinae* Makarova and several other marine to brackish diatoms were present in smaller numbers. Following the closure of the connection with the North Sea, the newly formed IJsselmeer became less saline within a few years, which is clearly visible in the completely changed diatom flora. The dominating diatom in the upper 42 cm of core IJ6 is *Staurosirella pinnata* (Ehrenberg) Williams et Round together with *Staurosira construens* Ehrenberg s.l. (Fig. 3). Other frequently occurring planktonic or tychoplanktonic diatoms include several taxa of the genera *Aulacoseira* and *Stephanodiscus*, *Cyclostephanos dubius* (Fricke) Round, *Diatoma tenuis* Agardh, *Nitzschia angustatula* Lange-Bertalot and *Nitzschia palea* (Kützing) W. Smith.

Trophic history of IJsselmeer

Trophic state indicator values

The downcore distribution of the most prominent diatom species (Fig. 3) is reproduced in Fig. 4 as trophic state categories, the compilation of which is based on species' trophic state indicator values according to Van Dam et al. (1994). The major part of the diatom assemblages throughout core IJ6 consists of *Staurosirella pinnata*, which is known to be indifferent to changes of a lake's nutrient load

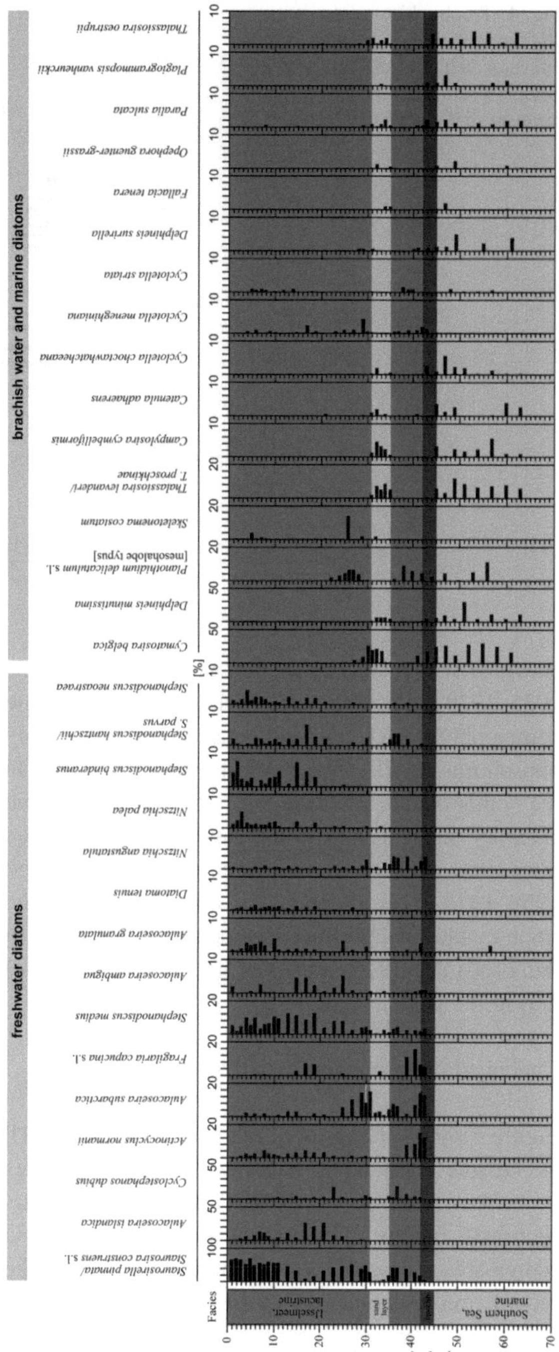

Fig. 3 Percentages of the most frequent freshwater and marine diatoms in core IJ6 from IJsselmeer

(Van Dam et al., 1994). Diatoms with eutrophic affinities are also very common throughout the past 70 years. However, Fig. 4 shows that the oligo-mesotrophic diatom *Aulacoseira subarctica* (O. Müller)

Fig. 4 Trophic history of IJsselmeer based on trophic indicator values (van Dam et al., 1994) of the most frequent diatoms

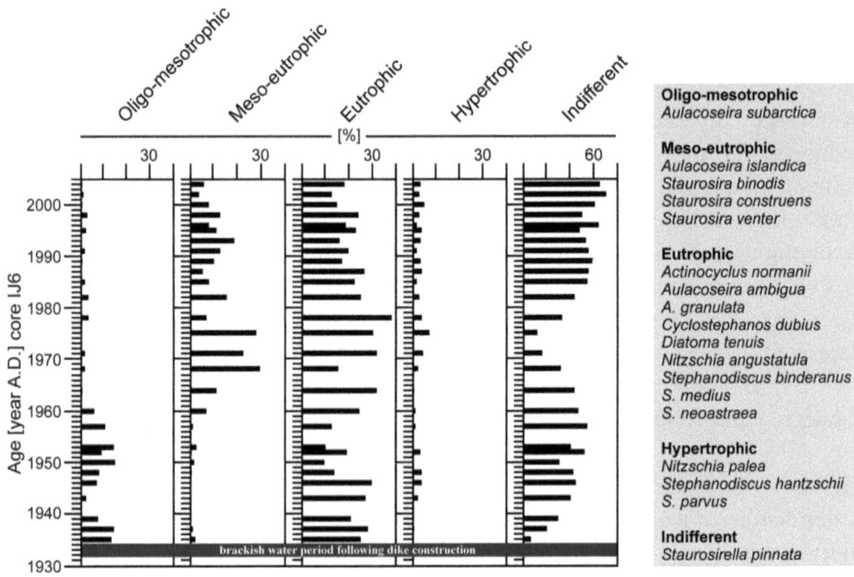

Haworth was mainly deposited during the intervals 1935–1939 and 1946–1960, which might indicate comparably lower nutrient concentrations, notably phosphorus, in IJsselmeer during these periods. On the other hand, diatoms of the meso-eutrophic category occur mainly in the assemblages deposited after 1960 and show highest frequencies in the period 1968–1975. This supports the assumption that the trophic state of IJsselmeer was already eutrophic in the 1960s and 1970s and even higher (hypertrophic) during the past 25 years.

Canonical correspondence analysis

Figure 5 shows the result of the partial CCA with TP as the sole constraining variable. The sample scores of the fossil diatom assemblages on the first CCA axis represent relative TP concentrations. The sample scores are comparably low in the earliest part of the record, representing the period 1935–1940 after which they increase considerably and generally remain with some fluctuations on a high level. Comparably lower sample scores and therefore lower TP concentrations are indicated for the periods 1948–1950, 1953 and 1960–1968. High sample scores, representing higher TP concentrations, are indicated for the period 1970–1978. Since the end-1970s, the high sample scores show a slightly declining, but fluctuating trend.

TP transfer function

The quantitative TP reconstruction in IJsselmeer since 1935 using the EDDI European training sets and a weighted MAT transfer function is shown in Fig. 6. The applied transfer function has a boot-strapped r^2 of 0.74 and a root mean squared error of prediction of 0.2955 log TP. Good modern analogues were defined as modern diatom samples having a chi-squared distance of less than the 5th percentile (Jones & Juggins, 1995). In the first years of the newly formed lake TP concentration increased rapidly towards the first maximum at ca. 100 μg l^{-1} in the early 1940s. Following this eutrophic period, the diatom-inferred TP concentration decreased to a value of ca. 60 μg l^{-1} in 1950, which was the lowest TP concentration ever calculated for IJsselmeer. Between 1950 and 1970 TP concentrations ranged between 60 and 100 μg l^{-1}. From the early 1970s, the onset of strong eutrophication is indicated by TP concentrations distinctly over 100 μg l^{-1}. The highest diatom-inferred TP concentration (156 μg l^{-1}) is calculated for 1987 indicating hypertrophic conditions in IJsselmeer at that time. A slight downward trend of the TP concentration in IJsselmeer is indicated since the 1990s but the diatom-inferred trophic state remains still in the eutrophic–hypertrophic range. Figure 6 also shows water column TP in IJsselmeer, which has been monitored since 1975.

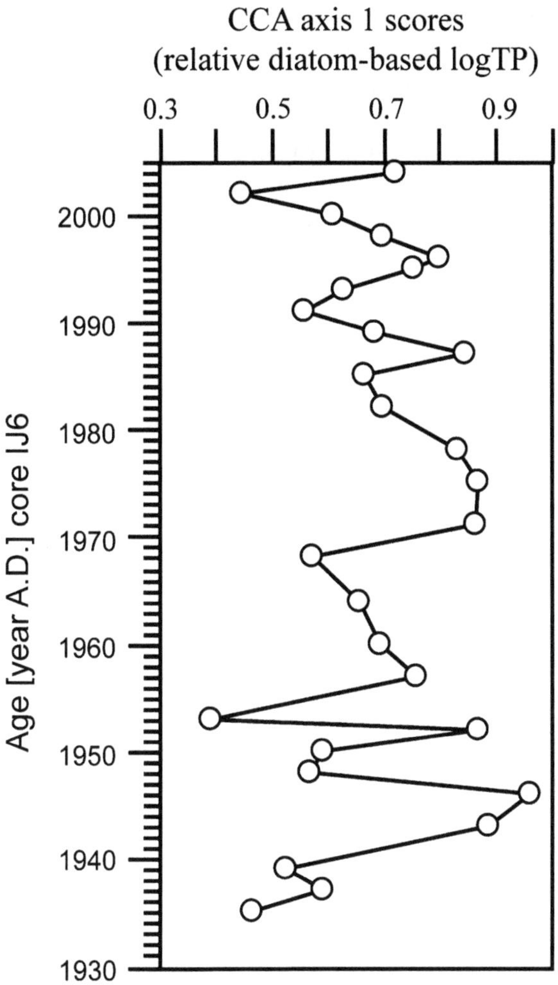

Fig. 5 Sample scores of the first CCA axis for core IJ6 from IJsselmeer. The scores on the ordination axis represent relative TP concentrations

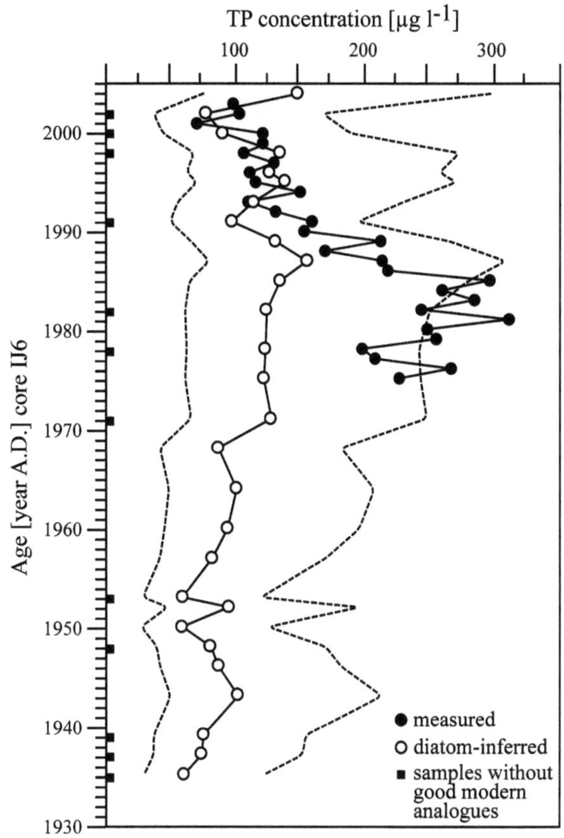

Fig. 6 Diatom-inferred (since 1935) and measured (since 1970) summer mean TP concentrations in IJsselmeer. The *dotted lines* represent the sample-specific standard errors of prediction and the *squares* represent samples without good modern analogues. The diatom-inferred TP reconstruction is based on the weighted Modern Analogue Technique using the EDDI TP diatom training set. Measured data from De Leeuw et al., (2006)

The recordings document highest TP concentrations between 1980 and 1985 (200–300 µg l^{-1}; Lammens et al., 2008). As a consequence of reduced nutrient load into IJsselmeer the measured TP concentrations decreased steadily since the mid-1980s and have reached present-day concentrations of slightly below 100 µg l^{-1}. This trend, however, is not as conspicuous in the diatom-based reconstruction (Fig. 6).

Discussion

The compilation of diatom-based trophic state classes provides first evidence of the development of the trophic history of IJsselmeer during the past 70 years. However, this qualitative approach only documents relative trends of trophic state changes. The multivariate approach using a direct gradient analysis (CCA) assumes a unimodal distribution of species (Ter Braak, 1986) along environmental gradients (here TP). As the CCA is constrained uniquely to TP, the CCA scores of the IJsselmeer samples (Fig. 5) show relative changes in TP concentration, and thus, represent a semiquantitative method. This approach is strongly recommended, if the general parameters of the studied fossil environment largely differ from those of the lakes contained in the modern training set (e.g. Bohncke et al., 2008). In this case, we are using several regional TP training sets that are based on diatom assemblages from surficial sediments from small lakes to infer past TP values in the artificially formed, large lake IJsselmeer. It is assumed here that the reconstructed TP curve reflects a realistic history of the trophic state of IJsselmeer for the past few decades. The curve (Fig. 6) basically shows the same trends as the CCA-based reconstruction (Fig. 5) of relative TP changes.

As 93.3% of the fossil diatom flora of IJsselmeer is represented in the EDDI training set, we feel confident that a transfer function approach such as MAT can be applied. However, in situ measurements made during the period of highest TP concentration show that diatom-inferred TP concentrations may underestimate actual TP values. Between 1975 and 1990, the average diatom-inferred TP concentration is distinctly lower than the measured average summer values of the same period (Fig. 6). The phenomenon of over- or underestimation of diatom-inferred compared to measured TP concentrations has been reported and discussed many times in the literature (e.g. Bennion et al., 1995; Lotter, 1998; Bradshaw & Anderson, 2001; Sayer, 2001). The reasons that might lead to such differences include training set- and model-inherent problems, e.g. the presence and number of low, medium and high TP sites in the training set or the fact that TP measurements document the situation during a single season (often summer), whereas the sedimentary diatom sample represents a larger period (one or several years). Another important factor might be the dominance of small, often benthic fragilarioid diatoms (*Staurosirella pinnata, Staurosira construens*; Figs. 3 and 4) in the sediment assemblages. These taxa have been described to react more sensitive to habitat availability than to TP concentration and therefore may have a distorting influence on TP reconstructions (Sayer, 2001). Furthermore, the variability of factors other than TP, including climate, light conditions, zooplankton grazing and competition between algal groups and genera, may also have a significant impact on the composition of fossil diatom assemblages and thus, indirectly have an influence on reconstructed TP concentrations. Spatial and temporal variabilities in the deposition of inorganic sediment and biogenic components, including microfossil remains, particularly in large lakes, may have an indirect influence on past TP estimates (Blom & Winkels, 1998). Taking all these factors, as well as the sample-specific standard errors of prediction into account, the reconstructed TP concentrations for IJsselmeer are within the measured TP ranges and therefore, can be regarded as reliable.

The diatom-inferred reconstruction of absolute TP concentrations shows two noticeable trends: first, the eutrophic maximum in the early 1940s, and second, the weakly declining TP concentrations since 1985 compared to the clearly reduced measured values (Fig. 6).

The true reasons for the first eutrophic period in IJsselmeer remain unexplained at the moment. It can be speculated that the construction of the Afsluitdijk caused hydrological, hydromorphological, geochemical and sedimentological conditions in the initial IJsselmeer that led to a massive release of phosphorus from the ancient sea bottom. Possible, even probable, triggers include changes in the redox potential and pH and/or increasing algal blooms (e.g. cyanobacteria) in the meanwhile freshened lake. These processes have been intensely studied and discussed in recent literature (e.g. Christophoridis & Fytianos, 2006; Xie et al., 2003 and references therein). Decrease of the TP concentration after 1945 might then be interpreted as a consequence of using up- and rebinding of phosphorus and reaching a more or less stable phosphorus budget in IJsselmeer.

There are also some possible reasons for the consistently relatively high reconstructed TP concentrations in the 1990s (TP 100–150 $\mu g\ l^{-1}$) compared to the previous interval. In contrast, the monitored measurements show a clear downward trend after 1985 (Fig. 6). The recovery of IJsselmeer from extreme phosphorus loads is seemingly not (or not yet) evident from the diatom data. A probable reason for this observation is that at hypertrophic conditions (TP $\geq$100 $\mu g\ l^{-1}$) phosphorus is not the limiting factor for diatom growth and particularly the composition of the diatom flora. Such overestimation of diatom-inferred TP concentrations has been also reported in other studies (e.g. Bradshaw & Anderson, 2001) and indicates that the decreased input of a single nutrient (e.g. phosphorus) does not necessarily lead directly to observable changes of the algal community, which might be a reflection of hysteresis effects. Moreover, model-inherent factors, such as the length of the phosphorus gradient covered by the calibration dataset, have also an effect on the reconstructed TP. The application of datasets that are not designed for the region in which it is used in many cases yields reconstructions that tend to either over- or underestimate the monitored phosphorus data (Bradshaw & Anderson, 2001). This does not necessarily mean, however, that such reconstructions cannot reveal valuable information on the trophic history of lakes.

The present study demonstrates the value of palaeolimnological approaches for assessing trophic histories of artificial lakes. As evidenced by diatom analysis, IJsselmeer was already meso-trophic within a few years of its formation in the early 1930s and underwent the initial eutrophic period in the early 1940s. Since the late 1950s, conditions in IJsselmeer were strongly eutrophic and increased to hypertrophy in the early 1970s. The continuous improvement in the trophic state since 1985 is not yet reflected in the diatom assemblages and the diatom-inferred TP values. This suggests that diatom-inferred nutrient reconstructions should preferably be validated by long-term monitoring data to reduce misinterpretations.

References

Battarbee, R. W., 1973. A new method for estimation of absolute microfossil numbers, with reference especially to diatoms. Limnology and Oceanography 18: 647–653.

Bennion, H. & R. W. Battarbee, 2007. The European Union Water Framework Directive: opportunities for palaeolimnology. Journal of Paleolimnology 38: 285–295.

Bennion, H., S. Wunsam & R. Schmidt, 1995. The validation of diatom-phosphorus transfer functions: an example from Mondsee, Austria. Freshwater Biology 34: 271–283.

Blom, G. & H. J. Winkels, 1998. Modelling sediment accumulation and dispersion of contaminants in Lake IJsselmeer (The Netherlands). Water Science and Technology 37: 17–24.

Bohncke, S. J. P., J. A. A. Bos, S. Engels, O. Heiri & C. Kasse, 2008. Rapid climate events as recorded in Middle Weichselian thermokarst lake sediments. Quaternary Science Reviews 27: 162–174.

Bradshaw, E. G. & N. J. Anderson, 2001. Validation of a diatom–phosphorus calibration set for Sweden. Freshwater Biology 46: 1035–1048.

Christophoridis, C. & K. Fytianos, 2006. Conditions affecting the release of phosphorus from surface lake sediments. Journal of Environmental Quality 35: 1181–1192.

Cremer, H. & F. P. M. Bunnik, 2006. Paleowaterkwaliteit van het IJsselmeer sinds de aanleg van de Afsluitdijk in 1932. [Paleo-water quality of Dutch lakes: reconstruction of the development of the phosphorus content of IJsselmeer base don fossil diatom assemblages.] TNO report 2006-U-R0133-B, Utrecht, 45 pp (in Dutch).

De Leeuw, J. J., C. Deerenberg, W. Dekker, R. van Hal & H. Jansen, 2006. Veranderingen in de visstand van het IJsselmeergebied: trends en oorzaken. Nederlands Instituut voor Visserijonderzoek [Paleo-water quality of Dutch lakes: reconstruction of the development of the phosphorus content of IJsselmeer base don fossil diatom assemblages.] TNO report C022-06, IJmuiden, 33 pp (in Dutch).

European Union, 2000. Directive 2000/60/EC of the European Parliament and of the Council of 23 October 2000 establishing a framework for Community action in the field of water policy. Official Journal of the European Communities L327: 1–73.

Jones, V. J. & S. Juggins, 1995. The construction of a diatom-based nutrient transfer function and its application at three lakes on Signy Island (maritime Antarctic) subject to differing degrees of nutrient enrichment. Freshwater Biology 34: 433–445.

Juggins, S., 2007. C2 Version 1.5 User guide. Software for ecological and palaeoecological data analysis and visualisation. Newcastle University, Newcastle upon Tyne, UK: 73.

Lammens, E., F. van Luijn, Y. Wessels, H. Bouwhuis, R. Noordhuis, R. Portielje & D. van der Molen, 2008. Towards ecological goals for the heavily modified lakes in the IJsselmeer area, The Netherlands. Hydrobiologia 599: 239–247.

Lotter, A. F., 1998. The recent eutrophication of Baldeggersee (Switzerland) as assessed by fossil diatom assemblages. The Holocene 8: 395–405.

Sayer, C. D., 2001. Problems with the application of diatom-total phosphorus transfer functions: examples from a shallow English lake. Freshwater Biology 46: 743–757.

Taylor, D., C. Dalton, M. Leira, P. Jordan, G. Cheng, L. Leon-Vintro, K. Irvine, H. Bennion & T. Nolan, 2006. Recent histories of six productive lakes in the Irish Ecoregion based on multiproxi palaeolimnological evidence. Hydrobiologia 571: 237–259.

Ter Braak, C. J. F., 1986. Canonical correspondence analysis: a new eigenvector technique for multivariate direct gradient analysis. Ecology 67: 1167–1179.

Ter Braak, C. J. F. & P. Šmilauer, 1998. CANOCO reference manual and user's guide to CANOCO for Windows: software for canonical community ordination (version 4). Microcomputer Power Ithaca, New York: 351.

Van Dam, H., A. Mertens & J. Sinkeldam, 1994. A coded checklist and ecological indicator values of freshwater diatoms from The Netherlands. Netherlands Journal of Aquatic Ecology 28: 117–133.

Xie, L. Q., P. Xie & H. J. Tang, 2003. Enhancement of dissolved phosphorus release from sediment to lake water by *Microcystis* blooms—an enclosure experiment in a hyper-eutrophic, subtropical Chinese lake. Environmental Pollution 122: 391–399.

Palaeolimnology of Lake Hess (Patagonia, Argentina): multi-proxy analyses of short sediment cores

Piero Guilizzoni · Julieta Massaferro · Andrea Lami · Eduardo Luis Piovano · Sergio Ribeiro Guevara · Stella Maris Formica · Romina Daga · Andrea Rizzo · Stefano Gerli

Originally published in the journal Hydrobiologia, Volume 631, No. 1, 289–302.
DOI: 10.1007/s10750-009-9818-5 © Springer Science+Business Media B.V. 2009

Abstract In contrast with the extensive palaeolimnological studies carried out in North America and Europe, relatively few studies have described the anthropogenic and/or climate impacts in Patagonian lakes. We addressed these issues by analysing geochemistry, lithology, pigments and chironomid remains from sediment cores collected from Lake Hess (41°22′20″S, 71°44′0″W) located in the Nahuel Huapi National Park in northern Patagonia. The aim of this study is to provide a palaeoenvironmental and climate reconstruction of the past ca. three centuries for this cold oligotrophic, quasi-pristine lake which receives meltwaters from the Tronador ice cap. Chronology was based on ^{137}Cs and ^{210}Pb measurements of the upper sediments, and the inferred sedimentation rate of 23.2 mg cm^{-2} y^{-1} (0.15 cm y^{-1}) was consistent with both sets of measurements. The sediment from Lake Hess was rich in tephra deposits particularly evident in the lower part of the cores. Tephras are valuable to use for core correlation and can be traced through peaks in the magnetic susceptibility (MS) profiles. Results from the multiproxy analyses in the longest core (83 cm) identify three main phases of change. From the bottom up to 42 cm (ca. AD 1800), the sediment is composed of light-grey organically rich clays. Both pigments and chironomids suggest variable trends in productivity and precipitation regime. At the end of the Little Ice Age chronozone (AD 1770–1850), pigment concentrations were very low. From 42 cm to ca. 25 cm (AD 1800–1940), the sedimentary record is composed of alternating black and dark organic-matter rich mud with variable amounts of macrophyte remains. Pigment concentrations and chironomid head capsule counts were also very low. These facies are composed of very fine plastic sediments with some faintly laminated intervals and an organic matter composition gradually decreasing towards the top of the zone. A sharp change occurs at 25 cm (ca. AD 1940) showing a strong increase in organic matter content,

Guest editors: K. Buczkó, J. Korponai, J. Padisák & S. W. Starratt
Palaeolimnological Proxies as Tools of Environmental Reconstruction in Fresh Water

P. Guilizzoni (✉) · A. Lami · S. Gerli
Consiglio Nazionale delle Ricerche – Istituto per lo Studio degli Ecosistemi, Largo Tonolli 50, 28922 Verbania Pallanza, Italy
e-mail: p.guilizzoni@ise.cnr.it

J. Massaferro
Laboratorio de Biodiversidad Darwin, Instituto de Investigaciones en Biodiversidad y Medioambiente/ Consejo Nacional de Investigaciones Científicas y Técnicas, Quintral 1250, 8400 Bariloche, Argentina

E. L. Piovano · S. M. Formica
The Centro de Investigaciones en Ciencias de la Tierra, Universidad de Córdoba, Av Velez Sarsfield 1611, 5016 Córdoba, Argentina

S. R. Guevara · R. Daga · A. Rizzo
Laboratorio de Análisis por Activación Neutrónica, Comisión Nacional de Energía Atómica, Av Bustillo km 9.5, 8400 Bariloche, Argentina

algal nutrients and plant pigments together with a change in the chironomid assemblages. This might document a change in the trophic condition of the lake associated with changes in erosion/deposition rates. Although there are records of human impact in the area studied, involving the use of fires, most of the observed chemical and biological changes in Lake Hess sediment sequence were interpreted in terms of climate changes, especially to changes in moisture balance brought about by variations in the strength of the westerlies.

Keywords Palaeolimnology · Pigments · Chironomids · Patagonia · South America

Introduction

Palaeolimnological research has shown that many remote areas (e.g., high mountain lakes) have experienced considerable biotic and sedimentary changes in the recent decades (e.g. Lami et al., 2000a; Lotter & Birks, 2003), but few studies have quantified the anthropogenic and climatic impacts in South America using Patagonian lakes (Ariztegui et al., 1997, 2007; Massaferro et al., 2005; Villarosa et al., 2006). Hydrological changes since the Last Glacial Maximum up to the Little Ice Age in Patagonia have been described from lake sediment cores on both sides of the Southern Andes in Lago Puyehue (40°S) in the Chilean Lake District (Chapron et al., 2006; Moernaut et al., 2007; Bertrand et al., 2008; Charlet et al., 2008; De Batist et al., 2008) and Laguna Frias (Ariztegui et al., 2007) in the Argentinean side. In the Patagonian Plateau, the record of Lago Cardiel (49°S; Gilli, 2003; Gilli et al., 2005; Markgraf et al., 2003) and Laguna Potrok Aike (52°S; Haberzettl et al., 2006; 2007a, b; Mayr et al., 2005) contain a detailed climate archive of the environmental variability since the late Pleistocene.

A multi-proxy approach is important in palaeolimnological studies to get an overview of the natural and anthropogenic events registered in the lake sediments. Northern Patagonia is a particularly interesting area for these studies because it is home to many different aquatic environments which are in turn controlled by regional and local climate effects (Villalba, 2007). For instance, the presence of the westerlies at the latitude of 40° S is a source of change in the atmospheric moisture coming from the Pacific Ocean (Ariztegui et al., 2007). Patagonia is particularly sensitive to seasonal atmospheric air-humidity changes driven by the Southern westerlies and the polar front position in winter (Prohaska, 1976). Temporal changes in the north–south precipitation gradient are controlled by the latitudinal position of the westerlies, which in turn is regulated by the strength and latitudinal position of the sub-tropical anticyclone in the southeast eastern Pacific and the circum-Antarctic low-pressure belt (Markgraf et al., 2000). The present-day northern limit of westerlies' influence along the Chilean Pacific coast lies at ca. 27°S. South of 38°S, the influence of westerlies is permanent resulting in high precipitations throughout the year, whereas at more northern latitudes, mean annual precipitation is comparatively low and is highest in winter (Veit, 1996).

Lake Hess is a small lake, which receives discharge from the River Manso Medio carrying meltwaters directly from the Tronador Glacier. Therefore, this lake is an interesting site, well placed to record variations in moisture regime and/or changes in the status of the glacier. Previous studies in the region have shown that the Tronador ice cap has reacted to climate change during distinct episodes such as the Late-Glacial–Holocene transition (Ariztegui et al., 1997; Hajdas et al., 2003), the Medieval Climatic Anomaly and the Little Ice Age with well-identified glacial advances between AD 1800–1850, and recent push-moraines (Rabassa & Clapperton, 1990; Villalba et al., 1997).

In this article, the issues of climate and palaeoenvironmental changes in a more recent period (the last ca. three centuries) have been addressed by analysing organic matter (by Loss on Ignition, LOI), nutrients (carbon and nitrogen), photosynthetic pigments and chironomid remains from three lithologically described sediment cores collected from Lake Hess located in the Nahuel Huapi National Park in northern Patagonia.

Records of organic matter, nutrients and fossil pigments in lake sediments reflect the environmental conditions in a lake and its catchment at the time of deposition (Guilizzoni et al., 2006). Their sedimentary chemical composition and algal abundance can record historical changes in lake trophic status, anthropogenic and natural (climate) change, adding greatly to our understanding of the development and functioning of lakes.

In addition, the use of chironomid remains (Diptera: Chironomidae) has, due to their stenotopic response to climate, productivity and water depth changes, and their high resistance to degradation, become widespread in palaeoecological studies (Brundin, 1958, 1966; Brooks & Birks, 2004).

The aims of this article are the following: (1) to provide additional paleolimnological information from northern Patagonian lake sediments during the last ca. 300 years, and (2) to establish the degree of human and climate impact within a quasi-pristine lake ecosystem.

Study area and sampling

Lake Hess (41°21′20″S; 71°44′0″W, 735 m a.s.l.) is a small, shallow (max depth 8.3 m), glacial lake located 100 km southeast of San Carlos de Bariloche. It receives meltwaters from Tronador ice cap via the Rio Manso Medio (Fig. 1). Water level fluctuations are quite common, given the small lake area and the significant water discharge by the outlet. The area is located far from urbanised regions, and no direct anthropogenic impacts on the lake are expected though records of extended fires in the region for the past 10,000 years have been reported in Kitzberger et al. (1997) and Veblen et al. (2003).

The climate of the area is predominantly cold temperate, and the vegetation is composed of mixed north Patagonian rainforest communities with mainly evergreen deciduous *Nothofagus* spp. At this latitude, the Patagonian region is dominated by air masses coming from the Pacific Ocean and strong constant west winds (westerlies) are dominant across the region.

In December 2005, using an inflatable boat, three short sediment cores named Hess 05-1 (83 cm long), Hess 05-2 (70 cm long) and Hess 05-3 (65 cm long) were sampled from the same site using a gravity corer (Ø PVC liner 6.3 cm).

Methods

In the laboratory, all cores were split lengthways, opened, photographed, visually inspected, and sub-sampled every 1 cm. Each sub-sample was freeze-dried and homogenised. Cores were correlated using organic matter, dry weight content, MS, tephra markers

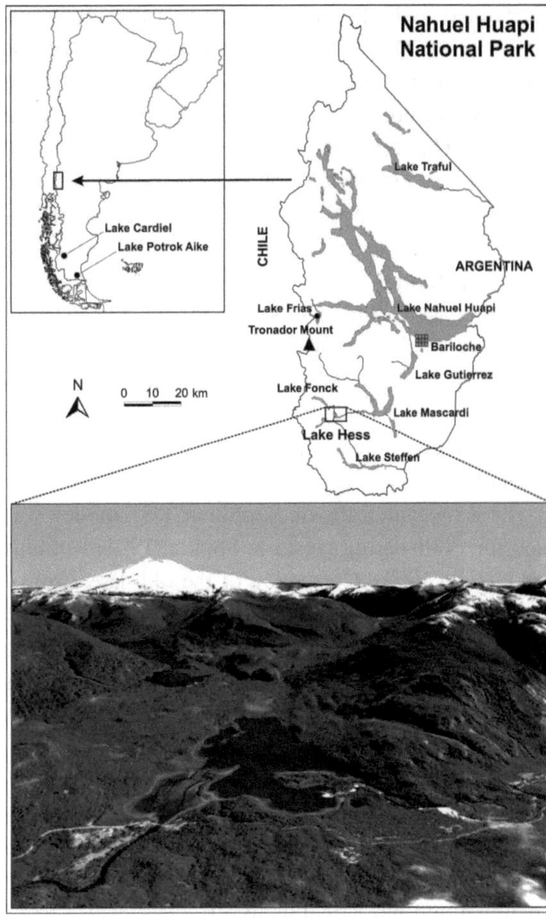

Fig. 1 Map showing the location of the study Lake Hess (Northern Patagonia, Argentina). An aerial view of Lake Hess showing the Tronador Mount and other investigated lakes mentioned in the text are also shown

and lithological features. MS was measured at 1 cm increments using a Bartington MS IB instrument prior to cutting the core. The sediment density was determined by computing the ratio between the dry weight of each sub-sampled sediment layer and its volume.

Core Hess 05-3 was used for radiometric dating at Laboratorio de Análisis por Activación Neutrónica, Comisión Nacional de Energía Atómica (LAAN CNEA), while cores Hess 05-1 and Hess 05-2 were studied for palaeolimnological indicators.

Chronology

Core dating was performed by ^{210}Pb and ^{137}Cs assay. ^{210}Pb, ^{226}Ra (in secular equilibrium with supported ^{210}Pb), and ^{137}Cs specific activity profiles were determined by high-resolution gamma ray spectrometry on core Hess 05-3. The measurements of the upper 8 cm are reported for ^{210}Pb and ^{137}Cs dating; below 8 cm, no significant radiometric values were obtained. The topmost samples (slices of 1 cm) were placed in a cylindrical plastic container (46 mm diameter and 8 mm height), sealed and counted in a close sample-detector geometry. ^{210}Pb and ^{226}Ra specific activities were obtained by measuring the 46 keV (from ^{210}Pb), 295 keV and 351.9 keV (from the decay of the ^{226}Ra short-lived daughters) emissions with a planar HPGe (LO-AX) detector. ^{137}Cs specific activity was determined by measuring the 661.7 keV emissions using an HPGe detector with 30 % relative efficiency.

The Constant Rate of Supply (CRS) model was used for ^{210}Pb dating (Joshi & Shukla, 1991; Robbins and Herche 1993). Correction of the old date error of the CRS model (Binford, 1990) was implemented by logarithmic extrapolation of the measurements to complement integration to infinite depth. For ^{137}Cs dating, the specific activity profiles were compared with the fallout sequence determined in this region, associated mainly with South Pacific nuclear tests from 1966 to 1974 (Ribeiro Guevara & Arribére, 2002).

Chemical and biological analyses

On core Hess 05-1, dry mass and organic matter were measured by LOI at 70 and 550°C, respectively. Total carbon, sulphur and total nitrogen were obtained using a Carbon Hydrogen Sulphur analyzer (NA 1500 Fisons). C/N has been based on an atomic ratio. All the chemical analyses were performed at CNR—Istituto per lo Studio degli Ecosistemi.

Photosynthetic pigments were extracted from ca. 1 g of wet sediment (large macrophyte remains free) in a 5 ml of 90% acetone, overnight in the dark, under nitrogen. The extract obtained was used to quantify both the chlorophylls and their derivatives (Chlorophyll Derivatives Units, CD) and total carotenoids (TC) by spectrophotometer (for details see Lami et al., 2000b). Individual carotenoids were detected by reversed phase high-performance liquid chromatography using a Dionex SUMMIT with a DAD-UV detector (Lami et al., 2000b). Carotenoid concentrations were expressed in nanomoles per gram of organic matter (nmol g^{-1} LOI; Züllig, 1986) and chlorophyll derivates in units per gram of organic matter (U g^{-1} LOI; Guilizzoni et al., 1982, 1983). Guilizzoni et al. (1992), Lami et al. (2000b), Guilizzoni & Lami (2002) and Lami et al. (2009) provide a detailed account of the spectrophotometric and chromatographic analyses of the plant pigments, their identification, quantification and interpretation.

Chironomid remains were studied in 82 samples at 1-cm intervals, throughout the core Hess 05-1. Subsamples of 3–4 g wet weight were deflocculated in 10% KOH, heated to 70°C for 20 min and subsequently sieved on a nested pair of 212- and 95-μm mesh sieves. Head capsules were picked from the wet residue in a Bogorov sorting tray under a stereo microscope at 25–40× magnification. Larval head capsules were mounted in Euparal and then identified with reference to Brundin (1966) and Cranston (1997). Chironomid and geochemical data were plotted using C2 software (Juggins, 2003). Data are expressed as percentage of relative abundance and as total concentration of heads capsules (HC) (number g d.w.$^{-1}$).

Data analysis

Profiles from each dataset were split into zones using CONISS (CONstrained Incremental Sum of Squared cluster analysis) with square root transformation to optimise the signal to noise ratio. Dendrograms are not shown to cause the self-evident large changes. Zones were constructed independently for each proxy using variables indicated in each figure. Samples with missing data values for some variables were excluded for the purposes of zonation.

Interpretation of pigment data

In low productivity lakes such as the alpine and remote sites like Lake Hess, a large proportion of the sedimentary organic matter is allochthonous, and pigments are poorly preserved in the terrestrial detritus, owing to its longer exposure to oxidation at the soil surface (Sanger, 1988). Pigments reach the surface of a lake as particulate matter from autochthonous planktonic and littoral organisms. Each population contains pigments that are distinctive and group specific that can be used in palaeoecological studies to indicate, for example, changes in the nature of the physical and chemical environments. The validity of using sedimentary pigments as an index of past events or processes such as trophic conditions and primary productivity depends mainly on the preservation of these relatively labile organic compounds, the extent of differential degradation during and after sedimentation, and the extent of allochthonous sources of sedimentary pigment products (Leavitt, 1993; Lami et al., 2000b).

Chlorophyll *a* is found in all photosynthetic aquatic organisms, with the exception of some species of autotrophic bacteria. Its degradation products (phaeophytin *a*, phaeophorbide *a*, chlorophyllide *a*) are relatively well preserved in lake sediments. Carotenoids are also widely distributed in all photosynthetic organisms. The principal carotene is β-carotene, which is present in all algal taxa. Lutein is present in Chlorophyta and macrophytes. In addition to these in Lake Hess, there are other specific xanthophylls, such as fucoxanthin and diatoxanthin, present in the diatoms, whereas echinenone and in part zeaxanthin are present in the cyanobacteria. Finally, alloxanthin is a carotenoid characteristic of strictly planktonic cryptophytes and could be used in alpine or arctic lakes to identify the degree of thaw experienced. Enhanced alloxanthin concentrations should indicate increased ice-free periods. The variation of alloxanthin (as well as diatoxanthin; Buchaca & Catalan, 2007; Lami et al. 2009) is also used to infer water level changes (Züllig, 1982; Pienitz et al., 1992; Leavitt et al., 1994). Phaeophorbide *a* is typically a pigment related to zooplankton and is produced by zooplankton during grazing (e.g., Leavitt et al., 1994; Buchaca & Catalan, 2007).

Results

Chronology

^{210}Pb-, ^{226}Ra-, and ^{137}Cs-specific activity profiles from Hess 05-3 are shown in Fig. 2. The total ^{210}Pb-specific activity (supported ^{210}Pb) was significantly higher than ^{226}Ra only in the five topmost slices. Therefore, a ^{210}Pb-based estimate of the sedimentation rate was only possible in the upper core section. Since there were few unsupported ^{210}Pb determinations, with highly associated uncertainties, the measured values of the CRS model were fitted to a logarithmic scale, and interpolated values were used. The estimation of the sedimentation rate obtained is 23.2 mg cm^{-2} y^{-1} (0.15 cm y^{-1}). Since Lake Hess is a geomorphologically active environment, approximate extrapolated age values have been calculated only for the core section 8–45 cm using the mean post 1950 sedimentation rates of 23.2 mg cm^{-2} y^{-1}. According to the extrapolation of the sedimentation rate estimated by ^{210}Pb, the sample at 5–6 cm corresponds approximately to 1965, which agrees well with the 1963 fallout peak estimated by ^{137}Cs dating (Fig. 2). From 45 cm downwards, the continuity in sedimentation rate cannot be assumed due to the strong changes in sediment density. No terrestrial plant remains suitable for ^{14}C analysis were found.

Although core Hess 05-3 was taken close to the other cores, the presence of coarser lenticular bodies interbedded in the sedimentary sequence suggests that different processes had affected normal sedimentation in the lake perhaps caused by irregularities in the bottom at the site of extraction.

Lithology and core-to-core correlation

The lithology of the cores shows a generally common pattern; the upper part is a unit composed of black silty organic matter-rich mud with a high abundance of macrophyte remains (Fig. 3). Although this unit has a variable thickness in the different cores (from 8 to 22 cm), it shows similar low values of dry weight (density is only measured in one core) (Figs. 3, 4). Below this unit, the sequence is composed of homogeneous dark grey-brownish organic-rich silty clay sediment, with some scattered macrophyte remains. The other remarkable feature is the presence of tephra layers intercalated in the sequence. Magnetic susceptibility in core 05-2, as well as density and LOI clearly identify these ash layers (Fig. 4). Based on several parameters (dry weight, LOI, tephras and lithology), used independently or in combination, a good correlation of cores Hess 05-1, Hess 05-2 and Hess 05-3 was established (Figs. 3, 4). Core Hess 05-1 is longer than cores Hess 05-2 and Hess 05-3, and shows some differences in lithology. Cores Hess 05-2 and Hess 05-3 appear to have more closely comparable lithologies.

We chose the longest core (Hess 05-1) for a detailed lithological description. This sedimentary sequence consists of three different lithological units: (I) light grey muds with tephra layers at the top; (II) dark grey muds which occur cyclically in the upper part of core and in the bottom of core; (III) black

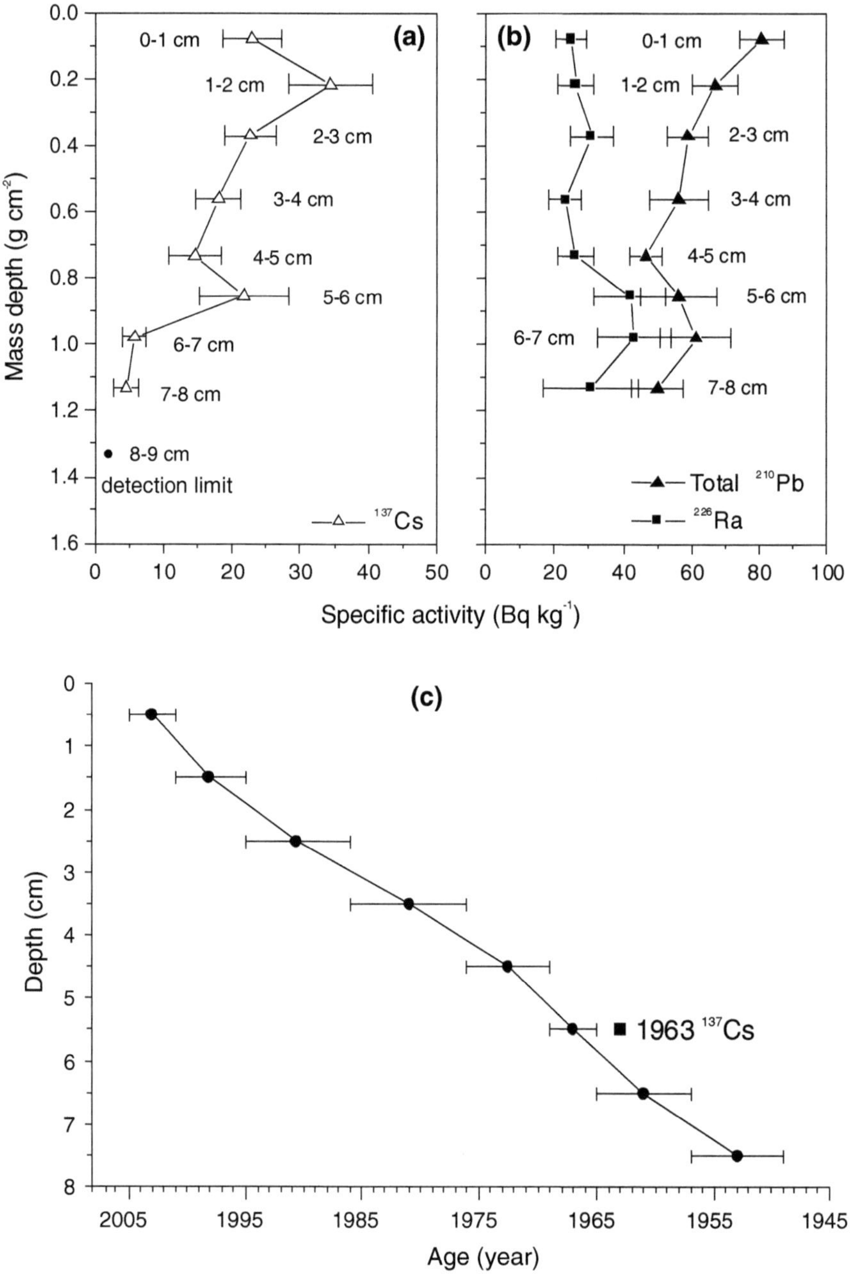

Fig. 2 Profiles of ^{137}Cs-specific activity (**a**); **b** ^{210}Pb and ^{226}Ra (in secular equilibrium with supported ^{210}Pb) specific activity versus cumulative dry mass (g cm^{-2}) in core Hess 05-3; **c** the plot of ages versus sediment depth is based on calculations using the CRS model applied to excess ^{210}Pb data; the depth of the 1963 ^{137}Cs peak is also shown

massive muds with abundant macrophyte remains (Fig. 3).

Abundant remains of macrophytes (mosses) are present between 0 and 6 cm, 12 and 15.5 cm, and 18 and 22 cm. The grey muds are composed of very fine, plastic sediment with faint laminations. This lithological feature is present below 22 cm and includes reworked tephra layers at 25–26 cm, 32–38 cm, 55–56 cm and 58–59 cm.

Tephra layers are located at different depths in the three cores (Fig. 3). The one located at 24–26 cm depth in core Hess 05-1 is characterised by brownish glassy shards, dark and grey fragments (scoria and rock fragments), plant remains and diatom frustules, among other organic materials, included in a silty clay matrix. This layer was related to the tephra located below 19–21 cm in core Hess 05-3 (Fig. 3). According to this characterisation, this ash could be

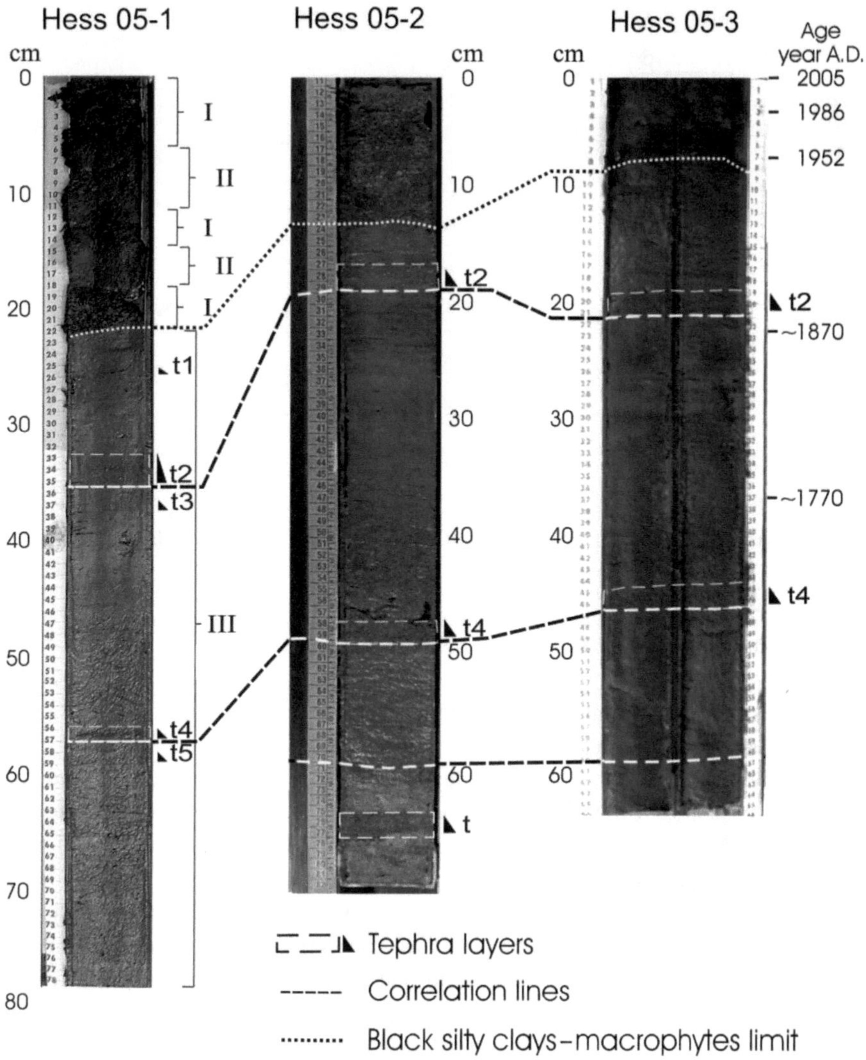

Fig. 3 Core correlations based on lithology and tephra layers (t) from Lake Hess. *Dotted line* correlates the top layer rich in macrophytes. *Dashed lines* indicate both the correlation of the main tephra layers and the thickness of the tephras (cf., the legend). Thickness of the tephras is also indicated by a *triangle*. Sedimentological units (I–III) are also indicated. *Note*: the top of core 05-2 starts at 10 cm (equivalent to 0 cm) because the photograph included the 'green foam' material used to store the sediment core after collection

correlated among cores, and associated with the same volcanic event (Fig. 3). A detailed microscopic and geochemical characterisation of the material is needed to achieve a better understanding of these events. Regarding the tephra layer identified at 16–17 cm in Hess 05-2 (located at 13–14 cm in Hess 05-3), additional analyses would be necessary before we could confidently conclude that they reflect the same event.

Additional tephra layers were identified in Hess 05-1 at the time of sub-sampling at 32–35 cm, 36.5–38 cm and 58–59 cm depths. These were not identifiable in core Hess-05-3. Instead in core Hess 05-3, a coarse tephra was visible at 46–47 cm depth (Fig. 3). In core Hess 05-2, four strong peaks of MS, between 60 and 70 cm, could indicate different volcanic events (Fig. 4).

Magnetic susceptibility and geochemistry

Magnetic susceptibility and dry weight profiles are often indicative of the presence of tephra layers or clastic materials derived, e.g. from episodes of high soil erosion or floods (Thompson & Oldfield, 1986). In our case, dry weight fluctuations are strongly related to tephra layers: three high peaks in core Hess 05-1 are shown at ca. 57, 43 and 25 cm core depth (Fig. 4). Dry weight is negatively correlated with organic matter (LOI). The upper 10–20 cm layer is rich in plant remains.

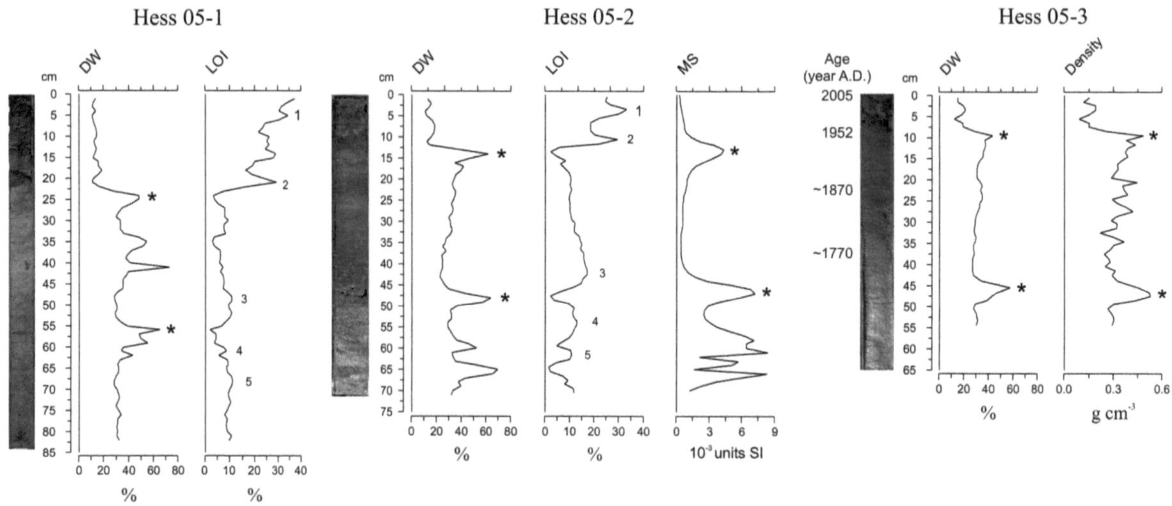

Fig. 4 Lake Hess, cores 05-1, 05-2 and 05-3. Depth distribution of dry weight (DW, % w.w.), density (g m^{-3}), magnetic susceptibility (MS, in SI 10^{-3}) and organic matter content (LOI, % d.w.). Core correlation between cores Hess 05-1 and Hess 05-2, based on LOI, is shown by *numbers*. *Asterisks* show the correlation between cores Hess 05-3 and Hess 05-2, also based on dry weight (d.w.). Additional information from the three cores and their correlation based on lithology and tephra layers is also shown in Fig. 3

Figures 3 and 4 summarise the data for core-to-core correlation and effects of the tephra deposits on organic matter. The same effect of tephra and the sharp increase in LOI in the upper part is also seen in core Hess 05-2 (Fig. 4).

In Lake Hess from ca. 25 cm upwards (zone III), a remarkable increase in LOI (from ca. 8–9% d.w. to ca. 40% d.w.) and plant pigments (see below) document a change in the trophic condition of the lake (Fig. 5). Those levels with LOI higher than the 20% correspond to gyttja. However, this term has been misused so frequently as to have been rendered meaningless (Schnurrenberger et al., 2003). Total carbon and nitrogen contents concentrations show the same recent increase. Sulphur, after a similar increase, decreases in the most recent sediments.

From core bottom to 25 cm (zones I and II), C and N concentrations (Fig. 5) are relatively constant with a few fluctuations, in part related to the impact of volcanic tephra deposits. From the C:N ratio, we infer that the proportion of allochthonous/autochthonous input of organic matter changed over time with a sharp increase from 25 cm onwards associated with the evident input of allochthonous materials (C:N > 10; Meyers & Ishiwatari, 1995). The core section between 50 and 25 cm points to a more autochthonous origin for the organic matter (C:N values around 9–10; Wetzel, 1983).

Fossil pigments

From the total and specific pigment profiles (Figs. 5, 6) we can recognise three main zones in core Hess 05-1: zone I (85–42 cm), zone II (42–25 cm) and zone III (25–0 cm).

The oldest part of the core (zone I) has relatively high algal pigment concentrations, which suggests a period of relatively high algal biomass (Fig. 5) (concentrations of β-carotene are also high). Pigments deceased at the end of this zone. According to the specific pigments (Fig. 6), the phytoplankton composition in zone I consisted mainly of cyanobacteria (echinenone), Chlorophyta (lutein) and siliceous algae (fucoxanthin and diatoxanthin). During this period, there was also a short phase (45–55 cm) where alloxanthin, a pigment specific to cryptophytes, showed relatively high values. Concentrations of phaeophorbides are low.

In zone II (42–25 cm), total and specific pigments (Figs. 5, 6) were barely detectable and indicate a phase of very low algal productivity probably associated to high flux of meltwater since this coincides with the low-productive phase observed among the chironomid remains. Inputs of clastic materials may dilute the pigment concentrations although the dry weight profiles generally do not change markedly except at the level of the tephra layers (cf. Fig. 4).

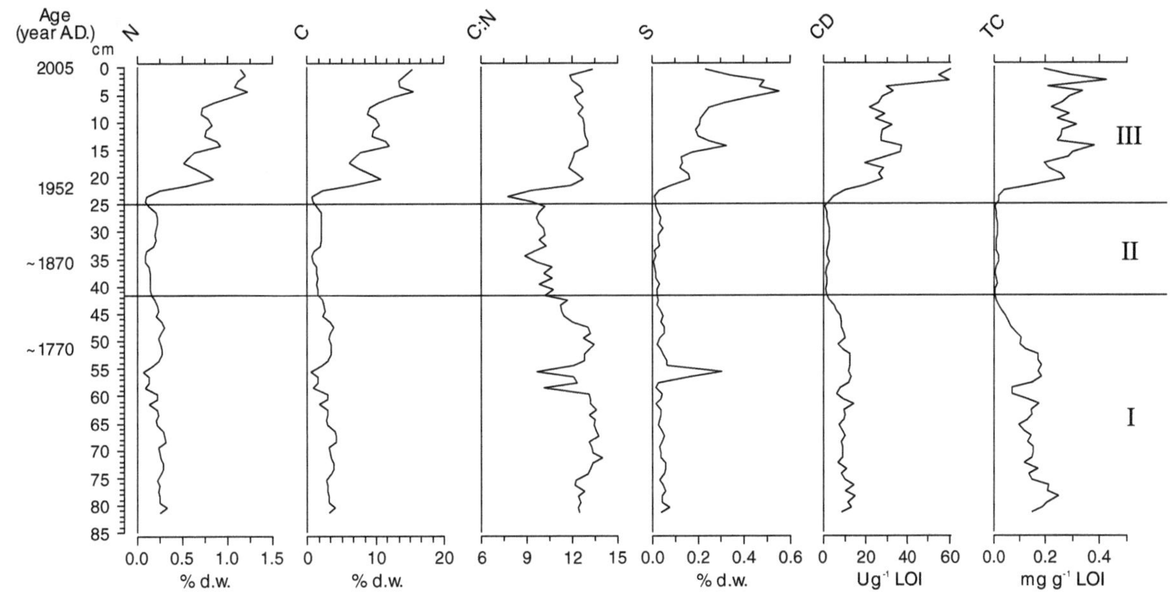

Fig. 5 Total nitrogen (N), total carbon (C), C:N ratio, sulphur (S) in per cent dry weight, total chlorophyll derivatives (CD, Units g^{-1} LOI), total carotenoids (TC; mg g^{-1} LOI) in core Hess 05-1 from Lake Hess. Zones I–III are also indicated

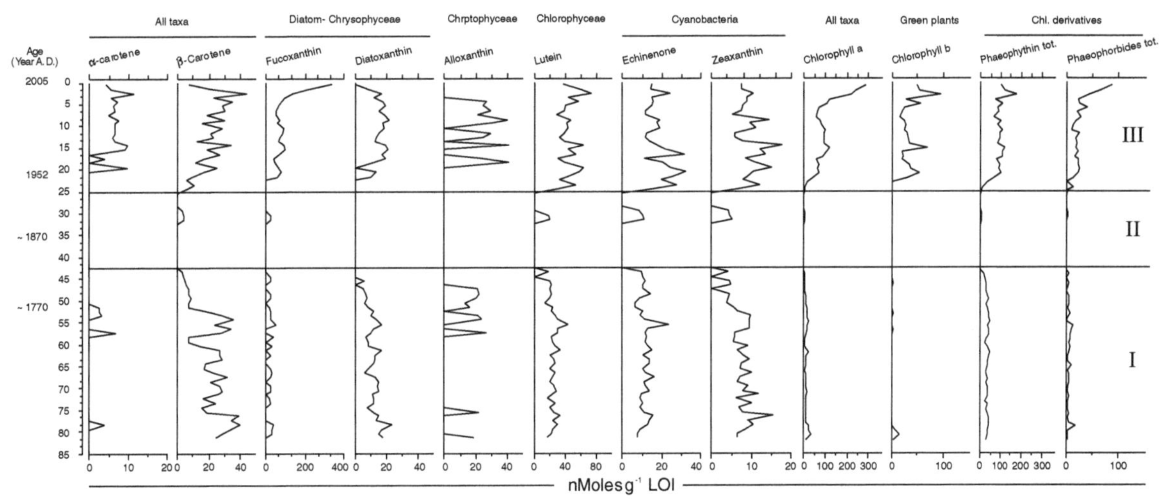

Fig. 6 Selected specific carotenoids and chlorophylls in core Hess 05-1 of Lake Hess. Zones I–III are also indicated

Among the specific carotenoids, the only distinctive episode was a very short-lived event around ca. 30 cm dominated by the presence of green algae and terrestrial vegetation (lutein) and cyanobacteria (echinenone and zeaxanthin).

Zone III (25 cm-top). The recent phytoplankton biomass, as detected by the pigment analysis (Fig. 5, 6), shows higher concentration values compared with the older periods (80–25 cm) and is dominated by siliceous algae (high concentrations of fucoxanthin, diatoxanthin). The carotenoid composition of this zone is also characterised by the presence of planktonic algae (alloxanthin), particularly abundant between 5 and 20 cm, and lutein that show the highest value in this zone. Lutein, as well as chlorophyll *b*, are specific pigments not only in algae, but they are also present in terrestrial vegetation and aquatic macrophytes (mosses remains in our cores). Therefore, their concentration peaks are also very likely associated with the presence of these organisms. Grazing on algae leads to the production of high concentrations of phaeophorbides, suggesting that grazing is an important factor in this zone.

Chironomids

The Lake Hess chironomid record (Fig. 7) is represented by a blend of profundal and littoral/lotic taxa (Brundin, 1958; Saether, 1979). Profundal taxa such as the Tribe Tanytarsini are poorly represented throughout the sequence. A CONISS cluster analysis (not shown) allowed us to divide the assemblages into three distinct zones described and interpreted as follows.

Zone I (from 85 to ~40 cm) is characterised by the coexistence of upper-littoral, littoral and lotic elements with profundal taxa. Although profundal taxa are present in high abundance (between 20 and 40%), the presence of littoral *Eukiefferiella* and *Demicryptochironomus*, a predator not only found in bottom sediments but also associated with upper littoral zones (Brodersen et al., 1998) together with other littoral elements, generally indicates low water levels. The occurrence of a mixed littoral/profundal chironomid assemblage could thus be related to a decrease in precipitation. As a consequence, as the lake dries, it develops a more fluvial character more strongly influenced by the melting waters of the Tronador ice cap. The presence of *Eukieferiella* and upper littoral chironomid remains could be linked to this fluvial system (Fig. 7).

In Zone II (from 40 to 25 cm), the abundance of head capsules drops sharply. The dominant and almost sole taxa present in this zone are *Ablabesmyia* and *Parapsectrocladius* showing abundance values close to 80%. Littoral and lotic elements disappear. A long period of low productivity (zone II) may correlate in part with a generally cold and wet period. During this period, chironomid assemblages are represented by profundal taxa, indicating an increase in precipitation, an increase in the lake water level and therefore, a more lotic regime. Another notable aspect of zone II is the decrease in the concentration of head capsules (total HC; Fig. 7) just after the tephra deposition, coinciding with the decrease in all biotic parameters studied (Figs. 5, 6). Three tephra layers were identified in this section, recording volcanic events that may strongly affect lake dynamics. Only *Ablabesmyia* and *Parapsectrocladius* relative abundance increased in zone II, while other taxa decreased drastically, revealing of the way in which these two genera are favoured by this kind of sediment, as observed by Urrutia et al. (2007) in Lake Galletué (Chilean Andes).

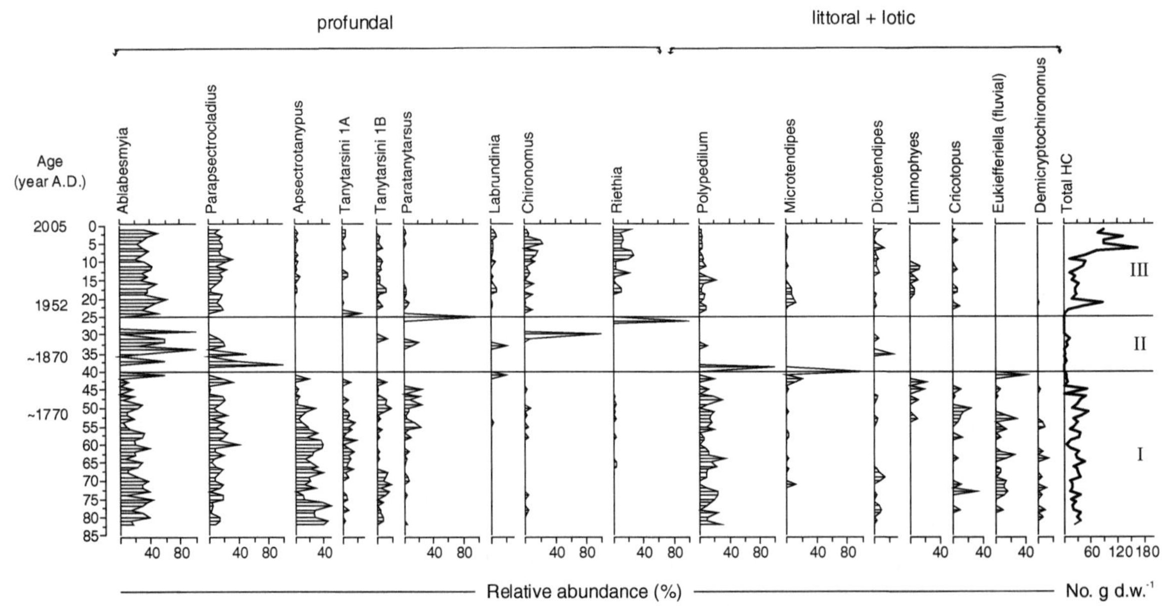

Fig. 7 Chironomid percentage diagram showing the most abundant taxa found in the Lake Hess sediment core (Hess 05-1). The taxa are grouped according to profundal and littoral plus lotic habitats. Zones I–III are also indicated. HC = head capsules

The uppermost zone III (25–0 cm) is characterised by the increase in profundal *Chironomus* and *Riethia* (Massaferro et al., 2008) and the appearance of littoral taxa such as *Limnophyes* and *Cricotopus*. It is interesting to note the total disappearance of *Apsectrotanypus* and *Eukiefferiella* which were very well represented in zone I.

Discussion

Multi-cores, multi-proxy analyses at Lake Hess show a number of palaeolimnological changes which, according to our extrapolated sedimentation rates, should represent ca. three centuries of lake and catchment history.

The ^{137}Cs fallout sequence in the sediments of some lakes from the Nahuel Huapi National Park area has been determined previously (Ribeiro Guevara & Arribére, 2002; Ribeiro Guevara et al., 2003). The most intense fallout peaks were registered in 1964–1966, as a consequence of the combination of stratospheric and tropospheric fallout, with the tropospheric contributions coming from the South Pacific nuclear tests. Another ^{137}Cs peak, associated mainly with tropospheric fallout, was detected in 1970–1972, with a secondary peak in 1974. No other relevant ^{137}Cs fallout was evident—not even the 1986 Chernobyl nuclear power plant accident that produced little stratospheric ^{137}Cs fallout in 1987 and 1988 and was not relevant here for dating purposes. The ^{137}Cs specific activity profile for the topmost 8 cm of our core did not follow the fallout sequence, and the reasons could be caesium mobilisation after deposition and/or sediment mixing.

Lake Hess sediments are represented by a dominance of organic matter-rich clay and are rich in tephra deposits particularly in the lower part of the cores. The lithology, tephras and geochemical parameters have been used for core correlation (Figs. 3, 4). Previous studies performed on sedimentary sequences from water bodies from Nahuel Huapi National Park demonstrated that the region was impacted by frequent volcanic eruptions from the Calbuco volcano and Puyehue-Cordón Caulle volcanic system with the Calbuco volcano eruptions being the most important affecting the south of the park (Daga et al., 2006, 2008a). However, it is not possible to assign dates to the eruptions without a better geochemical characterisation of the tephras providing a basis for correlation with potential source eruptions (Daga et al., 2008a). For the present, the sequence of tephras in the lower part of the cores cannot be used to strengthen the age model suggested here.

Although tephra layers from volcanic eruption events deposited in the lake may have had an impact on the biota, the main biological changes may have been in response to hydrological-related environmental changes in the catchment area.

Based on algal biomass (as inferred from the photosynthetic pigments), nutrient and chironomid assemblage analyses, three main intervals were evident along core Hess 05-1: at 85–40 cm (zone I; pre-AD 1800), 42–25 cm (zone II; ca. AD 1800–1940) and 25- top (zone III; last 50 years).

Zone I is characterised by low water level and low precipitation. In fact, the phaeophorbides (a marker carotenoid of zooplankton), as well as alloxanthin, are associated with a more lentic conditions (Buchaca & Catalan, 2007). In this zone, the sporadic presence of alloxanthin (cryptophytes, a truly planktonic algal group) is usually associated with high lake water levels (Pienitz et al., 1992). In general, however, zone I is characterised by a low water level and lower grazing pressure as shown by the generally low concentrations of phaeophorbides (Leavitt et al., 1994) compared with the rest of the core (Fig. 6).

Zone II shows intermediate environmental conditions, likely associated with relatively cold temperatures. From the organic matter, pigment contents and chironomid profiles, a long period of very low productivity is inferred. This is in agreement with a palaeoenvironmental study of the nearby Lake Frias (Ariztegui et al., 2007).

In zone III, the littoral/profundal chironomid fauna again indicates a lentic environment and a decrease in precipitation. From pigment (Fig. 6) and chironomid total abundance profiles (Fig. 7), it is possible to distinguish further trophic and hydrological changes in the topmost 10 cm which could be linked to the temperature increase in the recent times. A strong increase in nutrient (C, N) and sulphur concentrations (Fig. 5) occurs synchronously with these biotic changes. Also, the presence of macrophyte remains in the topmost 10 cm can be related to a decrease of water level and dry periods which, in turn, might suggest an increase in productivity. We cannot exclude the possibility that these changes during the last century are caused by an increased flux of inorganic sediments via the meltwater input. We believe that these changes are strongly affected by the physical environment such as water level fluctuations and the lotic/lentic regime of the lake, associated with dry/wet periods and incoming water from the glaciers. Chironomid assemblages, showing changes in profundal/littoral and lotic elements, confirm these repeated dry/wet alternations.

Another study based on a sedimentary sequence collected in Lake Toncek (1,700 m a.s.l.) from Nahuel Huapi National Park shows that both natural and anthropogenic events (long-range atmospheric transport of pollutants) occurred in the region during the last 900 years (Rizzo et al., 2007; Daga et al., 2008b; Ribeiro Guevara et al., 2009). The study of sub-fossil chironomid assemblages showed the predominance of *Pseudosmittia* and cold-stenothermic Podonominae from eleventh to seventeenth centuries, which decrease in the upper layers (eighteenth century to present) being replaced by the warm-adapted Tanypodinae. These observations together with a continuous increase of abundance and diversity indices during the twentieth century (Rizzo et al., 2007) are consistent with the observations in the sediment record from Lake Hess.

In summary, remote oligotrophic lakes Hess and Toncek both show biological and geological changes mainly related to climate and regional environmental variations. In contrast, at Lake Morenito (a small closed basin close to Lake Hess), located in an anthropogenically disturbed environment, the changes recorded were mainly related to productivity change due to human impact over the last 100 years (Massaferro et al., 2005).

Biological and geochemical profiles from Lake Hess sediment cores confirm a varying pattern of climate change. Chironomid assemblages show profundal, littoral, sub-littoral, terrestrial and fluvial elements, which may indicate alternating dry/wet periods. Whether the environmental variations are the result of the varying dominance of moisture and/or temperature controlling the Tronador Glacier mass balance remain an elusive question (Ariztegui et al. 2007).

In conclusion, owing to the remote location of this lake, this high-resolution short-core multiproxy study has added some valuable information on changes probably related to the westerlies influence and possibly ENSO-related variability (Jenny et al., 2002), and the occurrence of volcanic events in the area. Further research is needed in Patagonia, particularly to asses the extent to which climate changes over the last millennium have occurred in response to the El Niño Southern Oscillation (ENSO), Little Ice Age (LIA) and other short-term climate events in southern South America.

Acknowledgments This study was supported by the bilateral agreement between CNR (Italy) and CONICET (Argentina). We thank the three anonymous reviewers for their very helpful comments on the manuscript.We also like to acknowledge Krisztina Buczkó for providing assistance and encouragement.

References

Ariztegui, D., M. M. Bianchi, J. Massaferro, F. Lafargue & F. Niessen, 1997. Interhemispheric synchrony of late-glacial climatic instability as recorded in proglacial Lake Mascardi, Argentina. Journal Quaternary Science 12: 333–338.

Ariztegui, D., P. Bösch & E. Davaud, 2007. Dominant ENSO frequencies during the Little Ice Age in Northern Patagonia: The varved record of proglacial Lago Frìas, Argentina. Quaternary International 161: 46–55.

Bertrand, S., F. Charlet, B. Charlier, V. Renson & N. Fagel, 2008. Climate variability of southern Chile since the Last Glacial Maximum: a continuous sedimentological record from Lago Puyehue (40°S). Journal of Paleolimnology 39: 179–195.

Binford, M. W., 1990. Calculation and uncertainty analysis of ^{210}Pb date for PIRLA project lake sediment cores. Journal of Paleolimnology 3: 253–267.

Brodersen, K., P. Dall & C. Lindegaard, 1998. The fauna in the upper stony littoral of Danish lakes: macroinvertebrates as trophic indicators. Freshwater Biology 39: 577–592.

Brooks, S. J. & H. J. B. Birks, 2004. The dynamics of Chironomidae (Insecta: Diptera) assemblages in response to environmental change during the past 700 years on Svalbard. Journal of Paleolimnology 31: 483–498.

Brundin, L., 1958. The bottom faunistical lake type system and its application to the southern hemisphere. Moreover a theory of glacial erosion as a factor of productivity in lakes and oceans. Verhandlungen Internationale Vereinigung für Theoretische und Angewandte Limnologie 13: 288–297.

Brundin, L., 1966. Transantarctic relationships and their significance, as evidenced by chironomid midges, with a monograph of the subfamilies Potonominae and Aphroteniinae and the austral Heptagyiae. Kungliga Svenska Vetenskapsakademiens Handlingar 11: 1–472.

Buchaca, T. & J. Catalan, 2007. Factors influencing the variability of pigments in the surface sediments of mountain lakes. Freshwater Biology 52: 1365–1379.

Chapron, E., D. Ariztegui, S. Mulsow, G. Villarosa, M. Pino, V. Outes, E. Juvignie & E. Crivelli, 2006. Impact of 1960 major subduction earthquake in Northern Patagonia (Chile-Argentina). Quaternary International 158: 58–71.

Charlet, F., M. De Batist, E. Chapron, S. Bertrand, M. Pino & R. Urrutia, 2008. Seismic stratigraphy of Lago Puyehue (Chilean Lake District): new views on its deglacial and Holocene evolution. Journal of Paleolimnology 39: 163–177.

Cranston, P. S., 1997. Identification Guide to the Chironomidae of New South Wales. Australian Water Technologies Pty Ltd, West Ryde.

Daga, R., S. Ribeiro Guevara, M. L. Sánchez & M. Arribére, 2006. Geochemical characterization of volcanic ashes from recent events in Northern Patagonia Andean Range by INAA. Journal of Radioanalytical and Nuclear Chemistry 270: 677–694.

Daga, R., S. Ribeiro Guevara, M. L. Sánchez & M. Arribére, 2008a. Source identification of volcanic ashes by geochemical analysis of well preserved lacustrine tephras in Nahuel Huapi National Park. Applied Radiation and Isotopes 66: 1325–1336.

Daga, R., S. Ribeiro Guevara, M. L. Sánchez & M. Arribére, 2008b. Tephrochronology of recent events in Northern Patagonia Andean Range: spatial distribution and provenance of lacustrine ash layers in Nahuel Huapi National Park. Journal of Quaternary Science (Submitted).

De Batist, M., N. Fagel, M. F. Loutre & E. Chapron, 2008. A 17,900-year multi-proxy lacustrine record of Lago Puyehue (Chilean Lake District): introduction. Journal of Paleolimnology 39: 151–161.

Gilli, A., 2003. Tracking late Quaternary environmental change in southernmost South America using lake sediments of Lago Cardiel (49°S), Patagonia, Argentina. Ph.D. Thesis, ETH-Zurich, Switzerland: 172 pp.

Gilli, A., D. Ariztegui, F. S. Anselmetti, J. A. McKenzie, V. Markgraf, I. Hajdas & R. D. McCulloch, 2005. Mid-Holocene strengthening of the Southern Westerlies in South America—sedimentological evidences from Lago Cardiel, Argentina (49 S). Global and Planetary Changes 49: 75–93.

Guilizzoni, P. & A. Lami, 2002. Paleolimnology: use of algal pigments as indicators. In Bitton, G. (ed.), The Encyclopedia of Environmental Microbiology. Wiley & Sons, New York: 2306–2317.

Guilizzoni, P., G. Bonomi, G. Galanti & D. Ruggiu, 1982. Basic trophic status and recent development of some Italian lakes as revealed by plant pigments and other chemical components in sediment cores. Memorie dell'Istituto Italiano di Idrobiologia 40: 79–98.

Guilizzoni, P., G. Bonomi, G. Galanti & D. Ruggiu, 1983. Relationship between sedimentary pigments and primary production: evidence from core analyses of twelve Italian lakes. Hydrobiologia 103: 103–106.

Guilizzoni, P., A. Lami & A. Marchetto, 1992. Plant pigment ratios from lake sediments as indicators of recent acidification in alpine lakes. Limnology & Oceanography 37: 1565–1569.

Guilizzoni, P., A. Lami, M. Manca, S. Musazzi & A. Marchetto, 2006. Palaeoenvironmental changes inferred from biological remains in short lake sediment cores from the Central Alps and Dolomites. Hydrobiologia 562: 167–191.

Haberzettl, T., M. Wille, M. Fey, S. Janssen, A. Lücke, C. Mayr, C. Ohlendorf, F. Schäbitz, G. H. Schleser & B. Zolitschka, 2006. Environmental change and fire history of southern Patagonia (Argentina) during the last five centuries. Quaternary International 158: 72–82.

Haberzettl, T., H. Corbella, M. Fey, S. Janssen, A. Lücke, C. Mayr, C. Ohlendorf, F. Schäbitz, G. H. Schleser, M. Wille, S. Wulf & B. Zolitschka, 2007a. Late Glacial and Holocene wet-dry cycles in southern Patagonia: chronology, sedimentology and geochemistry of a lacustrine record from Laguna Potrok Aike, Argentina. The Holocene 17: 297–310.

Haberzettl, T., C. Mayr & B. Zolitschka, 2007b. Linkages between southern hemisphere westerlies and hydrological changes in semi-arid Patagonia during the last 16,000 years. PAGES News 15: 22–23.

Hajdas, I., G. Bonani, P. I. Moreno & D. Ariztegui, 2003. Precise radiocarbon dating of Late-Glacial cooling in mid-latitude South America. Quaternary Research 59: 70–78.

Jenny, B., B. L. Valero-Garcés, R. Urrutia, K. Kelts, H. Veit, P. G. Appleby & M. Geyh, 2002. Moisture changes and fluctuations of the Westerlies in Mediterranean Central Chile during the last 2000 years: the Laguna Aculeo record (33°50′S). Quaternary International 87: 3–18.

Joshi, S. R. & B. S. Shukla, 1991. Ab initio derivation of formulations for ^{210}Pb dating of sediments. Journal of Radioanalytical and Nuclear Chemistry 148: 73–79.

Juggins, S., 2003. C2 A Program for Analysing and Visualising Palaeoenvironmental Data, Version 1.4.3. University of Newcastle, Newcastle upon Tyne, UK. [http://www.staff.ncl.ac.uk/stephen.juggins].

Kitzberger, T., O. Veblen & R. Villalba, 1997. Climatic influences on fire regimes along a rain forest-to-xeric woodland gradient in northern Patagonia, Argentina. Journal of Biogeography 24: 35–47.

Lami, A., N. Cameron & A. Korhola (eds), 2000a. Paleolimnology and ecosystem dynamics of remote European Alpine lakes. Journal of Limnology 59(Suppl. 1): 119 pp.

Lami, A., P. Guilizzoni & A. Marchetto, 2000b. High resolution analysis of fossil pigments, carbon, nitrogen and sulphur in the sediments of eight European Alpine lakes: the MOLAR project. In Lami, A., N. Cameron & A. Korhola (eds), Paleolimnology and Ecosystem Dynamics at Remote European Alpine Lakes. Journal of Limnology 59 (Suppl. 1): 15–28.

Lami, A., S. Musazzi, A. Marchetto, T. Buchaca, M. Kernan, E. Jeppesen & P. Guilizzoni, 2009. Sedimentary pigments in 297 alpine lakes and their relation to environmental gradients. Archiv für Hydrobiologie (in press).

Leavitt, P. R., 1993. A review of factors that regulate carotenoid and chlorophyll deposition and fossil pigment abundance. Journal of Paleolimnology 9: 109–127.

Leavitt, P. R., P. R. Sanford, S. R. Carpenter & J. K. Kitchell, 1994. An annual fossil record of production, planktivory and piscivory during whole-lake manipulation. Journal of Paleolimnology 11: 133–149.

Lotter, A. F. & H. J. B. Birks (eds), 2003. The Holocene palaeolimnology of Sägistalsee and its environmental history. Journal of Paleolimnology 30: 342 pp.

Markgraf, V., T. R. Baumgartner, J. P. Bradbury, H. F. Diaz, R. B. Dunbar, B. H. Luckman, G. O. Seltzer, T. W. Swetnam &

R. Villalba, 2000. Paleoclimate reconstruction along the Pole-Equator-Pole transect of the Americas (PEP 1). Quaternary Science Reviews 19: 125–140.

Markgraf, V., J. P. Bradbury, A. Schwalb, S. J. Burns, C. Stern, D. Ariztegui, A. Gilli, F. S. Anselmetti, S. Stine & N. Maidana, 2003. Holocene palaeoclimates of southern Patagonia: limnological and environmental history of Lago Cardiel, Argentina. The Holocene 13: 581–591.

Massaferro, J., S. Ribeiro Guevara, A. Rizzo & M. Arribére, 2005. Short-term environmental changes in Lake Morenito (41°S, 71°W, Patagonia, Argentina) from the analysis of sub-fossil chironomids. Aquatic Conservation: Marine and Freshwater Ecosystems 15: 23–30.

Massaferro, J., P. Moreno, G. Denton, M. Vandergoes & A. Dieffenbacher-Krall, 2008. Chironomid and pollen evidence for climate fluctuations during the Last Glacial Termination in NW Patagonia. Quaternary Science Reviews (in press, available online).

Mayr, C., M. Fey, T. Haberzettl, S. Janssen, A. Lücke, N. I. Maidana, C. Ohlendorf, C. Schäbitz, G. H. Schleser, U. Struck, M. Wille & B. Zolitschka, 2005. Palaeoenvironmental changes in southern Patagonia during the last millennium recorded in lake sediments from Laguna Azul (Argentina). Palaeogeography, Palaeoclimatology, Palaeoecology 228: 203–227.

Meyers, P. A. & R. Ishiwatari, 1995. Organic matter accumulation records in lake sediments. In Lerman, A., D. Imboden & J. Gat (eds), Physics and Chemistry of Lakes. Springer-Verlag, Berlin, Heidelberg, New York: 279–328.

Moernaut, J., M. De Batist, F. Charlet, K. Heirman, E. Chapron, M. Pino, R. Brümmer & R. Urrutia, 2007. Giant earthquakes in South-Central Chile revealed by Holocene mass-wasting events in Lake Puyehue. Sediment Geology 195: 239–256.

Pienitz, R., I. R. Walker, B. A. Zeeb, J. P. Smol & P. R. Leavitt, 1992. Biomonitoring past salinity changes in an athalassic subarctic lake. International Journal of Salt Lake Research 1: 91–123.

Prohaska, F., 1976. The climate of Argentina, Paraguay and Uruguay. In Schwerdtfeger, W. (ed.), Climates of Central and South America. World Survey of Climatology. Elsevier, Amsterdam: 13–73.

Rabassa, J. & C. M. Clapperton, 1990. Quaternary glaciations of the southern Andes. Quaternary Science Reviews 9: 153–174.

Ribeiro Guevara, S. & M. Arribére, 2002. ^{137}Cs dating of lake cores from the Nahuel Huapi National Park, Patagonia, Argentina: Historical records and profile measurements. Journal of Radioanalytical and Nuclear Chemistry 252: 37–45.

Ribeiro Guevara, S., A. Rizzo, R. Sánchez & M. Arribére, 2003. ^{210}Pb fluxes in sediment layers sampled from Northern Patagonia lakes. Journal of Radioanalytical and Nuclear Chemistry 258: 583–595.

Ribeiro Guevara, S., M. Meili, A. Rizzo, R. Daga & M. Arribére, 2009. Highly variable mercury inputs to mountain lakes in Patagonia (Argentina) during the past millennium. 9th International Conference on Mercury as a Global Pollutant, June 7–12, 2009, Guiyang, China.

Rizzo, A., S. Ribeiro Guevara, M. Arribére & J. Massaferro, 2007. Study of subfossil chironomid assemblages of the last 900 yr recorded in a sediment sequence from a high altitude lake of Northern Patagonia (Argentina). 4th International Limnogeology Congress, July 11–14, 2007, Barcelona, Spain.

Robbins, J. A. & L. R. Herche, 1993. Models and uncertainty in ^{210}Pb dating of sediments. Radiochemical Limnology 25: 217–222.

Saether, O. A., 1979. Chironomid communities as water quality indicators. Ecography 2: 65–74.

Sanger, J. E., 1988. Fossil pigments in paleoecology and paleolimnology. Palaeogegraphy, Palaeoclimatology, Palaeolimnology 62: 343–359.

Schnurrenberger, D., J. Russell & K. Kelts, 2003. Classification of lacustrine sediments based on sedimentary components. Journal of Paleolimnology 29: 141–154.

Thompson, R. & F. Oldfield, 1986. Environmental Magnetism. Allen & Unwin, New York: 227 pp.

Urrutia, R., A. Araneda, F. Cruces, L. Torres, L. Chirinos, H. C. Treutler, N. Fagel, S. Bertrand, I. Alvial, R. Barra & E. Charpon, 2007. Changes in diatom, pollen, and chironomid assemblages in response to recent volcanic event in Lake Galletué (Chilean Andes). Limnologica 37: 49–62.

Veblen, T. T., T. Kitzberger, E. Raffaele & D. C. Lorenz, 2003. Fire history and vegetation changes in northern Patagonia, Argentina. In Veblen, T. T., W. L. Baker, G. Montenegro & W. T. Swetnam (eds), Fire and Climatic Change in Temperate Ecosystems of the Western Americas. Springer, New York, USA: 267–295.

Veit, H., 1996. Southern Westerlies during the Holocene deduced from geomorphological and pedological studies in the Norte Chico, Northern Chile (27–33jS). Palaeogeography Palaeoclimatology, Palaeoecology 123: 107–119.

Villalba, R., 2007. Tree-ring evidence for tropical-extratropical influences on climate variability along the Andes in South America. PAGES News 15: 23–25.

Villalba, R., J. Boninsegna, T. Veblen, A. Schmelter & S. Rubulis, 1997. Recent trends in tree-ring records from high elevation sites in the Andes of Northern Patagonia. Climatic Change 36: 425–454.

Villarosa, G., V. Outes, A. Hajduk, E. Crivelli Montero, D. Sellés, M. Fernández & E. Crivelli, 2006. Explosive volcanism during the Holocene in the Upper Limay River Basin: the effects of ashfalls on human societies, Northern Patagonia, Argentina. Quaternary International 158: 44–57.

Wetzel, R. G., 1983. Limnology. Saunders College Publisher, Philadelphia: 850 pp.

Züllig, H., 1982. Untersuchungen über die Stratigraphie von Carotinoiden im geschichteten Sediment von 10 Schweizer Seen zur Erkundung früherer Phytoplankton-Enfaltungen. Schweizerische Zeitschrift fur Hydrologie 47: 87–126.

Züllig, H., 1986. Carotenoids from plankton and photosynthetic bacteria in sediments as indicators of trophic changes in Lake Lobsigen during the last 14000 years. Hydrobiologia 143: 315–319.

A multi-proxy paleolimnological reconstruction of trophic state reference conditions for stratified carbonate-rich lakes in northern Germany

Thomas Hübener · Sven Adler · Petra Werner · Manuela Schult · Helmut Erlenkeuser · Hinrich Meyer · Mandy Bahnwart

Originally published in the journal Hydrobiologia, Volume 631, No. 1, 303–327.
DOI: 10.1007/s10750-009-9819-4

Abstract This study aims to identify reference conditions (nutrient status and diatom assemblages) as required by the European Water Framework Directive (WFD) for stratified, carbonate-rich lowland lakes with a large watershed area (watershed area to lake volume ratio (WV) > 1.5 km^2 10^{-6} m^{-3}) and a retention time (RT) from 0.1 to 10 years (Central Baltic Lake-Type 1, German Lake-Type 10) in European ecoregion 14. Diatoms, pollen and geochemistry were analysed from sediment cores of six lakes from northern Germany representing different subtypes of Lake-Type 10 (varying WV and RT) and covering the past 290–1,750 years. Historic total phosphorus levels were inferred using diatom-based transfer functions selected from a merged European data set and from optimised data sets identified with the moving-window approach. Pollen and geochemical proxies were used to identify occurrence and intensity of anthropogenic catchment usage. Lake trophic state reference conditions and associated diatom assemblages were identified for three of the six study lakes. In contrast, according to fossil pollen assemblages, two lakes were already strongly impacted by intensive catchment usage when the oldest investigated sediments were laid down. Thus, reference conditions of these already eutrophic lakes could not be identified. Similarly, the lowermost

Guest editors: K. Buczkó, J. Korponai, J. Padisák & S. W. Starratt
Palaeolimnological Proxies as Tools of Environmental Reconstruction in Fresh Water

T. Hübener (✉) · S. Adler
Institute of Biosciences, University of Rostock, Wismarsche Str. 8, 18051 Rostock, Germany
e-mail: thomas.huebener@uni-rostock.de

P. Werner
Institute of Geography, University of Osnabrück, Seminarstr. 19, 49069 Osnabrück, Germany

M. Schult
Institute of Botany, University of Greifswald, Grimmer Str. 88, 17487 Greifswald, Germany

H. Erlenkeuser
Leibniz-Laboratory for Radiometric Dating and Stable Isotope Research, University of Kiel, Max-Eyth-Str. 11–13, 24118 Kiel, Germany

H. Meyer
Institute of Geography & Geology, University of Greifswald, Jahnstr. 16, 17489 Greifswald, Germany

M. Bahnwart
State Agency for Nature and Environment Schleswig-Holstein, Hamburger Chaussee 25, 24220 Flintbek, Germany

samples of a core from the sixth lake showed signs of impact, and it remains unclear whether the identified dystrophic conditions occurred naturally or if they were due to the drainage of wetlands in Medieval times. Lakes with a relatively small WV (1.5–5.0 $km^2\ 10^{-6}\ m^{-3}$) and RT > 1 year were naturally oligotrophic to low mesotrophic and a typical, representative diatom assemblage was identified. In contrast, typical reference conditions or diatom assemblages for lakes with higher WV (5–18.6 $km^2\ 10^{-6}\ m^{-3}$) and RT < 1 year could not be identified as chemical precipitation and upstream lakes (nutrient sinks or sources) additionally influenced natural nutrient levels. Therefore, the reference situation of both trophic state and diatom assemblages in a lake may be strongly influenced by other modifying, limnological processes in addition to WV and RT. Overall, this study helps to implement the WFD by identifying reference conditions and by discussing the level of differentiation of lake types required to set reference conditions.

Keywords WFD · Reference values · Paleolimnology · Diatoms

Introduction

The identification of natural background conditions is often essential for conducting informed lake management programmes (Smol, 2008). The European Water Framework Directive (WFD; EU, 2000) requires that all European lakes >50 ha should be in "high" or "good" ecological conditions by 2015, as identified with bio-indicator assemblages (fish, macroinvertebrates, macrophytes, phytobenthos and phytoplankton). The ecological status of lakes and rivers is defined on the basis of the degree of deviation from type-specific minimally impaired reference conditions (EU-Intercalibration Technical Report, 2006). However, today undisturbed or minimally impaired systems are rare or absent due to widespread eutrophication (Bennion et al., 2004; Bennion & Battarbee, 2007). Hence, studies of Europe and North America have identified natural or reference trophic state or pH conditions using paleolimnological techniques (e.g. Ekdahl et al., 2004, 2007; Battarbee et al., 2005; Simpson et al., 2005; Werner et al., 2005).

In several European regions the human impact on trophic status of lakes and rivers has been detected since Subboreal to early Subatlantic Times (Beales, 1980; Günter, 1983; Dapples et al., 2002; Lotter & Birks, 2003), since late Medieval Times (Ohle, 1973; van Geel et al. 1994; Hübener & Dörfler, 2004; Bradshaw et al., 2005a, b; Sander-Johansson et al., 2005) or since the industrial developments during the nineteenth century (Barker et al., 2005; Taylor et al., 2006). The occurrence of naturally eutrophic (shallow) lakes has also been reported (Räsänen et al., 2006). However, some European lakes have been misclassified as "naturally eutrophic", due to an unexpected long history of anthropogenic eutrophication (Anderson, 1995). Furthermore, the magnitude of human impact at the local and regional scale has varied in both space and time since the earliest disturbances (e.g. Birks et al., 1988). Site- and region-specific factors that interfere with, or alter basic eutrophication processes include depth and quality of soil, degree and type of industrialisation and urbanisation, geology, climate, and other limnological features, natural trophic status, and the magnitude and duration of past pollution (Hall & Smol, 1999).

For the WFD, reference conditions must be defined on a lake type basis (EU, 2000). According to Carvalho et al. (2008) there are three lake types in the European Central-Baltic Lowlands (CB 1–3; ecoregion 14). For Germany, Mathes et al. (2002) identified 14 coarse lake-types based on ecoregions (Alps and foreland, central low mountains, lowlands), geologic basic reaction (carbonate-rich or carbonate-poor) and morphometric characteristics of the lakes (stratification pattern, ratio of watershed area versus lake volume: WV in $km^2\ 10^{-6}\ m^{-3}$, water retention time: RT). In Germany, trophic state reference conditions were modeled for each lake type according to overall TP input, which was calculated using the Vollenweider equation (Vollenweider, 1979; Vollenweider & Kerekes, 1982) and which thus takes into account catchment size, WV ratio, dominating soil type, TP export coefficients, water retention time and stratification patterns (polymictic or dimictic).

This study investigates German Lake-Type 10: carbonate-rich, dimictic lowland lakes from northern Germany (European ecoregion 14: Central plains) with a large watershed area to lake volume ratio (WV > 1.5 $km^2\ 10^{-6}\ m^{-3}$) and with a retention time of >0.1 year (Mathes et al., 2002). These lakes are associated with lake-type CB 1 according to Carvalho et al. (2008). This lake-type and the German Lake-Type 10 incorporate a wide range of lake characteristics. Accordingly, German Lake-Type 10 has currently been further divided into subtypes 10.1 and 10.2 (Mischke et al., 2007; Schaumburg et al., 2007). Mischke et al. (2007) base the subdivision on WV size (10.1: >1.5–15 km^2 10^{-6} m^{-3}; 10.2: >15 km^2 10^{-6} m^{-3}) as suggested by phytoplankton analyses. In contrast, Schaumburg et al. (2007) base the subdivision on retention time (10.1: 1–10 years; 10.2: 0.1–1 year) according to diatom analyses.

The aim of this study is to (1) reconstruct the trophic state development, including the trajectories of changes, for six carbonate-rich dimictic lakes with a large catchment area, (2) identify and possibly verify modelled trophic state reference conditions for German Lake-Type 10, (3) compare reference conditions with recent anthropogenic affected conditions and (4) test whether German Lake-Type 10 (WV > 1.5 km^2 10^{-6} m^{-3}, RT > 0.1 year) needs to be distinguished into sub-types 10.1 and 10.2, because of differing reference conditions.

Methods

Study sites

The six study lakes are situated in eastern Schleswig-Holstein, Germany (Fig. 1). The morphological and limnological key characteristics are listed in Table 1. All the study lakes are stratified, carbonate-rich lowland lakes with a large watershed area to lake volume ratio (WV, WV > 1.5 km^2 10^{-6} m^{-3}), i.e. they are classified as German Lake-Type 10 (Mathes et al., 2002). Mischke et al. (2007) suggested the subdivision of Lake-Type 10 into subcategories 10.1 and 10.2 based on differing WVs. Lake Bistensee (BIS), Lake Kellersee (KES), Lake Stolper See (STO) and Lake Tresdorfer See (TRE) represent subtype 10.1 (WV: >1.5–15 km^2 10^{-6} m^{-3}), while Lake Kleiner Plöner See (KPS) and Lake Gudower See (GUD) represent subtype 10.2 (>15 km^2 10^{-6} m^{-3}) (Table 1). In contrast, according to Schaumburg et al. (2007), Lakes STO and TRE also belong to subtype 10.2, as their subdivision is based on differing RT (10.1: 1–10 years; 10.2: 0.1–1 year) (Table 1). In addition, Lake BIS (10.1) is an isolated lake, i.e. there are no other lakes in its catchment area. In contrast, Lake KES (10.1 too) is located within a river–lake system, directly downstream of the recently hypereutrophic, shallow Lake Großer Eutiner See (Fig. 1). Lake STO and Lake TRE (both 10.1 and 10.2) have very similar morphometric features (Fig. 1). Lake STO is recently influenced by the upstream eutrophic dimictic Lake Belauer See, whereas Lake TRE has no connection with another lake (Fig. 1). Finally, Lake KPS (10.2) is part of a complex river–lake system, directly downstream of the dimictic Lake Großer Plöner See. Lake GUD (10.2) is an isolated, dystrophic lake (DOC = 21.0 μg/l, 2003, $n = 15$, LANU, 2003) with high proportions of fens and raised bogs in its catchment area.

Coring, subsampling

A freeze-corer (UWITEC, Mondsee, Austria; www.uwitec.at/html/frame.html; freeze corer for soft sediments, type 3) was used to retrieve 160–202-cm-long lake sediment cores in June 2005 (Table 2). If a steep bathymetry was present at the deepest water depth of a lake, cores were taken in a shallower location with flatter bathymetry to avoid areas with disturbed sedimentation (slumping) and high sedimentation rates. Thus, chances of undisturbed sediments and time represented by the cores were maximised. Coring positions and core lengths are given in Table 2.

The frozen sediment cores were sawed into 2.5-cm-thick slices, thawed and subsampled for diatoms, pollen, chemistry and chronology (^{210}Pb, ^{137}Cs and ^{14}C). Pollen, chemistry and diatoms were analysed at every 10 cm, with the exception of the diatoms of Lake Kellersee, which were investigated at every 5 cm. The activities of ^{210}Pb and ^{137}Cs activities were measured every 2.5 cm in the upper (most recent) 50 cm of each core. Two or three terrestrial macrofossils were taken from sediments below 80-cm sediment depth for ^{14}C analyses (Table 4).

^{210}Pb, ^{137}Cs and ^{14}C chronology

Radiocarbon dating was performed through AMS techniques (Nadeau et al., 1998) on macro-remains with typically 1 mg of sample carbon (Table 4). Sample pre-treatment involved removal of carbonates by hydrochloric acid. Contamination by infiltrated secondary humic acids of different ^{14}C-age appears negligible, as the samples were well sheltered within the sediment pile. ^{14}C-age calibration is based on the

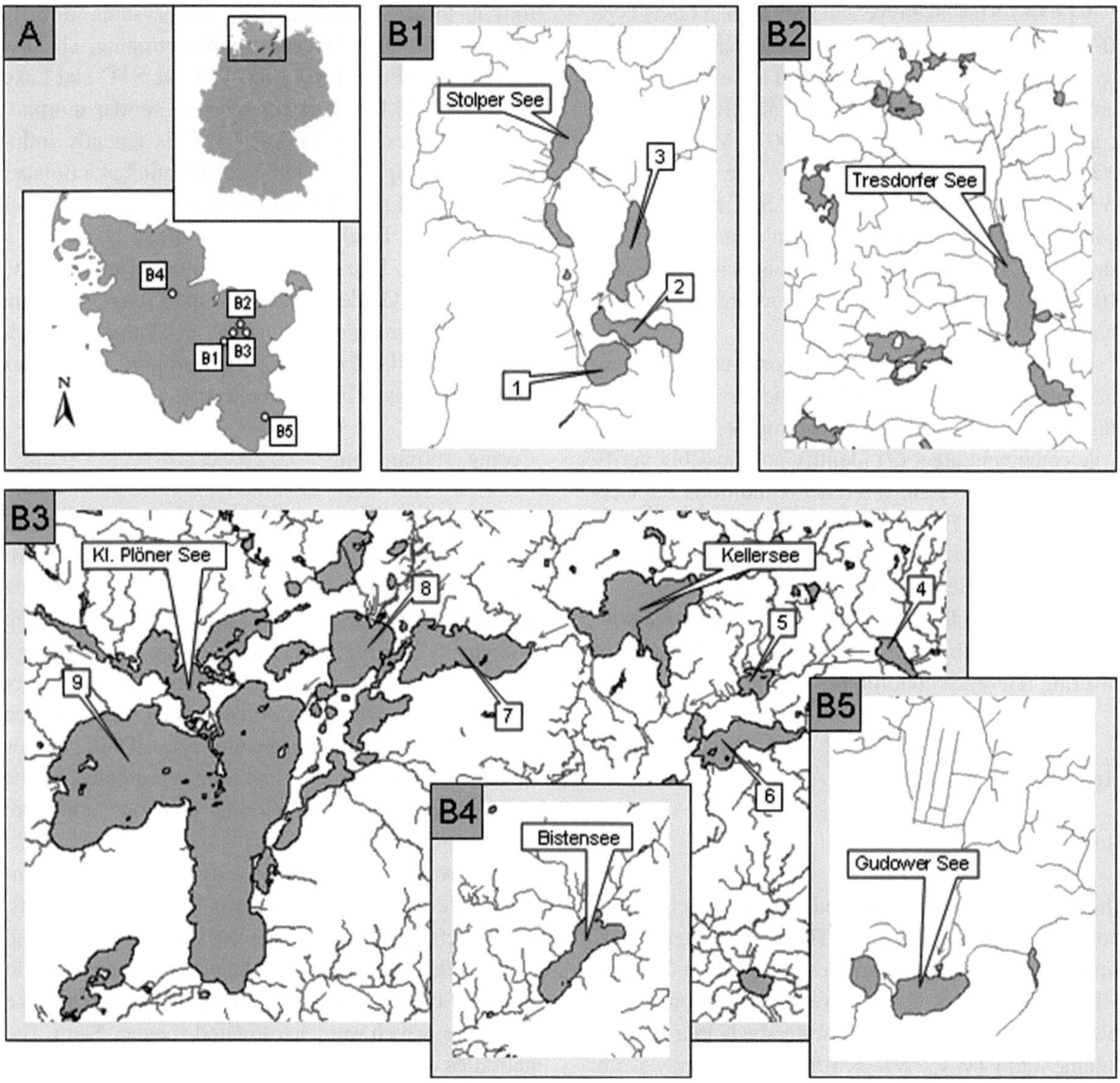

Fig. 1 **A** Location of study lakes in Schleswig-Holstein (northern Germany). **B1–B5** Detailed maps of the catchment area of the six investigated lakes. Three lakes are connected to river–lake systems. The Alte Schwentine river flows through Lake Stolper See (STO; **B1**). Lake STO is located downstream of the lakes Bornhöveder See (*1*), Schmalensee (*2*) and Belauer See (*3*). Lakes Kellersee (KES, **B3**) and Kleiner Plöner See (KPS; **B3**) are located in the river–lake system of the Schwentine river. They are connected with upstream lakes Stendorfer See (*4*), Sibbersdorfer See (*5*), Gr. Eutiner See (*6*; upstream KES and KPS), KES, Dieksee (*7*), Behler See (*8*) and Gr. Plöner See (*9*; upstream KPS)

CALIB rev 4.3 software (Stuiver et al., 1998) and the calibration Data Set 2. Results (Table 4) are given as calibrated years (AD).

The ^{210}Pb and ^{137}Cs values were quantified by measuring their gamma-radiation (46.5 keV and 661 keV, respectively) through a coaxial closed-ended reverse-electrode high-purity Ge-detector (40% efficiency) in a low-level arrangement. The arrangement and measuring time (2 days) is calibrated for varying filling heights of the beaker. Detection limits are 12 and 1.1 Bq/kg for ^{210}Pb and ^{137}Cs, respectively, for a typical sample weight of 10 g. The ^{210}Pb ages are calculated applying the Constant Initial Concentration Model (Robbins & Edgington, 1975). Supported ^{210}Pb was estimated from the ^{214}Pb level (352 keV) in the core. Chronology was established by referring to the known fallout history of ^{137}Cs (Table 3). The key events are (1) the Chernobyl accident 1986, (2) the

Table 1 Limnological key characteristics of the investigated lakes Bistensee (BIS), Kellersee (KES), Stolper See (STO), Tresdorfer See (TRE), Kleiner Ploner See (KPS) and Gudower See (GUD)

	BIS	KES	STO	TRE	KPS	GUD
Maximum depth (m)	14.7	25.8	14.6	14.7	31.4	9.7
Mean depth (m)	7.5	11.7	6.9	7.6	9.0	4.9
Lake area (km^2)	1.44	5.50	1.40	1.17	2.39	0.70
F (gradient of depth)	2.14	3.13	2.18	2.30	4.40	1.64
WV ($km^2\ 10^{-6}\ m^{-3}$)/subtype$_M$	2.1/10.1	2.3/10.1	6.2/10.1	7.3/10.1	18.6/10.2	17.3/10.2
Retention time (d)/subtype$_S$	544/10.1	438/10.1	192/10.2	146/10.2	73/10.2	183/10.2
Tropic index/classification (LAWA)	3.08/ve	2.81/e	2.76/e	2.85/e	2.88/e	3.52/h
Based on *n* analyses; year	6/2005	4/2002	4/2001	6/2005	11/2002	5/2003
TN (mg/l)	1.73	1.33	1.00	2.99	0.84	3.43
TP (μg/l)	47.0	63.5	58.2	44.0	85.2	91.2
Chla (μg/l)	22.96	13.81	16.85	18.44	13.14	49.53

F (gradient of depth) = maximum depth (m)/4.785(effective length (km) + effective width (km)^{028} according to LAWA (1999)

WV watershed to volume ratio, subtype$_M$ according to Mischke et al. (2007), subtype$_S$ according to Schaumburg et al. (2007) (for more details see text), trophic index, and classification according to LAWA (1999), *lm* low mesotrophic, *m* mesotrophic, *e* eutrophic, *ve* very eutrophic, *h* hypereutrophic

Table 2 Abbreviation (abv) of lake names, coring position, water depth of coring site and core length of the six study lakes from northern Germany

Lake	Abv.	Coring position E	N	Water depth (m)	Core length (cm)
Bistensee	BIS	9°41′ 55.85″	54°23′ 64.43″	11.2	150
Kellersee	KES	10°35′ 41.09″	54°10′ 42.49″	16.0	195
Stolper See	STO	10°13′ 96.03″	54°07′ 78.49″	11.1	167
Tresdorfer See	TRE	10°28′ 21.04″	54°13′ 59.81″	14.0	202
Gudower See	GUD	10°46′ 02.04″	53°32′ 99.75″	9.2	191
Kleiner Ploner See	KPS	10°23′ 18.44″	54°09′ 95.41″	17.8	197

Table 3 Sediment depth (cm) of ^{137}Cs peaks of known fallout-events from the six study lakes

Year of characteristic fallout events	BIS	KES	STO	TRE	KPS	GUD
2005	0	0	0	0	0	0
1986	s.d.	8.75	11.25	18.75	36.25	23.75
1963	s.d.	16.30	32.50	38.75	51.25	41.25
1954	s.d.	30.00	47.50	48.75	58.75	51.25
1945	s.d.	n.d.	n.d.	n.d.	68.75	68.75

For lake abbreviations see Table 2 (Year of characteristic fallout events)

n.d. not detectable, *s.d.* surface sediments disturbed

maximum of fallout deposition from atmospheric nuclear weapon testing (1963), and, at a lower level, of 1959 prior to the test ban, (3) the beginning of the large-scale fallout at about 1954 with the start of the fusion bomb testing and possibly (4) the faint traces of fallout starting in 1945 (UNSCEAR, 2000).

Chemical analysis

Before beginning with the analyses, 20 cm^3 of thawed sediment sample were freeze-dried, milled in an Ultra Centrifugal Mill ZM 100 (Retsch GmbH) to particle sizes <50 μm, dried at 105°C for 5 h and stored in an exsiccator. Loss-on-ignition (550°C for 2 h: LOI_{550}) was considered a measure of organic matter content. SiO_2 (μg/g) was determined using XRF-analyses of fused glass discs of Li-borate after combusting samples at 1100°C. The LOI_{550} may indicate lake productivity, and SiO_2 may indicate autochthonous productivity (diatoms and chrysophytes) as well as allochthonous influxes into the lake; the latter especially during times when SiO_2 is not correlated to diatom concentration.

Pollen analysis

For pollen analysis, 0.5 cm^3 sediment per sample was prepared according to standard procedures: boiling in HCl (25%), KOH (10%), sieving (meshes: 120 μm), treatment with HF (38–40%) (if necessary), acetolysis, preparation with ethyl alcohol and butanol, and mounting in silicon oil (Faegri & Iversen, 1975). Analysis of pollen samples was carried out with a Zeiss JenaMed Variant and Axiolab light microscope with a magnification of 400× and enumerated to a sum of at least 600 pollens. The pollen sum includes arboreal pollen and upland non-arboreal pollen. The relative pollen abundances were calculated to this pollen sum. Before begiining the pollen preparation, *Lycopodium clavatum*-pills (BATCH-number 938934, 10679 spores, Laboratory of Quaternary Geology University Lund, Sweden) were added as an exotic marker (Stockmarr, 1971). Identification and nomenclature of pollen types is performed as per Beug (2004) = b, Fægri (1993) = f, Moore et al. (1991) = m, Punt (1976), Punt & Clarke (1980, 1981, 1984), Punt & Blackmoore (1991) and Punt et al. (1988, 1995) = p. The abbreviations are added to the pollen types in the figures as suggested by Joosten & de Klerk (2002). Pollen assemblage zones (PZ) were identified using local pollen and the supra-regional pollen zonation according to Overbeck (1975) and according to Wiethold (1998).

Diatom analysis

Diatom sample preparation was based on Kalbe & Werner (1974) and Battarbee & Kneen (1982) using 0.1 g sediment, dried at 105°C. Samples were treated with 10 ml of 30% HCl, washed twice with distilled water, followed by centrifugation (12 min, 4000 revolutions min^{-1}) and treated with 10 ml H_2SO_4, 1 ml saturated $KMnO_4$ and a few drops of oxalic acid. The samples were rinsed again prior to adding microspheres (from V. Jones, University College London) and embedding diatoms into Naphrax®. Concurrent to counting a minimum of 500 valves, the microspheres were enumerated. Diatom concentration as '*n* valves/g dry weight' was calculated using the number of enumerated microspheres according to Battarbee & Kneen (1982). The identification of diatoms was based on Krammer & Lange-Bertalot (1986, 1988, 1991a, b); Lange-Bertalot & Moser (1994) and Krammer (1997a, b, 2000, 2002, 2003).

Diatom-based statistical analyses

A merged European training set was used for the reconstruction of diatom-inferred total phosphorous (DI-TP) levels of the pelagial, based on a local data set from northern Germany (MV, $n = 84$, Adler & Hübener, unpublished) and the EDDI-combined TP data set (European Diatom Database Initiative) (Battarbee et al., 2000, current data at http://craticula.nc.ac.uk/eddi/jsp/). DI-TP levels were reconstructed using the moving-window approach (MW). The MW approach (Hübener et al., 2008) is based on the identification of a certain number of the nearest neighbours between core- and training set diatom assemblages. For each fossil sample, the nearest neighbours were identified using Euclidean Distances derived from the sample scores of the first three DCA-axes (CANOCO 4.5, ter Braak and Smilauer 2002) of the merged MV-EDDI training set. Based on these distances, different "window sizes" (40, 60, …160 nearest samples) were identified for each fossil sample. These sample-specific training sets were compared based on error statistics (R^2, average bias, maximum bias and RMSEP—all error parameters are based on tenfold cross validation) of two related transfer functions (Weighted averaging Partial least squares, WA-PLS and Maximum Likelihood, ML) in each case. The training set and the transfer function with the best statistics were chosen for the TP reconstruction of each fossil sample from the six study lakes using an automated selection and the computer program R 2.4.1. (Hübener et al., 2008; Adler, 2008). Sample-specific training set performances were: mean $R^2 = 0.651–0.817$ and mean $R^2_{xval} = 0.462 - 0.491$. For the TP reconstructions of Lake GUD, the MW was run without *Aulacoseira subarctica* (O. Müller) Haworth, owing to the special dystrophic conditions of the lake (see results Lake GUD and discussion 1.2).

Significant zones of different diatom assemblages over time (diatom zones; DZ) were identified for the core samples of each lake using the cluster analysis-constrained incremental sum of squares (CONISS, Grimm, 1987). Significant differences ($P < 0.05$, Leps & Smilauer, 2003) of the DZ were tested using Monte-Carlo permutation test.

The combination of pollen zones (PZ), diatom zones (DZ) and DI-TP levels identified overall zones (OZ) of significant overall changes over time.

Identification of typical reference and recent diatom assemblages

A principal component analysis (PCA) was used to identify the diatom taxa of each core that best and typically represent (1) trophic state start (possibly reference-) conditions and (2) recent conditions (which mainly differed from background conditions). An example of the method is given in Fig. 2 using Lake Stolper See (STO). The 15 most important taxa of the core (according to total inertia explained) were plotted along axes one and two. In Lake STO, the 15 taxa presented in the PCA explain 60.4% total inertia of the 122 taxa identified from the whole core. The length and direction of the taxon vector identifies whether the taxon dominated the start (reference) or recent assemblages. Otherwise, the long vector lengths and the direction of the arrows (pointing between start and recent conditions) of *Cyclostephanos dubius* (Fricke) Round, *Aulacoseira granulata* var. *granulata* (Ehr.) Simonsen, *A. granulata* var. *angustissima* (O. Müller) Simonsen and *Fragilaria crotonensis* Kitton suggest that a strong overall presence of these taxa and that they are dominating the assemblages during both times (Fig. 2). Accordingly, dominant and sub-dominant taxa of start (reference) and recent assemblages, respectively, were identified for all the six lakes.

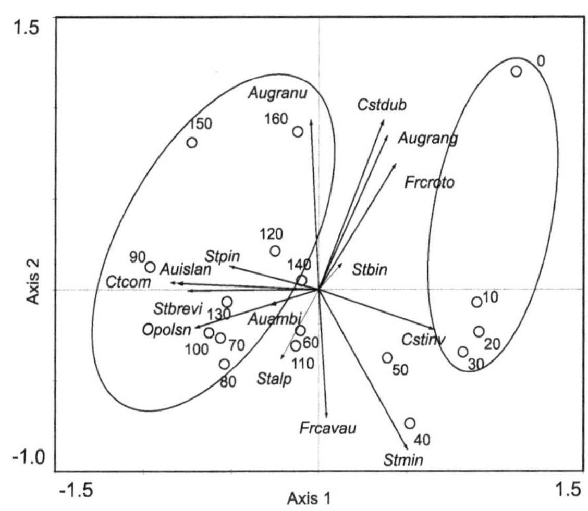

Fig. 2 Principal component analysis of 17 fossil core samples (0–160 cm) of Lake Stolper See. Also listed are the 15 taxa that explain most of the variation in the whole core (total inertia). Ellipses circle the samples representing starting conditions (160–70, except 110 cm) and recent conditions (30–0 cm), respectively. Auambi: *Aulacoseira ambigua*, Augranu: *A. granulata* var. *granulata*, Augrang: *A. granulata* var. *angustissima*, Auislan: *A. islandica*, Cstdub: *Cyclostephanos dubius*, Cstinv: *C. invisitatus*, Ctcom: *Cyclotella comensis*, Stbrevi: Staurosira *brevistriata*, Stpin: S. *pinnata*, Stbin: S. *binodis*, Frcavau: *F. capucina* var. *vaucheriae*, Frcroto: *F. crotonensis*, Opolsn: *Opephora olsenii*, Stalp: *Stephanodiscus alpinus*, Stmin: *S. minutulus*

Results

^{210}Pb, ^{137}Cs and ^{14}C chronology

The ^{137}Cs and ^{210}Pb peaks both suggest an undisturbed sedimentation of the most recent ~45 cm in five of the six study lakes. The ^{137}Cs peaks were distinct and excess ^{210}Pb followed a more or less regular exponential decrease with sediment depth. In contrast, ^{210}Pb and ^{137}Cs activities suggest some major disturbance in the upper sediments of Lake BIS.

Chronologies were established using the sediment depth of ^{137}Cs peaks and known ^{137}Cs fallout-events (Table 3). Decisions about macrofossil ages that hit the atmospheric ^{14}C-plateau (Table 4) were made in accordance with typical pollen events known from other regional and dated pollen stratigraphies (Averdieck, 1974, 1978; Behre, 1976; Wiethold, 1998), for example, the commencement (early fourteenth century) and the maximum (early eighteenth century) of *Fagopyrum* pollen, the occurrence of *Aesculus* pollen (first half of seventeenth century) or the increase of *Picea* pollen caused by cultivation of heathlands from both ~1750 to ~1850 onwards. The age of the oldest sediments of each core was extrapolated using sedimentation rates from the two oldest known dates. Thus, the oldest sediments from the six cores represent from ~1760 BP (KES) to ~290 BP (KPS). There are, however, large uncertainties associated with the extrapolations. Overall sedimentation rates of the whole cores ranged from ~0.11 cm/y (KES) to ~0.66 cm/y (KPS).

Catchment developments and lake histories

Lake Bistensee

The sediment core of Lake Bistensee (150 cm, Table 2) represents the lake and catchment development since early Medieval Times (~AD 950, Fig. 3). The core was divided into three overall zones (OZ I–III) according to changes in pollen assemblages, diatom assemblages and diatom-inferred total phosphorus (DI-TP) levels (Fig. 3). The indicators suggest near-natural conditions in the catchment area during Early to Mid-Medieval Times (OZ I, ~AD 950–1375, 150–100 cm). High tree pollen abundances (91–95%) indicate low-to-absent catchment usage in the oldest represented time (~AD 950). Subsequently, settlement-associated pollen started to slightly increase in the early thirteenth century, which suggests the commencement of relatively low settlement activities in Medieval Times. High relative abundances of oligotrophic *Aulacoseira subarctica* and benthic *Staurosira* species, as well as-low to-medium mesotrophic nutrient levels according to reconstructed TP levels (21–37 μg/l TP, Fig. 3) further support low anthropogenic impact on the lake until ~AD 1375 (100-cm sediment depth).

Clearing of the catchment and increasing sediment input from the catchment area began during Overall Zone II (OZ II, ~AD 1400–1750, 90–40 cm) in late Medieval Times, as indicated by continuously decreasing relative abundances of tree pollen, increasing cereal pollen and rising SiO_2 levels in the sediments. At-AD 1450, there was the first, significant shift in the diatom assemblages from a planktonic community dominated by *Aulacoseira subarctica* to higher abundances of *A. ambigua* (Grunow) Simonsen and *A. granulata* (Ehrenberg) *Simonsen* (Fig. 3). Consequently, DI-TP increased from low mesotrophic to mesotrophic nutrient levels.

During the last zone (OZ III, since ~AD 1825, 30–0 cm) human impact has intensified strongly, which has led to eutrophic nutrient levels since ~AD 1875 (20 cm). In the diatom assemblages, the formerly dominating oligotrophic to mesotrophic *Aulacoseira subarctica* is no longer detectable and is replaced by eutrophic *A. granulata*. In the most recent 10 cm (~AD 1950–2005), another significant shift took place in the diatom assemblages (CONISS, Fig. 3), as the relative abundances of eutrophic species such as *Stephanodiscus minutulus* (Kützing) Cleve & Müller, *Cyclostephanos dubius* and *Asterionella formosa* Hassall (Table 5) increased, indicating a further intensification of land use in the catchment.

Lake Kellersee

The sediment core of Lake Kellersee (140 cm) represents the last ~1615 years. Four overall zones (OZ I–IV) could be identified. The indicators suggest relatively low to only minimally anthropogenic impact in OZ I and II (~AD 390–1525) concurrent with oligo- to low mesotrophic conditions, and high-to-very high anthropogenic impact in OZ III and IV (~AD 1550–2005) concurrent with eutrophic to hypertrophic conditions (Fig. 4).

At the beginning of OZ I, the tree pollen (up to 97%) and low SiO_2 input from the catchment area suggest low settlement activities and a reforestation of once agriculturally used areas during the migration period. The dominant diatoms (*Aulacoseira subarctica* and *Cyclotella comensis* Grunow in Van Heurck) and DI-TP levels suggest oligo- to low mesotrophic conditions (2–23 μg TP/l) (Fig. 4). Decreasing *Fagus* abundances from ~700 to 1100 AD indicate continuous woodland clearing around Lake Kellersee. Concurrently, decreasing abundances of *Aulacoseira subarctica*, and increased abundances of *Cyclotella comensis*, *Aulacoseira islandica* (Müller) Simonsen and *Cyclostephanos dubius* indicate slightly increasing nutrient levels (Fig. 4).

The following OZ II (~AD 1100–1525, 100–75 cm) is characterised by a rise of indicators that suggest increased but still moderate settlement activities (see cereals, hemp and other settlement-associated pollen) compared to OZ I. Continued relatively high abundances of the diatoms *Cyclotella comensis* and *C. comta* (Ehrenberg) Kützing, suggests oligo-mesotrophic lake water conditions, although eutrophic taxa such as *Stephanodiscus minutulus* also

Table 4 Sediment depth and age of terrestrial macrofossils of the six study lakes determined with radiocarbon dating

Lake	Sediment depth (cm)	Lab-code	Selected material	Radiocarb age (^{14}C years BP)	Calibrated age (cal years)	Two sigma range (cal years AD)	Probability (%)
BIS	80–82.5	KIA 28008	Leaf fragment, bud-scale	456 ± 23	1439	**1418–1472**	95.4
	115	KIA 28007	Leaf fragment, bud-scale	828 ± 32	1219	**1160–1277**	95.4
KES	85	KIA 28014	Leaf fragment, bud-scale, nut Alnus	498 ± 21	1427	**1408–1440**	95.4
	147.5	KIA 28013	Leaf fragment, bud-scale, nut Alnus	1773 ± 49	244	**131–384**	95.4
STO	117.5–120	KIA 28019	Leaf fragment	128 ± 29	1690, 1728, 1811, 1921	1906–1940	14.3
						1801–1894	42.9
						1675–1776	38.2
	130	KIA 28020	Leaf fragment	213 ± 22	1664, 1784, 1787	1938–1955	19.1
						1761–1803	43.9
						1648–1677	32.4
	152.5	KIA 28021	Leaf fragment	178 ± 45	1674, 1777, 1800, 1941, 1954	1912–1955	17.3
						1826–1885	12.5
						1718–1823	46.3
						1650–1708	19.3
TRE	100–102.5	KIA 28017	Leaf fragment, bud-scale, moss	105 ± 38	1705/1818/1829/1882/1913	**1803–1938**	63.9
						1677–1761	31.5
	145	KIA 28016	Leaf fragment	252 ± 24	1652	1945–1955	5.7
						1780–1797	12.4
						1634–1670	71.6
						1530–1547	5.7
	195	KIA 28015	Leaf fragment	447± 24	1441	**1422–1478**	95.4
KPS	80–82.5	KIA 28004	Leaf fragment, bud-scale	91± 36	1890, 1909	**1804–1935**	67.7
						1679–1739	27.7
	122.5	KIA 28003	Wood	94 ± 22	1888, 1910	**1811–1921**	69.6
						1690–1728	25.8
	185	KIA 28005	Leaf fragment	55 ± 22	Intercept point n.d.	1875–1917	80.1
						1813–1847	15.3
GUD	90–92.5	KIA 28002	Leaf fragment, bud-scale	161 ± 23	1678/1742/1751/1757/1804/1936/1954	1917–1955	19.1
						1834–1877	6.7
						1791–1814	11.4
						1725–1783	42.9
						1666–1696	15.3
	135–137.5	KIA 28001	Leaf fragment, bud-scale, nut Betula	180 ± 27	1673/1777/1800/1942/1954	1919–1955	20.0
						1727–1812	57.2
						1656–1693	18.1
	182.5	KIA 28000	Leaf fragment	417 ± 21	1450	**1436–1489**	95.4

Bold letters: selected interval for dating. For lake abbreviations see Table 2

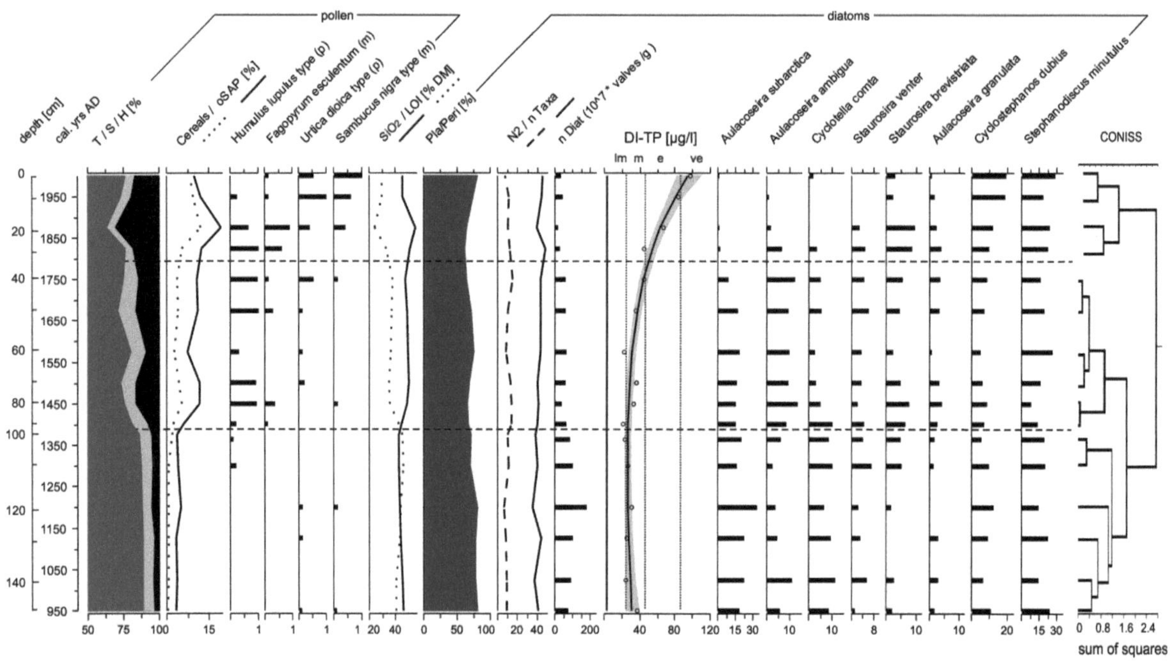

Fig. 3 Stratigraphy of sediments of Lake Bistensee: relation of pollen sum of trees (T), shrubs (S) and herbs (H); cereals (*dotted line*, %) and other settlement associated pollen (oSAP; *solid line*, %); relative abundance of indicative fossil pollen types; sedimentary SiO_2% DM (dry matter, *solid line*) and loss-on-ignition (LOI_{550}% DM; *dotted line*); relation of planktonic versus periphytic diatoms (Pla/Peri); number of detected diatom-taxa (*solid line*) and Hills-N2-diversity (*dashed line*); diatom concentration (*n* Diat/g DM); diatom-inferred total phosphorus levels (DI-TP) based on the moving-window approach, with single values (*open circles*), trend (*line*) and 95% confidence values (*grey areas*), *o* oligotrophic, *lm* low mesotrophic, *m* mesotrophic, *e* eutrophic, *ve* very eutrophic, *h* hypereutrophic: limits of trophic classification according to LAWA (1999) and Schaumburg et al. (2007); relative abundance of important diatom taxa (please note the different scales of *x*-axes) and zonation of diatom associations in the sediment core (CONISS) since AD 950. The *horizontal dashed lines* separated the overall zones OZ I (below), OZ II and OZ III

increase. Thus, DI-TP levels increased slightly but remained mesotrophic.

During OZ III settlement, activities expanded and forest clearance increased, while cereal and other settlement-associated pollen abundances were high. In the diatom assemblages, *Cyclotella comensis* was replaced by eutrophic taxa, *Aulacoseira granulata* and *Stephanodiscus minutulus*. The DI-TP levels increased from moderately eutrophic conditions at the beginning of OZ III to eutrophic conditions (78 μg/l) at the end of this period (~AD 1900; Fig. 4).

During the last period (OZ IV), all the indicators suggest a period of intense eutrophication, followed by a slight decline of anthropogenic impact. *Stephanodiscus minutulus* dominates throughout this zone. DI-TP levels reach a core maximum (~1970, 118 μm/l) at 20 cm, corresponding closely with measured epilimnetic TP values (LAWK, 1993). The uppermost diatom assemblages indicate a slight decrease in TP levels during the most recent times (5 and 0 cm).

Lake Stolper See (STO)

The sediment core (160 cm) represents the last ~370 years. Three overall zones could be differentiated. Anthropogenic impact was already distinct in OZ I, intensified in OZ II and remained very high in OZ III (Fig. 5).

Anthropogenic activities were already evident at the beginning of OZ I (~AD 1630–1900, 160–70 cm). The relative abundance of cereals was high and tree pollen abundances were low (65%), decreasing to 60% by the end of the eighteenth century and correspond with massive deforestations for glass foundries, charcoal burning and brick factories (Hase, 1983).

Table 5 List of the 15 diatom taxa that explain most of the total inertia (TI_{15}) in a principal component analysis (an example is given in Fig. 2) of the diatom assemblages of each study lake, listed in order of importance

Lake Main characteristics	Starting (reference) communities Dominant species	Recent communities Dominant species	Lake Main characteristics	Starting (reference) communities Dominant species	Recent communities Dominant species
BIS	*Aulacoseira subarctica*	*Aulacoseira islandica*	KES	*Aulacoseira subarctica*	*Stephanodiscus minutulus*
Subtype$_{MS}$ 10.1	*Cyclotella comta*	*Nitzschia palea*	Subtype$_{MS}$ 10.1	*Cyclotella comensis*	*Stephanodiscus alpinus*
Isolated	*Cyclotella comensis*	*Stephanodiscus minutulus*	Upstream lakes	*Aulacoseira islandica*	*Cyclostephanos dubius*
TI_{15} 60.6%	*Aulacoseira ambigua*	*Cyclostephanos dubius*	TI_{15} 75.3%	*Cyclotella comta*	*Aulacoseira granulata* v. *granulata*
	Staurosira construens	*Stephanodiscus alpinus*		*Stephanodiscus alpinus*	*Aulacoseira granulata* v. *angust.*
	Staurosira venter	*Fragilaria capucina* v. *vaucheriae*		*Cyclostephanos dubius*	*Aste formosa*
	Stephanodiscus minutulus	*Asterionella formosa*			*Aulacoseira ambigua*
	Cyclostephanos dubius				
	Staurosira binodis				
	Fragilaria capucina v. *vaucheriae*				
	Asterionella formosa				
STO	*Aulacoseira islandica*	*Stephanodiscus minutulus*	TRE	*Cyclotella comensis*	*Cyclostephanos invisitatus*
Subtype$_{M}$ 10.1	*Cyclostephanos dubius*	*Cyclostephanos dubius*	Subtype$_{M}$ 10.1	*Aulacoseira subarctica*	*Aulacoseira granulata* v. *angustissima*
Subtype$_{S}$ 10.2	*Staurosira brevistriata*	*Cyclostephanos invisitatus*	SubtypeS 10.2	*Cyclotella comta*	*Asterionella formosa*
Lake influenced	*Fragilaria crotonensis*	*Aulacoseira granulata* v. *granulata*	Isolated	*Asterionella formosa*	*Cyclostephanos dubius*
TI_{15} 60.5%	*Stephanodiscus minutulus*	*Aulacoseira granulata* v. *angust.*	TI_{15} 73.2%	*Fragilaria crotonensis*	*Stephanodiscus minutulus*
	Aulacoseira granulata v. *granulata*			*Tabellaria floculosa*	*Stephanodiscus hantzschii*
	Cyclotella comensis			*Stephanodiscus minutulus*	*Nitzschia palea*
	Aulacoseira ambigua			*Stepanodiscus alpinus*	
	Aulacoseira granulata v. *angust*				
	Opephora olsenii				
	Staurosira pinnata s.l.				

Table 5 continued

Lake Main characteristics	Starting (reference) communities Dominant species	Recent communities Dominant species	Lake Main characteristics	Starting (reference) communities Dominant species	Recent communities Dominant species
KPS	*Cyclotella comensis*	*Stephanodiscus minutulus*	GUD	*Aulacoseira subarctica*	*Aulacoseira subarctica*
Subtype$_{MS}$ 10.2	*Aulacoseira islandica*	*Aulacoseira islandica*	Subtype$_{MS}$ 10.2	*Stephanodiscus minutulus*	*Stephanodiscus minutulus*
Upstream lakes	*Staurosira brevistriata*	*Aulacoseira ambigua*	Isolated, dystrophic	*Aulacoseira ambigua*	*Aulacoseira granulata* v. *angust.*
TI$_{15}$ 65.9%	*Aulacoseira ambigua*	*Denticula tenuis*	TI$_{15}$ 56.2%	*Staurosira brevistriata*	*Fragilaria crotonensis*
	Stephanodiscus binderanus	*Aulacoseira granulata* v. *angustissima*		*Staurosira construens*	*Cyclostephanos dubius*
	Denticula tenuis	*Staurosira pinnata* s.l.		*Staurosira venter*	*Nitzschia fonticula*
	Cyclostephanos dubius	*Staurosira venter*		*Opephora olsenii*	*Cyclostephanos invisitatus*
	Staurosira venter	*Cyclostephanos invisitatus*		*Navicula scutellata*	
	Staurosira harrisonii	*Fragilaria crotonensis*		*Achnanthes joursacense*	
	Opephora olsenii			*Staurosira binodis*	
				Staurosira pinnata s.l.	

Given taxa are typical of starting or recent (2005) diatom assemblages

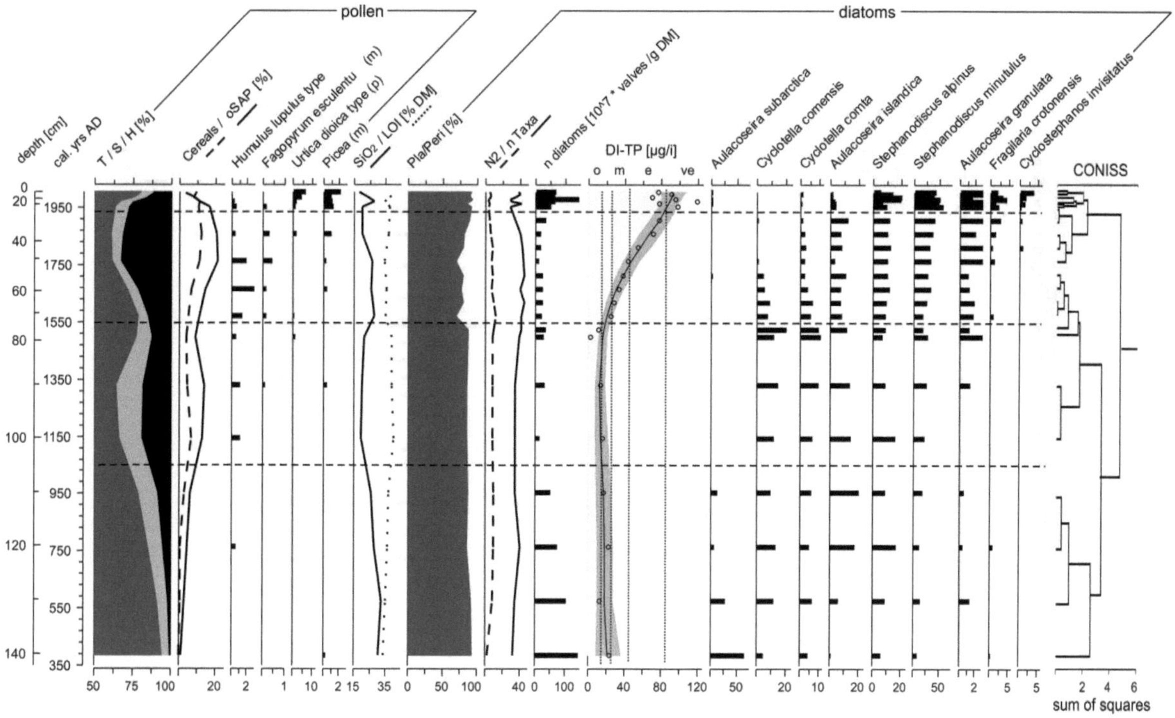

Fig. 4 Stratigraphy of sediments of Lake Kellersee since AD 390. Selected parameters from palynological, geochemical and diatom investigations. The *horizontal dashed lines* separated the overall zones OZ I (below), OZ II, OZ III and OZ IV. For abbreviations see Fig. 3

Reforestation of the catchment area commenced in ~AD 1800 (Hase, 1983) and is evident by increased relative abundances of coniferous trees (*Pinus*) and decreasing erosion (Fig. 5). The diatom assemblages of OZ I are characterised by high abundances of eutrophic plankton, such as *Stephanodiscus minutulus*, *Cyclostephanos dubius* and *Aulacoseira granulata.* Oligotrophic to mesotrophic taxa occurred only in low abundances. Thus, DI-TP values indicate mesotrophic to moderately eutrophic conditions (mean 53.9 μg/l) for this zone. The overall increase of terrestrial nutrient levels in OZ II (~AD 1910–1960, 60–40 cm) is indicated by increased abundances of nitrogen-loving *Sambucus nigra.* The oligo-mesotrophic diatoms *C. comensis* and *A. islandica* decreased markedly (Fig. 5). Thus, DI-TP levels increased significantly to eutrophic levels. The recent OZ III (since ~AD 1965, 30–0 cm) reflects very intense catchment usage, indicated by distinct increase of nitrogen loving species, such as *Urtica dioica* and *Sambucus nigra*. DI-TP values increased to very eutrophic conditions (>85 μg/l) with a maximum since ~AD 1965 (>100 μg/l, 30 cm, Fig. 5).

Lake Tresdorfer See (TRE)

The sediment core (202 cm) represents three overall zones during the last ~550 years with moderate anthropogenic impact in OZ I, increasing impact in OZ II and very high impact in OZ III (Fig. 6). OZ I (from AD 1450 to ~1750, 200–130 cm) is characterised by a distinct increase of catchment activity from moderate in Late Medieval Times to more intensive usage starting in the early eighteenth century (Fig 6). The DI-TP levels indicate stable, oligotrophic to slightly mesotrophic nutrient levels (mean DI-TP 200–140 cm: 12.7 μg/l, $n = 7$) (Fig. 6), as oligotrophic to mesotrophic *Cyclotella comensis* (e.g. Werner & Smol, 2006) dominated the assemblages.

In OZ II (~AD 1775–1900, 120–80 cm) pollen assemblages indicate a further increase of anthropogenic activities in the catchment area. Moderately eutrophic to eutrophic taxa such as *Stephanodiscus neoastraea* Hakansson & Hickel replaced *Cyclotella comensis* in the diatom plankton. Subsequently, *Cyclotella comensis* occurred only in very low

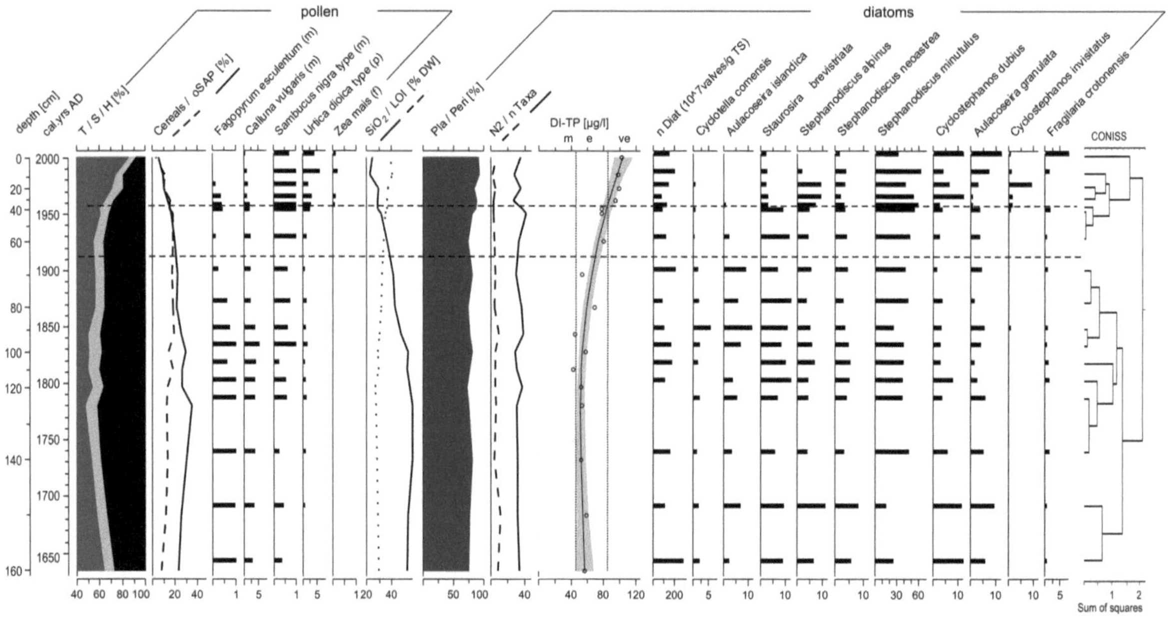

Fig. 5 Stratigraphy of sediments of Lake Stolper See since AD 1600. Selected parameters from palynological, geochemical and diatom investigations. The *horizontal dashed lines* separated the overall zones OZ I (below), OZ II and OZ III. For abbreviations, see Fig. 3

abundances in OZ II and was not evident in OZ III. Accordingly, trophic state increased to moderately eutrophic conditions by ~AD 1900 (DI-TP: 48 µg/l, 80 cm) (Fig. 6). During OZ III (since ~AD 1910, 70–0 cm), nitrogen-rich soil-preferring plants such as *Urtica dioica*, increased significantly and indicate an intensification of catchment usage. The diatom assemblages are dominated by planktonic, eutrophic *Stephanodiscus hantzschii* Grunow in Cleve & Grunow and *Cyclostephanos invisitatus* (Hohn & Hellerman) Theriot, Stoermer & Hakansson (Fig. 6). However, ^{210}Pb activities of the most recent 12.5–0 cm suggest possible mixing of these sediments. Therefore, increased DI-TP levels in the surface sediments should be interpreted with caution. Nevertheless, the overall DI-TP levels compare well with measured epilimnetic TP levels in 1999 ($n = 4$, mean of 53.7 µg/l; LANU, 2002) and in 2005 ($n = 6$, mean 44.5 µg/l, Table 1). Both measured TP means were clearly lower than expected according to estimations of 80 µg/l using catchment input calculations (LAWA, 1999).

Lake Kleiner Plöner See (KPS)

The 190-cm-long core only covers the time period since the early eighteenth century. Four overall zones (OZ I–IV) could be identified (Fig. 7). Anthropogenic impact was high-to-very high throughout the four zones. Decreasing *Fagus* and *Quercus*, high abundances of crop pollen, other settlement-associated pollen and the nitrogen-loving *Sambucus nigra* indicate a high and increasing usage of the catchment area since the beginning of OZ I (~AD 1720–1760, 190–160 cm). In the diatom assemblages, both oligotrophic to mesotrophic taxa (*Aulacoseira islandica* and *Cyclotella comensis*) and eutrophic taxa (*Cyclostephanos dubius*, *Aulacoseira ambigua*) dominated the assemblages (Fig. 7).

During OZ II (~AD 1775–1900, 150–80 cm), two important events influenced the lake. First, the reforestation with conifer trees (Hase, 1972) and second, the human-induced stepwise (1830 and 1881; Ohle, 1979) decrease of the water level by altogether ~1.5 m of upstream Lake Großer Plöner See, as documented by increased SiO_2 content of the sediments during these periods (Fig. 7). However, trophic state remained at mesotrophic levels until OZ III (~AD 1910–1960, 70–50 cm), when moderately eutrophic conditions were reached as a consequence of a more intense usage of the catchment area. The hypereutrophic *S. minutulus* occurred with more than 50% relative abundance in the diatom assemblages. Accordingly, DI-TP levels

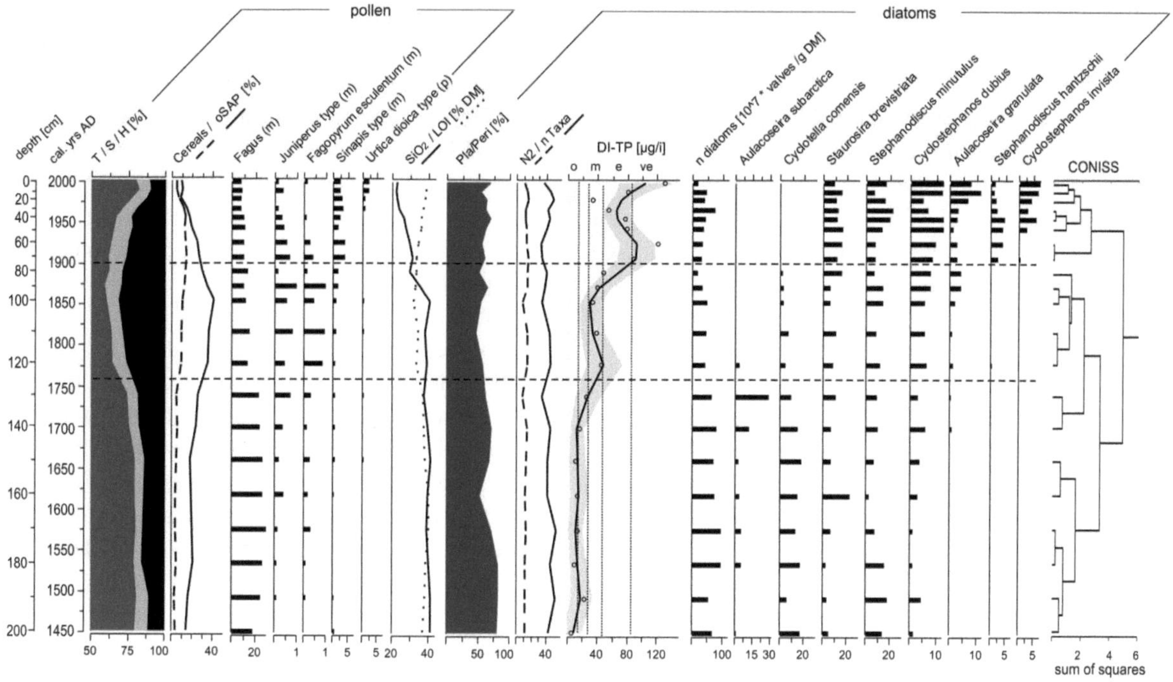

Fig. 6 Stratigraphy of sediments of Lake Tresdorfer See since AD 1450. Selected parameters from palynological, geochemical and diatom investigations. The *horizontal dashed lines* separated the overall zones OZ I (below), OZ II and OZ III. For abbreviations see Fig. 3

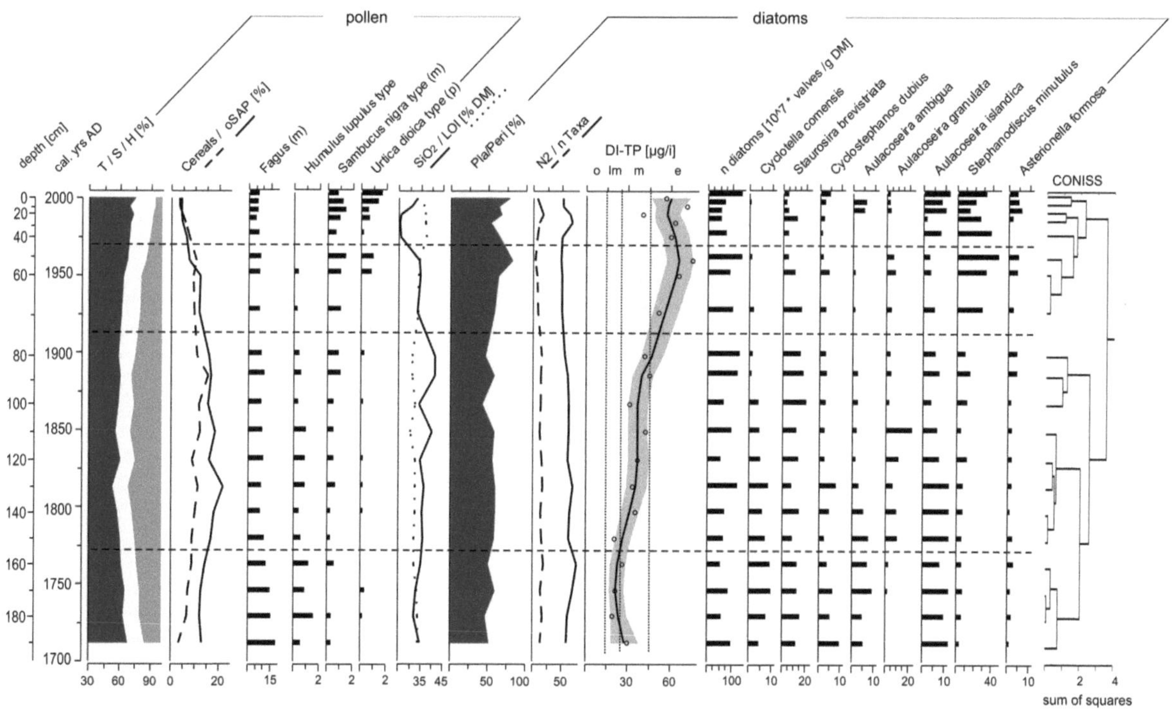

Fig. 7 Stratigraphy of sediments of Lake Kleiner Plöner See since AD 1700. Selected parameters from palynological, geochemical and diatom investigations. The horizontal dashed lines separated the overall zones OZ I (below), OZ II, OZ III and OZ IV. For abbreviations, see Fig. 3

reached their core maximum by ~AD 1960 (50 cm; 75 μg/l; Fig. 7). The most recent zone (OZ IV, since ~AD 1965, 40–0 cm) is characterised by increased abundances of moderately eutrophic taxa, such as *S. alpinus*, *Asterionella formosa*, *Aulacoseira islandica* and in the benthos *Staurosira venter* (Ehrenberg) Cleve & Möller. Thus, DI-TP levels decreased slightly in this zone, but remained at moderately eutrophic levels (~44–72 μg TP/l).

Lake Gudower See

The sediment core of Lake GUD (189 cm, Table 2) represents the history of the catchment area and the lake since Late Medieval Times (~AD 1410) (Fig. 8). According to pollen assemblages, diatom assemblages and DI-TP levels, three overall zones (OZ I–III) could be identified. Anthropogenic activities in the catchment area increased from moderate to very high during this time.

Lake GUD is dystrophic due to high inputs of humic substances (HS) from the catchment area (mean DOC 21.0 mg/l in 2003, $n = 15$, LANU, 2003). Humic substances may influence the light penetration and composition, photosynthetic electron transport and phosphate availability (Nürnberg & Shaw, 1999; Uhlmann & Horn, 2001). Thus, a given set of HS will induce a specific set of primary producers that are tolerant to the exposure of certain qualities and quantities of these HS (Steinberg et al., 2006). Therefore, the lake response to eutrophication and typical diatom taxa differ significantly to the other study lakes.

At the beginning of OZ I (~AD 1410 to ~1830, 187–100 cm), the fossil pollen indicate that logging activities already occurred in the catchment area. The increase of *Humulus lupulus*-type—pollen document—suggested that Lake GUD was used for hemp retting to accelerate breakdown processes for the fibre production (Dörfler, 1990). This practice leads to considerable eutrophication in the corresponding lakes (Beales, 1980; Müller & Kleinmann, 1998; Schwarz, 2006) and was also evident in Lake GUD by a concurrent increase of DI-TP levels from moderately eutrophic to eutrophic levels at the end of OZ I (Fig. 8). Nonetheless, the diatom assemblages of OZ I

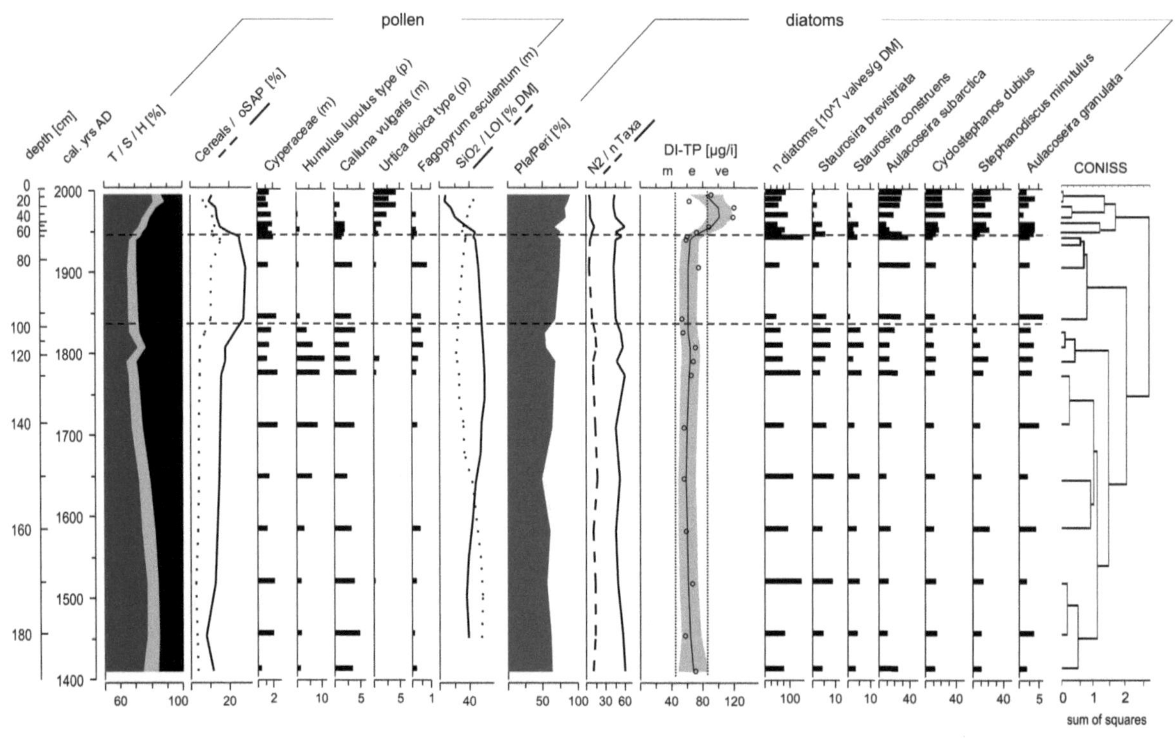

Fig. 8 Stratigraphy of sediments of Lake Gudower See since AD 1410. Selected parameters from palynological, geochemical and diatom investigations. The horizontal dashed lines separated the overall zones OZ I (below), OZ II and OZ III. For abbreviations, see Fig. 3

were dominated by the nordic-alpine *Aulacoseira subarctica*, which usually occurs in oligotrophic waters in areas with cold winters (Hustedt, 1930; Cleve-Euler, 1951).

Zone OZ II (~AD 1850–1950, 90–60 cm) was characterised by a reforestation with fast-growing coniferous trees, the cultivation of corn and decreased cultivation of buckwheat. Phosphorus (DI-TP) levels indicate relatively stable, moderately eutrophic conditions throughout OZ II. Since ~AD 1955 (OZ III, 50 cm upward), littoral diatoms (*Staurosira construens, S. brevistriata*) decreased and thus suggest decreased light levels in the littoral, possibly caused by increasing input of HS or maybe by increased plankton abundances of all algae (not just diatoms) or cyanobacteria. Moderately eutrophic DI-TP levels were lower than measured TP, probably due to lower availability of phosphate for algae in dystrophic lakes. Measured epilimnetic TP concentrations indicate eutrophic conditions with 154 μg TP/l in 1987/1988 ($n = 14$) and 91.2 μg/l (mean, $n = 5$, Table 1) in 2003 (LANU, 2005).

Characterisation of reference and recent diatom assemblages

According to the fossil pollen assemblages, anthropogenic catchment usage was already moderate (Lake TRE) to high (lakes STO, KPS, GUD) in the earliest time period represented by the sediments of the study lakes. The oldest sediments of Lake BIS and KES seem to represent natural background conditions, with the typically oligo-mesotrophic *Cyclotella comensis* characterising the earliest assemblages of all the study lakes. In addition, *Aulacoseira subarctica*, *Cyclotella comta*, *Staurosira construens* and S. *venter* are typical diatoms in the oldest sediments of lakes BIS, KES, TRE and GUD (Table 5). Important subdominant taxa of these lakes were *Encyonopsis microcephala* (Grunow) Krammer and *Caloneis schumanniana* (Grunow in Van Heurck) Cleve. Thus, *Cyclotella comensis* was also present in the impacted lakes STO and KPS, with minimal abundances (1.3–2.3%) in lake STO and with more importance (4.5–10.3%) in lake KPS. Taxa with higher TP optima dominate the oldest assemblages in the impacted lakes STO, KPS and GUD (*Aulacoseira islandica*, *A. ambigua*, *Cyclostephanos dubius*, *Stephanodiscus binderanus* (Kützing) Krieger and *S. minutulus*, Table 5).

The recent diatom assemblages in the six study lakes were dominated by eutrophic taxa, namely *Stephanodiscus minutulus*, *Aulacoseira granulata*, *A. islandica*, *A. ambigua*, *Discostella pseudostelligera* (Hustedt) Houk & Klee, *Cyclostephanos dubius*, *Asterionella formosa* and *Fragilaria crotonensis*. Important eutrophic subdominant taxa were *Cocconeis placentula* Ehrenberg, *Amphora inariensis Krammer* and *Diatoma vulgare* Bory (Table 5).

Generally, *Cyclostephanos dubius* is an indicator of relatively high trophic states (Bennion et al., 1996; Schönfelder et al., 2002; Bradshaw et al., 2002; Bradshaw & Anderson, 2003). This taxon was abundant in mesotrophic to eutrophic conditions in both reference and recent diatom assemblages.

Discussion

Reference conditions

The paleolimnological indicators in this study demonstrate that the point in time of significant and continuous eutrophication may vary from AD 1400 to 1700 for the six study lakes representing German Lake-Type 10, part of European Ecoregion 14 (Table 6). In contrast, reference conditions were provisorily assumed to be reached prior to AD 1850 in Great Britain (European Ecoregion 18; EU, 2000; REFCOND, 2003; Bennion et al., 2004). However, this study confirms the findings from other lakes in the European central plains that suggest significant, continuous and increasing anthropogenic eutrophication since Medieval Times (Ohle, 1973; Hübener & Dörfler, 2004; Bradshaw et al., 2005b). Similarly 17 out of 21 investigated lakes in Denmark had already been eutrophic in AD 1850 (Bjerring et al., 2008), while reference conditions were given at oligotrophic (deep lakes) or mesotrophic (shallow lakes) levels (Søndergaard et al. (2005). Thus, it might be necessary to go back to early Medieval Times or even to the migration period to identify reference conditions for lakes in European Ecoregion 14.

Table 6 summarises the historic proposed and inferred trophic classification of the six study lakes from Schleswig-Holstein, Germany. The required type-specific undisturbed to only minimally impaired reference conditions (EU-Intercalibration Technical Report, 2006) and associated diatom assemblages

Table 6 Anticipated and diatom-inferred trophic state conditions in the oldest sediments of the study lakes (starting conditions)

Name of the lake	Subtype/anticipated reference acc. Schaumburg et al. (2007) –Mischke et al. (2007)	Starting conditions inferred/until
Bistensee	10.1/m–10.1/lm	**lm/1375**
Kellersee	10.1/m–10.1/lm	**o–lm/1525**
Stolper See	10.2/e–10.1/lm	e/1960
Tresdorfer See	10.2/e–10.1/lm	**o/1700**
Kleiner Plöner See	10.2/e–10.2/m	m/1900
Gudower See	10.2/e–10.2/m	e/1950

Subtypes (10.1 or 10.2) and anticipated reference conditions are given according (acc.) to Schaumburg et al. (2007) and according to Mischke et al. (2007). Diatom-inferred trophic state conditions/points in time, when starting conditions changed towards a higher trophic state

Bold letters: starting conditions correspond to reference conditions (see text). Trophic classification according to LAWA (1999)

o oligotrophic, *lm* low mesotrophic, *m* mesotrophic, *e* eutrophic

were surely identified for two lakes (BIS, KES). The four remaining lakes had already moderately (TRE) to relatively high (STO, KPS, GUD) intensities of catchment usage at the beginning of the examined time period according to the pollen assemblages. Thus, undisturbed to only minimally impaired conditions in the catchment and trophic conditions for these lakes remain unknown, but were likely of lower trophic status than the levels inferred from the oldest sediments (starting conditions in Table 6).

Diatom-inferred total phosphorus (DI-TP) levels

All the six study lakes belong to lake-type CB 1 (Carvalho et al., 2008) and are part of European ecoregion 14. The TP reference value for lake type CB 1 is 24 μg TP/l (low mesotrophic conditions) (EU-Intercalibration Technical Report, 2006). Denmark and the Netherlands also belong to European ecoregion 14. Here, the boundary between high and good ecological status for stratified lakes is set at 12.5 μg TP/l (Denmark; Søndergaard et al., 2005) and 20 μg TP/l (the Netherlands; lake type M20; STOWA, 2007), respectively. Similarly, this study suggests 15–25 μg/l for German Lake-Type 10.1. Mischke et al. (2007) and Schaumburg et al. (2007) subdivided the German Lake-Type 10 according to hydrological key characteristics and based on different TP-input and assumed graduated differences in reference conditions (Table 6). In Denmark, Søndergaard et al. (2005) further differentiate lakes categorised as CB 1 only into shallow (mean depth <3 m) and deep (stratified, mean depth >3 m) lakes. In the Netherlands lakes with a surface area >100 km^2 belong to type M21 with a TP reference of 40 μg/l (STOWA, 2007).

Theoretically, lakes BIS and KES have favourable hydrologic conditions for a naturally low TP-input from the catchment area and were thus classified as Subtype 10.1 according to Schaumburg et al. (2007) (high retention time; RT) and Mischke et al. (2007) (low watershed to volume ratio; WV). In this study, the oldest sediments very likely reflect undisturbed background conditions in both lakes. The DI-TP levels indicate oligotrophic to low mesotrophic (KES) and low mesotrophic (BIS) reference conditions. These reference conditions were in accordance with the adoptions of the EU-Intercalibration Technical Report (2006) and were slightly better than the proposed mesotrophic (Schaumburg et al., 2007) and low mesotrophic (Mischke et al., 2007) conditions for German Lake-Type 10.1 (Table 6). In other lakes in northern Germany, characterised by even better hydrological parameters (lower WV-ratio), this low mesotrophic reference value was confirmed (Dreßler et al., 2006; Schwarz, 2006).

In Lake BIS, DI-TP reference conditions lasted throughout OZ I (Fig. 3). Conditions then changed in OZ II through increased anthropogenic settlement activities (according to fossil pollen assemblages) to higher trophic state levels in ~AD 1450 (Fig. 3). In Lake KES, trophic state levels left the oligotrophic to low mesotrophic reference conditions in ~AD 1525 (Fig. 4) due to intensive settlement activities.

Lakes STO and TRE had a higher theoretical natural TP-input from the catchment area than the previous two study lakes (BIS and KES), due to a higher WV and lower RT. Accordingly, lakes STO and TRE are classified as Subtype 10.2 according to Schaumburg et al. (2007) (relatively low RT), but still as 10.1 according to Mischke et al. (2007) (still relatively low WV) (Table 1). Thus, Schaumburg et al. (2007) suggest eutrophic reference conditions, while Mischke et al. (2007) still assume low mesotrophic reference conditions for lakes STO and TRE. In both catchments, pollen assemblages suggest already moderate (TRE) to intense (STO) usage of the catchment area in the oldest represented time period of the cores (~AD 1450 and ~AD 1600, respectively). The moderate eutrophic TP values and associated diatom assemblages in Lake STO reflect these usages of the catchment and suggest that the proposed eutrophic reference conditions from Schaumburg et al. (2007) for German Lake-Type Subtype 10.2 are probably too high. Despite moderate anthropogenic impact in the catchment area of Lake TRE, DI-TP levels suggest an oligotrophic state with ~7 μg TP/l during the oldest represented time period. Reference conditions must also be oligotrophic, as this is the 'best' category available for trophic states. Thus, the moderate usage of the catchment may have had only a very minor or no effect on the lakes trophic state.

The remaining two study lakes (GUD, KPS) have the most unfavourable hydrologic conditions for naturally low TP input and were thus classified as Subtype 10.2 according to both Schaumburg et al. (2007) (low RT) and Mischke et al. (2007) (high WV). Schaumburg et al. (2007) assume eutrophic natural conditions, while Mischke et al. (2007) propose mesotrophic conditions for Lake-Type 10.2. According to pollen analyses, the oldest sediments very likely reflect already disturbed conditions in both lakes. Thus, natural reference conditions may be lower than the diatom-inferred mesotrophic (KPS) and eutrophic (GUD) TP levels from the oldest sediments. For Lake GUD, it is unknown whether the inflow "Stichelsbach" is a natural creek or anthropogenically built for draining the catchment, as the name 'Stichelsbach' suggests. In the latter case, the natural trophic state condition was very likely lower than the inferred eutrophic conditions. Thus, a further investigation that pre-dates the first settlement activities in Medieval Times is needed to identify natural trophic state conditions for Lake GUD.

Diatom assemblages

According to anthropogenic usage of the catchment area (identified by fossil pollen), trophic state reference conditions and thus also associated diatom assemblages were only identified for lakes BIS and KES. Although moderate settlement activities were already evident in the catchment area of Lake TRE, similar diatom taxa to those in lakes BIS and KES dominated the assemblages in the oldest sediments of Lake TRE, indicating oligotrophic to mesotrophic DI-TP levels. However, the subdominance of eutrophic taxa (*Asterionella formosa*, *Fragilaria crotonensis*, *Stephanodiscus minutulus*, Table 5) indicate some degree of anthropogenic impact during this time. Lakes STO, KPS and GUD had already relatively high intensities of catchment usage at the beginning of the examined time period. Thus, natural diatom assemblages remain unknown for these already impacted lakes that mainly represent Lake-Type 10.2.

Diatom assemblages have a natural variability and thus reference diatom assemblages for Central Baltic lakes (CB 1) and German Lake-Type 10 may incorporate several diatom assemblages. These assemblages may be present during phases of moderate catchment usage (indicated by the pollen) and may include slight changes in the diatom assemblages, as we have shown for Lake TRE. In the following discussion, we represent the diatom assemblages from all the periods of similar TP levels as reference assemblages for Lake-Type 10.

For the implementation of the WFD, the diatom assemblages identified here as 'reference assemblages' cannot necessarily be used directly, unless the lakes are examined using paleolimnology. In Germany, the ecological status of lakes according to the WFD is identified using either pelagic algal assemblages or benthic diatom assemblages from the shoreline (Mischke et al., 2007; Schaumburg et al., 2007). Thus, this study has identified diatoms typical of reference assemblages, but the percentages or dominances of each taxon may vary depending on the method used to identify the current ecological status.

In lakes BIS and KES (both Sub-type 10.1), the reference conditions were typically dominated by the diatoms *Aulacoseira subarctica*, *Cyclotella comta* and *Cyclotella comensis*, which were replaced by *Aulacoseira granulata* with increasing anthropogenic impact (Figs. 3 and 4, Table 5).

In Lake KES, *Aulacoseira subarctica* dominated during OZ I but was replaced by diatom taxa already present (mainly *Cyclotella comta* and *Stephanodiscus minutulus*) during OZ II (and DZ II), when anthropogenic impact increased due to moderate settlement activities. However, inferred trophic state levels remained at low mesotrophic level. Thus, this shift could have been due to factors other than anthropogenic impact, such as increased temperatures. For example, *Aulacoseira* spp. have been shown to decrease at the expense of *Cyclotella* spp. with increased climate warming (Rühland et al. in press). Similarly, *A. subarctica* is known for preferring cold waters. Thus, this decrease of *A. subarctica* could have been due to the warmer period (climatic optimum) during Medieval Times and therefore, both assemblages (OZ I and OZ II) may represent reference diatom assemblages.

In Lake TRE, diatom assemblages changed during DZ II with *Aulacoseira subarctica* (plankton) and *Staurosira construens* (periphyton) replacing *Cyclotella comensis* (plankton) and *S. brevistriata* (periphyton) and then dominating the assemblages until the middle of the eighteenth century (Fig. 6). Although pollen taxa indicative of settlement slightly increased and diatom assemblages changed, the lake remained at low mesotrophic levels and thus the diatoms may still represent reference conditions.

Overall, reference diatom assemblages could only be identified for German Lake-Type 10.1 (WV: 1.5–5.0; RT > 1 year), but not for Lake-Type 10.2 (Table 4). With increasing WV and decreasing RT, the complexity of factors influencing (reference) diatom assemblages also increases. Thus, several diatom assemblages may represent reference conditions for Lake-Type 10.2., including taxa typical for oligo-mesotrophic conditions (e.g. *Cyclotella comensis*) to eutrophic conditions (e.g. *Asterionella formosa*, *Fragilaria crotonensis*, *Stephanodiscus minutulus*; Lake TRE).

Unfortunately, the two lakes with very high WV (KPS and GUD) differed greatly from each other and thus, provide little evidence for typical diatom assemblages or general trends. Lake KPS is located at the end of a long river–lake system and thus, the high WV is not necessarily reflected by high TP input from the catchment area. Accordingly, the oldest assemblages of the KPS core also contain oligo-mesotrophic taxa, such as dominant *Cyclotella comensis* or co-dominant *Encyonopsis microcephala* and *Caloneis schumanniana*. These taxa possibly do not represent reference conditions of lakes with a high WV, as the up-stream lakes may act as nutrient sinks (see below). On the other hand, the oldest assemblages of Lake KPS do differ from the mesotrophic reference assemblages of lakes BIS and KEL in that eutrophic taxa such as *Aulacoseira ambigua* and *Stephanodiscus binderanus* were present.

In Lake GUD different factors interact to influence the diatom assemblages throughout the core compared to the other study lakes, as Lake GUD is dystrophic. Thus, these assemblages cannot be compared to the other lakes. The assemblages of Lake GUD were dominated by taxa indicating very eutrophic conditions (*Discostella pseudostelligera*, *Aulacoseira granulata*, *Cocconeis placentula*) and taxa typical of mesotrophic conditions (*A. subarctica*, *Staurosira venter*). These assemblages confirm that high levels of humic substances favour humic-substance tolerating taxa, i.e. that humic substances may determine the taxa occurring in the diatom assemblages (see also Steinberg et al., 2006).

Influence of other lakes in the catchment area

One isolated (BIS, TRE, GUD) and one lake within a river–lake system (KES, STO, KPS) always had a similar sized catchment area, i.e. in this study three 'pairs' of lakes were investigated (Table 1). Interestingly, sometimes the isolated lake had a higher inferred trophic state reference condition than the lake of a river–lake system (e.g. low mesotrophic conditions for isolated Lake BIS versus oligotrophic to low mesotrophic Lake KES) and sometimes vice versa (e.g. oligotrophic to low mesotrophic Lake TRE versus probably less than eutrophic Lake STO).

Lake KES is located downstream of the lakes Großer Eutiner See and Sibbersdorfer See (part of the Schwentiner lake chain; Fig. 1). As expected, these lakes functioned as a nutrient sink for nutrients from the upstream catchment area during pre-impact times and also recently. In contrast, the upstream lakes from Lake STO, especially the hypereutrophic Lake Belauer See, function recently as a nutrient and biomass source and may thus explain the lagging reaction of the diatom assemblages to reduced nutrient loading from the catchment in Lake STO (Fig. 1). Although most plankton (including diatoms) may be filtered in the stream that connects Lake STO and Lake Belauer See (Bahnwart et al., 1999; Bahnwart, 2001), Lake STO seems to be inoculated with high nutrient levels indicating diatoms from Lake Belauer See. Therefore, recently measured nutrient levels (Table 1) differ greatly from diatom-inferred TP levels (Fig. 5), most likely because some diatoms originate from hypereutrophic Lake Belauer See Lake.

Lake KPS (low RT, high WV) is located at the end of the Schwentiner lake-river system (Fig. 1), with Lake Großer Plöner See (30 km^2 surface area) draining directly into Lake KPS (2 km^2 surface area) via a short stream. Thus, water from 97.2% of the catchment area flows through Lake Großer Plöner See into Lake KPS, i.e. lake conditions of Lake Großer Plöner See will likely have a high impact on Lake KPS. Accordingly, the anthropogenic induced increased nutrient input during the last century led to increased planktonic primary production in Lake Großer Plöner See. The outflow of such lakes typically contains distinctly higher plankton concentrations than their inflow (Moss et al., 1989; Köhler, 1993), i.e. the increased nutrient levels are mainly transformed into planktonic biomass. Almost the entire biomass may then sink to the lake sediment, may be filtered by sessile mollusks in the lake or may leave in the outflow (Schmitt, 2000; Bahnwart, 2001). Thus, during reference conditions and probably also during moderately eutrophic conditions, most nutrients will already be transformed into biomass in Lake Großer Plöner See. However, during periods of 'nutrient overload' Lake Großer Plöner See may functionally change from a nutrient and biomass sink to a nutrient and biomass source, as was probably the case during the two human-induced water level reductions of Lake Großer Plöner See in ~AD 1830 and ~1881. This water level reduction of Lake Großer Plöner See is reflected by increased DI-TP levels in Lake KPS (Fig. 7). Unfortunately, the core only dates back to ~AD 1730. The first water level increase took place ~AD 1250 and led to a massive eutrophication of Lake Großer Plöner See (Ohle, 1973). Therefore, this study can not identify the direct impact of this first major water level rise and the subsequent eutrophication of Lake Großer Plöner See on Lake KPS. Hence, reference conditions (before ~AD 1250) of Lake KPS remain unknown but were very likely lower than the inferred mesotrophic conditions in ~AD 1730.

Unexpected stable TP periods

During some relatively long periods, DI-TP levels were comparatively stable, despite minor-to-moderate and increasing anthropogenic settlement activities in the catchment area (according to pollen assemblages) of several lakes (BIS, KES, TRE). In Lake BIS, DI-TP levels were constant during moderate settlement activities throughout OZ II, i.e. for 150 years (Fig. 3). Similarly, in Lake KES DI-TP levels were constant during moderate settlement activities throughout OZ II, i.e. for ~425 years (Fig. 4), while in Lake TRE DI-TP levels were lower than expected from the distinct settlement activities during the entire OZ I (Fig. 6).

One explanation for the unexpected stable TP periods may be a TP-Fe precipitation within the lakes. For example, a palaeolimnological study of the history from Lake Dudinghausener See, (northern Germany) identified a masking of the settlement activities on DI-TP levels from Slavic to Late Medieval Times (Dreßler et al., 2006). A constant inflow of iron-rich groundwater led to TP-Fe precipitation, and thus, the increased TP from the settlement activities were not available for the phytoplankton of Lake Dudinghausener See. DI-TP levels remained within the mesotrophic level (Dreßler et al., 2006). Current measurements indicate low Fe and high TP levels in the groundwater around the study lakes as historic Fe-levels were not identified causes for unexpected (according to fossil pollen assemblages) periods of constant DI-TP levels remain unknown.

Conclusions

This study confirms the value of diatom analyses to infer and identify historic reference conditions for carbonate-rich, stratified lakes with a large catchment area from North German lowlands (part of European ecoregion 14: Central plain). The studied lake sediment cores covered the past 290–1760 years until 2005. Similar to other lakes in the central European lowlands, several study lakes were already impacted in Medieval Times. For three out of the six study lakes (BIS, KES, TRE), historic trophic state reference conditions in the lake were identified, as well as the corresponding natural (BIS, KES) or minimally impacted (TRE) diatom assemblages. The inferred TP-levels (10–25 μg TP/l) are in accordance with estimated reference values of such lakes in European Ecoregion 14 (12.5–24 μg TP/l; Søndergaard et al. 2005; EU-Intercalibration Technical Report, 2006; STOWA, 2007). The remaining three study lakes (STO, KPS, and GUD) had already high settlement activities in their catchment at the beginning of the studied time period. Thus, total phosphorus levels were probably lower than the inferred mesotrophic to eutrophic conditions of the oldest represented time.

Oligotrophic to low mesotrophic reference conditions were indicated for lakes in German Lake-Type 10 with a relatively small catchment area and high retention time. Oligotrophic reference conditions were also indicated for Lake TRE with a higher WV and lower RT (WV: 7.3; RT: 143 d). Local characteristics, such as TP–Fe precipitation through naturally high Fe availability, may be responsible for the low phosphate levels in this lake.

Lakes with yet higher WV and lower RT compared to Lake TRE may have naturally mesotrophic levels, i.e. >25 μg TP/l, due to a higher nutrient input from the catchment. Hence, according to our results, the eutrophic reference conditions suggested by Schaumburg et al. (2007) are possibly not realistic for these lakes. Further research is needed to clarify and typify the reference conditions of these lakes.

Overall, the differing reference conditions of the subtypes of Lake-Type 10 are probably not solely due to differences in water exchange of the lake with the catchment area (WV and RT), but due to more complex mechanisms, such as (i) upstream lakes in the catchment area that can function as a nutrient sink or (ii) as a nutrient and phytoplankton source or (iii) within-lake processes (TP precipitation through high iron availability, decreased TP availability or light limitation through abundant humic substances). Thus, assumed trophic state reference conditions (or TP levels) may be too high (ii) or too low (i, iii) compared to actual reference conditions.

Thus, this study confirms that a further differentiation of Central Baltic Lake-Type 1 and German Lake-Type 10 is needed, when identifying the reference conditions of trophic state or diatom assemblages, but that other factors in addition to WV and RT also need to be considered.

Acknowledgements We gratefully acknowledge the assistance provided by and the conducting of the coring from R. Niederreiter (UWITEC, Mondsee, Austria), and the company "AquaSed" (Rostock-Warnemünde, Germany). We would also like to thank Mrs. Paschen for her technical assistance in the laboratory, an anonymous reviewer for the constructive comments and all the contributors of EDDI-subsets for allowing the use of their data. This project was funded by the State Agency for Nature and Environment Schleswig-Holstein, Germany.

References

Adler, S., 2008. New package paltran 1.0: WA, WA-PLS, (POM) and MW for paleolimnology in R 2.6.1. http://www.r-project.org.

Anderson, N. J., 1995. Naturally eutrophic lakes: reality, myth or myophia? Trends in Ecology and Evolution 10: 137–138.

Averdieck, F. R., 1974. Zur Vegetations-, Siedlungs- und Seegeschichte. In Hinz, H., H. E. Nellissen, H. Helmuth, W. Prange & F. R. Averdieck (eds), Bosau. Untersuchung einer Siedlungskammer in Ostholstein 1. Offa 31: 150–169.

Averdieck, F. R., 1978. Palynologischer Beitrag zur Entwicklungsgeschichte des Grossen Plöner Sees und der Vegetation seiner Umgebung. Archiv für Hydrobiologie 85: 1–46.

Bahnwart, M., 2001. Ökophysiologische Untersuchungen zu den steuernden Faktoren der Phytoplanktonsukzession im Verlauf der Fließstrecke der Warnow. PhD Theses Univ. Rostock, Berlin, Weissensee-V: 1–141.

Bahnwart, M., T. Hübener & H. Schubert, 1999. Downstream changes in phytoplankton composition and biomass in a lowland river-lake system (Warnow River, Germany). Hydrobiologia 391: 99–111.

Barker, P. A., J. M. Pates, R. J. Payne & R. M. Healey, 2005. Changing nutrient levels in Grasmere, English Lake District, during recent centuries. Freshwater Biology 50: 1971–1981.

Battarbee, R. W. & M. J. Kneen, 1982. The use of electronically counted microspheres in absolute diatom analysis. Limnology and Oceanography 27: 184–188.

Battarbee, R. W., S. Juggins, F. Gasse, N. J. Anderson, H. Bennion & N. G. Cameron, 2000. European Diatom Database (EDDI). An information system for paleoenvironmental reconstruction. European Climate Science Conference, European Commission, Vienna, Austria 1998: 1–10.

Battarbee, R. W., D. T. Monteith, S. Juggins, C. D. Evans, A. Jenkins & G. L. Simpson, 2005. Reconstructing pre-acidification pH for an acidified Scottish loch: a comparison of palaeolimnological and modeling approaches. Environmental Pollution 137: 135–149.

Beales, P. W., 1980. The late Devensian and Flandrian vegetational history of Crose Mere, Shropshire. New Phytologist 85: 133–161.

Behre, K. E., 1976. Pollenanalytische Untersuchungen zur Vegetations- und Siedlungsgeschichte bei Flögeln und im

Ahlenmoor (Elbe-Weser-Winkel). Probleme der Küstenforschung im südlichen Nordseegebiet 11: 101–118.

Bennion, H. & R. Battarbee, 2007. The European Union Water Framework Directive: opportunities for palaeolimnology. Journal of Paleolimnology 38: 285–295.

Bennion, H., S. Juggins & N. J. Anderson, 1996. Predicting epilimnetic phosphorus concentrations using an improved diatom-based transfer function, and its application to lake eutrophication management. Environmental Science Technology 30: 2004–2007.

Bennion, H., J. Fluin & G. L. Simpson, 2004. Assessing eutrophication and reference conditions for Scottish freshwater lochs using subfossil diatoms. Journal of applied Ecology 41: 124–138.

Beug, H. J., 2004. Leitfaden der Pollenbestimmung für Mitteleuropa und angrenzende Gebiete. Pfeil, München.

Birks, H. H., H. J. B. Birks, P. E. Kaland & D. Moe, 1988. The Cultural Landscape—Past, Present and Future. Cambridge University Press, Cambridge.

Bjerring, R., E. G. Bradshaw, S. L. Amsinck, L. S. Johansson, B. V. Odgaard, A. B. Nielsen & E. Jeppsen, 2008. Inferring recent changes in the ecological state of 21 Danish candidate reference lakes (EU Water Framework Directive) using palaeolimnology. Journal of Applied Ecology 45: 1566–1575.

Bradshaw, E. G. & N. J. Anderson, 2003. Environmental factors that control the abundance of *Cyclostephanos dubius* (Bacillariophyceae) in Danish lakes, from seasonal to century scale. European Journal of Phycology 38: 265–276.

Bradshaw, E. G., N. J. Anderson, J. P. Jensen & E. Jeppesen, 2002. Phosphorus dynamics in Danish lakes and the implications for diatom ecology and palaeoecology. Freshwater Biology 47: 1963–1975.

Bradshaw, E. G., P. Rasmussen & B. V. Odgaard, 2005a. Mid- to late holocene land-use change and lake development at Dallund So, Denmark: synthesis of multiproxy data, linking land and lake. Holocene 15: 1152–1262.

Bradshaw, E. G., P. Rasmussen, H. Nielsen & N. J. Anderson, 2005b. Mid- to late holocene land-use change and lake development at Dallund So, Denmark: trends in lake primary production as reflected by algal and macrophyte remains. Holocene 15: 1130–1242.

Carvalho, L., A. Solomini, G. Phillips, M. van den Berg, O.-P. Pietiläinen, A. Lynche Solheim & U. Mischke, 2008. Chlorophyll reference conditions for European lake types used for intercalibration of ecological status. Aquatic Ecology 42: 203–211.

Cleve-Euler, A., 1951. Die Diatomeen von Schweden und Finnland. Kungl. Svenska Vetenskapsakademiens Handlingar, Fjärde Serien, Band 2, No. 1. Almquist & Wiksells Boktryckeri, Stockholm.

Dapples, F., A. F. Lotter, F. N. van Leewen, W. O. van der Knaap, S. Dimitriadis & D. Oswald, 2002. Paleolimnological evidence for increased landslide activity due to forest clearing and land-use since 3600 cal BP in the western Swiss Alps. Journal of Paleolimnology 27: 239–248.

Dörfler, W., 1990. Die Geschichte des Hanfanbaus in Mitteleuropa aufgrund palynologischer Untersuchungen und von Greßrestnachweisen. Prähistorische Zeitschrift 65: 218–244.

Dreßler, M., U. Selig, W. Dörfler, S. Adler, H. Schubert & Th. Hübener, 2006. Environmental changes and the migration period in Europe by the example of Lake Dudinghausen (northern Germany). Quarternary Research 66: 25–37.

EDDI (European Diatom Database Initiative). http://craticula.nc.ac.uk/eddi/jsp.

Ekdahl, E. J., J. L. Teranes, T. P. Guilderson, C. A. Wittkop, C. F. Turton, J. H. McAndrews & E. F. Stoermer, 2004. A prehistorical record of cultural eutrophication from Crawford Lake, Ontario. Geology 32: 745–748.

Ekdahl, E. J., J. L. Teranes, C. A. Wittkop, E. F. Stoermer, E. D. Reavie & J. P. Smol, 2007. Diatom assemblage response to Iroquoian and Euro-Canadian eutrophication of Crawford Lake, Ontario, Canada. Journal of Paleolimnology 37: 233–246.

EU-Intercalibration Technical Report, 2006. European environmental NGO technical review of the water framework directive intercalibration process. http://www.eeb.org/activities/water/200609_NGO-review-WFD-Intercalibration.

European Union, 2000. Direktive 2000/60/EC of the European Parliament and the Council of 23.10.2000 establishing a framework for Community action in the field of water policy. Official Journal of the EC L327: 1–72.

Fægri, K., 1993. Bestimmungsschlüssel für die nordwesteuropäische Pollenflora. Fischer, Jena.

Fægri, K., & J. Iversen, 1975. Textbook of pollen analysis. Munksgaard. Kopenhagen: 101–117.

Grimm, E. C., 1987. Coniss: a fortran 77 program for stratigraphically constrained cluster analysis by the method of incremental sum of squares. Computers & Geosciences 13: 13–35.

Günter, J., 1983. Development of Grossensee (Holstein, Germany): variations in trophic status from the analysis of subfossil microfauna. Hydrobiologia 103: 231–234.

Hall, R. I. & J. P. Smol, 1999. Diatoms as indicators of lake eutrophication. In Stoermer, E. F. & J. P. Smol (eds), The Diatoms: Applications for the Environmental and Earth Sciences. Cambridge University Press, Cambridge: 128–168.

Hase, W., 1972. Anbau und Bewirtschaftung des Nadelholzes in Schleswig-Holstein zu preußischer Zeit (1864–1945). Forstarchiv 43: 71–74.

Hase, W., 1983. Abriss der Wald- und Forstgeschichte Schleswig-Holstein im letzten Jahrtausend. Schriftenreihe des Naturwissenschaftlichen Vereins Schleswig-Holstein 53: 83–124.

Hübener, Th. & W. Dörfler, 2004. Reconstruction of the trophic development of the Lake Krakower Obersee (Mecklenburg, Germany) by means of sediment-diatom- and pollen-analysis. Studia Quaternaria 21: 101–108.

Hübener, Th., M. Dreßler, A. Schwarz, K. Langner & S. Adler, 2008. Dynamic adjustment of training sets ('moving-window' reconstruction) by using transfer functions in paleolimnology—a new approach. Journal of Paleolimnology 40: 79–95.

Hustedt, F., 1930. Die Kieselalgen Deutschlands, Österreichs und der Schweiz. In L. Rabenhorst's Kryptogamen-Flora, Band VII, 1. Teil, reprint Koelz, 1991.

Joosten, H. & P. de Klerk, 2002. What's in a name? Some thoughts on pollen classification, identification, and nomenclature in Quarternary palynology. Review of Palaeobotany & Palynology 122: 29–45.

Kalbe, L. & H. Werner, 1974. Das Sediment des Kummerower Sees. Untersuchungen des Chemismus und der Diatomeenflora. Internationale Revue der Gesamten Hydrobiologie 596: 755–782.

Köhler, J., 1993. Growth, production and losses of phytoplankton in the lowland River Spree. 1. Population dynamics. Journal of Plankton Research 15: 335–349.

Krammer, K., 1997a. Die cymbelloiden Diatomeen. Eine Monographie der weltweit bekannten Taxa, Teil 1 Allgemeines und Encyonema Part. Bibliotheca Diatomologica, Vol. 36. Cramer Berlin, Stuttgart.

Krammer, K., 1997b. Die cymbelloiden Diatomeen, Eine Monographie der weltweit bekannten Taxa, Teil 2 Encyonema part., Encyonopsis and Cymbellopsis. Bibliotheca Diatomologica, Vol. 37. Cramer, Berlin, Stuttgart.

Krammer, K., 2000. The Genus Pinnularia. In Lange-Bertalot, H. (ed.), Diatoms of Europe, Vol. 1. A.R.G. Gantner Verlag, Ruggel.

Krammer, K., 2002. Cymbella. In Lange-Bertalot, H. (ed.), Diatoms of Europe, Vol. 3. A.R.G. Gantner Verlag, Ruggel.

Krammer, K., 2003. Cymbopleura, Delicata, Navicymbula, Gophocymbelloides, Afrocymbella. In Lange-Bertalot, H. (ed.), Diatoms of Europe, Vol. 4. A.R.G. Gantner Verlag, Ruggel.

Krammer, K. & H. Lange-Bertalot, 1986. Süßwasserflora von Mitteleuropa, Bacillariophyceae 2/1. In Ettl, H., J. Gerloff, H. Heinig & D. Mollenhauer (eds), Süsswasserflora von Mitteleuropa, Vol. 2, no. 1. Gustav Fisher Verlag, Stuttgart.

Krammer, K. & H. Lange-Bertalot, 1988. Süßwasserflora von Mitteleuropa, Bacillariophyceae 2/2. In Ettl, H., J. Gerloff, H. Heinig & D. Mollenhauer (eds), Süsswasserflora von Mitteleuropa, Vol. 2, no. 2. Gustav Fisher Verlag, Stuttgart.

Krammer, K. & H. Lange-Bertalot, 1991a. Süßwasserflora von Mitteleuropa, Bacillariophyceae 2/3. In Ettl, H., J. Gerloff, H. Heinig & D. Mollenhauer (eds), Süsswasserflora von Mitteleuropa, Vol. 2, no. 3. Gustav Fisher Verlag, Stuttgart.

Krammer, K. & H. Lange-Bertalot, 1991b. Süßwasserflora von Mitteleuropa, Bacillariophyceae 2/4. In Ettl, H., J. Gerloff, H. Heinig & D. Mollenhauer (eds), Süsswasserflora von Mitteleuropa, Vol. 2, no. 4. Gustav Fisher Verlag, Stuttgart.

Lange-Bertalot, H. & G. Moser, 1994. Brachysira, Monographie der Gattung. Bibliotheca Diatomologica, Vol. 29. Cramer, Berlin, Stuttgart.

LANU, 2002. Measurement report. Landesamt für Umwelt und Natur des Landes Schleswig Holstein, unpublished.

LANU, 2003. Measurement report. Landesamt für Umwelt und Natur des Landes Schleswig Holstein, unpublished.

LANU, 2005. Measurement report. Landesamt für Umwelt und Natur des Landes Schleswig Holstein, unpublished.

LAWA, 1999. Gewässerbewertung – stehende Gewässer. Länderarbeitsgemeinschaft Wasser, Kulturbuchverlag, Berlin.

LAWK, 1993. Die Seen der oberen Schwentine (Krs. Ostholstein), Bericht des Ministeriums für Umwelt des Landes Schleswig-Holstein.

Leps, J. & P. Smilauer, 2003. Multivariate Analysis of Ecological Data Using CANOCO. Cambridge University Press, Cambridge.

Lotter, A. F. & H. J. B. Birks, 2003. The Holocene palaeolimnology of Sägistalsee and its environmental history a synthesis. Journal of Paleolimnology 30: 333–342.

Mathes, J., G. Plambeck & J. Schaumburg, 2002. Das Typisierungssystem für stehende Gewässer in Deutschland mit Wasserflächen ab 0.5 km^2 zur Umsetzung der Wasserrahmenrichtlinie. Aktuelle Reihe Brandenburgische Technische Universität Cottbus 5: 15–23.

Mischke, U., U. Riedmüller, E. Hoehn & B. Nixdorf, 2007. Praxistest Phytoplankton in Seen. Endbericht zum LAWA Projekt (O 5.05). Berlin, Freiburg, Bad Saarow.

Moore, P. D., J. A. Webb & M. E. Collinson, 1991. Pollen Analysis. Blackwell Science, Oxford.

Moss, B., I. Booker, H. Balls & K. Manson, 1989. Phytoplankton distribution in a temperate flood-plain lake and river system. 1. Hydrology, nutrient sources and phytoplankton biomass. Journal of Plankton Research 11: 813–838.

Müller, H. & A. Kleinmann, 1998. Palynologische Untersuchung eines Sedimentprofiles aus dem Wollingster See. In Schutz und Erhaltung nährstoffarmer Stillgewässer am Beispiel des Wollingster Sees. Mitteilungen der Arbeitsgruppe Geobotanik in Schleswig-Holstein und Hamburg 57: 44–52.

Nadeau, M. J., P. M. Grootes, M. Schleicher, P. Hasselberg, A. Rieck & M. Bitterling, 1998. Sample throughput and data quality at the Leibniz-Labor AMS facility. Radiocarbon 40: 239–245.

Nürnberg, G. K. & M. Shaw, 1999. Productivity of clear and humic lakes: nutrients, phytoplankton, bacteria. Hydrobiologia 382: 97–112.

Ohle, W., 1973. Rasante Eutrophierung des Großen Plöner Sees in frühgeschichtlicher Zeit. Naturwissenschaften 60: 47.

Ohle, W., 1979. Ontogeny of the Lake Grosser Plöner See. In Horie, S. (ed.), Paleolimnology of Lake Biwa and the Japanese pleistocene, Vol. 7. Kyoto University, Otsu: 3–33.

Overbeck, F., 1975. Botanisch-geologische Moorkunde. Wachholz Verlag, Neumünster.

Punt, W. (ed.), 1976. The Northwest European Pollen Flora, Vol. 1. Elsevier, Amsterdam.

Punt, W. & S. Blackmoore (eds), 1991. The Northwest European Pollen Flora, Vol. 6. Elsevier, Amsterdam.

Punt, W. & G. C. S. Clarke (eds), 1980. The Northwest European Pollen Flora, Vol. 2. Elsevier, Amsterdam.

Punt, W. & G. C. S. Clarke (eds), 1981. The Northwest European Pollen Flora, Vol. 3. Elsevier, Amsterdam.

Punt, W. & G. C. S. Clarke (eds), 1984. The Northwest European Pollen Flora, Vol. 4. Elsevier, Amsterdam.

Punt, W., S. Blackmoore & G. C. S. Clarke (eds), 1988. The Northwest European Pollen Flora, Vol. 5. Elsevier, Amsterdam.

Punt, W., P. P. Hoen & S. Blackmoore (eds), 1995. The Northwest European Pollen Flora, Vol. 7. Elsevier, Amsterdam.

Räsänen, J., T. Kaupilla & V. P. Salonen, 2006. Sediment-based investigation of naturally or historically eutrophic lakes-implications for lake management. Journal of Lake Management 79: 253–265.

REFCOND, 2003. Final guidance on establishing reference conditions and ecological status class boundaries from inland surface waters. Common Implementation Strategy (CIS) Working Group 2.3, Version 7.0, 5.3.2003.

Robbins, A. & D. N. Edgington, 1975. Determination of recent sedimentation rates in Lake Michigan using Pb-210 and Cs-137. Geochimica et Cosmochimica Acta 39: 285–304.

Rühland, K., A. M. Paterson & J. P. Smol, in press. Hemispheric-scale patterns of climate-related shifts in planktonic diatoms from North American and European lakes. Global Change Biology 14: 2740–2754.

Sander-Johansson, L., S. L. Amsinnck, R. Bjerring & E. Jeppesen, 2005. Mid- to late Holocene land-use change and lake development at Dallund So, Denmark: trophic structure inferred from cladoceran subfossils. Holocene 15: 1143–1151.

Schaumburg, J., Ch. Schranz, D. Stelzer & G. Hofmann, 2007. Handlungsanweisung für die ökologische Bewertung von Seen zur Umsetzung der EU-Wasserrahmenrichtlinie:Makrophyten und Phytobenthos. Bayerisches Landesamt für Umwelt.

Schmitt, M., 2000. Partikuläres Material (POM) in einem seebürtigen Flachlandfließgewässer: Eintrags- und Retentionsdynamik unter besonderer Berücksichtigung des Phytoplanktons. PhD Theses, Brandenburgigische Technische Universität, Berlin, Weißensee-Verlag, Berlin.

Schönfelder, I., J. Gelbrecht, J. Schönfelder & C. E. W. Steinberg, 2002. Relationship between littoral diatoms and their chemical environment in northeastern German lakes and rivers. Journal of Phycology 38: 66–82.

Schwarz, A., 2006. Rekonstruktion der Entwicklung des Schulzensees und des Tiefen Sees (Mecklenburg-Vorpommern) seit dem Spätglazial mittels Diatomeenanalyse unter besonderer Berücksichtigung der Trophiegeschichte. Greifswalder Geographische Arbeiten 41: 1–166.

Simpson, G. L., E. M. Shilland, J. M. Winterbottom & J. Keay, 2005. Defining reference conditions for acidified waters using a modern analogue approach. Environmental Pollution 137: 119–133.

Smol, J. P., 2008. Pollution of Lakes and Rivers—A Paleoenvironmental Perspective. Blackwell Publishing, Oxford.

Søndergaard, M., E. Jeppesen, J. P. Jensen & S. L. Amsinck, 2005. Water Framework Directive: ecological classification of Danish lakes. Journal of Applied Ecology 42: 616–629.

Steinberg, C. E. W., S. Kamara, V. Yu. Prokhotskaya, L. Manusadzianas, T. A. Karasyova, M. A. Timofeyev, Z. Jie, A. Paul, T. Meinelt, V. F. Farjalla, A. Y. O. Matsuo, B. K. Burnison & R. Menzel, 2006. Dissolved humic substances—ecological driving forces from the individual to the ecosystem level? Freshwater Biology 51: 1189–1210.

Stockmarr, J., 1971. Tablets with spores used in absolute pollen analysis. Pollen et Spores 13: 615–621.

STOWA, 2007. Referenties en Maatlatten voor natuurlijke Watertypen voor de Kaderrichtlijn Water. Rap. 32, 375 pp. (www.stowa.nl).

Stuiver, M., P. J. Reimer, E. Bard, J. W. Beck, G. S. Burr, K. A. Hughen, B. Kromer, G. McCormac, J. Van der Pflicht & M. Spurkm, 1998. INTCAL98 radiocarbon age calibration, 24.000-0 cal BP. Radiocarbon 40: 1041–1083.

Taylor, D., C. Dalton, M. Leira, P. Jordan, G. Chen, L. Leòn-Vintró, K. Irvine, H. Bennion & T. Nolan, 2006. Recent histories of six productive lakes in the Irish Ecoregion based on multiproxy palaeolimnological evidence. Hydrobiologia 571: 237–259.

ter Braak, C. J. F. & P. Śmilauer, 2002. CANOCO Reference Manual and CanoDraw for Windows User's Guide. Software for Canonical Community Ordination (version 4.5). Microcomputer Power, Ithaca.

Uhlmann, D. & H. Horn, 2001. Hydrobiologie der Binnengewässer. Ulmer Verlag, Stuttgart.

UNSCEAR, 2000. Sources and effects of ionizing radiation, in UNSCEAR 2000 Report to the General Assembly, Technical Paper, U.N. Science Commission. On the Effects of Atom. Radiation, Vienna. Vol. I: Sources, Annex C: Exposures from man-made sources of radiation.

UWITEC: According to 192–193. http://www.uwitec.at/html/frame.html.

van Geel, B., L. A. Mur, M. Ralskajasiewiczowa & T. Goslar, 1994. Fossil akinetes of Aphanizomenon and Anabaena as indicators for medieval phosphate-eutrophication of Lake Gosciaz (Central Poland). Review of Palaeobotany and Palynology 83: 97–105.

Vollenweider, R. A., 1979. Das Nährstoffbelastungskonzept als Grundlage für den externen Eingriff in den Eutrophierungsprozeß stehender Gewässer und Talsperren. Zeitschrift für Wasser- und Abwasserforschung 12: 46–56.

Vollenweider, R. & J. Kerekes, 1982. Eutrophication of Waters—Monitoring, Assessment and Control. OECD, Paris.

Werner, P. & J. P. Smol, 2006. Diatom-environmental relationships and nutrient transfer functions from contrasting shallow and deep limestone lakes in Ontario, Canada. Hydrobiologia 533: 145–173.

Werner, P., M. Chaisson & J. P. Smol, 2005. Long-term limnological changes in six lakes with differing human impacts from a limestone region in southwestern Ontario, Canada. Lake and Reservoir Management 21: 436–452.

Wiethold, J., 1998. Studien zur jüngeren postglazialen vegetations- und Siedlungsgeschichte im östlichen Schleswig-Holstein. Universitätsforschung zur prähistorischen Archeologie, Vol. 45. Habelt, Bonn.

FSC
www.fsc.org
MIX
Papier aus verantwortungsvollen Quellen
Paper from responsible sources
FSC® C105338

If you have any concerns about our products,
you can contact us on
ProductSafety@springernature.com

In case Publisher is established outside the EU,
the EU authorized representative is:
Springer Nature Customer Service Center GmbH
Europaplatz 3, 69115 Heidelberg, Germany

Printed by Libri Plureos GmbH
in Hamburg, Germany